Cytogenomics

Cytogenomics

Edited by

Thomas Liehr
Jena University Hospital, Friedrich Schiller University,
Institute of Human Genetics, Jena, Germany
http://cs-tl.de/

ELSEVIER

ACADEMIC PRESS
An imprint of Elsevier

Academic Press is an imprint of Elsevier
125 London Wall, London EC2Y 5AS, United Kingdom
525 B Street, Suite 1650, San Diego, CA 92101, United States
50 Hampshire Street, 5th Floor, Cambridge, MA 02139, United States
The Boulevard, Langford Lane, Kidlington, Oxford OX5 1GB, United Kingdom

Notices
Knowledge and best practice in this field are constantly changing. As new research and experience
broaden our understanding, changes in research methods, professional practices, or medical treatment
may become necessary.

Practitioners and researchers must always rely on their own experience and knowledge in evaluating
and using any information, methods, compounds, or experiments described herein. In using such
information or methods they should be mindful of their own safety and the safety of others, including
parties for whom they have a professional responsibility.

To the fullest extent of the law, neither the Publisher nor the authors, contributors, or editors, assume
any liability for any injury and/or damage to persons or property as a matter of products liability,
negligence or otherwise, or from any use or operation of any methods, products, instructions, or ideas
contained in the material herein.

Library of Congress Cataloging-in-Publication Data
A catalog record for this book is available from the Library of Congress

British Library Cataloguing-in-Publication Data
A catalogue record for this book is available from the British Library

ISBN : 978-0-12-823579-9

For information on all Academic Press publications
visit our website at https://www.elsevier.com/books-and-journals

Publisher: Andre Gerhard Wolff
Acquisitions Editor: Peter Linsley
Editorial Project Manager: Sara Valentino
Production Project Manager: Punithavathy Govindaradjane
Cover Designer: Victoria Pearson

Typeset by SPi Global, India

Working together
to grow libraries in
developing countries
www.elsevier.com • www.bookaid.org

Contents

CHAPTER 9   Application of CRISPR/Cas9 to visualize defined genomic sequences in fixed chromosomes and nuclei

Takayoshi Ishii, Kiyotaka Nagaki, and Andreas Houben

CHAPTER 10 Approaches for studying epigenetic aspects of the human genome

Tigran Harutyunyan and Galina Hovhannisyan

SECTION 2 Current cytogenomic research

Contributors

Mario Davide Maria Avarello
Genomic Vision, Bagneux, France

Vladislav S. Baranov
D.O. Ott Research Institute of Obstetrics, Gynecology and Reproductology, Saint Petersburg, Russia

Hayk Barseghyan
Bionano Genomics, San Diego, CA; Children's National Research Institute, Washington, DC, United States

Prakhar Bisht
Genomic Vision, Bagneux, France

Sven Bocklandt
Bionano Genomics, San Diego, CA, United States

Alka Chaubey
Bionano Genomics, San Diego, CA, United States

Joan-Ramon Daban
Department of Biochemistry and Molecular Biology, Faculty of Biosciences, Autonomous University of Barcelona, Barcelona, Spain

Yannick Delpu
Bionano Genomics, San Diego, CA, United States

Thomas Eggermann
Uniklinik RWTH Aachen, Institut für Humangenetik, Aachen, Germany

Jean-Baptiste Gaillard
Unit of Chromosomal Genetics and Research Platform Chromostem, Department of Medical Genetics, Arnaud de Villeneuve Hospital, Montpellier, France

Vincent Gatinois
Unit of Chromosomal Genetics and Research Platform Chromostem, Department of Medical Genetics, Arnaud de Villeneuve Hospital; INSERM 1183 Unit "Genome and Stem Cell Plasticity in Development and Aging" Institute of Regenerative Medicine and Biotherapies, St Eloi Hospital, Montpellier, France

Tigran Harutyunyan
Department of Genetics and Cytology, Yerevan State University, Yerevan, Armenia

Alex Hastie
Bionano Genomics, San Diego, CA, United States

Andreas Houben
Leibniz Institute of Plant Genetics and Crop Plant Research (IPK), Seeland, Germany

Galina Hovhannisyan
Department of Genetics and Cytology, Yerevan State University, Yerevan, Armenia

Yuichi Ichikawa
Division of Cancer Biology, The Cancer Institute of JFCR, Tokyo, Japan

Ivan Y. Iourov
Yurov's Laboratory of Molecular Genetics and Cytogenomics of the Brain, Mental Health Research Center; Laboratory of Molecular Cytogenetics of Neuropsychiatric Diseases, Veltischev Research and Clinical Institute for Pediatrics of the Pirogov Russian National Research Medical University; Russian Medical Academy of Continuous Postgraduate Education, Moscow, Russia

Takayoshi Ishii
Arid Land Research Center (ALRC), Tottori University, Tottori, Japan

Tatiana V. Kuznetzova
D.O. Ott Research Institute of Obstetrics, Gynecology and Reproductology, Saint Petersburg, Russia

Thomas Liehr
Jena University Hospital, Friedrich Schiller University, Institute of Human Genetics, Jena, Germany

Kiyotaka Nagaki
Institute of Plant Science and Resources, Okayama University, Kurashiki, Japan

Franck Pellestor
Unit of Chromosomal Genetics and Research Platform Chromostem, Department of Medical Genetics, Arnaud de Villeneuve Hospital; INSERM 1183 Unit "Genome and Stem Cell Plasticity in Development and Aging" Institute of Regenerative Medicine and Biotherapies, St Eloi Hospital, Montpellier, France

Jacques Puechberty
Unit of Chromosomal Genetics and Research Platform Chromostem, Department of Medical Genetics, Arnaud de Villeneuve Hospital, Montpellier, France

Noriko Saitoh
Division of Cancer Biology, The Cancer Institute of JFCR, Tokyo, Japan

Anouck Schneider
Unit of Chromosomal Genetics and Research Platform Chromostem, Department of Medical Genetics, Arnaud de Villeneuve Hospital, Montpellier, France

Malte Spielmann
Institute of Human Genetics, University of Lübeck, Lübeck; Human Molecular Genomics Group, Max Planck Institute for Molecular Genetics, Berlin, Germany

Uirá Souto Melo
Human Molecular Genomics Group, Max Planck Institute for Molecular Genetics, Berlin, Germany

Martin Ungelenk
Jena University Hospital, Friedrich Schiller University, Institute of Human Genetics, Jena, Germany

Svetlana G. Vorsanova
Yurov's Laboratory of Molecular Genetics and Cytogenomics of the Brain, Mental Health Research Center; Laboratory of Molecular Cytogenetics of Neuropsychiatric Diseases, Veltischev Research and Clinical Institute for Pediatrics of the Pirogov Russian National Research Medical University, Moscow, Russia

Anja Weise
Jena University Hospital, Friedrich Schiller University, Institute of Human Genetics, Jena, Germany

Veronica Yumiceba
Institute of Human Genetics, University of Lübeck, Lübeck, Germany

Yuri B. Yurov
Yurov's Laboratory of Molecular Genetics and Cytogenomics of the Brain, Mental Health Research Center; Laboratory of Molecular Cytogenetics of Neuropsychiatric Diseases, Veltischev Research and Clinical Institute for Pediatrics of the Pirogov Russian National Research Medical University, Moscow, Russia

A definition for cytogenomics - Which also may be called chromosomics

1

Thomas Liehr

Jena University Hospital, Friedrich Schiller University, Institute of Human Genetics, Jena, Germany

Chapter outline

From cytogenetics to cytogenomics

Research and diagnostics in human genetics, with the chromosome in focus, were originally designated as "cytogenetics" (Liehr & Claussen, 2002). Working with human chromosomes rather than DNA and sequences was very popular in the 1970s and 1980s. However, since the beginning of the 1990s, mainstream human geneticists have looked at people dealing with chromosomes as something like outdated "fossils" (Liehr et al., 2017; Salman et al., 2004). Interestingly, this attitude was never justified by any real evidence. This way of thinking may arise from the impression that the higher the resolution - down to the base pair level, as made possible by Sanger sequencing and faster by modern next-generation sequencing approaches (Ungelenk, 2021) - the more informative and meaningful genetic research and diagnostics may be (Roberts, 2015). Still, it is important to understand that all the yet available techniques to study the human genome, at different levels of resolutions, and at the level of the single cell or by approaching millions of cells at a time, complement rather than work against each other (Liehr et al., 2017; Ungelenk, 2021; Weise et al., 2019).

Nonetheless, cytogeneticists in particular seem to feel driven by the pressure from molecular geneticists, and have (over)reacted in part by changing well-established

Cytogenomics. https://doi.org/10.1016/B978-0-12-823579-9.00001-1

designations from "cytogenetics" to "cytogenomics," to appear more "modern," maybe fancier and/or attractive, as the following examples show:

- For decades, the American Cytogenetic Association has held a conference on chromosomes every 2 years, which was renamed from the "American Cytogenetics Conference" to the "American Cytogenomics Conference" in 2018 (American Cytogenomics Conference, 2020).
- The well-established "European Cytogenetic Association" also did the same, i.e., their biennial conference, called the "European Cytogenetics Conference" until 2017, has been known as the "European Cytogenomics Conference" only since 2019 (ECA - European Cytogeneticists Association, 2020).
- The "bible" of human chromosome nomenclature, the ISCN, standing for "international system of cytogenetic nomenclature" since 1978 (Shaffer et al., 2013), was renamed the "international system of cytogenomic nomenclature" in 2016 (McGowan-Jordan et al., 2016). Even though it is explained by a half sentence why this change was made ("to reflect the changes in technology under its purview"), no definition is given for cytogenomics in this booklet. Astonishingly, in the latest version of the ISCN (McGowan-Jordan et al., 2020) it was even decided to show no chromosomes at all on the title page; instead 14 lanes with 26 different-colored horizontal bars are shown, most likely representing sequencing results of 364 base pairs. However, instead of four different colors, as might be expected for four different nucleotides (guanine, adenine, cytosine, and thymine), bars comprised of five different colors are depicted. The fifth color may be methylated cytosine, but it cannot be visualized in a sequencing run along with unmethylated cytosine (apart from nanopore technique). In addition, the publishing house Karger uses the same five colors in its logo.

The designation "cytogenomics" appeared in 1999 and has since then mainly been used as an alternative to "cytogenetics and genomics approaches" (Liehr, 2021a). Only in 2019 was it suggested by Russian colleagues that the term cytogenomics could, in retrospect, be deduced as well from the word "cyto(post)genomics" (Iourov, 2019).

A definition of cytogenomics

In this book, we also use the neologism "cytogenomics," but with the goal to paraphrase a new field of research in genomics and diagnostics in human genetics, which has a comprehensive and integrative view on the field. Cytogenomics, as we understand it here, is not at all restricted to diagnostics; such a definition for cytogenomics was given by Silva et al. (2019), where it is "a general term that encompasses conventional, as well as molecular cytogenetics (fluorescence in situ hybridization (FISH), microarrays) and molecular-based techniques." Here "cytogenomics" is understood as equivalent wording under the definition used for chromosomics by Prof. Uwe Claussen (Jena, Germany) in 2005 (Claussen, 2005) and outlined in Fig. 1.1 and in the next paragraph (Chromosomics, 2020; Liehr, 2019).

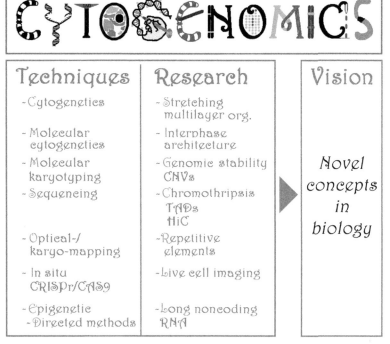

FIG. 1.1

Schematic overview on what cytogenomics comprises. In the upper box the word cytogenomics is written as a pictogram: "C" is depicted like GTG-banded chromosomes; "Y" represents a DNA forming a replication fork; "T" has as *red* upper part standing for a topologically associating domain; "O" is an interphase nucleus, showing the interphase architecture; "G" highlights the epigenetic changes as posttranslational methylation (*green* M) of cytosine; "E" stands for a chromosome, which is microdissected by a glass needle; "N" includes left hand side a chromosome with one fragile site in the long arm and a so-called gap in the short arm; "O" represents the molecular karyotyping approach; "M" is depicted intentionally in *green* and *red*, representing a rearrangement between to nonhomologous chromosomes stained by two-color fluorescence in-situ hybridization; "I" represents a chromosome in *blue* counterstain DAPI with a *green* and a *red* locus-specific probe as applied in fluorescence in-situ hybridization; "C" includes a pipette, which drops cytogenetically worked-up cell suspension on a slide; "S" is shown as a chromosome 1 after C-banding. In the two left-hand lower boxes some representative approaches/techniques used and actual research applications in cytogenomics are listed; all of them are treated in the following chapters of this book. The right-hand lower box shows that the term cytogenomics is meant to be integrative and include the vision to create novel concepts in biology and medicine.

Prof. Claussen stated that the term "chromosomics," here used as equivalent to the term "cytogenomics," "was introduced to draw attention to the three-dimensional morphological changes in chromosomes that are essential elements in gene regulation" (Claussen, 2005). His idea was to subsume under such a term all chromosome-related research with the goal to lead us to novel concepts in biology. This contrasts with other omics-designations, which mainly aim to show the importance of their specific field by separating it from others. Thus, chromosomics/cytogenomics includes the following kinds of studies/fields:

- research on plasticity of chromosomes in relation to the three-dimensional positions of genes, which affect cell function in a developmental and tissue-specific manner during the cell cycle. Here, my own studies on chromosome structure in meta- and interphase, as well as studies by Thomas Cremer and others, can be mentioned (Liehr, 2021b), which use three-dimensional-FISH as well as HiC-analyses (Ungelenk, 2021; Weise & Liehr, 2021a). Overall, it has already been suggested that gene expression is dependent on and regulated by chromosome structure in interphase. Thus, new ways of thinking are already on the way to being integrated into transcriptomic research, with chromatin modifications being considered more than they were previously (Ichikawa & Saitoh, 2021; Weise & Liehr, 2021a; Yumiceba et al., 2021), as well as some other ways of looking at epigenetic factors influencing gene expression (Eggermann, 2021; Harutyunyan & Hovhannisyan, 2021).
- research into chromatin-modification-mediated changes in the architecture of chromosomes, which may influence the functions and lifespans of cells, tissues, organs, and individuals. Getting more insights into the flexible, three-dimensional structures of metaphase chromosomes (Daban, 2021) may also help us to understand the influence of aforementioned "positional effects" on cells (Weise & Liehr, 2021a; Yumiceba et al., 2021) at different stages of their development (Baranov & Kuznetzova, 2021; Pellestor et al., 2021). One thing necessary to consider here in more detail is the recent observation that each cell of the (human) body remembers which of each homologous chromosome set derives from the mother and which from the father of the individual (Weise et al., 2016). Additionally, effects of copy number alterations appearing during aging and their effects on nuclear architecture are not completely established yet (Mkrtchyan et al., 2010).
- Research on species-specific differences in the architecture of chromosomes, which has been overlooked in the past. In that sense, the use of the word cytogenomics by the aforementioned Russian colleagues was correct (Iourov, 2019). However, also here the construction of interphase and the effects of this architecture are the focus of research covered by the term cytogenomics/chromosomics in evolution-focused research. It is, for example, still unknown if conserved genes in mammalians keep their position always in the same kind of chromosomal band (Giemsa-dark or -light) during evolution (Kosyakova et al., 2009) and, if it changes position, when this leads to differential

expression. Here molecular cytogenetics is the approach to be applied in future studies (Bisht & Avarello, 2021; Liehr, 2021c), and perhaps also optical mapping and karyo-mapping (Delpu et al., 2021).

- research about the occurrence and prevalence of chromosomal gaps and breaks and interchanges, to be studied in the first place by banding cytogenetics (Weise & Liehr, 2021b) and molecular cytogenetics (Bisht & Avarello, 2021; Liehr, 2021c). Here especially fragile sites and their putative role as seeding points of (i) evolutionary conserved breakpoints, (ii) breakpoints observed in inherited, and (iii) acquired chromosomal aberrations in tumors are of interest. Recently, as suggested by Uwe Claussen, the fragile site-related breaks were attributed to chromosomes' three-dimensional structure and function rather than to DNA-sequence (Liehr, 2019).

Conclusion

Overall, this book summarizes, in Section 1, traditional and new approaches to study the human genome at different levels of resolution, like cytogenetics (Weise & Liehr, 2021b), molecular karyotyping (Weise & Liehr, 2021c), or application of CRISPR/Cas9 in cytogenomics (Ishii et al., 2021); Section 2 highlights current hot topics of cytogenomic research, e.g., chromothripsis (Pellestor et al., 2021), cytogenomic landscape of the human brain (Iourov et al., 2021), or copy number variants (Liehr, 2021d), without claiming to provide the complete possible content here.

Ultimately and primarily, this book is meant to focus on and bring the reader back to the necessity to relate all obtained genetic/genomic/cytogenomic results to the fact that human cells do not contain one single 2 m long DNA-string, but that this string is divided normally into 46 parts of significantly different lengths (Liehr, 2020, 2021a). Number of portions (i.e., if there is maybe an extra chromosome), arrangements of smaller and larger DNA-blocks, epigenetic DNA-changes, positioning in the nucleus, inter- and intrachromosomal interactions, number, and possible expression of repetitive elements (at least as RNA), all these and many more things to be discovered in future should be considered in a comprehensive cytogenomic analysis, which deserves such a designation.

References

American Cytogenomics Conference. (2020). https://chromophile.org/.

Baranov, V. S., & Kuznetzova, T. V. (2021). Nuclear stability in early embryo. Chromosomal aberrations. In T. Liehr (Ed.), *Cytogenomics* (pp. 307–325). Academic Press. Chapter 15 (in this book).

Bisht, P., & Avarello, M. D. M. (2021). Molecular combing solutions to characterize replication kinetics and genome rearrangements. In T. Liehr (Ed.), *Cytogenomics* (pp. 47–71). Academic Press. Chapter 5 (in this book).

Chromosomics. (2020). http://cs-tl.de/.

Claussen, U. (2005). Chromosomics. *Cytogenetic and Genome Research, 111*(2), 101–106. https://doi.org/10.1159/000086377.

Daban, J.-R. (2021). Multilayer organization of chromosomes. In T. Liehr (Ed.), *Cytogenomics* (pp. 267–296). Academic Press. Chapter 13 (in this book).

Delpu, Y., Barseghyan, H., Bocklandt, S., Hastie, A., & Chaubey, A. (2021). Next-generation cytogenomics: High-resolution structural variation detection by optical genome mapping. In T. Liehr (Ed.), *Cytogenomics* (pp. 123–146). Academic Press. Chapter 8 (in this book).

ECA—European Cytogeneticists Association. (2020). https://www.e-c-a.eu/en/.

Eggermann, T. (2021). Epigenetics. In T. Liehr (Ed.), *Cytogenomics* (pp. 389–401). Academic Press. Chapter 20 (in this book).

Harutyunyan, T., & Hovhannisyan, G. (2021). Approaches for studying epigenetic aspects of the human genome. In T. Liehr (Ed.), *Cytogenomics* (pp. 155–209). Academic Press. Chapter 10 (in this book).

Ichikawa, Y., & Saitoh, N. (2021). Shaping of genome by long noncoding RNAs. In T. Liehr (Ed.), *Cytogenomics* (pp. 357–372). Academic Press. Chapter 18 (in this book).

Iourov, I. Y. (2019). Cytopostgenomics: What is it and how does it work? *Current Genomics, 20*(2), 77–78. https://doi.org/10.2174/1389202920002190422120524.

Iourov, I. Y., Vorsanova, S. G., & Yurov, Y. B. (2021). Cytogenomic landscape of the human brain. In T. Liehr (Ed.), *Cytogenomics* (pp. 327–348). Academic Press. Chapter 16 (in this book).

Ishii, T., Nagaki, K., & Houben, A. (2021). Application of CRISPR/Cas9 to visualize defined genomic sequences in fixed chromosomes and nuclei. In T. Liehr (Ed.), *Cytogenomics* (pp. 147–153). Academic Press. Chapter 9 (in this book).

Kosyakova, N., Weise, A., Mrasek, K., Claussen, U., Liehr, T., & Nelle, H. (2009). The hierarchically organized splitting of chromosomal bands for all human chromosomes. *Molecular Cytogenetics, 2*, 4. https://doi.org/10.1186/1755-8166-2-4.

Liehr, T. (2019). From human cytogenetics to human chromosomics. *International Journal of Molecular Sciences, 20*, 826.

Liehr, T. (2020). *Human genetics—Edition 2020: A basic training package*. Epubli.

Liehr, T. (2021a). Overview of currently available approaches used in cytogenomics. In T. Liehr (Ed.), *Cytogenomics* (pp. 11–24). Academic Press. Chapter 2 (in this book).

Liehr, T. (2021b). Nuclear architecture. In T. Liehr (Ed.), *Cytogenomics* (pp. 297–305). Academic Press. Chapter 14 (in this book).

Liehr, T. (2021c). Molecular cytogenetics. In T. Liehr (Ed.), *Cytogenomics* (pp. 35–45). Academic Press. Chapter 4 (in this book).

Liehr, T. (2021d). Repetitive elements, heteromorphisms, and copy number variants. In T. Liehr (Ed.), *Cytogenomics* (pp. 373–388). Academic Press. Chapter 19 (in this book).

Liehr, T., & Claussen, U. (2002). Current developments in human molecular cytogenetic techniques. *Current Molecular Medicine, 2*(3), 283–297. https://doi.org/10.2174/1566524024605725.

Liehr, T., Mrasek, K., Klein, E., & Weise, A. (2017). Modern high throughput approaches are not meant to replace 'old fashioned' but robust techniques. *Journal of Genetics and Genomes, 1*(1), e101.

McGowan-Jordan, J., Hastings, R., & Moore, S. (Eds.). (2020). *ISCN 2020—An international system for human cytogenomic nomenclature (2020)* Karger.

McGowan-Jordan, J., Simons, J., & Schmid, M. (Eds.). (2016). *ISCN 2016—An international system for human cytogenomic nomenclature (2016)* Karger.

Mkrtchyan, H., Gross, M., Hinreiner, S., Polytiko, A., Manvelyan, M., Mrasek, K., … Weise, A. (2010). The human genome puzzle—The role of copy number variation in somatic mosaicism. *Current Genomics, 11*(6), 426–431. https://doi.org/10.2174/138920210793176047.

Pellestor, F., Gaillard, J.-B., Schneider, A., Puechberty, J., & Gatinois, V. (2021). Chromoanagenesis phenomena and their formation mechanisms. In T. Liehr (Ed.), *Cytogenomics* (pp. 213–245). Academic Press. Chapter 11 (in this book).

Roberts, J. P. (2015). *Molecular cytogenetics: Arrays, NGS and beyond.* Biocompare. https://www.biocompare.com/Editorial-Articles/175781-Molecular-Cytogenetics-Arrays-NGS-and-Beyond/.

Salman, M., Jhanwar, S. C., & Ostrer, H. (2004). Will the new cytogenetics replace the old cytogenetics? *Clinical Genetics*, *66*(4), 265–275. https://doi.org/10.1111/j.1399-0004.2004.00316.x.

Shaffer, L., McGowan-Jordan, J., & Schmid, M. (Eds.). (2013). *ISCN 2013—An international system for human cytogenetic nomenclature (2013)* Karger.

Silva, M., de Leeuw, N., Mann, K., Schuring-Blom, H., Morgan, S., Giardino, D., … Hastings, R. (2019). European guidelines for constitutional cytogenomic analysis. *European Journal of Human Genetics*, *27*(1), 1–16. https://doi.org/10.1038/s41431-018-0244-x.

Ungelenk, M. (2021). Sequencing approaches. In T. Liehr (Ed.), *Cytogenomics* (pp. 87–122). Academic Press. Chapter 7 (in this book).

Weise, A., Bhatt, S., Piaszinski, K., Kosyakova, N., Fan, X., Altendorf-Hofmann, A., … Chaudhuri, J.P. (2016). Chromosomes in a genome-wise order: Evidence for metaphase architecture. *Molecular Cytogenetics*, *9*(1), 36. https://doi.org/10.1186/s13039-016-0243-y.

Weise, A., & Liehr, T. (2021a). Interchromosomal interactions with meaning for disease. In T. Liehr (Ed.), *Cytogenomics* (pp. 349–356). Academic Press. Chapter 17 (in this book).

Weise, A., & Liehr, T. (2021b). Cytogenetics. In T. Liehr (Ed.), *Cytogenomics* (pp. 25–34). Academic Press. Chapter 3 (in this book).

Weise, A., & Liehr, T. (2021c). Molecular karyotyping. In T. Liehr (Ed.), *Cytogenomics* (pp. 73–85). Academic Press. Chapter 6 (in this book).

Weise, A., Mrasek, K., Pentzold, C., & Liehr, T. (2019). Chromosomes in the DNA era: Perspectives in diagnostics and research. *Medgen*, *31*(1), 8–19. https://doi.org/10.1007/s11825-019-0236-4.

Yumiceba, V., Souto Melo, U., & Spielmann, M. (2021). 3D cytogenomics: Structural variation in the three-dimensional genome. In T. Liehr (Ed.), *Cytogenomics* (pp. 247–266). Academic Press. Chapter 12 (in this book).

Technical aspects

Overview of currently available approaches used in cytogenomics

2

Thomas Liehr

Jena University Hospital, Friedrich Schiller University, Institute of Human Genetics,
Jena, Germany

Chapter outline

What is cytogenomics?

Cytogenomics is a designation introduced to describe the formidable technical developments in (human) genetics in the last decades. According to PubMed (PubMed, 2020), the term "cytogenomics" was in use as early as 1999 in a paper from France (Bernheim, 1999). Then, until 2012, PubMed lists only 1–13 publications per year applying this term. Between 2013 and 2018 it was mentioned in ~35 papers per annum. In 2019 and 2020, it was used almost 70 times each year (PubMed, 2020). In 2021, several established journals, such as *Frontiers in Genetics* or *International Journal of Molecular Sciences* (*MDPI*), published special issues on cytogenomics.

  A new designation like "cytogenomics" became necessary (see also Liehr, 2021a) due to a tremendous shift from traditional ways of looking on the genetic content of organisms, to more sophisticated ones, often called "high throughput approaches" (e.g., Haeri et al., 2016; Mitsuhashi & Matsumoto, 2020). These modern, fancy, and generally quite expensive approaches are only feasible in connection with high-tech apparatuses and/or bioinformatics, and they enable a high-resolution view on genomes, as well as creation of large data sets in short times (Liehr, 2017). This is in sharp contrast to traditional methods of genetic analyses, which is - in terms of molecular genetic approaches - concentrated on single genes or relatively small regions of DNA (i.e., small amounts of data), rather than a whole genome (Liehr, 2020). The most traditional way to analyze genomes is cytogenetics, which means studying whole genomes, but on the low-resolution level of whole chromosomes and, if available, their subbands (achievable resolution not more than 5–10 Mb). The specificity of cytogenetic analyses is that it focuses on the single cell level; this enables totally different insights into genomes than "high throughput approaches," meant in the first place to analyze millions of cells at a time (Liehr, 2020).

Overall, the introduction of the term "cytogenomics" was intended to create a designation comprising all the below-detailed approaches suited to the study of genomes. It is an integrative idea, which aims to collect all classical and new cytogenetic and molecular-genetic/-genomic and bioinformatics approaches under one roof. Interestingly, it is hard to find a definition for the term "cytogenomics" in the literature. In 2019, Ivan Iourov (author of a chapter in this book (Iourov et al., 2021)) stated that this term could be deduced from the word "cyto(post)genomics," and defined it as encompassing "all areas of chromosome biology addressed in the genomic context" (Iourov, 2019). Interestingly, as outlined in Chapter 1 of this book (Liehr, 2021a), the term cytogenomics is also quite similar, if not identical, to what Uwe Claussen introduced as "chromosomics" in 2005 (Claussen, 2005; Liehr, 2019). He argued that to restrict the definition of "chromosomics" = "cytogenomics" to the pure technical aspects is falling short and needs to be expanded by a vision. Prof. Claussen's vision was "developing of novel concepts in biology" (Claussen, 2005). Perhaps one can also put it that the goal of cytogenomics is to uncover the underlying principles of life by genetic and genomic research; this is what drives all researchers using cytogenomic approaches. Cytogenomics should never stand alone as "pure application of approaches," but should be used in context with visions, theories, and ideas of how to look at human or other genomes and try to understand nature, and to be a naturalist, which is to be a student of nature in the best sense.

Accordingly, this book is divided into two parts: a more technically oriented one, treating major approaches of cytogenomics, and a more application-oriented part, presenting selected current cytogenomic research. The central part of this actual chapter consists of Tables 2.1–2.4, which include the most important currently available cytogenomic approaches, and also list major achievements made possible by their dedicated use.

Cytogenomic approaches

Progress in the field that we now call cytogenomics was in most cases stimulated by technical developments from outside the genetics field. To give two examples:

Table 2.1 Cytogenomic technique cytogenetics.

Cytogenomic approach cytogenetics	Introduced in year	Milestones due to this approach in human genetics
Classical and banding cytogenetics	1879 by W. Flemming	– 1888: chromosomes are denominated by H.W. Waldeyer – 1902/3: chromosome theory of inheritance by W. Sutton and T. Boveri – 1956: J.H. Tijo and A. Levan publish correct human chromosome number as 46, thus enabling start of cytogenetic diagnostics – 1968–70: chromosome banding techniques established, e.g., by L. Zech, and since then applied in human genetic basic diagnostics

Information when a cytogenetic approach was introduced, together with some (subjectively selected) milestones achieved by its application in human genetics, is provided in Liehr (2020).

Table 2.2 Cytogenomic technique molecular genetics.

Cytogenomic approach molecular genetics	Introduced in year	Milestones due to this approach in human genetics
Basics	1905: the word "genetics" was introduced by W. Bateson	– 1941: one-gene-one enzyme hypothesis by E.L. Tatum and G.W. Beadle – 1944: DNA and not proteins are carrier of genetic information acc. to O.T. Avery – 1949: cytosine and guanine as well as adenine and thymidine are present in 1:1 ratio in DNA acc. to E. Chargaff – 1951: R. Franklin and M. Wilkins deliver X-ray depiction of DNA – 1953: DNA double-helix model by J. Watson and F. Crick – 1957: central dogma of molecular biology that information goes from DNA via RNA to protein by F. Crick – 1965: genetic triplet code deciphered by M. Nierenberg
Classical techniques	1968–83: basic molecular genetic approaches are introduced	– 1968: first restriction enzymes are isolated; restriction enzymes were important tools for "cytogenomics" → see also paragraph (optical mapping - Bionano) in this table – 1973: first cloning of extrinsic DNA to a bacterium by S. Cohen and H. Boyer – 1975: blotting of DNA acc. to E.M. Southern – 1983: DNA-fingerprint analyses based on repetitive elements acc. to A. Jeffreys – 1983: polymerase chain reaction (PCR) acc. to K. Mullis

Continued

Table 2.2 Cytogenomic technique molecular genetics—cont'd

Cytogenomic approach molecular genetics	Introduced in year	Milestones due to this approach in human genetics
Sequencing	1975: F. Sanger develops sequencing of DNA	– Since 1975: classical sequencing used in multiple research and diagnostic settings – Since 1990s: development of multiple alternative sequencing approaches, mainly driven by human genome (HUGO)-project – 2001: first draft of human genome is published by C. Venter and HUGO – From 2000: next-generation sequencing (NGS - also second-generation sequencing) is introduced – 2011: chromothripsis is detected in cancer cells based on NGS – Since ~2010: efforts to establish third generation sequencing = long-read sequencing – 2014: topologically associating domains (TADs) are reported and start to play major roles in interpretation of HiC data and also in research and diagnostics – 2014: CRISPR/Cas9 system for genomic editing is reported by J.A. Doudna and E. Charpentier – 2020: multilayer organization of the chromosomes is suggested by J.R. Daban
Optical mapping (Bionano)	~2010	– Optical mapping based on high molecular weight DNA and sequences related to restriction enzyme recognition sites – At present, application of optical mapping in diagnostics and research

Information on when a molecular genetic approach was introduced, together with some (subjectively selected) milestones achieved by its application in human genetics is provided (Daban, 2021; Delpu et al., 2021; Ishii et al., 2021; Liehr, 2020; Pellestor et al., 2021).

1. Banding cytogenetics (Weise & Liehr, 2021a) could only be established after standardized, high-quality microscopes became available, which was not before Carl Zeiss, Ernst Abbe, and Otto Schott started to work on that in the 1880s in Jena, Germany (Carl Zeiss Biography, 2020). Continuous improvements of microscopy techniques led to characterization of correct chromosome numbers in humans in 1956 (Tijo & Levan, 1956). Later, establishment of the whole field of molecular cytogenetics (Liehr, 2021b) was dependent on corresponding reliable fluorescence microscopes (Liehr, 2019).

2. Similarly, "high throughput approaches" (Ichikawa & Saitoh, 2021; Pellestor et al., 2021; Ungelenk, 2021; Weise & Liehr, 2021b, 2021c; Yumiceba et al., 2021) only became realistic after miniaturization of machines in terms of microchip technology (McGlennen, 2001) and/or management of tremendous amounts of data had been achieved and established, mainly in fields outside

Table 2.3 Cytogenomic technique molecular cytogenetics.

Cytogenomic approach molecular cytogenetics	Introduced in year	Milestones due to this approach in human genetics
General	1986 by D. Pinkel et al. for human chromosomes based on radioactive in-situ hybridization developed in 1969 by M.L. Pardue and J.G. Gall	– Since 1986: insights into interphase architecture – Since 1986: application in diagnostics – Since 1991: application of repetitive probes by V.R. Babu and A. Wiktor – 1992: comparative genomic hybridization (CGH) by Kallioniemi et al.; since 2000s as array-CGH (aCGH or molecular karyotyping) applied – 1992: glass-needle-based microdissection is used for the first time to create probes for fluorescence in-situ hybridization (FISH) by Meltzer et al. – 1996 to present: establishment of diverse multicolor FISH (mFISH) probe sets, according to questions to be studied – Since ~2000: genomic landscape of the brain is studied – Since ~2010: intrachromosomal interactions in human diseases come into focus – 2011: chromothripsis is detected in cancer cells and first studied by FISH in 2013 by Mackinnon et al. – 2012: chromothripsis and its meaning for nuclear stability of human embryo is discovered – 2014: CRISPR-mediated live imaging is developed – 2020: multilayer organization of the chromosomes is suggested by J.R. Daban
Molecular combing	1994 Bensimon et al. and in 1995 by Florijn et al. published as fiber-FISH	– Since 1995: insights into high-resolution stretched DNA fibers – Since ~2011: application in diagnostics
Molecular karyotyping	Since ~1998, based on CGH (1992) different aCGH approaches were developed	– aCGH was developed for use in human genetics diagnostics – Since ~1998: aCGH based on bacterial artificial chromosome probes (BACs) – Since ~2006: aCGH based on oligonucleotide probes – Since ~2010: aCGH based on single nucleotide polymorphisms (SNPs) enabling besides detection of copy number variations also detection of epigenetic changes (uniparental disomy) in trio settings

Information on when a molecular genetic approach was introduced, together with some (subjectively selected) milestones achieved by its application in human genetics, is provided (Bisht & Avarello, 2021; Daban, 2021; Liehr, 2020, 2021b, 2021c; Pellestor et al., 2021; Weise & Liehr, 2021b).

Table 2.4 Cytogenomic techniques in connection with epigenetics.

Cytogenomic approach in connection with epigenetics	Introduced in year	Milestones due to this approach in human genetics
Epigenetic research methods come from molecular (cyto)genetics	1956: the term epigenetics was introduced by C.H. Waddington	– 1980: concept of uniparental disomy introduced by Eric Engel – Since 1986: insights into interphase architecture – ~2010: influence of environmental factors on offspring was acc. to L.H. Lumey – 2014: topologically associating domains (TADs) are reported and start to play major roles also in connection with epigenetics and (long) noncoding RNAs

Information on when a molecular genetic approach was introduced, together with some (subjectively selected) milestones achieved by its application in human genetics, is provided (Eggermann, 2021; Harutyunyan & Hovhannisyan, 2021; Ishii et al., 2021; Liehr, 2020; Ichikawa & Saitoh, 2021; Yumiceba et al., 2021).

genetics. Application of each technique in genetic/genomic research undoubtedly also led to improvements in the corresponding machines and tools (Wooley et al., 2005).

Separate chapters in this book are devoted to most of the cytogenomic approaches listed below and in Tables 2.1–2.4. Here, these techniques are just put together in terms of temporal context; in addition, some (subjectively selected) milestones are provided, which were achieved by those cytogenomic technologies in human genetics.

Before the word "genetics" was defined

The word "genetics" was introduced by William Bateson in 1905 (William Bateson, 2020). Before this could take place, groundbreaking work and other basic insights into nature were necessary. Those were provided by many scientists, and it is impossible to mention them all here. However, it was thought-leaders like Gregor Mendel, Charles Darwin, Walther Flemming, Thomas Hunt Morgan, and others who provided seminal input here (Liehr, 2020).

Cytogenetics

Even though around the year 1600 the first microscopes had made it possible to see and denominate cells in plants for the first time (done by Robert Hook), it was not until 1879 that Walther Flemming had microscopes available to visualize and document the first chromosomes as such. Thus, he is nowadays called the "founder of cytogenetics" (Liehr, 2020). Among the milestones listed in Table 2.1

for cytogenetics, the most important one for human genetics was the publication of the correct modal human chromosome number as 46 in 1956 (Tijo & Levan, 1956). This was the starting point of human genetic diagnostics based on cytogenetics and for comparative genetics. For further details, see Chapter 3.

Molecular genetics

Molecular genetics developed from chemistry, physics, biochemistry, and other disciplines including biology and medicine. Thus, in Table 2.2 for all listed achievements of molecular genetics between the years 1941 and 1965, no specific "cytogenomic" approaches can be given.

Important specific approaches that may be included as "cytogenomic ones" are listed in Table 2.2 from the late 1960s onwards, as (i) use of restriction enzymes (Meselson & Yuan, 1968; Roberts, 2005) (see also Table 1 - optical mapping - Bionano), (ii) cloning of extrinsic DNA, (iii) blotting, (iv) DNA-fingerprint analyses, and (v) polymerase chain reaction (PCR) (Liehr, 2020); (vi) sequencing, introduced basically by Frederick Sanger in 1975, is listed separately in Table 2.2, and is also an important cytogenomic tool. All mentioned approaches had and/or have importance in human genetic diagnostics and research.

Sequencing had, after 1975, a specific historical evolution; in connection with sequencing of the human genome (human genome = HUGO project), NGS, or second-generation sequencing, was introduced in the 1990s to 2000s, followed by third-generation, long-range sequencing in the 2010s. Besides the fact that sequencing identified a lot of disease-causing mutations, an important breakthrough for cytogenomics based on sequencing was the detection of chromothripsis in cancer in 2011 (Baranov & Kuznetzova, 2021; Colnaghi et al., 2011; Pellestor et al., 2021), even though it had been seen before, it had not been much recognized in cytogenetics (Houge et al., 2003). While detection of chromothripsis was based on NGS, long-range sequencing identified the "topologically associating domains" (TADs) in 2014 (Shibayama et al., 2014), providing elementary new insights into three-dimensional organization of the interphase nucleus and also genetically caused diseases (Ungelenk, 2021; Weise & Liehr, 2021c; Yumiceba et al., 2021). In the same year, the CRISPR/Cas9 system for genomic editing was reported by Doudna and Charpentier (2014), which also enables new cytogenomic research methods (Ishii et al., 2021).

An approach with high potential is optical mapping (Bionano), which is based on combined use of high molecular weight DNA and restriction enzyme recognition sites with specific miniaturized tools and bioinformatics (Reisner et al., 2010). The approach is discussed by Delpu et al. (2021) in this book and has started to be applied in diagnostics and research (Young et al., 2020; Zook et al., 2020).

Molecular cytogenetics

From 1986 onwards, molecular cytogenetics based on fluorescence in-situ hybridization (FISH) was established and refined. It was deduced from radioactive in-situ hybridization, invented in 1969 (Liehr, 2020) (Table 2.3).

From the very beginning, FISH was applied in human genetic diagnostics and research. In particular, Thomas Cremer (Munich, Germany) used this approach intensely to study the three-dimensional structure of the interphase (Cremer et al., 2020; Thomas Cremer, 2020); however, other groups were also working in this field (Lemke et al., 2002; Liehr, 2021c; Rada-Iglesias et al., 2018; Weise et al., 2002; Yu & Ren, 2017; Yurov et al., 2001).

Repetitive regions of genomes are best studied by FISH approaches (Liehr, 2021d), and thus, multiple probe sets for heterochromatic and euchromatic regions of the human genome were developed (Babu & Wiktor, 1991; Liehr, 2020). Among other techniques, glass-needle-based microdissection is applied to establish FISH-probes (Al-Rikabi et al., 2019). Major achievements in cytogenomics have been made by combination of FISH and NGS, and FISH and long-range-sequencing approaches, since the 2010s (Liehr, 2020; Zlotina et al., 2020). Intra- and interchromosomal interactions in human diseases have come into focus (Meguro-Horike et al., 2011; Weise & Liehr, 2021c; Yumiceba et al., 2021), the presence of chromothripsis was confirmed in cancer cells (MacKinnon & Campbell, 2013), and also its natural occurrence in human embryos was discovered (Baranov & Kuznetzova, 2021; Pellestor, 2014; Pellestor et al., 2014, 2021). In addition, recently, the multilayer organization of the chromosomes was suggested (Daban, 2020, 2021). Due to the CRISPR/Cas9 system for genomic editing, in 2014, CRISPR-mediated live cell imaging was developed (Anton et al., 2014), leading to similar results to FISH, but in the living cell (Ishii et al., 2021).

Molecular combing (Bensimon et al., 1994) or fiber-FISH (Florijn et al., 1995) is another molecular cytogenetic approach, developed in 1994/1995 and published since then in various forms (Duell et al., 1997; Rautenstrauss et al., 1997). When doing FISH, the lowest resolution between two clearly separated DNA-level probes is achievable on metaphase chromosomes. Due to decondensation in interphase, a slightly higher resolution may be obtained using those (Lemke et al., 2002). In fiber-FISH, the DNA is stretched artificially and, according to stretching, even probes in the kilobase range may be resolved, depicted, and studied concerning their order, orientation, and copy numbers. Since ~2011, molecular combing has even been applied in diagnostics (Bisht & Avarello, 2021; Bensimon et al., 1994; Nguyen et al., 2019).

Overall, FISH is a field with all its possibilities still not explored; multicolor-FISH approaches are summarized elsewhere (Liehr, 2020a) and can, just to mention one example, also be used to visualize cDNA on lampbrush chromosomes (Zamariolli et al., 2020; Zlotina et al., 2020).

It is a matter of unresolvable discussion if molecular karyotyping is a molecular cytogenetic or molecular genetic approach. What is certain is that in 1992, comparative genomic hybridization (CGH) on chromosomes was established by Kallioniemi et al. as a FISH-approach (Liehr, 2020; Weise & Liehr, 2021b). Chromosome-based CGH still plays a major role in comparative cytogenomic studies among different but closely related species (de Moraes et al., 2019; Spangenberg et al., 2020).

In the late 1990s, CGH was translated to a chromosome-free variant, first called array-CGH (aCGH) and later designated as molecular karyotyping. CGH and aCGH work according to the same principle, i.e., samples of two whole genomes - a normal (labeled, e.g., in green) and a potentially abnormal one (labeled, e.g., in red) - are hybridized against normal DNA. By CGH/aCGH, one can detect gains and losses in the potentially abnormal DNA probe; both methods have the same restrictions: they cannot detect low-level mosaics, they are blind to heterochromatin, and they cannot detect balanced aberrations. Between 2000 and the 2010s, only copy number changes could be detected (Pinkel et al., 1998; Snijders et al., 2001; Stankiewicz & Beaudet, 2007); since ~2010, molecular karyotyping based on single nucleotide polymorphisms (SNPs) has also enabled detection of epigenetic changes (i.e., uniparental isodisomy) in trio settings (Papenhausen et al., 2011).

Besides DNA-based aCGH, there are other more or less related biochip technologies, nicely summarized by Dr. O.P. Kallioniemi in 2001 as follows: besides detection of "(1) disease predisposition by using single-nucleotide polymorphism (SNP) microarrays," there are also essays for detection of "(2) global gene expression patterns by cDNA microarrays, (3) concentrations, functional activities or interactions of proteins with proteomic biochips," and also such for "(4) cell types or tissues as well as clinical endpoints associated with molecular targets by using tissue microarrays" (Kallioniemi, 2001).

Epigenetics

"The term 'epigenetics' was originally used to denote the poorly understood processes by which a fertilized zygote developed into a mature, complex organism. With the understanding that all cells of an organism carry the same DNA, and with increased knowledge of mechanisms of gene expression, the definition was changed to focus on ways in which heritable traits can be associated not with changes in nucleotide sequence, but with chemical modifications of DNA, or of the structural and regulatory proteins bound to it" (Felsenfeld, 2014). Accordingly, the field of epigenetics (term introduced in 1956; see Waddington, 1956) deals with topics like uniparental disomy (Engel, 1980; Liehr, 2014), imprinting (Monk et al., 2019), and other different kinds of monoallelic gene expression (Liehr, 2020); however, the interphase-architecture can also be seen as an epigenetic operative influence (Cremer et al., 2020; Liehr, 2021c), together with DNA-methylation and chromatin modifying influences (Table 2.4).

In addition, nonprotein-coding DNA, previously regarded as "junk-DNA" is nowadays seen as an important influencer and potential epigenetic regulator, and as being the source of micro-RNA, long-noncoding RNA, etc. (Liehr, 2020; Noordermeer & Feil, 2020; Ichikawa & Saitoh, 2021; Shibayama et al., 2014; Yamamoto & Saitoh, 2019). Finally, the opponent of Charles Darwin, Jean-Baptiste de Lamarck, has enjoyed some rehabilitation during recent decades, as it has turned out that environmental (i.e., epigenetic) factors, such as nutrition or smoking, may have influences on offspring (Lumey et al., 2011).

Conclusion

Overall, the range of approaches applicable in cytogenomics is unrestricted and unpredictable. For example, a decade ago, the cytogenomics importance of an approach accessing high molecular weight DNA like in the Bionano optical mapping approach would have been unimaginable (Delpu et al., 2021).

References

Al-Rikabi, A. B. H., Cioffi, M. D. B., & Liehr, T. (2019). Chromosome microdissection on semi-archived material. *Cytometry Part A*, *95*(12), 1285–1288.

Anton, T., Bultmann, S., Leonhardt, H., & Markaki, Y. (2014). Visualization of specific DNA sequences in living mouse embryonic stem cells with a programmable fluorescent CRISPR/Cas system. *Nucleus*, *5*(2), 163–172.

Babu, V. R., & Wiktor, A. (1991). A fluorescence in situ hybridization technique for retrospective cytogenetic analysis. *Cytogenetics and Cell Genetics*, *57*(1), 16–17.

Baranov, V. S., & Kuznetzova, T. V. (2021). Nuclear stability in early embryo. Chromosomal aberrations. In T. Liehr (Ed.), *Cytogenomics* (pp. 307–325). Academic Press. Chapter 15 (in this book).

Bensimon, A., Simon, A., Chiffaudel, A., Croquette, V., Heslot, F., & Bensimon, D. (1994). Alignment and sensitive detection of DNA by a moving interface. *Science*, *265*(5181), 2096–2098. https://doi.org/10.1126/science.7522347.

Bernheim, A. (1999). Exploration du génome dans les proliférations malignes: de la cytogénétique à la cytogénomique. *Annales de Pathologie*, *19*, S1–S3.

Bisht, P., & Avarello, M. D. M. (2021). Molecular combing solutions to characterize replication kinetics and genome rearrangements. In T. Liehr (Ed.), *Cytogenomics* (pp. 47–71). Academic Press. Chapter 5 (in this book).

Carl Zeiss Biography. (2020). https://www.schott.com/english/company/corporate_history/biography/carl_zeiss.html.

Claussen, U. (2005). Chromosomics. *Cytogenetic and Genome Research*, *111*(2), 101–106. https://doi.org/10.1159/000086377.

Colnaghi, R., Carpenter, G., Volker, M., & O'Driscoll, M. (2011). The consequences of structural genomic alterations in humans: Genomic disorders, genomic instability and cancer. *Seminars in Cell and Developmental Biology*, *22*(8), 875–885. https://doi.org/10.1016/j.semcdb.2011.07.010.

Cremer, T., Cremer, M., Hübner, B., Silahtaroglu, A., Hendzel, M., Lanctôt, C., … Cremer. (2020). The interchromatin compartment participates in the structural and functional organization of the cell nucleus. *BioEssays*, *42*, e1900132.

Daban, J. (2020). Supramolecular multilayer organization of chromosomes: Possible functional roles of planar chromatin in gene expression and DNA. *FEBS Letters*, *594*, 395–411.

Daban, J.-R. (2021). Multilayer organization of chromosomes. In T. Liehr (Ed.), *Cytogenomics* (pp. 267–296). Academic Press. Chapter 13 (in this book).

de Moraes, R., Sember, A., Bertollo, L. A., De Oliveira, E. A., Rab, P., Hatanaka, T., … Cioffi, M.D. (2019). Comparative cytogenetics and neo-Y formation in small-sized fish species of the genus Pyrrhulina (Characiformes, Lebiasinidae). *Frontiers in Genetics*, *10*, 678. https://doi.org/10.3389/fgene.2019.00678.

Delpu, Y., Barseghyan, H., Bocklandt, S., Hastie, A., & Chaubey, A. (2021). Next-generation cytogenomics: High-resolution structural variation detection by optical genome mapping. In T. Liehr (Ed.), *Cytogenomics* (pp. 123–146). Academic Press. Chapter 8 (in this book).

Doudna, J., & Charpentier, E. (2014). Genome editing. The new frontier of genome engineering with CRISPR-Cas9. *Science, 346*, 1258096.

Duell, T., Nielsen, L. B., Jones, A., Young, S. G., & Weier, H. U. G. (1997). Construction of two near-kilobase resolution restriction maps of the 5′ regulatory region of the human apolipoprotein B gene by quantitative DNA fiber mapping (QDFM). *Cytogenetic and Genome Research, 79*(1–2), 64–70. https://doi.org/10.1159/000134685.

Eggermann, T. (2021). Epigenetics. In T. Liehr (Ed.), *Cytogenomics* (pp. 389–401). Academic Press. Chapter 20 (in this book).

Engel, E. (1980). A new genetic concept: Uniparental disomy and its potential effect, isodisomy. *American Journal of Medical Genetics, 6*(2), 137–143. https://doi.org/10.1002/ajmg.1320060207.

Felsenfeld, G. (2014). A brief history of epigenetics. *Cold Spring Harbor Perspectives in Biology, 6*, a018200.

Florijn, R. J., Bonden, A. J., Vrolijk, H., Wiegant, J., Vaandrager, J. W., Bass, F., ... Raap, A.K. (1995). High-resolution DNA fiber-fish for genomic DNA mapping and colour barcoding of large genes. *Human Molecular Genetics, 4*(5), 831–836. https://doi.org/10.1093/hmg/4.5.831.

Haeri, M., Gelowani, V., & Beaudet, A. L. (2016). Chromosomal microarray analysis, or comparative genomic hybridization: A high throughput approach. *MethodsX, 3*, 8–18. https://doi.org/10.1016/j.mex.2015.11.005.

Harutyunyan, T., & Hovhannisyan, G. (2021). Approaches for studying epigenetic aspects of the human genome. In T. Liehr (Ed.), *Cytogenomics* (pp. 155–209). Academic Press. Chapter 10 (in this book).

Houge, G., Liehr, T., Schoumans, J., Ness, G. O., Solland, K., Starke, H., ... Vermeulen, S. (2003). Ten years follow up of a boy with a complex chromosomal rearrangement: Going from a >5 to 15-breakpoint CCR. *American Journal of Medical Genetics, 118*(3), 235–240. https://doi.org/10.1002/ajmg.a.10106.

Ichikawa, Y., & Saitoh, N. (2021). Shaping of genome by long noncoding RNAs. In T. Liehr (Ed.), *Cytogenomics* (pp. 357–372). Academic Press. Chapter 18 (in this book).

Iourov, I. Y. (2019). Cytopostgenomics: What is it and how does it work? *Current Genomics, 20*(2), 77–78. https://doi.org/10.2174/1389202920021904221520524.

Iourov, I. Y., Vorsanova, S. G., & Yurov, Y. B. (2021). Cytogenomic landscape of the human brain. In T. Liehr (Ed.), *Cytogenomics* (pp. 327–348). Academic Press. Chapter 16 (in this book).

Ishii, T., Nagaki, K., & Houben, A. (2021). Application of CRISPR/Cas9 to visualize defined genomic sequences in fixed chromosomes and nuclei. In T. Liehr (Ed.), *Cytogenomics* (pp. 147–153). Academic Press. Chapter 9 (in this book).

Kallioniemi, O. P. (2001). Biochip technologies in cancer research. *Annals of Medicine, 33*(2), 142–147. https://doi.org/10.3109/07853890109002069.

Lemke, J., Claussen, J., Michel, S., Chudoba, I., Mühlig, P., Westermann, M., ... Claussen, U. (2002). The DNA-based structure of human chromosome 5 in interphase. *American Journal of Human Genetics, 71*(5), 1051–1059. https://doi.org/10.1086/344286.

Liehr, T. (2014). *Uniparental disomy (UPD) in clinical genetics. A guide for clinicians and patients*. Springer.

Liehr, T. (2017). *What about the real costs of next generation sequencing (NGS) in human genetic diagnostics?* http://atlasofscience.org/what-about-the-real-costs-of-next-generation-sequencing-ngs-in-human-genetic-diagnostics.

Liehr, T. (2019). From human cytogenetics to human chromosomics. *International Journal of Molecular Sciences, 20*(4), 826. https://doi.org/10.3390/ijms20040826.

Liehr, T. (2020). *Human genetics—Edition 2020: A basic training package.* Epubli.

Liehr, T. (2020a). *Basics and literature on multicolor fluorescence in situ hybridization application.* http://cs-tl.de/DB/TC/mFISH/0-Start.html.

Liehr, T. (2021a). A definition for cytogenomics - Which also may be called chromosomics. In T. Liehr (Ed.), *Cytogenomics* (pp. 1–7). Academic Press. Chapter 1 (in this book).

Liehr, T. (2021b). Molecular cytogenetics. In T. Liehr (Ed.), *Cytogenomics* (pp. 35–45). Academic Press. Chapter 4 (in this book).

Liehr, T. (2021c). Nuclear architecture. In T. Liehr (Ed.), *Cytogenomics* (pp. 297–305). Academic Press. Chapter 14 (in this book).

Liehr, T. (2021d). Repetitive elements, heteromorphisms, and copy number variants. In T. Liehr (Ed.), *Cytogenomics* (pp. 373–388). Academic Press. Chapter 19 (in this book).

Lumey, L. H., Stein, A. D., & Susser, E. (2011). Prenatal famine and adult health. *Annual Review of Public Health, 32,* 237–262. https://doi.org/10.1146/annurev-publhealth-031210-101230.

MacKinnon, R. N., & Campbell, L. J. (2013). Chromothripsis under the microscope: A cytogenetic perspective of two cases of AML with catastrophic chromosome rearrangement. *Cancer Genetics, 206*(6), 238–251. https://doi.org/10.1016/j.cancergen.2013.05.021.

McGlennen, R. C. (2001). Miniaturization technologies for molecular diagnostics. *Clinical Chemistry, 47*(3), 393–402. https://doi.org/10.1093/clinchem/47.3.393.

Meguro-Horike, M., Yasui, D. H., Powell, W., Schroeder, D. I., Oshimura, M., LaSalle, J. M., & Horike, S. (2011). Neuron-specific impairment of inter-chromosomal pairing and transcription in a novel model of human 15q-duplication syndrome. *Human Molecular Genetics, 20*(19), 3798–3810. https://doi.org/10.1093/hmg/ddr298.

Meselson, M., & Yuan, R. (1968). DNA restriction enzyme from E. coli. *Nature, 217*(5134), 1110–1114. https://doi.org/10.1038/2171110a0.

Mitsuhashi, S., & Matsumoto, N. (2020). Long-read sequencing for rare human genetic diseases. *Journal of Human Genetics, 65*(1), 11–19. https://doi.org/10.1038/s10038-019-0671-8.

Monk, D., Mackay, D. J. G., Eggermann, T., Maher, E. R., & Riccio, A. (2019). Genomic imprinting disorders: Lessons on how genome, epigenome and environment interact. *Nature Reviews Genetics, 20*(4), 235–248. https://doi.org/10.1038/s41576-018-0092-0.

Nguyen, K., Broucqsault, N., Chaix, C., Roche, S., Robin, J., Vovan, C., … Magdinier, F. (2019). Deciphering the complexity of the 4q and 10q subtelomeres by molecular combing in healthy individuals and patients with facioscapulohumeral dystrophy. *Journal of Medical Genetics, 56*(9), 590–601. https://doi.org/10.1136/jmedgenet-2018-105949.

Noordermeer, D., & Feil, R. (2020). Differential 3D chromatin organization and gene activity in genomic imprinting. *Current Opinion in Genetics and Development, 61,* 17–24. https://doi.org/10.1016/j.gde.2020.03.004.

Papenhausen, P., Schwartz, S., Risheg, H., Keitges, E., Gadi, I., Burnside, R. D., … Tepperberg, J. (2011). UPD detection using homozygosity profiling with a SNP genotyping microarray. *American Journal of Medical Genetics. Part A, 155*(4), 757–768. https://doi.org/10.1002/ajmg.a.33939.

Pellestor, F. (2014). Chromothripsis: How does such a catastrophic event impact human reproduction? *Human Reproduction, 29*(3), 388–393. https://doi.org/10.1093/humrep/deu003.

Pellestor, F., Gaillard, J.-B., Schneider, A., Puechberty, J., & Gatinois, V. (2021). Chromoanagenesis phenomena and their formation mechanisms. In T. Liehr (Ed.), *Cytogenomics* (pp. 213–245). Academic Press. Chapter 11 (in this book).

Pellestor, F., Gatinois, V., Puechberty, J., Geneviève, D., & Lefort, G. (2014). Chromothripsis: Potential origin in gametogenesis and preimplantation cell divisions. A review. *Fertility and Sterility*, *102*(6), 1785–1796. https://doi.org/10.1016/j.fertnstert.2014.09.006.

Pinkel, D., Segraves, R., Sudar, D., Clark, S., Poole, I., Kowbel, D., … Albertson, D.G. (1998). High resolution analysis of DNA copy number variation using comparative genomic hybridization to microarrays. *Nature Genetics*, *20*(2), 207–211. https://doi.org/10.1038/2524.

PubMed. (2020). https://pubmed.ncbi.nlm.nih.gov/.

Rada-Iglesias, A., Grosveld, F. G., & Papantonis, A. (2018). Forces driving the three-dimensional folding of eukaryotic genomes. *Molecular Systems Biology*, *14*(6). https://doi.org/10.15252/msb.20188214, e8214.

Rautenstrauss, B., Fuchs, C., Liehr, T., Grehl, H., Murakami, T., & Lupski, J. R. (1997). Visualization of the CMT1A duplication and HNPP deletion by FISH on stretched chromosome fibers. *Journal of the Peripheral Nervous System*, *2*(4), 319–322.

Reisner, W., Larsen, N. B., Silahtaroglu, A., Kristensen, A., Tommerup, N., Tegenfeldt, J. O., & Flyvbjerg, H. (2010). Single-molecule denaturation mapping of DNA in nanofluidic channels. *Proceedings of the National Academy of Sciences of the United States of America*, *107*(30), 13294–13299. https://doi.org/10.1073/pnas.1007081107.

Roberts, R. J. (2005). How restriction enzymes became the workhorses of molecular biology. *Proceedings of the National Academy of Sciences of the United States of America*, *102*(17), 5905–5908. https://doi.org/10.1073/pnas.0500923102.

Shibayama, Y., Fanucchi, S., Magagula, L., & Mhlanga, M. M. (2014). lncRNA and gene looping: What's the connection? *Transcription*, *5*, e28658.

Snijders, A., Nowak, N., Segraves, R., Blackwood, S., Brown, N., Conroy, J., … Albertson, D.G. (2001). Assembly of microarrays for genome-wide measurement of DNA copy number. *Nature Genetics*, *29*(3), 263–264. https://doi.org/10.1038/ng754.

Spangenberg, V., Arakelyan, M., Cioffi, M. D. B., Liehr, T., Al-Rikabi, A., Martynova, E., … Kolomiets, O. (2020). Cytogenetic mechanisms of unisexuality in rock lizards. *Scientific Reports*, *10*, 8697. https://doi.org/10.1038/s41598-020-65686-7.

Stankiewicz, P., & Beaudet, A. L. (2007). Use of array CGH in the evaluation of dysmorphology, malformations, developmental delay, and idiopathic mental retardation. *Current Opinion in Genetics and Development*, *17*(3), 182–192. https://doi.org/10.1016/j.gde.2007.04.009.

Thomas Cremer. (2020). https://en.wikipedia.org/wiki/Thomas_Cremer.

Tijo, J. H., & Levan, A. (1956). The chromosome number of man. *Hereditas*, *42*, 1–6. https://doi.org/10.1111/j.1601-5223.1956.tb03010.x.

Ungelenk, M. (2021). Sequencing approaches. In T. Liehr (Ed.), *Cytogenomics* (pp. 87–122). Academic Press. Chapter 7 (in this book).

Waddington, C. H. (1956). Embryology, epigenetics and biogenetics. *Nature*, *177*(4522), 1241. https://doi.org/10.1038/1771241a0.

Weise, A., & Liehr, T. (2021a). Cytogenetics. In T. Liehr (Ed.), *Cytogenomics* (pp. 25–34). Academic Press. Chapter 3 (in this book).

Weise, A., & Liehr, T. (2021b). Molecular karyotyping. In T. Liehr (Ed.), *Cytogenomics* (pp. 73–85). Academic Press. Chapter 6 (in this book).

Weise, A., & Liehr, T. (2021c). Interchromosomal interactions with meaning for disease. In T. Liehr (Ed.), *Cytogenomics* (pp. 349–356). Academic Press. Chapter 17 (in this book).

Weise, A., Starke, H., Heller, A., Uwe, C., & Liehr, T. (2002). Evidence for interphase DNA decondensation transverse to the chromosome axis: A multicolor banding analysis. *International Journal of Molecular Medicine, 9*(4), 359–361. https://doi.org/10.3892/ijmm.9.4.359.

William Bateson. (2020). https://en.wikipedia.org/wiki/William_Bateson.

Wooley, J., Lin, H., & National Research Council (US) Committee on Frontiers at the Interface of Computing and Biology (Eds.). (2005). *Catalyzing inquiry at the interface of computing and biology* National Academies Press.

Yamamoto, T., & Saitoh, N. (2019). Non-coding RNAs and chromatin domains. *Current Opinion in Cell Biology, 58*, 26–33. https://doi.org/10.1016/j.ceb.2018.12.005.

Young, E., Abid, H. Z., Kwok, P. Y., Riethman, H., & Xiao, M. (2020). Comprehensive analysis of human subtelomeres by whole genome mapping. *PLoS Genetics, 16*(1). https://doi.org/10.1371/journal.pgen.1008347, e1008347.

Yu, M., & Ren, B. (2017). The three-dimensional organization of mammalian genomes. *Annual Review of Cell and Developmental Biology, 33*, 265–289. https://doi.org/10.1146/annurev-cellbio-100616-060531.

Yumiceba, V., Souto Melo, U., & Spielmann, M. (2021). 3D cytogenomics: structural variation in the three-dimensional genome. In T. Liehr (Ed.), *Cytogenomics* (pp. 247–266). Academic Press. Chapter 12 (in this volume).

Yurov, Y. B., Vostrikov, V. M., Vorsanova, S. G., Monakhov, V. V., & Iourov, I. Y. (2001). Multicolor fluorescent in situ hybridization on post-mortem brain in schizophrenia as an approach for identification of low-level chromosomal aneuploidy in neuropsychiatric diseases. *Brain and Development, 23*(1), S186–S190. https://doi.org/10.1016/S0387-7604(01)00363-1.

Zamariolli, M., Di-Battista, A., Moysés-Oliveira, M., de Mello, C. B., de Paula Ramos, M. A., Liehr, T., & Melaragno, M. I. (2020). Disruption of PCDH10 and TNRC18 genes due to a balanced translocation. *Cytogenetic and Genome Research, 160*(6), 321–328. https://doi.org/10.1159/000508820.

Zlotina, A., Maslova, A., Pavlova, O., Kosyakova, N., Al-Rikabi, A., Liehr, T., & Krasikova, A. (2020). New insights into chromomere organization provided by lampbrush chromosome microdissection and high-throughput sequencing. *Frontiers in Genetics, 11*(1), 57. https://doi.org/10.3389/fgene.2020.00057.

Zook, J., Hansen, N., Olson, N., Chapman, L., Mullikin, J., Xiao, C., … Salit, M. (2020). A robust benchmark for detection of germline large deletions and insertions. *Nature Biotechnology, 38*(1), 1347–1355. https://doi.org/10.1038/s41587-020-0538-8.

Cytogenetics

Anja Weise and Thomas Liehr

*Jena University Hospital, Friedrich Schiller University, Institute of Human Genetics,
Jena, Germany*

Chapter outline

What is cytogenetics?

Cytogenetics is the oldest and most basic technique used to perform cytogenomic studies. It is far from being outdated and offers unique possibilities to approach the whole genome of a given species at once; cytogenetics takes advantage of the possibility of making a genome visible to the human eye under a microscopic lens (Liehr, 2020, 2021a, 2021b).

Cytogenetics is defined slightly differently, according to the following three references:

- Cytogenetics is "the study of chromosomes, which are long strands of DNA and protein that contain most of the genetic information in a cell. Cytogenetics involves testing samples of tissue, blood, or bone marrow in a laboratory to look for changes in chromosomes, including broken, missing, rearranged, or extra chromosomes. Changes in certain chromosomes may be a sign of a genetic disease or condition or some types of cancer. Cytogenetics may be used to help

Cytogenomics. https://doi.org/10.1016/B978-0-12-823579-9.00003-5

diagnose a disease or condition, plan treatment, or find out how well treatment is working" (Cytogenetics - Cancer-Dictionary, 2020).

- "Cytogenetics is essentially a branch of genetics but is also a part of cell biology/cytology (a subdivision of human anatomy), that is concerned with how the chromosomes relate to cell behavior, particularly to their behavior during mitosis and meiosis. Techniques used include karyotyping, analysis of G-banded chromosomes and other cytogenetic banding techniques" (Cytogenetics - Wikipedia, 2020).
- Cytogenetics is "the branch of genetics that is concerned primarily in cellular components, especially chromosomes, in relation to heredity, genetic anomalies, and pathologic conditions. It is the branch of genetics that deal at the cellular level" (Cytogenetics - BiologyOnline, 2020).

Thus, cytogenetics, as discussed here, is about studying chromosomes by the classical methods outlined later. It is single-cell oriented, and studies chromosomes for research and diagnostic purposes. In research, it provides basic information about genomes, including necessary insights into chromosome numbers, shape, and structures, like telomeres, centromeres, chromosome-arms, or banding pattern. In addition, it is of interest that mitotic as well as meiotic chromosomes can be accessed. Overall, cytogenetics is the "entry method," in case genetic diagnostics is required in pre-, postnatal, or leukemia diagnostics, but also in studies on karyotypes and genomes in evolution research (Liehr, 2020).

One major issue of cytogenetics is the nomenclature of the chromosomes. The latter is not only the basis for description of inborn or acquired chromosomal rearrangements in humans, but also essential for evolutionary studies of different species. Furthermore, chromosomes are the backbone along which all molecular genetic information is aligned, such as in genomic browsers (e.g., UCSC Genome Browser, n.d.). Thus, there are international systems for uniform nomenclature guidelines published and regularly updated for humans (McGowan-Jordan et al., 2020) and animals (DiBerardino et al., 1990). To the best of our knowledge there is nothing similar available for plants overall, but there are efforts to provide some standards for heavily studied species like *Arabidopsis* (Arabidopsis-nomenclature, 2020). In Fig. 3.1, Giemsa-based banding pattern of human chromosome 1 and the corresponding idiogram with nomenclature according to ISCN (= international system for human cytogenomic nomenclature) (McGowan-Jordan et al., 2020) is depicted. The ISCN is the best-elaborated cytogenetic nomenclature system, as it is also the most applied one to describe in detail all possible changes appearing along chromosomes that are associated with human diseases. Some examples of what the nomenclature looks like are given in Table 3.1, providing an impression of nomenclature's complexity. An idea is also provided that karyotype formulas enable relatively short descriptions about complicated events going on in the human genome. However, the nomenclature does not reflect real biological band splitting (see later), or real sizes of human chromosomes. It needs to be highlighted in this context, that the indeed smallest human chromosome is that which is denominated as #21; even though it is known since

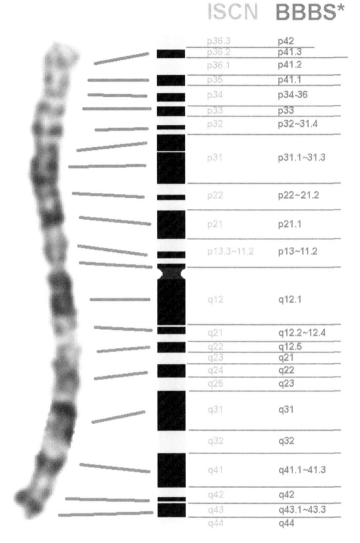

ISCN   BBBS*

ISCN	BBBS*
p36.3	p42
p36.2	p41.3
p36.1	p41.2
p35	p41.1
p34	p34-36
p33	p33
p32	p32~31.4
p31	p31.1~31.3
p22	p22~21.2
p21	p21.1
p13.3~11.2	p13~11.2
q12	q12.1
q21	q12.2~12.4
q22	q12.5
q23	q21
q24	q22
q25	q23
q31	q31
q32	q32
q41	q41.1~41.3
q42	q42
q43	q43.1~43.3
q44	q44

GTG   IDIOGRAM

CHROMOSOME 1

*biological based band splitting

FIG. 3.1

Chromosome 1 after GTG-banding together with an idiogram. Here, a GTG-banded human chromosome 1 is depicted; at its right side an idiogram is shown. The *black and white* bands are numbered according to an international consensus (the ISCN nomenclature) as shown in the *light gray* caption. The ISCN-based nomenclature implies to describe the band splitting from shorter to larger chromosomes. However, the biologically based band splitting (BBBS) is different and the according reality is shown in the *dark gray* caption.

Table 3.1 Examples of cytogenetic nomenclature.

Cytogenetic karyotype	Abbreviations/symbols used	Explanation
46,XX	46 = number of chromosomes , = used to separate from next information X = X-chromosome	Overall, 46 chromosomes were seen in cytogenetic analyses; the number 46 includes two X-chromosomes. Accordingly, this is a normal female karyotype
47,XY,+21	47 = number of chromosomes , = used to separate from next information X = X-chromosome Y = Y-chromosome + = an additional chromosome is present compared to the normal constitutional chromosome-condition 21 = chromosome 21	Overall, 47 chromosomes were seen in cytogenetic analyses; the number 47 includes one X- and one Y chromosome, and a supernumerary chromosome 21. Accordingly, this is a karyotype of a male Down-syndrome patient
46,XX,t(9;22)(q34;q11.2)	46 = number of chromosomes , = used to separate from next information X = X-chromosome t = translocation 9 = chromosome 9 22 = chromosome 22 q = long arm of a chromosome q34 = subband q34 in chromosome 9 were a break and fusion appeared q11.2 = subband q11.2 in chromosome 22 were a break and fusion appeared ; = a break had happened	Overall, 46 chromosomes were seen in cytogenetic analyses; the number 46 includes two X-chromosomes There is also a balanced translocation between chromosomes 9 and 22 with breakpoints in 9q34 and 22q11.2 Accordingly, this is a female karyotype with BCR-ABL1-rearrangement being typical for chronic myelogenous leukemia

Three examples of karyotype formulas are given and abbreviations, symbols used, and overall meaning are explained.

decades to be smaller than what we call chromosome #22, and it is suggested to order chromosomes of a species by size, this was not ever changed.

A short history of cytogenetics

As summarized elsewhere (Liehr, 2020, 2021b), cytogenetics started with visualization of chromosomes by Walther Flemming in 1879, who also introduced the words "chromatin" and "mitosis," based on his microscopic findings. In 1888, Heinrich Wilhelm Waldeyer introduced the name "chromosome" (meaning "stained body") and Walter Sutton and Theodor Boveri suggested in 1902/1903 the chromosome theory of inheritance. They postulated that Mendel's "hereditary factors" are localized on chromosome pairs. This was further worked out by Thomas Hunt Morgan (1910), who could show that genes are aligned like pearls on a necklace along the chromosome. The human cytogenetics discipline, in particular, "went through many different steps of developments - each of them providing some more possibilities for the characterization of structurally abnormal and/or supernumerary chromosomes of unknown origin, which can be found in karyotypes of cancer patients and patients with constitutional genetic disorders. The era of reliable identification of human chromosomes started with the invention of the banding method by Dr. Lore Zech (Upsala) in 1970. Further techniques, like C-banding or silver staining of the nucleolus organizing regions followed in 1971 and 1976, respectively, and completed the cytogenetic set of methods for the next decade. Nowadays, still GTG-banding (G-bands by Trypsin using Giemsa) is the gold standard against which all rising molecular (cyto)genetic techniques are measured" (Liehr & Claussen, 2002).

Material and methods in cytogenetics
Material applied in cytogenetics

Pure cytogenetics, as it is understood in this chapter, can only work with highly condensed chromosomes and not with interphase nuclei. Thus, biological material used must either comprise a substantial part of cells in metaphase stage or be stimulatable to enter metaphase in cell culture. In addition, gamete cells can be used to prepare meiotic chromosomes from them. Thus, an important feature of cytogenetics is that it relies on living and dividing cells. Accordingly, each cytogenetic working laboratory needs to have suitable cell culture facilities and personnel able to handle them. Protocols for how to cultivate germinal cells, blood-, skin fibroblast-, spleen- (in animal studies), amnion-, and chorion- or bone marrow cells can be found in the literature (e.g., Verma & Babu, 1995).

Cytogenetic methods

Initially, between the years ~1880 and ~1970, chromosomes were studied under the microscope without and/or with many different staining approaches. The most often used - and still applied in species whose chromosomes cannot be banded - were Giemsa and Orcein for so-called "solid staining." This kind of staining makes it possible to visualize clearly chromosomes, their size, their shape, and their centromeric position (Sassi et al., 2020). Based on chromosomes stained in this way, the first chromosomal syndromes were identified in the 1960s (Liehr, 2020).

As mentioned above, Dr. Lore Zech (Upsala, Sweden), found that human chromosomes show a banding pattern, if specifically treated (Schlegelberger, 2013).

Besides, this quinacrine based Q-bands, nowadays it is GTG- and R-banding (R = reverse) which are mainly used to produce a protein-mediated black and white banding along chromosomes (Claussen, 2005; Claussen et al., 2002), as shown in Fig. 3.1. In case of doubtful heterochromatic regions, C-banding (CBG) or silver staining of the nucleolus organizing regions can also be done (Liehr, 2020; Liehr & Claussen, 2002). Important improvements were also achieved for chromosome preparation techniques such as: (i) arrest of growing cells in the metaphase stage by adding colchicine to the cell culture, some hours before preparation; (ii) use of hypotonic solution to induce cell swelling and chromosome separation; and (iii) adaptation of humidity during chromosome spreading for optimal band splitting. Details on corresponding protocols are reported elsewhere (Verma & Babu, 1995; Weise & Liehr, 2017).

Advantages and restrictions

Cytogenetics has benefits and shortcomings, as do all approaches. On the plus side:

- there is a whole genomic view;
- there is a single-cell view, enabling detection of low-grade mosaics;
- it is a long-standing (experienced since the 1970s), inexpensive approach;
- aberrations like Robertsonian translocations or heterochromatic supernumerary marker chromosomes can only be picked up by cytogenetics; and
- cytogenetic data are a prerequisite to interpret correctly high-throughput or complex data on DNA sequences.

On the other hand, there are the following points to consider:

- resolution is normally not below 5–8 megabases;
- there may be cell cultural artifacts;
- cell cultivation is time consuming;
- there is a need for trained personnel (however, this is also a problem in other cytogenomic approaches, like next-generation sequencing (Ungelenk, 2021) or molecular karyotyping (Weise & Liehr, 2021); and
- interphase cells cannot be studied, thus, cytogenetics is restricted to cultivable cell types.

Overall, cytogenetics is one approach among many available ones in the field of cytogenomic diagnostics and research. Its appropriate use must be planned according to the questions to be studied. Nowadays, it is imperative to expand cytogenetic studies where possible and necessary to molecular cytogenetic ones (Liehr, 2021c; Weise & Liehr, 2021), overcoming in this way practically all mentioned disadvantages of pure cytogenetics (Claussen, 2005; Claussen et al., 2002; Kumar & Eng, 2014; Liehr, 2020; Liehr & Claussen, 2002; Verma & Babu, 1995; Weise & Liehr, 2017). An overview on presently available cytogenomic approaches and their spectrum of resolution is given in Fig. 3.2.

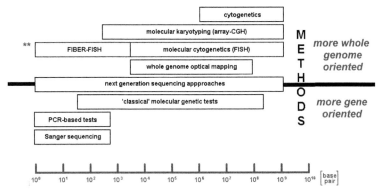

FIG. 3.2

Schematic depiction of the range of resolution of different cytogenomic approaches. Here, the power of resolution in terms of approximate base pair resolution in a logarithmic scheme for major cytogenomic approaches is visualized. Even though cytogenetics provides the lowest resolution, its single-cell and whole genomic view makes this approach irreplaceable in the concert of cytogenomic possibilities to study genomes. In addition, it is highlighted by a horizontal bar which methods are more whole genome and which are more (single) gene oriented; as FISH and FIBER-FISH can also go toward (sub)gene-level, they are marked with "**" to highlight this fact. *Array-CGH*, array-comparative genomic hybridization; also denominated as "chromosome microarray = CMA"; *FISH*, fluorescence in-situ hybridization; *PCR*, polymerase chain reaction.

Applications
Human genetic diagnostics

Diagnostics in human genetics includes three major branches, based on (i) prenatal, (ii) postnatal, and (iii) tumor cytogenetic indications (Liehr, 2020; Liehr & Claussen, 2002; Liehr et al., 2017). Hundreds of thousands of cytogenetic studies are done each year in Germany alone. Tissues like amnion, chorion, blood, fibroblast, and/or bone marrow cells are examined with indications such as sonographic abnormalities of an unborn child, dysmorphic signs in a newborn, infertility in couples, or suspicion of a leukemic disease. Numerous inborn syndromes based on trisomies of chromosomes 13, 18, and 21, numerical aberrations of gonosomes, or structural chromosomal rearrangements can be diagnosed just by banding cytogenetics. In addition, cytogenetic detectable chromosomal markers are imperative for leukemia diagnostics, which naturally appear as mosaics and can only be reliably grasped by a single-cell directed approach (Cytogenetics - Cancer-Dictionary, 2020; Liehr, 2020; Verma & Babu, 1995).

Cytogenetic-based research

Cytogenetic-based research is most often done nowadays in combination with other approaches - specifically molecular cytogenetic ones. This is valid for chromosome-oriented research in plants (e.g., Desta et al., 2019), as in animals (e.g., Sassi et al., 2020). In the latter studies, determination of correct chromosome numbers and shapes by cytogenetics is especially important for species characterization and often detection of cryptic species or subpopulations with specific chromosomal contents (e.g., De Oliveira et al., 2015). In addition, based on cytogenetics and correct chromosome numbers alone, next-generation sequencing data can be interpreted correctly (Reichwald et al., 2015).

For cytogenetic research, another method must be mentioned, too: glass-needle-based chromosome microdissection is an approach that not only enables highly specific collection of chromosomes and chromosomal subregions in picogram-levels (Al-Rikabi et al., 2019), but also is used for evolutionary studies in many kinds of chromosomes, including also lampbruchs chromosomes (Zlotina et al., 2016). This kind of experimental setting, i.e., glass-needle-based chromosome microdissection, can also be used for chromosome stretching (Claussen et al., 1994). Based on this, Uwe Claussen showed that human chromosomal subbands split in a specific manner, and that only GTG-dark bands further split into subbands and GTG-light ones never do (Kuechler et al., 2001). Building on this spadework, later on, the biologically based band splitting for all human chromosomes could be established (Kosyakova et al., 2009); in Fig. 3.1, this is shown on the example of chromosome 1. However, this insight is still not adequately recognized in the interpretation of cytogenomic data. Interestingly, recent research by Dr. J.-R. Daban revealed that the banding pattern can be understood also on an electron-microscopic level (Daban, 2020, 2021).

As the mainstream of cytogenomic research favors the application of new and "presently fancy" approaches, only little cytogenetic (or cytogenetic combined with molecular cytogenetic) research is done in human or other chromosomes. Unanswered questions relating to these approaches, and especially to chromosome stretching or also molecular combing (Bisht & Avarello, 2021), include the following:

- Do bands split differently in different tissues?
- How is band splitting in human compared to that of other great or lower apes?
- How many subbands are there?
- What is the biological function of the bands?
- How is band splitting altered in case of translocations, inversions, or other rearrangements?

Conclusion

Cytogenomics is built on insights into genomes gained elementarily by cytogenetic studies. While cytogenetics is a well-recognized field in botany and zoology, in human genetics, it has been regarded as dead for decades (Liehr, Mrasek, et al., 2017). However, due to its indubitable advantages and benefits, it is an approach which was, is, and will be necessary in cytogenomics, as in other approaches summarized in this book and in Fig. 3.2.

References

Al-Rikabi, A. B. H., Cioffi, M. D. B., & Liehr, T. (2019). Chromosome microdissection on semi-archived material. *Cytometry Part A*, *95*(12), 1285–1288. https://doi.org/10.1002/cyto.a.23896.

Arabidopsis-nomenclature. (2020). https://www.arabidopsis.org/portals/nomenclature/nomenclature.jsp.

Bisht, P., & Avarello, M. D. M. (2021). Molecular combing solutions to characterize replication kinetics and genome rearrangements. In T. Liehr (Ed.), *Cytogenomics* (pp. 47–71). Academic Press. Chapter 5 (in this book).

Claussen, U. (2005). Chromosomics. *Cytogenetic and Genome Research*, *111*(2), 101–106. https://doi.org/10.1159/000086377.

Claussen, U., Mazur, A., & Rubtsov, N. (1994). Chromosomes are highly elastic and can be stretched. *Cytogenetics and Cell Genetics*, *66*(2), 120–125. https://doi.org/10.1159/000133681.

Claussen, U., Michel, S., Mühlig, P., Westermann, M., Grummt, U., Kromeyer-Hauschild, K., & Liehr, T. (2002). Demystifying chromosome preparation and the implications for the concept of chromosome condensation during mitosis. *Cytogenetic and Genome Research*, *98*(2–3), 136–146. https://doi.org/10.1159/000069817.

Cytogenetics—BiologyOnline. (2020). https://www.biologyonline.com/dictionary/cytogenetics.

Cytogenetics—Cancer-Dictionary. (2020). https://www.cancer.gov/publications/dictionaries/cancer-terms/def/cytogenetics.

Cytogenetics—Wikipedia. (2020). https://en.wikipedia.org/wiki/Cytogenetics.

Daban, J. (2020). Supramolecular multilayer organization of chromosomes: Possible functional roles of planar chromatin in gene expression and DNA. *FEBS Letters*, *594*, 395–411.

Daban, J.-R. (2021). Multilayer organization of chromosomes. In T. Liehr (Ed.), *Cytogenomics* (pp. 267–296). Academic Press. Chapter 13 (in this book).

De Oliveira, E. A., Bertollo, L. A. C., Yano, C. F., Liehr, T., & Cioffi, M. D. B. (2015). Comparative cytogenetics in the genus Hoplias (Characiformes, Erythrinidae) highlights contrasting karyotype evolution among congeneric species. *Molecular Cytogenetics*, *8*(1), 56. https://doi.org/10.1186/s13039-015-0161-4.

Desta, Z. A., Kolano, B., Shamim, Z., Armstrong, S. J., Rewers, M., Sliwinska, E., … de Koning, D.J. (2019). Field cress genome mapping: Integrating linkage and comparative maps with cytogenetic analysis for rDNA carrying chromosomes. *Scientific Reports*, *9*(1), 17028. https://doi.org/10.1038/s41598-019-53320-0.

DiBerardino, D., Hayes, H., Fries, R., & Long, S. (Eds.). (1990). *International system for cytogenetic nomenclature of domestic animals (1989)* Karger.

Kosyakova, N., Weise, A., Mrasek, K., Claussen, U., Liehr, T., & Nelle, H. (2009). The hierarchically organized splitting of chromosomal bands for all human chromosomes. *Molecular Cytogenetics*, *2*(1), 4. https://doi.org/10.1186/1755-8166-2-4.

Kuechler, A., Mueller, C. R., Liehr, T., & Claussen, U. (2001). Detection of microdeletions in the short arm of the X chromosome by chromosome stretching. *Cytogenetics and Cell Genetics*, *95*(1–2), 12–16. https://doi.org/10.1159/000057010.

Kumar, D., & Eng, C. (2014). Oxford monographs on medical genetics. In *Genomic medicine—Principles and practice* Oxford University Press.

Liehr, T. (2020). *Human genetics—Edition 2020: A basic training package* (1st ed.). Neopubli.

Liehr, T. (2021a). A definition for cytogenomics - Which also may be called chromosomics. In T. Liehr (Ed.), *Cytogenomics* (pp. 1–7). Academic Press. Chapter 1 (in this book).

Liehr, T. (2021b). Overview of currently available approaches used in cytogenomics. In T. Liehr (Ed.), *Cytogenomics* (pp. 11–24). Academic Press. Chapter 2 (in this book).

Liehr, T. (2021c). Molecular cytogenetics. In T. Liehr (Ed.), *Cytogenomics* (pp. 35–45). Academic Press. Chapter 4 (in this book).

Liehr, T., & Claussen, U. (2002). Multicolor-FISH approaches for the characterization of human chromosomes in clinical genetics and tumor cytogenetics. *Current Genomics*, *3*(3), 213–235. https://doi.org/10.2174/1389202023350525.

Liehr, T., Mrasek, K., Klein, E., & Weise, A. (2017). Modern high throughput approaches are not meant to replace 'old fashioned' but robust techniques. *Journal of Genetics and Genomes*, *1*(1), e101.

Liehr, T., Weise, A., & Schreyer, I. (2017). Humangenetische Diagnostik – muss es immer NGS sein? *BIOspektrum*, *03*(17), 350–351.

McGowan-Jordan, J., Hastings, R., & Moore, S. (Eds.). (2020). *ISCN 2020—An international system for human cytogenomic nomenclature (2020)* Karger.

Reichwald, K., Petzold, A., Koch, P., Downie, B. R., Hartmann, N., Pietsch, S., … Platzer, M. (2015). Insights into sex chromosome evolution and aging from the genome of a short-lived fish. *Cell*, *163*(6), 1527–1538. https://doi.org/10.1016/j.cell.2015.10.071.

Sassi, F. D. M. C., Deon, G. A., Moreira-Filho, O., Vicari, M. R., Bertollo, L. A. C., Liehr, T., … Cioffi, M.B. (2020). Multiple sex chromosomes and evolutionary relationships in amazonian catfishes: The outstanding model of the genus harttia (siluriformes: Loricariidae). *Genes*, *11*(10), 1–16. https://doi.org/10.3390/genes11101179.

Schlegelberger, B. (2013). In memoriam: Prof. Dr. rer. nat. Dr. med. h.c. Lore Zech; 24.9.1923–13.3.2013: Honorary member of the European Society of Human Genetics, Honorary member of the German Society of Human Genetics, Doctor laureate, the University of Kiel, Germany. *Molecular Cytogenetics*, *20*. https://doi.org/10.1186/1755-8166-6-20.

UCSC Genome Browser;(n.d.) genome.ucsc.edu.

Ungelenk, M. (2021). Sequencing approaches. In T. Liehr (Ed.), *Cytogenomics* (pp. 87–122). Academic Press. Chapter 7 (in this book).

Verma, R., & Babu, A. (1995). *Human chromosomes: Manual of basic techniques* (2nd ed.). McGraw-Hill Inc.

Weise, A., & Liehr, T. (2017). Pre- and postnatal diagnostics and research on peripheral blood, bone marrow chorion, amniocytes, and fibroblasts. In T. Liehr (Ed.), *Fluorescence in situ hybridization (FISH)—Application guide* (2nd ed., pp. 171–180). Springer.

Weise, A., & Liehr, T. (2021). Molecular karyotyping. In T. Liehr (Ed.), *Cytogenomics* (pp. 73–85). Academic Press. Chapter 6 (in this book).

Zlotina, A., Kulikova, T., Kosyakova, N., Liehr, T., & Krasikova, A. (2016). Microdissection of lampbrush chromosomes as an approach for generation of locus-specific FISH-probes and samples for high-throughput sequencing. *BMC Genomics*, *17*(1), 126. https://doi.org/10.1186/s12864-016-2437-4.

Molecular cytogenetics

4

Thomas Liehr

Jena University Hospital, Friedrich Schiller University, Institute of Human Genetics,
Jena, Germany

Chapter outline

What is molecular cytogenetics?

Molecular cytogenetics developed from cytogenetics (Weise & Liehr, 2021a). Similar to cytogenetics, molecular cytogenetics is mainly single-cell oriented, and is able to study the whole genome, as well as single gene-loci or repetitive elements in the genomic content; in other words, the morphologically preserved chromosome can be studied (Liehr, 2020a, 2020b; Liehr & Claussen, 2002). Nowadays, molecular cytogenetics is a cytogenomic approach (Liehr, 2021a: Liehr, 2021b), which relates only to one single technique, i.e., the fluorescence in-situ hybridization (FISH) approach. However, there is a second technique, which was originally considered to be a molecular cytogenetic basic application; this is the so-called primed in-situ hybridization (PRINS) (Liehr, 2017), established in 1989 (Koch et al., 1989). The latter is an in-situ polymerase chain reaction (PCR), which leads to comparable results to those obtained by FISH. As PRINS is only suited for detection of repetitive DNA in the genome (Pellestor, 2006), and its main advantage, that it could lead to results within 2 h, was recently also trumped by FISH (Brockhoff et al., 2016), PRINS is only performed in a few specialized laboratories worldwide (Pellestor, 2006).

The in situ hybridization approach, allowing nucleic acid sequences (DNA and RNA) to be examined inside cells or on chromosomes, was first described in 1969. It was available initially exclusively as a radioactive variant (Gall & Pardue, 1969) until nonradioactive probe labeling was reported in 1981. Biotin was the first nonradioactive hapten, which could be detected by avidin, coupled to a fluorochrome (Langer et al., 1981). Even though restricted by the physics of visible light wavelengths, and the possibilities to produce filters enabling separate excitation and detection of fluorochromes, nowadays, up to eight flurochromes may be applied simultaneously in multicolor-FISH (mFISH) experiments (Kytölä et al., 2000; Liehr & Claussen, 2002). The FISH technique has been continuously developed and improved, and it is now available in a variety of applications in research and diagnostics. It is possible to detect DNA and RNA in FISH (Liehr, 2020a); however, here only DNA-based FISH is treated.

The principle of FISH is shown in Fig. 4.1. Basically, a (i) target DNA is required in order to hybridize a (ii) probe DNA to it. Target DNA can be introduced in a FISH experiment as native cells, nuclei, tissue sections, metaphase chromosomes, or pure DNA. Probe-DNA is much more diverse and is discussed separately below; more technical details can be found in Liehr (2017).

Probes applied in molecular cytogenetics and how to get them

There are two different kinds of probes available for FISH. Most of those applied nowadays in human genetics are commercially available probes, which are ready to use and have been labeled by the supplying companies with corresponding fluorophores or nonfluorescent haptens. Others, and in most cases much more interesting for cytogenomic research, are homemade probes. The latter need to be labeled before use; as described elsewhere (Liehr, 2017), this can be either PCR-based (Telenius et al., 1992) or done by Nick-translation (Rigby et al., 1977); another approach for labeling, not often used in research labs is the so-called Universal Linkage System (ULS) (Wiegant et al., 1999).

Irrespective of whether they are commercial or homemade probes, here we distinguish five basic types of probe-DNA that are suitable and used for FISH.

Locus-specific probes

Locus-specific probes (LSPs) are, in most cases, derived from molecular genetic cloning experiments. Thus, all different kinds of genetic vectors, like plasmids, cosmids, bacterial artificial chromosomes, yeast artificial chromosomes, and others can be used, as long as they contain the desired insert of species-specific DNA, which is

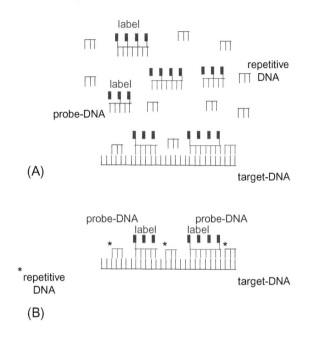

FIG. 4.1

Principle of fluorescence in-situ hybridization (FISH).

(A) After denaturation of target- and probe-DNA, both are incubated in hybridization solution at 37°C for several hours. In a normal setting, it is necessary to block repetitive DNA sequences in the target DNA by cohybridizing unlabeled repetitive DNA. Thus, background signals are avoided. (B) After hybridization is finished, target-DNA with probe-DNA and repetitive DNA form a hybrid. If the label is a ligand that is already emitting fluorescence on excitation, the result can be evaluated under the fluorescence microscope. Otherwise, the hapten (like biotin or digoxigenin) needs to be detected by corresponding fluorochrome-labeled antibodies. Signal amplification is also possible here.

targeted by FISH (Liehr, 2017). All the above mentioned vectors able to carry such inserts of at least 12 kb in size can be used (Liehr et al., 1995). Thus, if plasmid-based clones are used, it will be necessary to apply contiguous probes (Smith et al., 1997). Otherwise, the signals obtained with smaller probes will not be present in each metaphase and/or on each of both the homologous chromosomes. For mapping of probes in exceptional cases, 0.44 kb-sized cDNA probes can be applied successfully (Pröls et al., 1997). But to answer scientific or diagnostic questions needing reliable and reproducible numbers of FISH-signals per cell, normally LSPs of 20 kb and larger should be used. Accordingly, all attempts to introduce oligonucleotide or peptide nucleic acid (PNA)-based LSPs in routine FISH have not been particularly successful as yet (Cartwright et al., 2019).

Repetitive probes

In genomes, there are, in general, more repetitive (also in parts called heterochromatic) DNA sequences than single or low copy (euchromatic) DNA (Liehr, 2014). Repetitive DNA has the advantage that it can be easily visualized in FISH. Repeats detected can be limited to two to 200 base pairs in length. Thus, they are definitely below what can be detected for LSPs. However, as such repeats span tens to thousands of kilo base pairs, if tracked by corresponding FISH-probes, they result in strong and easily evaluable signals (Liehr, 2014).

Repetitive DNA is undergoing rapid evolution; thus, it is an optimal target to be studied by FISH, and widely applied in corresponding basic research in comparative cytogenomics, e.g., in plants (Jiang, 2019) and animals, like fish (Cioffi & Bertollo, 2012), mollusks (Gallardo-Escárate et al., 2005), amphibians (Schmid & Steinlein, 2015), and others (Spangenberg et al., 2020). Besides more or less randomly selected 2n to 6n repeats, telomeric repeats (6n), centromeric repeats (~170n), and repeats being present in nucleolus organizing regions (NOR) are also used as FISH-probes (e.g., Xu et al., 2019). Interestingly, centromeric repeats (also called satellite DNA) can, but must not be, chromosome-specific. In humans, almost each chromosomic centromere has its private satellite-DNA sequence; however, e.g., in mice, this is not the case, and all centromeres have one sequence in common (Liehr, 2014). Interestingly, at least one repetitive DNA, D4Z4 in localized in 4q24, has some meaning in human genetic diagnostics and can be traced by molecular combing (Bisht & Avarello, 2021; Nguyen et al., 2019).

Partial chromosome paints

To understand better chromosomal rearrangements happening during evolution, as well as in human acquired or inherited diseases, FISH can be used. LSPs, repetitive DNA-probes (see above), and painting probes can be applied (Liehr, 2020a). A probe is denominated as a painting probe as soon as it simultaneously stains at least one or two euchromatic chromosomal subbands and not only a locus or a repetitive (heterochromatic) block.

However, only glass needle-based chromosome microdissection (midi) is suited to establish small, chromosome-region-specific painting probes - so-called partial chromosome paints, or PCPs (Al-Rikabi et al., 2019). Normally, the largest PCPs are those covering one arm of a meta- or submetacentric chromosome (Kytölä et al., 2000). In addition, there are PCPs that are specific for a part of, to several subbands of, a chromosome (Liehr et al., 2002; Senger et al., 1990). Finally, very small PCPs, being similar in size to an LSP, can be produced in animals presenting extremely large chromosomes, i.e., vertebrates with lampbrush chromosomes (Zlotina et al., 2016), insects with polytene chromosomes (Scalenghe et al., 1981), and principally also plants with polytene chromosomes (Guerra, 2001); however, to the best of our knowledge the latter has not been done yet.

Finally, radiation hybrid cells can also be taken advantage of for setting up whole and partial chromosome paints; however, this cannot be done in a directed way and is nowadays rarely applied (Guyon et al., 2012).

Whole chromosome paints

DNA-probes, which are used in FISH and stain a whole chromosome, are called whole chromosome paints (WCPs). They are routinely either produced by midi (Al-Rikabi et al., 2019; Liehr, 2017), as mentioned above for PCPs, or by chromosome flow sorting (Ferguson-Smith et al., 2005; Liehr, 2017). In addition, similar to radiation hybrids, interspecies hybrids (e.g., mouse/human somatic cell hybrid) can be used as a source of species-specific WCP probes (Sabile et al., 1997).

Whole genome

Interestingly, it also makes sense to use whole genomic DNA in research and diagnostic settings as probes in FISH. Certainly, when just hybridizing whole genomic DNA of a healthy human on chromosomes of another healthy human, only a meaningless whole genome stain can be expected. However, when whole genomic tumor DNA labeled in red and whole genomic peripheral blood DNA of a healthy person labeled in green are cohybridized to normal human blood metaphases, one gets informative results; i.e., gains and losses in a tumor, compared to normal DNA, become visible. This technique is called comparative genomic hybridization (CGH) (Kallioniemi et al., 1992) and was further developed to molecular karyotyping, or array-CGH (aCGH), in human genetic diagnostics (Pinkel et al., 1998; Weise & Liehr, 2021b). Finally, CGH can be used for comparative cytogenomics in evolution research (Xu et al., 2019).

Probe-sets applied in molecular cytogenetics

The abovementioned FISH-probes can be combined in two- to multicolor-FISH experiment settings (Liehr, 2017); there are endless possibilities, summarized in Liehr (2020a). Just to mention a few for humans: there are mFISH probe sets combining all human WCP probes (Schröck et al., 1996; Speicher et al., 1996), or all centromeric probes (Nietzel et al., 2001); there are cancer-disease specific (Nagai et al., 2019) or probe sets specific for prenatal testing (Eiben et al., 1999); and there are those directed against specific regions in the genome, like subtelomeres (Brown et al., 2000) or subcentromeres (Starke et al., 2003). According to the diagnostic or scientific question, new probe sets are developed in humans, animals, or plants (Liehr, 2020a).

Different resolutions in molecular cytogenetics

FISH may be applied on each kind of tissue. Optimal for cytogenomic studies are metaphase chromosomes. However, metaphase chromosomes allow only for a resolution of several mega base pairs (Liehr et al., 2018); also, only a few cell types are suited for chromosome preparation (Weise & Liehr, 2021a). Thus, many FISH-studies are restricted to interphase cells (e.g., studies in the brain; see Iourov et al., 2021). Nonetheless, using interphase nuclei has a major advantage. On the one hand it was shown that in interphase, chromosomes have approximately the same lengths as in metaphase. However, in interphase chromosomes decondense square to chromosome axis (Lemke et al., 2002). Thus, interphase allows for a higher resolution of two closely located FISH-probes than metaphase does (Weise et al., 2002). Accordingly, the highest resolution may be achieved when using DNA fibers in molecular combing (Bisht & Avarello, 2021; Nguyen et al., 2019).

Cytogenomic applications of molecular cytogenetics

The FISH-approach was and is still used for many cytogenomic research topics. The following is a list of what can be done using molecular cytogenetics, but without any claim of completeness:

- location and order of genes on chromosomes, i.e., gene mapping can be done by metaphase (Pröls et al., 1997) or fiber-FISH (Nguyen et al., 2019); microdissection and sequencing can help to elucidate otherwise not accessible genomic regions by sequencing (Zlotina et al., 2016);
- in human genetics, molecular cytogenetics is a major tool in genetic diagnostics of inherited (Liehr et al., 2013; Liehr & Claussen, 2002) and acquired diseases (Liehr et al., 2015; Liehr & Claussen, 2002);
- insights into interphase architecture can be gained on flattened and 3D-preserved interphase nuclei (Cremer et al., 2020; Lemke et al., 2002; Liehr, 2021c); data can be compiled with such obtained by other approaches (like electron microscopy) and lead to new insights of chromosome structure (Daban, 2020, 2021); here especially studies based on multicolor banding and the hierarchical nature of chromosome band splitting were important (Kosyakova et al., 2009);
- small and large repetitive probes, as well as CGH studies, can provide insights into speciation and can help to distinguish closely related species and identify sex-determination systems (Cioffi & Bertollo, 2012; Spangenberg et al., 2020; Xu et al., 2019);
- genomic damage after chemical toxification or radiation may be studied using FISH, applied on chromosomes, micronuclei, or in comet assay (Hovhannisyan, 2010); chromosome (in)stability may be determined in meta- and interphase of normal and cancerous tissues (Yurov et al., 2009), including phenomena like chromothripsis (Koltsova et al., 2019; Pellestor et al., 2021), chromosome pulverization (Nazaryan-Petersen et al., 2020), and gene amplification (Shimizu, 2009);
- uniparental disomy can be visualized in trios, taking advantage of copy number variants (Weise et al., 2015);

- FISH is applied in aging research, as it can uncover, e.g., cryptic mosaics (Mkrtchyan et al., 2010) or changes in telomere structure (Hagman et al., 2020) associated with getting older; and
- CRISPR-mediated live imaging allows insights into living cells (Anton et al., 2014), while other molecular cytogenetic approaches can visualize only dead cells (Liehr, 2017).

Conclusion

Overall, what Serakinci and Koelvraa stated in 2009 is still valid:

"FISH techniques were originally developed as extra tools in attempts to map genes and a number of advances were achieved with this new technique. However, it soon became apparent that the FISH concept offered promising possibilities also in a number of other areas in biology and its use spread into new areas of research and also into the area of clinical diagnosis. In very general terms the virtues of FISH are in two areas of biology, namely genome characterization and cellular organization, function and diversity... To what extent FISH technology will be further developed and applied in new areas of research in the future remains to be seen"

(Serakinci & Koelvraa, 2009)

References

Al-Rikabi, A. B. H., Cioffi, M. D. B., & Liehr, T. (2019). Chromosome microdissection on semi-archived material. *Cytometry Part A*, *95*(12), 1285–1288. https://doi.org/10.1002/cyto.a.23896.

Anton, T., Bultmann, S., Leonhardt, H., & Markaki, Y. (2014). Visualization of specific DNA sequences in living mouse embryonic stem cells with a programmable fluorescent CRISPR/Cas system. *Nucleus*, *5*(2), 163–172. https://doi.org/10.4161/nucl.28488.

Bisht, P., & Avarello, M. D. M. (2021). Molecular combing solutions to characterize replication kinetics and genome rearrangements. In T. Liehr (Ed.), *Cytogenomics* (pp. 47–71). Academic Press. Chapter 5 (in this book).

Brockhoff, G., Bock, M., Zeman, F., & Hauke, S. (2016). The FlexISH assay brings flexibility to cytogenetic HER2 testing. *Histopathology*, *69*(4), 635–646. https://doi.org/10.1111/his.12974.

Brown, J., Horsley, S., Jung, C., Saracoglu, K., Janssen, B., Brough, M., … Kearney, L. (2000). Identification of a subtle t(16;19)(p13.3;p13.3) in an infant with multiple congenital abnormalities using a 12-colour multiplex FISH telomere assay, M-TEL. *European Journal of Human Genetics*, *8*(12), 903–910. https://doi.org/10.1038/sj.ejhg.5200545.

Cartwright, I. M., Haskins, J. S., & Kato, T. A. (2019). PNA telomere and centromere FISH staining for accurate analysis of radiation-induced chromosomal aberrations. *Methods in Molecular Biology*, *1984*, 95–100. https://doi.org/10.1007/978-1-4939-9432-8_11.

Cioffi, M., & Bertollo, L. (2012). Chromosomal distribution and evolution of repetitive DNAs in fish. In M. Garrido-Ramo (Ed.), *Vol. 7. Genome dynamics* (1st ed.). Karger.

Cremer, T., Cremer, M., Hübner, B., Silahtaroglu, A., Hendzel, M., Lanctôt, C., … Cremer, C. (2020). The interchromatin compartment participates in the structural and functional organization of the cell nucleus. *BioEssays*, *42*, e1900132.

Daban, J. (2020). Supramolecular multilayer organization of chromosomes: Possible functional roles of planar chromatin in gene expression and DNA. *FEBS Letters*, *594*, 395–411.

Daban, J.-R. (2021). Multilayer organization of chromosomes. In T. Liehr (Ed.), *Cytogenomics* (pp. 267–296). Academic Press. Chapter 13 (in this book).

Eiben, B., Trawicki, W., Hammans, W., Goebel, R., Pruggmayer, M., & Epplen, J. (1999). Rapid prenatal diagnosis of aneuploidies in uncultured amniocytes by fluorescence in situ hybridization. Evaluation of >3,000 cases. *Fetal Diagnosis and Therapy*, *14*, 193–197.

Ferguson-Smith, M., Yang, F., Rens, W., & O'Brien, P. (2005). The impact of chromosome sorting and painting on the comparative analysis of primate genomes. *Cytogenetic and Genome Research*, *108*, 112–121.

Gall, J. G., & Pardue, M. L. (1969). Formation and detection of RNA-DNA hybrid molecules in cytological preparations. *Proceedings of the National Academy of Sciences of the United States of America*, *63*(2), 378–383. https://doi.org/10.1073/pnas.63.2.378.

Gallardo-Escárate, C., Álvarez-Borrego, J., Del Río-Portilla, M. A., Cross, I., Merlo, A., & Rebordinos, L. (2005). Fluorescence in situ hybridization of rDNA, telomeric (TTAGGG)n and (GATA)n repeats in the red abalone *Haliotis rufescens* (Archaeogastropoda: Haliotidae). *Hereditas*, *142*(2005), 73–79. https://doi.org/10.1111/j.1601-5223.2005.01909.x.

Guerra, M. (2001). Fluorescent in situ hybridization in plant polytene chromosomes. *Methods in Cell Science*, *23*, 133–138.

Guyon, R., Rakotomanga, M., Azzouzi, N., Coutanceau, J., Bonillo, C., D'Cotta, H., … Galibert, F. (2012). A high-resolution map of the Nile tilapia genome: A resource for studying cichlids and other percomorphs. *BMC Genomics*, *13*, 222.

Hagman, M., Werner, C., Kamp, K., Fristrup, B., Hornstrup, T., Meyer, T., … Krustrup, P. (2020). Reduced telomere shortening in lifelong trained male football players compared to age-matched inactive controls. *Progress in Cardiovascular Diseases*, *63*(6), 738–749.

Hovhannisyan, G. G. (2010). Fluorescence in situ hybridization in combination with the comet assay and micronucleus test in genetic toxicology. *Molecular Cytogenetics*, *3*(1). https://doi.org/10.1186/1755-8166-3-17.

Iourov, I. Y., Vorsanova, S. G., & Yurov, Y. B. (2021). Cytogenomic landscape of the human brain. In T. Liehr (Ed.), *Cytogenomics* (pp. 327–348). Academic Press. Chapter 16 (in this book).

Jiang, J. (2019). Fluorescence in situ hybridization in plants: Recent developments and future applications. *Chromosome Research*, *27*, 153–165.

Kallioniemi, A., Kallioniemi, O. P., Sudar, D., Rutovitz, D., Gray, J. W., Waldman, F., & Pinkel, D. (1992). Comparative genomic hybridization for molecular cytogenetic analysis of solid tumors. *Science*, *258*(5083), 818–821. https://doi.org/10.1126/science.1359641.

Koch, J. E., Kølvraa, S., Petersen, K. B., Gregersen, N., & Bolund, L. (1989). Oligonucleotide-priming methods for the chromosome-specific labelling of alpha satellite DNA in situ. *Chromosoma*, *98*(4), 259–265. https://doi.org/10.1007/BF00327311.

Koltsova, A., Pendina, A., Efimova, O., Chiryaeva, O., Kuznetzova, T., & Baranov, V. S. (2019). On the complexity of mechanisms and consequences of chromothripsis: An update. *Frontiers in Genetics*, *10*, 393.

Kosyakova, N., Weise, A., Mrasek, K., Claussen, U., Liehr, T., & Nelle, H. (2009). The hierarchically organized splitting of chromosomal bands for all human chromosomes. *Molecular Cytogenetics*, *2*, 4.

Kytölä, S., Rummukainen, J., Nordgren, A., Karhu, R., Farnebo, F., Isola, J., & Larsson, C. (2000). Chromosomal alterations in 15 breast cancer cell lines by comparative genomic hybridization and spectral karyotyping. *Genes, Chromosomes & Cancer, 28,* 308–317.

Langer, P., Waldrop, A., & Ward, D. (1981). Enzymatic synthesis of biotin-labeled olynucleotides: Novel nucleic acid affinity probes. *Proceedings of the National Academy of Sciences of the United States of America, 78,* 6633–6637.

Lemke, J., Claussen, J., Michel, S., Chudoba, I., Mühlig, P., Westermann, M., … Claussen, U. (2002). The DNA-based structure of human chromosome 5 in interphase. *American Journal of Human Genetics, 71*(5), 1051–1059. https://doi.org/10.1086/344286.

Liehr, T. (2014). *Benign & pathological chromosomal imbalances—Microscopic and submicroscopic copy number variations (CNVs) in genetics and counseling* (1st ed.). Academic Press.

Liehr, T. (Ed.). (2017). *Fluorescence in situ hybridization (FISH)—Application guide* (2nd ed.). Springer.

Liehr, T. (2020a). *Basics and literature on multicolor fluorescence in situ hybridization application.* http://cs-tl.de/DB/TC/mFISH/0-Start.html.

Liehr, T. (2020b). *Human genetics—Edition 2020: A basic training package* (1st ed.). Epubli.

Liehr, T. (2021a). A definition for cytogenomics - Which also may be called chromosomics. In T. Liehr (Ed.), *Cytogenomics* (pp. 1–7). Academic Press. Chapter 1 (in this book).

Liehr, T. (2021b). Overview of currently available approaches used in cytogenomics. In T. Liehr (Ed.), *Cytogenomics* (pp. 11–24). Academic Press. Chapter 2 (in this book).

Liehr, T. (2021c). Nuclear architecture. In T. Liehr (Ed.), *Cytogenomics* (pp. 297–305). Academic Press. Chapter 14 (in this book).

Liehr, T., & Claussen, U. (2002). Multicolor-FISH approaches for the characterization of human chromosomes in clinical genetics and tumor cytogenetics. *Current Genomics, 3*(3), 213–235. https://doi.org/10.2174/1389202023350525.

Liehr, T., Heller, A., Starke, H., Rubtsov, N., Trifonov, V., Mrasek, K., … Claussen, U. (2002). Microdissection based high resolution multicolor banding for all 24 human chromosomes. *International Journal of Molecular Medicine, 9*(4), 335–339. https://doi.org/10.3892/ijmm.9.4.335.

Liehr, T., Othman, M., Rittscher, K., & Alhourani, E. (2015). The current state of molecular cytogenetics in cancer diagnosis. *Expert Review of Molecular Diagnostics, 15,* 517–526.

Liehr, T., Schreyer, I., Kuechler, A., Manolakos, E., Singer, S., Dufke, A., … Weise, A. (2018). Parental origin of deletions and duplications—About the necessity to check for cryptic inversions. *Molecular Cytogenetics, 11,* 20.

Liehr, T., Thoma, K., Kammler, K., Gehring, C., Ekici, A., Bathke, K., … Rautenstrauss, B. (1995). Direct preparation of uncultured EDTA-treated or heparinized blood for interphase FISH analysis. *Applied Cytogenetics, 21,* 185–188.

Liehr, T., Weise, A., Hamid, A. B., Fan, X., Klein, E., Aust, N., … Kosyakova, N. (2013). Multicolor FISH methods in current clinical diagnostics. *Expert Review of Molecular Diagnostics, 13*(3), 251–255. https://doi.org/10.1586/erm.12.146.

Mkrtchyan, H., Gross, M., Hinreiner, S., Polytiko, A., Manvelyan, M., Mrasek, K., … Weise, A. (2010). The human genome puzzle—The role of copy number variation in somatic mosaicism. *Current Genomics, 11,* 426–431.

Nagai, T., Okamura, T., Yanase, T., Chaya, R., Moritoki, Y., Kobayashi, D., … Yasui, T. (2019). Examination of diagnostic accuracy of UroVysion fluorescence in situ hybridization for bladder cancer in a single community of Japanese hospital patients. *Asian Pacific Journal of Cancer Prevention, 20,* 1271–1273.

Nazaryan-Petersen, L., Bjerregaard, V., Nielsen, F., Tommerup, N., & Tümer, Z. (2020). Chromothripsis and DNA repair disorders. *Journal of Clinical Medicine*, *9*, 613.

Nguyen, K., Broucqsault, N., Chaix, C., Roche, S., Robin, J., Vovan, C., ... Magdinier, F. (2019). Deciphering the complexity of the 4q and 10q subtelomeres by molecular combing in healthy individuals and patients with facioscapulohumeral dystrophy. *Journal of Medical Genetics*, *56*(9), 590–601. https://doi.org/10.1136/jmedgenet-2018-105949.

Nietzel, A., Rocchi, M., Starke, H., Heller, A., Fiedler, W., Wlodarska, I., ... Liehr, T. (2001). A new multicolor-FISH approach for the characterization of marker chromosomes: Centromere-specific multicolor-FISH (cenM-FISH). *Human Genetics*, *108*(3), 199–204. https://doi.org/10.1007/s004390100459.

Pellestor, F. (Ed.). (2006). *PRINS and in situ PCR protocols* Humana Press.

Pellestor, F., Gaillard, J.-B., Schneider, A., Puechberty, J., & Gatinois, V. (2021). Chromoanagenesis phenomena and their formation mechanisms. In T. Liehr (Ed.), *Cytogenomics* (pp. 213–245). Academic Press. Chapter 11 (in this book).

Pinkel, D., Segraves, R., Sudar, D., Clark, S., Poole, I., Kowbel, D., ... Albertson, D.G. (1998). High resolution analysis of DNA copy number variation using comparative genomic hybridization to microarrays. *Nature Genetics*, *20*(2), 207–211. https://doi.org/10.1038/2524.

Pröls, F., Liehr, T., Rinke, R., & Rautenstrauss, B. (1997). Assignment of the microvascular endothelial differentiation gene 1 (MDG1) to human chromosome band 14q24.2→q24.3 by fluorescence in situ hybridization. *Cytogenetic and Genome Research*, *79*(1–2), 149–150. https://doi.org/10.1159/000134706.

Rigby, P. W. J., Dieckmann, M., Rhodes, C., & Berg, P. (1977). Labeling deoxyribonucleic acid to high specific activity in vitro by nick translation with DNA polymerase I. *Journal of Molecular Biology*, *113*(1), 237–251. https://doi.org/10.1016/0022-2836(77)90052-3.

Sabile, A., Poras, I., Cherif, D., Goodfellow, P., & Avner, P. (1997). Isolation of monochromosomal hybrids for mouse chromosomes 3, 6, 10, 12, 14, and 18. *Mammalian Genome*, *8*, 81–85.

Scalenghe, F., Turco, E., Edström, J. E., Pirrotta, V., & Melli, M. (1981). Microdissection and cloning of DNA from a specific region of *Drosophila melanogaster* polytene chromosomes. *Chromosoma*, *82*(2), 205–216. https://doi.org/10.1007/BF00286105.

Schmid, M., & Steinlein, C. (2015). Chromosome banding in Amphibia. XXXII. The genus Xenopus (Anura, Pipidae). *Cytogenetic and Genome Research*, *145*, 201–217.

Schröck, E., du Manoir, S., Veldman, T., Schoell, B., Wienberg, J., Ferguson-Smith, M., ... Ried, T. (1996). Multicolor spectral karyotyping of human chromosomes. *Science*, *273*, 494–497.

Senger, G., Lüdecke, H. J., Horsthemke, B., & Claussen, U. (1990). Microdissection of banded human chromosomes. *Human Genetics*, *84*(6), 507–511. https://doi.org/10.1007/BF00210799.

Serakinci, N., & Koelvraa, S. (2009). Molecular cytogenetic applications in diagnostics and research—An overview. In T. Liehr (Ed.), *Fluorescence in situ hybridization (FISH)—Application guide* (1st ed.). Springer.

Shimizu, N. (2009). Extrachromosomal double minutes and chromosomal homogeneously staining regions as probes for chromosome research. *Cytogenetic and Genome Research*, *124*(3–4), 312–326. https://doi.org/10.1159/000218135.

Smith, C., Ma, N., Nowak, N., Shows, T., & Gerhard, D. (1997). A 3-Mb contig from D11S987 to MLK3, a gene-rich region in 11q13. *Genome Research*, *7*, 835–842.

Spangenberg, V., Arakelyan, M., Cioffi, M., Liehr, T., Al-Rikabi, A., Martynova, E., ... Kolomiets, O. (2020). Cytogenetic mechanisms of unisexuality in rock lizards. *Scientific Reports*, *10*, 8697.

Speicher, M., Gwyn Ballard, S., & Ward, D. (1996). Karyotyping human chromosomes by combinatorial multi-fluor FISH. *Nature Genetics, 12*, 368–375.

Starke, H., Nietzel, A., Weise, A., Heller, A., Mrasek, K., Belitz, B., … Liehr, T. (2003). Small supernumerary marker chromosomes (SMCs): Genotype-phenotype correlation and classification. *Human Genetics, 114*(1), 51–67. https://doi.org/10.1007/s00439-003-1016-3.

Telenius, H., Carter, N. P., Bebb, C. E., Nordenskjöld, M., Ponder, B. A. J., & Tunnacliffe, A. (1992). Degenerate oligonucleotide-primed PCR: General amplification of target DNA by a single degenerate primer. *Genomics, 13*(3), 718–725. https://doi.org/10.1016/0888-7543(92)90147-K.

Weise, A., & Liehr, T. (2021a). Cytogenetics. In T. Liehr (Ed.), *Cytogenomics* (pp. 25–34). Academic Press. Chapter 3 (in this book).

Weise, A., & Liehr, T. (2021b). Molecular karyotyping. In T. Liehr (Ed.), *Cytogenomics* (pp. 73–85). Academic Press. Chapter 6 (in this book).

Weise, A., Othman, M. A. K., Bhatt, S., Löhmer, S., & Liehr, T. (2015). Application of BAC-probes to visualize copy number variants (CNVs). *Methods in Molecular Biology, 1227*, 299–307. https://doi.org/10.1007/978-1-4939-1652-8_16.

Weise, A., Starke, H., Heller, A., Uwe, C., & Liehr, T. (2002). Evidence for interphase DNA decondensation transverse to the chromosome axis: A multicolor banding analysis. *International Journal of Molecular Medicine, 9*(4), 359–361. https://doi.org/10.3892/ijmm.9.4.359.

Wiegant, J. C. A. G., Van Gijlswijk, R. P. M., Heetebrij, R. J., Bezrookove, V., Raap, A. K., & Tanke, H. J. (1999). ULS: A versatile method of labeling nucleic acids for FISH based on a monofunctional reaction of cisplatin derivatives with guanine moieties. *Cytogenetics and Cell Genetics, 87*(1–2), 47–52. https://doi.org/10.1159/000015390.

Xu, D., Sember, A., Zhu, Q., De Oliveira, E. A., Liehr, T., Al-Rikabi, A. B. H., … De Bello Cioffi, M. (2019). Deciphering the origin and evolution of the X1X2Y system in two closely-related Oplegnathus species (Oplegnathidae and Centrarchiformes). *International Journal of Molecular Sciences, 20*(14), 3571. https://doi.org/10.3390/ijms20143571.

Yurov, Y. B., Vorsanova, S. G., & Iourov, I. Y. (2009). GIN'n'CIN hypothesis of brain aging: Deciphering the role of somatic genetic instabilities and neural aneuploidy during ontogeny. *Molecular Cytogenetics, 2*(1), 23. https://doi.org/10.1186/1755-8166-2-23.

Zlotina, A., Kulikova, T., Kosyakova, N., Liehr, T., & Krasikova, A. (2016). Microdissection of lampbrush chromosomes as an approach for generation of locus-specific FISH-probes and samples for high-throughput sequencing. *BMC Genomics, 17*(1), 126. https://doi.org/10.1186/s12864-016-2437-4.

Molecular combing solutions to characterize replication kinetics and genome rearrangements

Prakhar Bisht* and Mario Davide Maria Avarello*

Genomic Vision, Bagneux, France

Chapter outline

Introduction

The previous century has been focused on discovering and deciphering the human genome. It is accurately mapped, and the organization of genomic functional sequences has also been discovered (Lander et al., 2001; Venter et al., 2001). Nonetheless, there are still unsolved questions regarding the genetic organization of the eukaryotic genome, on how it is regulated at the level of gene expression or even DNA replication; such studies are nowadays summarized as cytogenomic research (Liehr, 2021a, 2021b).

Since the beginning of the genome mapping era, scientists have developed more sophisticated and precise ways to uncover the unresolved secrets hidden in our genome. The initial method to study human genomes was cytogenetics (Weise & Liehr, 2021a), having a low resolution but has proven to be suitable as a basic way to validate genomic rearrangements caused by DNA damage or mutation (Gunderson et al., 2006; Hoogendoorn et al., 2000). Many other molecular genetic oriented cytogenetic

---

Cytogenomics. https://doi.org/10.1016/B978-0-12-823579-9.00005-9

approaches have been developed to study human and other genomes (Pinkel & Albertson, 2005; Spinner et al., 2013) as summarized in other chapters of this book (Daban, 2021; Delpu et al., 2021; Iourov et al., 2021; Ishii et al., 2021; Liehr, 2021c, 2021d; Weise & Liehr, 2021b). It is important here that such molecular approaches have resolutions between several mega base pair (Mbp) down to 1 base pair (bp) (e.g., in sequencing approaches) (Ungelenk, 2021). There are different levels of organization of the genome from a single base to genes and chromatin organization for their regulation. The investigation of a higher level of DNA organization reflects the understandings of the genome rearrangements, their involvement in disease, and the ability to predict the onset. Another method to visualize genomic regions at a higher resolution than cytogenetics is fluorescence in situ hybridization (FISH) (Liehr, 2021c; Reisinger et al., 2006). It allows a resolution of ~ 10 Mbp in metaphases and of ~ 1–2 Mbp in interphase (Fonseca et al., 2001; Jenkins et al., 2006; Liehr, 2021e). Overall, the genome can be accessed by the aforementioned approaches at all levels of resolution. During the years, numerous tentative ideas to fill the gap were not really successful due to a lack of repeatability and the robustness of the statistical analyses. A robust solution for this problem has been developed in 1994 with the setting up of a technique called molecular combing (Allemand et al., 1997; Bensimon et al., 1994).

Molecular combing, as the name suggests, refers to the physical combing of the DNA on a glass surface. DNA extraction is performed by a unique strategy where the raw material (cells) is embedded in agarose plugs (Fig. 5.1A and B). After several enzymatic treatments and washing steps, the removal of "dirty materials of the cells" such as proteins and lipids is carried out. Following the plugs are dissolved and very long DNA fibers, a necessary prerequisite for combing is obtained in the solution (Allemand et al., 1997). Long fibers are now taken up by a special chemically coated coverslip (silanized coverslips), to allow stretching of DNA fibers only from one end (Fig. 5.1C). Coverslip is then pulled out with a specific speed of 300 µm/s to have a constant stretching factor (1 mm = 2 kb) to establish reproducibility (Técher et al., 2013). As a meniscus is generated between air and liquid phases, the force generated on the DNA at the meniscus effectively stretches the DNA during the pulling out of the coverslip (Herrick & Bensimon, 1999; Kaykov et al., 2016).

Form the DNA solution, hundreds of genomes can be combed on a singular coverslip, thus the sampling representation per coverslips can be applied to robust statistic evaluation after performing, e.g., a FISH procedure on this material. Thereby, one can use FISH probes to visualize specific DNA sequences, and by a specific protocol, the Genomic Morse Code (GMC) can be applied to identify chromosomal rearrangements (Herrick et al., 2000a; Yadav & Sharma, 2019).

In fact, molecular combing has opened up an era of DNA investigation that leads to the deep understanding of replication kinetics in model systems, at genome-wide or locus-specific level (Fig. 5.1D and E). In addition, by molecular combing, it has been possible to see complex DNA rearrangement that is currently lost with other technologies: such as small deletions, amplifications, transpositions, and inversions

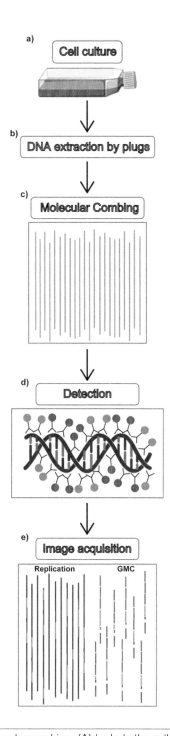

FIG. 5.1

Schematic workflow of molecular combing. (A) Include the cell culture and/or the sample preparation. (B) The DNA extractions are performed by the use of agarose plugs to prevent DNA breaks and degradation. (C) DNA is combed on a silanized coverslip, and the quality of the dsDNA is checked by using YOYO1 counterstaining. (D) The detection of targeted regions is carried out on unstained dsDNA by hybridization using probes in the case of GMC, or immunostaining in the case of RCA. (E) The coverslip is scanned and imaged. The DNA fibers are stitched together and the signals for RCA or GMC are visualized, and a report is generated in the end by the use of the assay-specific scan configuration software.

(Fig. 5.1D and E). Therefore, molecular combing represents a method to bring a high comprehension for studying many diseases such as cancer and other genetic and aging-related diseases.

Basic studies on DNA replication

Originally, investigations to understand DNA replication were developed in bacteria using autoradiography. These experiments were carried out by pulsing cells with two ^{3}H-thymidine analogs (low and high specificity) as isotope for the radiography and lysing cells to collect the DNA (Cairns, 1963). The first autoradiogram experiment was conducted on HeLa cells (Cairns, 1966). The DNA molecules appeared as venation tracks of silver grains and DNA-fork movements were deduced according to grain density (Huberman & Riggs, 1968; Huberman & Tsai, 1973). Thus, quantitative data on DNA replication kinetics of HeLa cells, such as the direction and speed of the forks and the distances between the origins of replication, could be obtained. Before these experiments, it was a matter of debate if replication fork was moving uni- or bidirectionally; until bidirectional movements of the forks along the DNA replication were proven and accepted (Hand & Tamm, 1973). Also, it was shown that a fork moves along DNA with a speed of 1.5–3.6 kb/min and a maximum rate of 7.5–15 kb/min. It is estimated further that a fork changes its movement rate throughout the different regions it is passing over or during the cycle (i.e., fork is threefolds faster at the end of replication). While there is no variation through the different S-phases (early, mid, or late replication), average size covered is about 90 kb/min (Housman & Huberman, 1975). It was also found that replicon sizes are different between species as well between different cells of a body. Finally, replications occur in specific spots within the nucleus, called foci, and each focus contains 10–100 replicons; replication is not simultaneous among the foci, rather there is a 20% of asynchrony between foci (Hand, 1975, 1978).

The experiment using ^{3}H-thymidine analogs revealed also a relation between DNA replication defects and genome instability, based on studies in Fanconi anemia and ataxia telangiectasia cells (Hand, 1977). In a key experiment, in cells derived from patients suffering from Werner and Bloom syndrome (common defect is a mutation in RecQ-related helicase), it could be shown that such cells cannot recover after replication stall; instead, the homologous recombination pathway was activated (Hanaoka et al., 1985; Hand & German, 1975). Activation of the latter was the underlying reason for damage caused during the fork stall or the excessive slowdown in the cells of affected patients.

Despite all the incredible pieces of evidence being achieved by autoradiography-based experiments, there were imprecision in measurements and doubt about the bias that they could comprise. Principal limitations were due to instability of DNA during extraction with the consequence to have in the end, less analizable signals. In this case, data cannot be reproducible as the success of the experiment is totally dependent on environmental conditions and/or chance. Furthermore, since the labeling method was on the use of ³H-thymidine only, it was difficult to give proper directions to the forks introducing a bias in the analyses. In addition, the ³H-thymidine analog is toxic for the cells, and its use in high concentrations made it difficult to prevent an influence/a bias on the replication process itself. Furthermore, unequal labeling of certain areas could bring to an underestimation of the origins of replication. Finally, it was impossible to identify termination of replication due to the fact that the label of the merging forks was visible as a longer fork. However, the ³H-thymidine analog protocol has provided a basic knowledge of the DNA replication process.

Replication kinetic studies by molecular combing

At the beginning of the 1990s, a method has been setup to do DNA replication studies based on DNA-fiber analysis - the molecular combing approach. The latter is more robust and free of radioactivity than the ³H-thymidine analog protocol. In molecular combing, pulses are provided to the cells by applying one or a combination of two different thymidine analogs such as CldU (5-chloro-2′-deoxyuridine), IdU (5-iodo-2′-deoxyuridine), BrdU (5-bromo-2′-deoxyuridine), and/or EdU (5-ethynyl-2′-deoxyuridine). After cells chasing with dNTPs, they are embedded in plugs to carry out DNA extraction (Herrick & Bensimon, 1999). The use of plugs allows to enrich and stabilize the DNA during protein and lipid digestion, in fact, releasing gently, very long double-stranded DNA (dsDNA) fibers into solution, once the plug is dissolved. The actual combing of the dsDNA is achieved by using special silanized coverslips to ameliorate DNA binding to the chemically coated glass surface. Then, there is a step of dipping the coverslip into DNA solution for a few minutes. Finally, pulling out of the coverslip creates a meniscus between the air and liquid phases, as mentioned earlier.

Molecular combing is a nontoxic system that can be used safely, as only nonradioactive material is used. Also, the nonradioactive nucleotides reduce the cytotoxic effects of pulsing steps and thus minimize a possible bias on replication analyses. Altogether, these advantages have brought molecular combing to be considered absolutely valuable method to study replication kinetics in more complex organisms such as eukaryotes.

Molecular combing was used to study the activation of origins and the forks of replication. Experiments carried out on *Saccharomyces cerevisiae* and *Saccharomyces pombe* uncovered some mechanisms of staggered origin activation on time, either within the same chromosome or between different chromosomes of the same organism (Czajkowsky et al., 2008; Patel et al., 2006; Rhind, 2006). A similar approach was done on *Xenopus laevis* to study stochasticity of activation in the origins of replication. This time, double labeling pulses were applied to visualize the newly incorporated nucleotides. They could then distinguish replicated DNA between and after the pulses to follow the replication itself. This approach unrevealed the dynamics of DNA molecule synthesis and duplication at the genome-wide level. Besides, it also showed the presence of spacing of 5–15 kb between the origins in *X. laevis*, and the fact that the frequency of origin activation increases is higher in late S-phase (Herrick et al., 2000b; Marheineke & Hyrien, 2001).

The combed DNA helped to explain increased forks rates and numbers at the end of the S-phase. In fact, two proteins were considered to have a crucial role in it: the ribonucleotide reductase (RNR) and CHK1 (Maya-Mendoza et al., 2007; Nordlund & Reichard, 2006). RNR is the protein that catalyzes the last step and promotes dNTPs synthesis, while CHK1 is a checkpoint surveillance protein that regulates the arrest of the DNA replication in case of issues and the holding of the activation of origins of replication. It was found that RNR activity is higher from the mid-S-phase on, thus it allows to have more dNTPs available for the fork at mid and late of S-phase (Barkley et al., 2007). Additionally, the action of CHK1 in promoting RNR activation upon DNA stress increased replication rate and kept off delay in replication timing due to fork stall. However, it was found in S. pombe that checkpoint regulation is not involved in the regulation of the origins of replication (Mickle et al., 2007). Thus, the interaction between RNR and CHK1 has not still understood completely. Moreover, DNA synthesis can be also affected independently of the rest of the genome in the mid- or late-S-phase by affecting a tumor suppressor PERP1 (PKNOX1). Indeed, molecular combing experiments revealed that PERP1 downregulation is associated to increase simultaneous origin firing, to affect DNA replication rate and fork asymmetry, and ultimately to lead to genomic instability (Palmigiano et al., 2018).

Molecular combing has made accessible the origins of replication for complex organisms to scientific studies. The number of origins of replication per chromosome in higher eukaryotes varies through the S-phase, and many factors, such as DNA replication stress, may lead to the activation of extra origins so-called dormant origins of replication. Then, having a formula that defines the threshold of activated origins per chromosome can represent a new mathematical way to help in the mapping of origins of replication for genomes, specific for every organism. The parameter out derivate from this formula has been called Minimum number of replication Origins (MO). The MO takes into account chromosome size, S-phase duration, and replication rate: the first MO was obtained from NCBI databanks, while the other two were acquired by using molecular combing (da Silva, 2020). It was also studied what is the efficiency of the origins of replication. This is due to the fact that normally among all the origins, only a few are active per each cell cycle. Specifically, it was found that in S. pombe, Hsk1-Dfp1 replication kinase plays a role in regulating the efficiency of origin activation. Researchers found that Hsk1-Dfp1 is dissolved in the nucleus, but they co-localize once the origins are accessible. Then, the efficiency could be controlled and so do the genome stability (Patel et al., 2008).

Application of molecular combing for investigations of origins of replication led to uncovering the mechanisms driving the DNA re-replication. In fact, this occurrence is due to the activation of an origin, and subsequently, another origin is fired within the previous one. Re-replication drives to genome instability, and it occurs in cancer cells with high frequencies. The DNA combing analyses have unveiled modifications on the origin initiation pathway which leads to two different effects: flexible overactivation of re-replicating origins, while replication stress in normal mitotic growth activates dormant origins (Fu et al., 2020).

Furthermore, molecular combing analyses have contributed to uncovering mechanistic processes and variations in the molecular pathways related to DNA replication and repair. Particularly, the attention has been directed to the genomic instability that is caused by the alteration of physiological replication. In fact, any DNA damage causes replication stress, and also more replication stress increases the presence of damages into the DNA, leading to the origination of cancerous cells (Gaillard et al., 2015). Many studies featuring the DNA damage have been focused on the different components of the molecular machinery for the DNA replication itself or on the many proteins that are involved with the regulation of the replication or maintenance of the DNA.

Along with the many proteins of the replication machinery, the DNA replication is the clamp proliferating cell nuclear antigen (PCNA) that plays a crucial role in assuring the processivity of the fork. It is known that PCNA is ubiquitylated on the residue of lysine 164 (K164) in the S-phase or in response to DNA replication stress. PCAN-K164 is, in fact, important for the maturation of the Okazaki fragments and in the activation of the DNA damage tolerance pathway (Chang & Cimprich, 2009; Thakar et al., 2020). Moreover, it has been found that mutation in the K164 residue of the PCNA has a negative impact on the origin licensing, and it is the cause of accumulation of single-stranded DNA gaps coming from the previous cell cycle and impairing the deposition of MCM2-7 helicase (Leung et al., 2020). Other studies have identified additional proteins associated with PCNA, and this was linked to increasing the opportunity of genome instability avoidance. This was the case with a PCNA accessory protein called SDE2. It was shown that loss of SDE2 effects caused a defeat of the maintenance of the replication and replication stress counteraction, through the impairment of the fork protection complex (Rageul et al., 2020). Molecular combing was used, also, to uncover the negative regulation of the checkpoint activation in humans, via the cyclin-dependent kinases (CDKs), polo-like kinase 1 (PLK1), and 9-1-1 (ortholog to PCNA). The researchers found that CDK phosphorylates a subunit of 9-1-1 called RAD9, which binds PLK1. In this way, the checkpoint was not activated, and DNA replication rates were maintained even in the presence of replication stress (Wakida et al., 2017). Thus, genotoxic stress causes the activation of the DNA damage checkpoint that leads cells to slow down the replication, but a too-slow replication is correlated with genomic instability. The performances of specific DNA polymerases have been disclosed thanks to the precision obtained by molecular combing. In a study, one specific DNA polymerase, called polymerase epsilon (PolE), was studied. More precisely, the investigators have generated a truncated form of the protein for the subunit 4 (POLE4 −/−) to correlate mechanistic, genetic, and physiological effects of such mutation. Thus, the POLE4 −/− led embryonic lethality in mice, while the lethality was suppressed when the *P53* was also mutated (Bellelli et al., 2018).

Molecular combing becomes a breakthrough method for the studies of DNA repair defects, replication stress, and their effect on genome instability. These investigations have unrevealed the crucial mechanisms of the replication underlying in tumors. Therefore, molecular combing has represented an efficient application to discover a potential therapeutic target and/or a compound. In fact, the effect of an inhibitor for the CDC7 has been approved. CDC7 is involved in the firing of origins of replication via the phosphorylation of one subunit of the replicative helicase called MCM2. This represents a biomarker that can be used as an anticancer target due to the fact that tumors show over numeracy of origins of replication (Iwai et al., 2019; Rainey et al., 2017). However, it was found a new role for CDC7 that was not connected to the origins. It has been shown that it can interplay with MRE11. Indeed, it has been shown that MRE11 localization at the forks, upon DNA damage, is lost as a consequence of CDC7 inhibition. In addition, CDC7 activity was found at the reversed forks, and its deregulating activity on MRE11 was necessary to onset the pathological behavior typical of the cells BRCA2 −/− (Rainey et al., 2020). Then, CDC7 has been identified as a biomarker for some anticancer treatments because of its roles: on the origins firing and, on the initiation, and elongation during the DNA replication.

Other biomarkers that have been studied by using molecular combing is BRCA1 and USP1. BRCA1 negative cells have been proved to be tumorigenic, and it is involved in the breast cancer onset. It is due to the increased genomic instability derived by the defects on the homologous recombination pathway (Gudmundsdottir & Ashworth, 2006). USP1 is a deubiquitinase protein, and it has been showing to have a synthetic lethal relationship with BRCA1 because it has been shown to be upregulated in BRCA1-deficient cells (Lim et al., 2018). Thus, BRCA1 negative cells have been shown to be sensitive to a PARP inhibitor, though some can acquire resistance (Farmer et al., 2005). Thus, USP1 inhibitor treatment can be used as an additional therapeutics strategy in combination with PARP inhibition.

Another interesting mention by the use of molecular combing is the DNA replication in telomere and the effect on senescence. Senescence is defined as the gradual decay of the physiological processes which contributes to the progressive and extreme slowdown of the cell cycle. The main characteristic of senescence is the shortening of the telomeres, which as a consequence leads to loss of doubling capacity from cells known as replicative senescence or as a protective method from oncogene overexpression. It has been shown that replicative senescence reduces the fork progression rate and causes the activation of the dormant origins. However, globally the replication time has not seemed to be affected. The replication timing does not derivate from senescence progression even though it is associated with accelerated aging. There is a relationship between senescence, aging, and DNA replication, but the replication timing is definitely unaffected by the replication stress (Rivera-Mulia et al., 2018). To assure that the senescence processes are not prematurely activated, cells need to control the DNA replication of the telomeres.

There are two proteins solving this role: telomerase and RTEL1. Telomerase is specifically able to perform DNA synthesis of the telomeric repetitive sequences (TTAGGG in vertebrates), while RTEL1 is needed to resolve the secondary structures of the telomere to promote the passage of the replisome. It has been proved that

RTEL1 negative cells increased the reverse forks at telomeres, and that telomerase binding to the reversed replicative forks inhibits the replication. The replicative problem of telomeres can be resolved by using PARP inhibitors or by depleting UBC13 or ZRANB3 to impede the accumulation of dysfunctional telomeres (Margalef et al., 2018). Thus, events of premature telomere shortening can be attributed to an incorrectly binding of telomerase to reverse forks, compromising, then, DNA replication by machinery falling. Thus, truncated replication of the telomere leads to the shortening of such chromosome regions.

Replication kinetic studies in locus-specific manner

The study of the DNA replication kinetics requires complicate analyses and knowledge of the DNA synthesis in detail, especially in complex organisms like eukaryotes. In addition, these investigations of DNA copy are made complicated by the intimate interplays between the DNA replication and all the DNA repair systems. Moreover, the synthesis of the DNA in higher eukaryotes is asynchronous, and the exact sequence of the origins is not known. Historically, only a few sequences have been identified in yeasts, so-called autonomously replicating sequences (ARS), but these are not present in higher eukaryotes, such as in the human being. Thus, it is not possible to predict where an origin is licensed and regulated. Replication kinetics studies require that the scientist should comprehend the functions of the origins of replication (either the replicative or the dormant), thus the origins densities in a specific moment and the fork rates. Even though some traditional methods, like the 2D gel, could identify some origins, their analyses could not be done at a genome-wide level. On contrary, molecular combing allows to investigate singular origins of replication as well as singular forks, on a genome-wide manner. Moreover, another interesting idea is to investigate the replication dynamics in a specific DNA sequence, by using Genome Morse Code assay (GMC) in combination with the Replication Combing Assay (RCA) protocol. The so-called RCA locus makes it possible then to compare the kinetics of DNA synthesis in a replicating sequence with the rest of the replicating genome.

The use of RCA locus was used on replicon to prove the previous observations of its size. By molecular combing, it was found that a replicon corresponds to two or more potential origins, and the average size identified was 25 kb (15–95 kb). This observation disagreed with the previous sizing of 50–300 kb, but it was explained with the hypothesis that a given cell would use only one origin per replicon at a given S-phase. Thus, the origins, within a replicon (or replicon clusters), are stochastically activated. Nevertheless, once one origin is activated, it interferes with the activation of another one located 10–25 kb from the original origin. This proved again that origins are stochastically chosen in higher eukaryotes, but the positioning does not become random. The origin activation was studied by using a molecular combing of singular molecule analysis of DNA replication on the mouse *IgH* locus. It was found that potential origins were 20 kb spaced, and in the normal S-phase, the initiation zones had 25-kb spacing with a minimal replicon size of 12 kb (Lebofsky et al., 2006; Taylor & Hozier, 1976). Change in the replicon size was due to the involvement of multiple replicon units (of 12 kb in a modular way). It could be caused by the contemporizing transcription activity in the *IgH* locus.

The RCA locus done via molecular combing (Fig. 5.2) is very useful to uncover some peculiar mechanism of "common deletion" happening in mitochondria (mt), which involves a 4977-bp region flanked by 13-bp repeats. In fact, mutations of the mtDNA are connected to several diseases related to aging and neurological disorders. Researchers have performed RCA locus to understand how the deletion occurred. A stalled fork was found at the point of the deletion. Thus, it was explained as an impaired replication that causes the formation of toxic lesions (Phillips et al., 2017). The theory was supported by the in vivo evidence of the asymmetric mode of replication in mtDNA synthesis.

RCA locus is considered to be particularly helpful to investigate how the DNA synthesis is carried out in DNA regions enriched with repetitive sequences, such as centromeres and telomeres. Fork progression can be hampered in these hard-to-replicate regions, as a consequence it causes replication stress and the possibility to trigger chromosome fragility, aging, and cancer. FRA6E is the most common chromosome fragile site (Mrasek et al., 2010), and it has been connected to many other genes. It is considered to be replicated in the mid-to-late S-phase. An experiment was conducted analyzing the replication kinetics of FRA6E and LMNB2 as control of early replicating locus. By molecular combing, it has been possible to confirm that fork rates were highly comparable to those detected at LMNB2. However, replication timing of PARK2, one gene connected to FRA6E, appeared to be fixed at early/late replication. Thus, DNA breaks do arise preferentially at early/late transition zones (Palumbo et al., 2010). RCA locus dedicated to replication process at the telomere has also brought some major discovery. DNA combing has led to understanding how TRF2 is assisting DNA synthesis at the pericentromeric region specifically, but not at the centromeric. Particularly, the N terminus of TRF2 was able to bind a specific pericentromeric sequence called Satellite III, but only during the S-phase. Such interaction has triggered the recruitment of RTEL1 to resolve the G-quadruplex present and to allow the replication machinery to pass through (Mendez-Bermudez et al., 2018). The TFR2 stabilizes the heterochromatin, but more studies are needed to uncover the correlation between heterochromatic replication and its relationship with aging and cancer. The telomere lengthening is always associated with diseases, such as tumor onset. A specific pathway has been discovered to be active on telomere, known as alternative lengthening of telomeres (ALT). ALT is an error-prone system, where newly generated mistakes can contribute to developing aggressive cancer type. Recently, an interplay of ALT and complexes of DNA repairs has been discovered. Specifically, ALT-mediated telomere synthesis requires the presence of composite proteins: the BLM-TOP3A-RMI (BTR) dissolvase complex. In the ALT-mediated telomere synthesis, a recombination intermediate is formed, and this point has been identified as requiring BRT complex to be successful. Finally, it causes telomere replication mediated by POLD3 (Sobinoff et al., 2017). Thus, the SLX4-SLX1-ERCC4

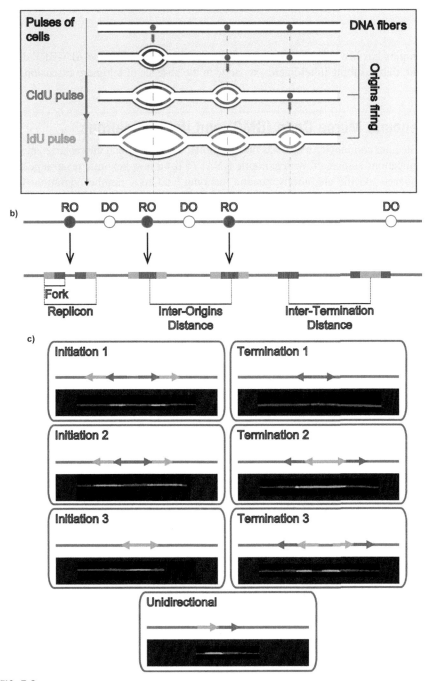

FIG. 5.2

Schematic representation of the Replication Combing Assay (RCA) and the interpretation of signals. (A) Cells are pulsed with the same concentration of CldU and IdU consecutively, the time of the pulse depends on the model system used; (B) schematic of replicative origins (RO) and dormant origins (DO) on the DNA fibers, and the consequence replication parameters are considered in RCA once the DNA is combed; (C) representation and interpretation of all the potential replicon signals that can be identified by RCA.

complex has been shown to be able to neutralize and resolve the ALT-BRT dependent, thus resulting in telomere exchange in the absence of telomere extension.

Genomic Morse Code (GMC) and its applications

Molecular combing as a tool has been exploited to identify and characterize multiple applications associated with genomic DNA. To list a few, genomic re-arrangements, sequence-specific alterations, genetic mapping, and copy number variation (CNV) studies are applications that utilize uniform stretched DNA on a solid surface to illustrate information using genetic material as a building block. The molecular combing allows to take standard physical measurements, to score for high genome copy number per glass coverslips, and to have statistical significance. Compared with other existing techniques, molecular combing technology has assisted in answering multiple scientific questions.

GMC designs for CNV studies

Over the past several years, researchers have found that structural variants, within specific repetitive regions within the human genomic DNA, are responsible for genomic diversity similar to single-nucleotide polymorphisms (SNPs) (Redon et al., 2006) which span from few base pairs (bp) to kilobases (kb). Such repetitive genetic alterations that contribute to human variability are called CNVs. CNVs are found in humans, as well as other mammals, and it is responsible for recurrent deletions and duplications which are to date linked mostly to neuropsychiatric phenotypes (Freeman et al., 2006; Weise & Liehr, 2021c). The variation in gene copy number is associated with diseases onset such as psoriasis (Hollox et al., 2008) and asthma (Handsaker et al., 2015). One of the most defined studies of small CNVs is on the trinucleotide repeats (TNRs), which comprise of expansion linked to three nucleotide tandem repeats; sometimes also defined as special kind of microsatellite repeats (Liehr, 2021e). TNR tandem repeats exhibit dynamic contractions and expansions and are linked with diseases such as Huntington's disease. Similarly, the pathological link of CNVs is also associated with susceptibility to systemic lupus erythematosus and schizophrenia (Sekar et al., 2016). Interestingly, the characterization of these CNVs is unobserved due to lack of technical tools to identify the correct measurements and thus limits the true identification of copy numbers by introducing a hereditary paradox linked to disease identification.

Molecular combing has proven to be a suitable technology for studies focusing on the characterization and identification of such disorders linked to CNV. In one study, human blood neutrophil peptides (HNP1-3) are studied to identify how copy number variants of alpha-defensins gene *DEFA1* and *DEFA3* link to the susceptibility of infections and auto immune disorders (Hughes et al., 2020). This study was performed on 2504 samples from 1000 genomes (G1K) project where the *DEFA1A3* locus CNV

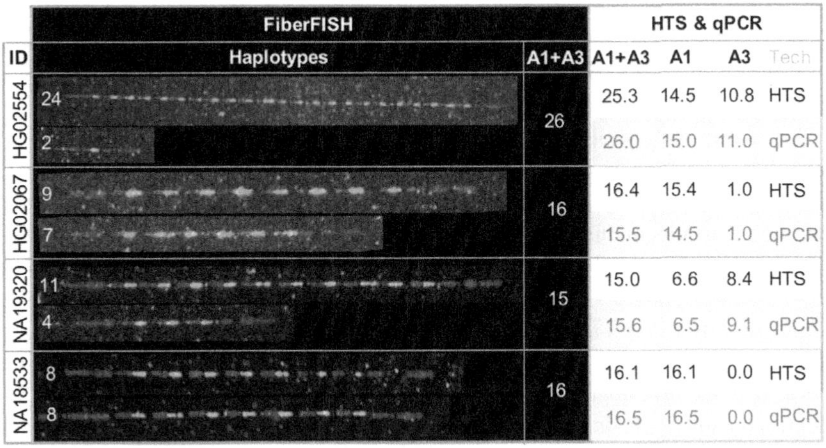

		FiberFISH		HTS & qPCR			
ID		**Haplotypes**	**A1+A3**	**A1+A3**	**A1**	**A3**	Tech
HG02554	24		26	25.3	14.5	10.8	HTS
	2			26.0	15.0	11.0	qPCR
HG02067	9		16	16.4	15.4	1.0	HTS
	7			15.5	14.5	1.0	qPCR
NA19320	11		15	15.0	6.6	8.4	HTS
	4			15.6	6.5	9.1	qPCR
NA18533	8		16	16.1	16.1	0.0	HTS
	8			16.5	16.5	0.0	qPCR

FIG. 5.3

The identification of copy number for DEAF1A3 haplotypes using molecular combing. The DEAF1A3 gene *(red)*, DEFT1P pseudogene *(green)*, left flanking region *(red and blue)* while right flanking region *(blue alone)* depicts the copy number identification. The four samples (HG02554, HG02067, NA19320, and NA18533) have high high-throughput sequencing (HTS) CN and are obtained from the NHGRI Repository at Coriell Institute of Medical Research. The HG02554 qPCR estimates match the molecular combing (MC) results as this sample's scaling/calibration of qPCR estimates was done using MC results (Hughes et al., 2020).

was re-estimated to be highest in the African Caribbean population (genotypes of 26 copy numbers), while previously the highest CNV detected genome-wide was 15 copy numbers. In this, molecular combing was used to visualize the haplotypes on selected samples (Fig. 5.3). Similarly, CNV of the human amylase gene clusters, i.e., *AMY1*, *AMY2A*, and *AMY2B* associated with obesity and dietary adaptation, was also defined by scoring with molecular combing (Shwan et al., 2017). The study demonstrated that the unexpected genomic rearrangements of amylase genes led to genomic instability involving five recurrent relocations of pancreatic amylase genes *AMY2A* and *AMY2B*. It was carried out on the sub-Saharan African population where copy number analysis obtained high-order expansions of all amylase genes, i.e., *AMY1*, *AMY2A*, and *AMY2B*.

GMC designs for sequence-specific alterations

Molecular combing was originally created to identify and score genetic diseases arising from deletions of gene sequences ranging from 50 to 200 kb (Michalet et al., 1997). It was later followed up by involving unique color-coding patterns (i.e., GMC) to specific sequence regions or repeat units to distinguish and detect micro-deletions and duplications. One study led to the identification of *BRCA* gene locus for loss of heterozygosity when compared with control and altered alleles. With the GMC design for the *BRCA1* gene, deletions of 24 kb and two 3 kb were scored followed by 6-kb duplication. In addition, an undocumented 17-kb duplication was also identified (Gad et al., 2001a, 2001b).

Later on, using molecular combing, a genetic test has been developed for the detection of structural variations and large rearrangements in the *BRCA1* and *BRCA2* genes for breast and ovarian cancer susceptibility (Cheeseman et al., 2012). Four of ten large rearrangements that had never previously found were characterized in this study. These included deletion within *BRCA1* of 11.4 kb in the exon 3, 5.8-kb deletion in exon 24, and 8.1-kb duplication of the exons 5–7, while for *BRCA2*, 14.8 kb duplication of exons 17–20 were found. The same study also focused on covering noncoding regions that included 5′-region of *BRCA1*, which had the *NBR2* gene, the ψ *BRCA1* pseudogene, and the *NBR1* gene. These regions are difficult to be identified using PCR or aCGH probes, due to rearrangement-prone genomic regions (Rouleau et al., 2007; Staaf et al., 2008). The existence of a 100-kb "sequencing gap" was also found in the genomic region upstream of the *BRCA1* gene, which supported the genomic structure proposed in GRCh37 genome.

Molecular combing successfully also characterized the genetic heterogeneity disorder called *familial adult myoclonic epilepsy* (FAME). It is an autosomal dominant syndrome where cortical hand tremors associated with myoclonic jerks, sometimes with generalized tonic-clonic seizures (GTCS) and more rarely focal seizures occur (van den Ende et al., 2018). Recently, it was found that the diseases onset is due to intronic TTTTA/TTTCA repeat expansions on chromosome 8q24 in *SAMD12* (*FAME1*), and it is major cause of the disorder especially in the Chinese and Japanese populations (Ishiura et al., 2018; Zeng et al., 2019). Similarly, a parallel study provided evidences to describe the causes of a different subtype of *FAME*, known as *FAME3*. FAME3 disorder onset is also caused by the same TTTTA/TTTCA repeat expansions, but it occurs in the intron 1 of a gene called *MARCH6* (Florian et al., 2019). Using the molecular combing technique, these genomic rearrangements were identified to vary from 3.3 to 14 kb on average. The link of these genomic expansions was confirmed to be correlated with the age at epilepsy onset. Also, these expansions at the *MARCH6* were directly associated with the TTTCA repeat units which contributed to the pathogenicity of FAME3.

Using molecular combing, another disorder that has been well characterized is facioscapulohumeral muscular dystrophy (FSHD). It is the third most prevalent muscular hereditary myopathy. The clinical phenotype of FSHD introduces atrophy and weakness in the face, scapulohumeral, and anterior foreleg muscles. Due to great variability in clinical severity, the disease varies from a chronic infantile form to individuals who remain asymptomatic throughout their lives. This hereditary autosomal dominant neuromuscular disorder is found to be affecting 1 in 8,300 to 1 in 20,000 people in different Western populations (Deenen et al., 2014; Mostacciuolo et al., 2009). The clinical symptoms are known to arise between the age of 20–40 years. There are two types of FSHD: FSHD Type 1 and FSHD Type 2 where the majority of patients (95%) are Type 1 while only a few (5%) are Type 2. The locus linked to the pathology of FSHD is located on the subtelomeric region of the 4q35 chromosome end. The disease is associated with contraction of the microsatellite tandem repeat array *D4Z4* of 3.3-kb length (Fig. 5.4). In a healthy population, the repeated units of *D4Z4* (3.3 kb) span from 11 to 150 copies, while in patients with FSHD1, it is 1–10 units (van Deutekom et al., 1993). The *D4Z4* pathogenicity is linked with the 4qA haplotype of the chromosome 4q35. Although the sequence similarity of haplotype 4qB is 92% homologous to 4qA and likewise 10q26 loci have 98% homology to 4qA polymorphism, the contractions of 10q array or 4qB haplotypes are not found to be associated with FSHD (Lemmers et al., 2007, 2010; van Geel et al., 2002). Even with such detailed insight into the genotype of the disease, it is a daunting task to extract the causative information using conventional diagnostic techniques, i.e., qPCR, Southern blot, pulse-field gel electrophoresis (PFGE), and/or linear gel electrophoresis (LGE). Additionally, these conventional methods, e.g., Southern blot apart from being labor-intensive fail to detect events such as somatic mosaicism and chromosomal rearrangements. However, PFGE could identify those noncanonical variants, but the setup requires a multistep cumbersome experimental plan. Thus, using of molecular combing-based strategy, the advantages were to be user-friendly, high thoroughput, and above all ability to directly visualize the loci at 4q35 vs 10q26 in the same field of view, facilitating diagnosis and interpretation of results. In the year 2013, Genomic Vision obtained the mark of CE-IVD (Conformitè Européenne In-Vitro Diagnostics) for the commercial application use as FSHD diagnostic assay.

GMC design for gene editing studies

Gene and cell therapies have surfaced as promising treatments for several diseases, including viral infections, hereditary disorders, and cancer. Recent advances in the development of gene-editing technologies, based on the CRISPR-Cas9 system, have enabled rapid and accessible genome engineering in eukaryotic cells and living organisms. Albeit these efforts, these methods can still result in a wide range of unintended modifications that hamper their full translation into the clinic. With the combination of molecular combing and GMC, the detection, characterization, and quantification of gene-editing events can be achieved. A study was performed in which by the CRISPR-Cas9 system, a 6.5-kb-deletion in the *BRCA1* gene was created using HEK293 cells (Fig. 5.5).

To detect and characterize the gene-editing events induced at the *BRCA1* locus, DNA extracted was combed from control and transfected HEK293 cells which were hybridized with *BRCA1*-specific Genomic Morse Code (GMC). No edited *BRCA1* gene in control cells was obtained, whereas efficiencies ranging between 6.5% and 11% were seen with different gRNA pairs (Nguyen et al., 2011). The presence of unexpected large rearrangements (>40 kb) was also detected (Barradeau et al., 2018). The results obtained by molecular combing were compared with those obtained by droplet digital PCR (ddPCR) and targeted sequencing on the same edited material. It was concluded that molecular combing is a unique tool that enables to visualize and

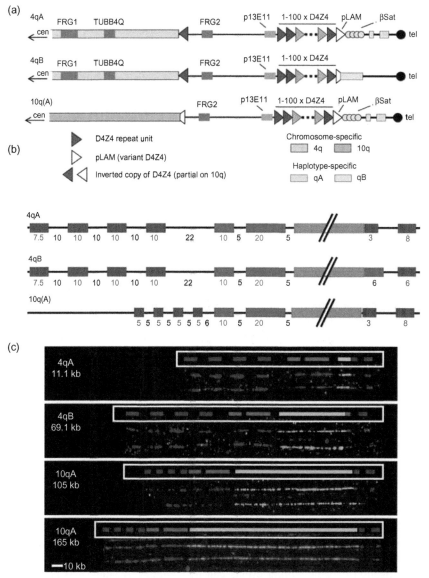

FIG. 5.4

The FSHD barcode design. (A) The schematic representation of the target haplotypes, i.e., 4qA, 4qB, and 10qA regions. (B) The representation of the three color-coding probe design used for distinguishing the 4qA, 4qB, and 10qA, with the lengths of each region mentioned below in kilobases, respectively. The *green color* probe annotates the D4Z4 repeats array. The *blue color* probe identifies the common proximal region of chromosomes 4 and 10 and distinguishes the 4qA from 4qB at the telomeric end of the bar code. The *red color* probe pattern identifies chromosome 4 to chromosome 10 and distinguishes the 4qA from 4qB at the telomeric end of the bar code. (C) The representation of validated signals obtained for each haplotype identified in patient samples is shown with the schematic representation on top and actual single molecule detection below in each panel. The variable length measurements of D4Z4 is indicated in kilobases for each allele (Nguyen et al., 2011).

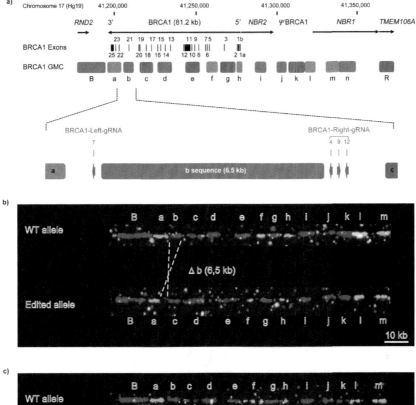

FIG. 5.5

Detection by GMC gene-editing events. (A) Illustration of GMC design for the target locus with nomenclature for each color code using alphabets. The schema also shows the gRNA target locus site and size with color coding and nomenclature to carry out the deletion using CRISPR-Cas9 system. (B) Example of detection of WT and edited BRCA1 GMCs from HEK293 cells transfected with the gRNA 7/4 pair. The deletion of BRCA1 genomic region covered by the 6.5-kb apparent *blue* "b" probe of the BRCA1 GMC is pointed out by the *white dash lines*. (C) Example of duplication/inversion in the BRCA1 gene induced by CRISPR-Cas9 system. The schematic representation of the hybridization patterns corresponding to the potential duplication/inversion of the BRCA1 gene is shown above in the fluorescent arrays (Barradeau et al., 2018).

quantify both specific and unexpected rearrangements in a single experiment, without amplification bias. Altogether, these results demonstrate that molecular combing, through the analysis of hundreds of single DNA molecules, provides an orthogonal method to NGS-based approaches and a promising assay to quantify the efficiency for precise novel gene therapies. Thus, with a similar approach, molecular combing can also serve as a tool for the characterization of bio-manufacturing applications. The use of adeno-associated virus integration site 1 (AAVS1), as the safe harbor locus for target gene insertion using the CRISPR-Cas9 system, is the novel approach exploited for multiple bio-production purposes. This approach can be highly informative for cell line genetic stability and homogeneity analysis of master cell banks, used in antibody production and cellular therapeutics industries.

Conclusion

The basis of human genetic diseases is mostly attributed to variations at the genomic level. The analyses of these genomic alterations are comprehensive only if the applied methods can provide an accurate indication of the genomic rearrangements occurring on genome wide and sequence-specific scale. Many of the setup methods can bring different pieces of information according to the level of complexity. Nowadays, they can have a resolution of a single base pair, but the big picture is lost. While some other techniques can see the bigger variation, but the sensitivity and the complexity are not accurate. To such extent, analyses, by molecular combing, have become a central technique to investigate the phenomena caused by genomic modifications.

The Replication Combing Assay (RCA) is the technique used to understand how the replication kinetics are changed upon a specific protein mutation or a treatment. The use of RCA has, indeed, facilitated the experimental performance and enlarged the number of investigations about the DNA replication. RCA has led to a higher comprehension of the variations of DNA synthesis genome-wide or sequence specific due to its extraordinary accuracy. In first, RCA has brought to the confirmation of the basic mechanisms of DNA replication in various organisms such as yeasts and human cells and also in symbiotic organelle such as mitochondria. Afterward, molecular combing has become a breakthrough assay to understand how a mutation affects DNA synthesis. By accounting for the numbers of the DNA replication kinetics, RCA is used to measure the effect of genome modifications on a protein and to screen the potential treatment actions. Nevertheless, performing RCA can bring to understand how a genomic variation affects the DNA replication kinetics, in the entire genome or in a specific locus. Ultimately, RCA by molecular combing gives insights into the impact of a genomic variation on an organism. Thus, RCA is a unique method to quantify the impact of genomic rearrangements on the vitality of the cells and or the organisms.

Similarly in cytogenomics, the genetic modifications can be categorized from small to large rearrangements: the range goes from one to few base pairs: single-nucleotide variation (SNV), small insertions and deletions (INDELs) to large structural variants (SVs) up to CNV. The use of multicolor-coding pattern by Genome Morse Code (GMC) application has proven to be a powerful yet elegant tool to answer multiple genetic complexities. Since the user has the power to design, create, and permutate varied color barcodes, the applicability of the technique is vast. The potential to score for genomic events at a minuscule resolution of 2 kilobases to a span of few megabases makes this technique potent to cover a huge spectrum of identification. The GMC assay allows physical mapping and comprehensive analysis of genetic modifications, such as amplifications, deletions, repeats, inversions, translocations, and complex rearrangements such as mosaicism. This attributes to directly link such rearrangements to genetic disorders that can help in genotypic mapping of familial syndromes and identify hard to score events. Due to physical stretching of DNA and characterization of region of interest (ROI), molecular combing has proven to recognize events which are even unaccounted for. Thus, in cytogenomics, molecular combing can prove to be a powerful tool to uncover and provide answers to chromosomal modifications linked to replication kinetics or genomic rearrangements and more. As quoted by Albert Szent-Györgyi, "Research is to see what everybody else has seen, and to think what nobody else has thought." It is critical to push the barriers on usage of existing techniques to answer the questions which still elude us. Thus, using molecular combing technology, it is envisioned to implement the same and find answers to what is still unknown in the field of cytogenomics.

References

Allemand, J. F., Bensimon, D., Jullien, L., Bensimon, A., & Croquette, V. (1997). pH-dependent specific binding and combing of DNA. *Biophysical Journal, 73*(1), 2064–2070. https://doi.org/10.1016/S0006-3495(97)78236-5.

Barkley, L. R., Ohmori, H., & Vaziri, C. (2007). Integrating S-phase checkpoint signaling with trans-lesion synthesis of bulky DNA adducts. *Cell Biochemistry and Biophysics, 47*(3), 392–408. https://doi.org/10.1007/s12013-007-0032-7.

Barradeau, S., Bensimon, A., & Cavarec, L. (2018). *Method for the monitoring of modified nucleases induced gene editing events by molecular combing.* US Patent App. 15/813,974, 2018.

Bellelli, R., Borel, V., Logan, C., Svendsen, J., Cox, D. E., Nye, E., … Boulton, S.J. (2018). Polε instability drives replication stress, abnormal development, and tumorigenesis. *Molecular Cell, 70*(4). https://doi.org/10.1016/j.molcel.2018.04.008. 707–721.e7.

Bensimon, A., Simon, A., Chiffaudel, A., Croquette, V., Heslot, F., & Bensimon, D. (1994). Alignment and sensitive detection of DNA by a moving interface. *Science, 265*(5181), 2096–2098. https://doi.org/10.1126/science.7522347.

Cairns, J. (1963). The bacterial chromosome and its manner of replication as seen by autoradiography. *Journal of Molecular Biology, 6*, 208–213. https://doi.org/10.1016/s0022-2836(63)80070-4.

Cairns, J. (1966). Autoradiography of HeLa cell DNA. *Journal of Molecular Biology, 15*(1), 372–373. https://doi.org/10.1016/s0022-2836(66)80233-4.

Chang, D. J., & Cimprich, K. A. (2009). DNA damage tolerance: When it's OK to make mistakes. *Nature Chemical Biology, 5*(2), 82–90. https://doi.org/10.1038/nchembio.139.

Cheeseman, K., Rouleau, E., Vannier, A., Thomas, A., Briaux, A., Lefol, C., … Ceppi, M. (2012). A diagnostic genetic test for the physical mapping of germline rearrangements in the susceptibility breast cancer genes BRCA1 and BRCA2. *Human Mutation, 33*(6), 998–1009. https://doi.org/10.1002/humu.22060.

Czajkowsky, D. M., Liu, J., Hamlin, J. L., & Shao, Z. (2008). DNA combing reveals intrinsic temporal disorder in the replication of yeast chromosome VI. *Journal of Molecular Biology, 375*(1), 12–19. https://doi.org/10.1016/j.jmb.2007.10.046.

da Silva, S. S. (2020). Estimation of the minimum number of replication origins per chromosome in any organism. *Bio-Protocol, 10*(20), e3798. https://cn.bio-protocol.org/e3798.

Daban, J.-R. (2021). Multilayer organization of chromosomes. In T. Liehr (Ed.), *Cytogenomics* (pp. 267–296). Academic Press. Chapter 13 (in this book).

Deenen, J. C., Arnts, H., van der Maarel, S. M., Padberg, G. W., Verschuuren, J. J., Bakker, E., … van Engelen, B.G. (2014). Population-based incidence and prevalence of facioscapulohumeral dystrophy. *Neurology, 83*(12), 1056–1059. https://doi.org/10.1212/WNL.0000000000000797.

Delpu, Y., Barseghyan, H., Bocklandt, S., Hastie, A., & Chaubey, A. (2021). Next-generation cytogenomics: High-resolution structural variation detection by optical genome mapping. In T. Liehr (Ed.), *Cytogenomics* (pp. 123–146). Academic Press. Chapter 8 (in this book).

Farmer, H., McCabe, N., Lord, C. J., Tutt, A. N., Johnson, D. A., Richardson, T. B., … Ashworth, A. (2005). Targeting the DNA repair defect in BRCA mutant cells as a therapeutic strategy. *Nature, 434*(7035), 917–921. https://doi.org/10.1038/nature03445.

Florian, R. T., Kraft, F., Leitão, E., Kaya, S., Klebe, S., Magnin, E., … Depienne, C. (2019). Unstable TTTTA/TTTCA expansions in MARCH6 are associated with familial adult myoclonic epilepsy type 3. *Nature Communications, 10*(1), 4919. https://doi.org/10.1038/s41467-019-12763-9.

Fonseca, R., Oken, M. M., Harrington, D., Bailey, R. J., Van Wier, S. A., Henderson, K. J., … Dewald, G.W. (2001). Deletions of chromosome 13 in multiple myeloma identified by interphase FISH usually denote large deletions of the q arm or monosomy. *Leukemia, 15*(6), 981–986. https://doi.org/10.1038/sj.leu.2402125.

Freeman, J. L., Perry, G. H., Feuk, L., Redon, R., McCarroll, S. A., Altshuler, D. M., … Lee, C. (2006). Copy number variation: New insights in genome diversity. *Genome Research, 16*(8), 949–961. https://doi.org/10.1101/gr.3677206.

Fu, H., Redon, C. E., Utani, K., Thakur, B. L., Jang, S., Marks, A. B., … Aladjem, M.I. (2020). Dynamics of replication origin over-activation. *BiorXiv*. https://doi.org/10.1101/2020.01.27.922211.

Gad, S., Aurias, A., Puget, N., Mairal, A., Schurra, C., Montagna, M., … Stoppa-Lyonnet, D. (2001a). Color bar coding the BRCA1 gene on combed DNA: A useful strategy for detecting large gene rearrangements. *Genes, Chromosomes & Cancer, 31*(1), 75–84. https://doi.org/10.1002/gcc.1120.

Gad, S., Scheuner, M. T., Pages-Berhouet, S., Caux-Moncoutier, V., Bensimon, A., Aurias, A., … Stoppa-Lyonnet, D. (2001b). Identification of a large rearrangement of the BRCA1 gene using colour bar code on combed DNA in an American breast/ovarian cancer family previously studied by direct sequencing. *Journal of Medical Genetics, 38*(6), 388–392. https://doi.org/10.1136/jmg.38.6.388.

Gaillard, H., García-Muse, T., & Aguilera, A. (2015). Replication stress and cancer. *Nature Reviews. Cancer, 15*(5), 276–289. https://doi.org/10.1038/nrc3916.

Gudmundsdottir, K., & Ashworth, A. (2006). The roles of BRCA1 and BRCA2 and associated proteins in the maintenance of genomic stability. *Oncogene, 25*(43), 5864–5874. https://doi.org/10.1038/sj.onc.1209874.

Gunderson, K. L., Steemers, F. J., Ren, H., Ng, P., Zhou, L., Tsan, C., … Shen, R. (2006). Whole-genome genotyping. *Methods in Enzymology, 410,* 359–376. https://doi.org/10.1016/S0076-6879(06)10017-8. 16938560.

Hanaoka, F., Yamada, M., Takeuchi, F., Goto, M., Miyamoto, T., & Hori, T. (1985). Autoradiographic studies of DNA replication in Werner's syndrome cells. *Advances in Experimental Medicine and Biology, 190,* 439–457. https://doi.org/10.1007/978-1-4684-7853-2_22.

Hand, R. (1975). Deoxyribonucleic acid fiber autoradiography as a technique for studying the replication of the mammalian chromosome. *The Journal of Histochemistry and Cytochemistry, 23*(7), 475–481. https://doi.org/10.1177/23.7.1095649.

Hand, R. (1977). Human DNA replication: Fiber autoradiographic analysis of diploid cells from normal adults and from Fanconi's anemia and ataxia telangiectasia. *Human Genetics, 37*(1), 55–64. https://doi.org/10.1007/BF00293772.

Hand, R. (1978). Eucaryotic DNA: Organization of the genome for replication. *Cell, 15*(2), 317–325. https://doi.org/10.1016/0092-8674(78)90001-6.

Hand, R., & German, J. (1975). A retarded rate of DNA chain growth in Bloom's syndrome. *Proceedings of the National Academy of Sciences of the United States of America, 72*(2), 758–762. https://doi.org/10.1073/pnas.72.2.758.

Hand, R., & Tamm, I. (1973). DNA replication: Direction and rate of chain growth in mammalian cells. *The Journal of Cell Biology, 58*(2), 410–418. https://doi.org/10.1083/jcb.58.2.410.

Handsaker, R. E., Van Doren, V., Berman, J. R., Genovese, G., Kashin, S., Boettger, L. M., & McCarroll, S. A. (2015). Large multiallelic copy number variations in humans. *Nature Genetics, 47*(3), 296–303. https://doi.org/10.1038/ng.3200.

Herrick, J., & Bensimon, A. (1999). Single molecule analysis of DNA replication. *Biochimie, 81*(8–9), 859–871. https://doi.org/10.1016/s0300-9084(99)00210-2.

Herrick, J., Michalet, X., Conti, C., Schurra, C., & Bensimon, A. (2000a). Quantifying single gene copy number by measuring fluorescent probe lengths on combed genomic DNA. *Proceedings of the National Academy of Sciences of the United States of America, 97*(1), 222–227. https://doi.org/10.1073/pnas.97.1.222.

Herrick, J., Stanislawski, P., Hyrien, O., & Bensimon, A. (2000b). Replication fork density increases during DNA synthesis in X. laevis egg extracts. *Journal of Molecular Biology, 300*(5), 1133–1142. https://doi.org/10.1006/jmbi.2000.3930.

Hollox, E. J., Huffmeier, U., Zeeuwen, P. L., Palla, R., Lascorz, J., Rodijk-Olthuis, D., … Schalkwijk, J. (2008). Psoriasis is associated with increased beta-defensin genomic copy number. *Nature Genetics, 40*(1), 23–25. https://doi.org/10.1038/ng.2007.48.

Hoogendoorn, B., Norton, N., Kirov, G., Williams, N., Hamshere, M. L., Spurlock, G., … O'Donovan, & M. C. (2000). Cheap, accurate and rapid allele frequency estimation of single nucleotide polymorphisms by primer extension and DHPLC in DNA pools. *Human Genetics, 107*(5), 488–493. https://doi.org/10.1007/s004390000397.

Housman, D., & Huberman, J. A. (1975). Changes in the rate of DNA replication fork movement during S phase in mammalian cells. *Journal of Molecular Biology, 94*(2), 173–181. https://doi.org/10.1016/0022-2836(75)90076-5.

Huberman, J. A., & Riggs, A. D. (1968). On the mechanism of DNA replication in mammalian chromosomes. *Journal of Molecular Biology, 32*(2), 327–341. https://doi.org/10.1016/0022-2836(68)90013-2.

Huberman, J. A., & Tsai, A. (1973). Direction of DNA replication in mammalian cells. *Journal of Molecular Biology, 75*(1), 5–12. https://doi.org/10.1016/0022-2836(73)90525-1.

Hughes, T., Hansson, L., Akkouh, I., Hajdarevic, R., Bringsli, J. S., Torsvik, A., … Djurovic, S. (2020). Runaway multi-allelic copy number variation at the α-defensin locus in African and Asian populations. *Scientific Reports, 10*(1), 9101. https://doi.org/10.1038/s41598-020-65675-w.

Iourov, I. Y., Vorsanova, S. G., & Yurov, Y. B. (2021). Cytogenomic landscape of the human brain. In T. Liehr (Ed.), *Cytogenomics* (pp. 327–348). Academic Press. Chapter 16 (in this book).

Ishii, T., Nagaki, K., & Houben, A. (2021). Application of CRISPR/Cas9 to visualize defined genomic sequences in fixed chromosomes and nuclei. In T. Liehr (Ed.), *Cytogenomics* (pp. 147–153). Academic Press. Chapter 9 (in this book).

Ishiura, H., Doi, K., Mitsui, J., Yoshimura, J., Matsukawa, M. K., Fujiyama, A., … Tsuji, S. (2018). Expansions of intronic TTTCA and TTTTA repeats in benign adult familial myoclonic epilepsy. *Nature Genetics, 50*(4), 581–590. https://doi.org/10.1038/s41588-018-0067-2.

Iwai, K., Nambu, T., Dairiki, R., Ohori, M., Yu, J., Burke, K., … Ohashi, A. (2019). Molecular mechanism and potential target indication of TAK-931, a novel CDC7-selective inhibitor. *Science Advances, 5*(5), eaav3660. https://doi.org/10.1126/sciadv.aav3660.

Jenkins, E. C., Velinov, M. T., Ye, L., Gu, H., Li, S., Jenkins, E. C., Jr., … Silverman, W.P. (2006). Telomere shortening in T lymphocytes of older individuals with Down syndrome and dementia. *Neurobiology of Aging, 27*(7), 941–945. https://doi.org/10.1016/j.neurobiolaging.2005.05.021.

Kaykov, A., Taillefumier, T., Bensimon, A., & Nurse, P. (2016). Molecular combing of single DNA molecules on the 10 megabase scale. *Scientific Reports, 6*, 19636. https://doi.org/10.1038/srep19636.

Lander, E. S., Linton, L. M., Birren, B., Nusbaum, C., Zody, M. C., Baldwin, J., … International Human Genome Sequencing Consortium. (2001). Initial sequencing and analysis of the human genome. *Nature, 409*(6822), 860–921. https://doi.org/10.1038/35057062.

Lebofsky, R., Heilig, R., Sonnleitner, M., Weissenbach, J., & Bensimon, A. (2006). DNA replication origin interference increases the spacing between initiation events in human cells. *Molecular Biology of the Cell, 17*(12), 5337–5345. https://doi.org/10.1091/mbc.e06-04-0298.

Lemmers, R. J., van der Vliet, P. J., Klooster, R., Sacconi, S., Camaño, P., Dauwerse, J. G., … van der Maarel, S.M. (2010). A unifying genetic model for facioscapulohumeral muscular dystrophy. *Science, 329*(5999), 1650–1653. https://doi.org/10.1126/science.1189044.

Lemmers, R. J., Wohlgemuth, M., van der Gaag, K. J., van der Vliet, P. J., van Teijlingen, C. M., de Knijff, P., … van der Maarel, S.M. (2007). Specific sequence variations within the 4q35 region are associated with facioscapulohumeral muscular dystrophy. *American Journal of Human Genetics, 81*(5), 884–894. https://doi.org/10.1086/521986.

Leung, W., Baxley, R. M., Thakar, T., Rogers, C. B., Buytendorp, J. P., Wang, L., … Bielinsky, A.-K. (2020). PCNA-K164 ubiquitination facilitates origin licensing and mitotic DNA synthesis. *BioXiv*. https://doi.org/10.1101/2020.06.25.172361.

Liehr, T. (2021a). A definition for cytogenomics - Which also may be called chromosomics. In T. Liehr (Ed.), *Cytogenomics* (pp. 1–7). Academic Press. Chapter 1 (in this book).

Liehr, T. (2021b). Overview of currently available approaches used in cytogenomics. In T. Liehr (Ed.), *Cytogenomics* (pp. 11–24). Academic Press. Chapter 2 (in this book).

Liehr, T. (2021c). Molecular cytogenetics. In T. Liehr (Ed.), *Cytogenomics* (pp. 35–45). Academic Press. Chapter 4 (in this book).

Liehr, T. (2021d). Nuclear architecture. In T. Liehr (Ed.), *Cytogenomics* (pp. 297–305). Academic Press. Chapter 14 (in this book).

Liehr, T. (2021e). Repetitive elements, heteromorphisms, and copy number variants. In T. Liehr (Ed.), *Cytogenomics* (pp. 373–388). Academic Press. Chapter 19 (in this book).

Lim, K. S., Li, H., Roberts, E. A., Gaudiano, E. F., Clairmont, C., Sambel, L. A., … D'Andrea, A.D. (2018). USP1 is required for replication fork protection in BRCA1-deficient tumors. *Molecular Cell*, 72(6). https://doi.org/10.1016/j.molcel.2018.10.045. 925–941.e4.

Margalef, P., Kotsantis, P., Borel, V., Bellelli, R., Panier, S., & Boulton, S. J. (2018). Stabilization of reversed replication forks by telomerase drives telomere catastrophe. *Cell*, 172(3). https://doi.org/10.1016/j.cell.2017.11.047. 439–453.e14.

Marheineke, K., & Hyrien, O. (2001). Aphidicolin triggers a block to replication origin firing in Xenopus egg extracts. *The Journal of Biological Chemistry*, 276(20), 17092–17100. https://doi.org/10.1074/jbc.M100271200.

Maya-Mendoza, A., Petermann, E., Gillespie, D. A., Caldecott, K. W., & Jackson, D. A. (2007). Chk1 regulates the density of active replication origins during the vertebrate S phase. *The EMBO Journal*, 26(11), 2719–2731. https://doi.org/10.1038/sj.emboj.7601714.

Mendez-Bermudez, A., Lototska, L., Bauwens, S., Giraud-Panis, M. J., Croce, O., Jamet, K., … Ye, J. (2018). Genome-wide control of heterochromatin replication by the telomere capping protein TRF2. *Molecular Cell*, 70(3). https://doi.org/10.1016/j.molcel.2018.03.036. 449–461.e5.

Michalet, X., Ekong, R., Fougerousse, F., Rousseaux, S., Schurra, C., Hornigold, N., … Bensimon, A. (1997). Dynamic molecular combing: Stretching the whole human genome for high-resolution studies. *Science*, 277(5331), 1518–1523. https://doi.org/10.1126/science.277.5331.1518.

Mickle, K. L., Ramanathan, S., Rosebrock, A., Oliva, A., Chaudari, A., Yompakdee, C., … Huberman, J.A. (2007). Checkpoint independence of most DNA replication origins in fission yeast. *BMC Molecular Biology*, 8, 112. https://doi.org/10.1186/1471-2199-8-112.

Mostacciuolo, M. L., Pastorello, E., Vazza, G., Miorin, M., Angelini, C., Tomelleri, G., … Trevisan, C.P. (2009). Facioscapulohumeral muscular dystrophy: Epidemiological and molecular study in a north-east Italian population sample. *Clinical Genetics*, 75(6), 550–555. https://doi.org/10.1111/j.1399-0004.2009.01158.x.

Mrasek, K., Schoder, C., Teichmann, A. C., Behr, K., Franze, B., Wilhelm, K., … Weise, A. (2010). Global screening and extended nomenclature for 230 aphidicolin-inducible fragile sites, including 61 yet unreported ones. *International Journal of Oncology*, 36(4), 929–940. https://doi.org/10.3892/ijo_00000572.

Nguyen, K., Walrafen, P., Bernard, R., Attarian, S., Chaix, C., Vovan, C., … Lévy, N. (2011). Molecular combing reveals allelic combinations in facioscapulohumeral dystrophy. *Annals of Neurology*, 70(4), 627–633. https://doi.org/10.1002/ana.22513.

Nordlund, P., & Reichard, P. (2006). Ribonucleotide reductases. *Annual Review of Biochemistry*, 75, 681–706. https://doi.org/10.1146/annurev.biochem.75.103004.142443.

Palmigiano, A., Santaniello, F., Cerutti, A., Penkov, D., Purushothaman, D., Makhija, E., … Blasi, F. (2018). PREP1 tumor suppressor protects the late-replicating DNA by controlling its replication timing and symmetry. *Scientific Reports*, 8(1), 3198. https://doi.org/10.1038/s41598-018-21363-4.

Palumbo, E., Matricardi, L., Tosoni, E., Bensimon, A., & Russo, A. (2010). Replication dynamics at common fragile site FRA6E. *Chromosoma*, 119(6), 575–587. https://doi.org/10.1007/s00412-010-0279-4.

Patel, P. K., Arcangioli, B., Baker, S. P., Bensimon, A., & Rhind, N. (2006). DNA replication origins fire stochastically in fission yeast. *Molecular Biology of the Cell*, *17*(1), 308–316. https://doi.org/10.1091/mbc.e05-07-0657.

Patel, P. K., Kommajosyula, N., Rosebrock, A., Bensimon, A., Leatherwood, J., Bechhoefer, J., & Rhind, N. (2008). The Hsk1(Cdc7) replication kinase regulates origin efficiency. *Molecular Biology of the Cell*, *19*(12), 5550–5558. https://doi.org/10.1091/mbc.e08-06-0645.

Phillips, A. F., Millet, A. R., Tigano, M., Dubois, S. M., Crimmins, H., Babin, L., … Sfeir, A. (2017). Single-molecule analysis of mtDNA replication uncovers the basis of the common deletion. *Molecular Cell*, *65*(3). https://doi.org/10.1016/j.molcel.2016.12.014. 527–538.e6.

Pinkel, D., & Albertson, D. G. (2005). Comparative genomic hybridization. *Annual Review of Genomics and Human Genetics*, *6*, 331–354. https://doi.org/10.1146/annurev. genom.6.080604.162140.

Rageul, J., Park, J. J., Zeng, P. P., Lee, E. A., Yang, J., Hwang, S., … Kim, H. (2020). SDE2 integrates into the TIMELESS-TIPIN complex to protect stalled replication forks. *Nature Communications*, *11*(1), 5495. https://doi.org/10.1038/s41467-020-19162-5.

Rainey, M. D., Quachthithu, H., Gaboriau, D., & Santocanale, C. (2017). DNA replication dynamics and cellular responses to ATP competitive CDC7 kinase inhibitors. *ACS Chemical Biology*, *12*(7), 1893–1902. https://doi.org/10.1021/acschembio.7b00117.

Rainey, M. D., Quinlan, A., Cazzaniga, C., Mijic, S., Martella, O., Krietsch, J., … Santocanale, C. (2020). CDC7 kinase promotes MRE11 fork processing, modulating fork speed and chromosomal breakage. *EMBO Reports*, *21*(8), e48920. https://doi.org/10.15252/ embr.201948920.

Redon, R., Ishikawa, S., Fitch, K. R., Feuk, L., Perry, G. H., Andrews, T. D., … Hurles, M.E. (2006). Global variation in copy number in the human genome. *Nature*, *444*(7118), 444–454. https://doi.org/10.1038/nature05329.

Reisinger, J., Rumpler, S., Lion, T., & Ambros, P. F. (2006). Visualization of episomal and integrated Epstein-Barr virus DNA by fiber fluorescence in situ hybridization. *International Journal of Cancer*, *118*(7), 1603–1608. https://doi.org/10.1002/ijc.21498.

Rhind, N. (2006). DNA replication timing: Random thoughts about origin firing. *Nature Cell Biology*, *8*(12), 1313–1316. https://doi.org/10.1038/ncb1206-1313.

Rivera-Mulia, J. C., Schwerer, H., Besnard, E., Desprat, R., Trevilla-Garcia, C., Sima, J., … Lemaitre, J.M. (2018). Cellular senescence induces replication stress with almost no affect on DNA replication timing. *Cell Cycle*, *17*(13), 1667–1681. https://doi.org/10.1080/15384 101.2018.1491235.

Rouleau, E., Lefol, C., Tozlu, S., Andrieu, C., Guy, C., Copigny, F., … Lidereau, R. (2007). High-resolution oligonucleotide array-CGH applied to the detection and characterization of large rearrangements in the hereditary breast cancer gene BRCA1. *Clinical Genetics*, *72*(3), 199–207. https://doi.org/10.1111/j.1399-0004.2007.00849.x.

Sekar, A., Bialas, A. R., de Rivera, H., Davis, A., Hammond, T. R., Kamitaki, N., … McCarroll, S.A. (2016). Schizophrenia risk from complex variation of complement component 4. *Nature*, *530*(7589), 177–183. https://doi.org/10.1038/nature16549.

Shwan, N. A. A., Louzada, S., Yang, F., & Armour, J. A. L. (2017). Recurrent rearrangements of human amylase genes create multiple independent CNV series. *Human Mutation*, *38*(5), 532–539. https://doi.org/10.1002/humu.23182.

Sobinoff, A. P., Allen, J. A., Neumann, A. A., Yang, S. F., Walsh, M. E., Henson, J. D., … Pickett, H.A. (2017). BLM and SLX4 play opposing roles in recombination-dependent replication at human telomeres. *The EMBO Journal*, *36*(19), 2907–2919. https://doi. org/10.15252/embj.201796889.

Spinner, N. B., Ferguson-Smith, M. A., & Ledbetter, D. H. (2013). Cytogenetic analysis. In D. Rimoin, R. Pyeritz, & B. Korf (Eds.), *Emery and Rimoin's principles and practice of medical genetics and genomics* (pp. 1–18). https://doi.org/10.1016/B978-0-12-383834-6.00029-X.

Staaf, J., Törngren, T., Rambech, E., Johansson, U., Persson, C., Sellberg, G., … Borg, A. (2008). Detection and precise mapping of germline rearrangements in BRCA1, BRCA2, MSH2, and MLH1 using zoom-in array comparative genomic hybridization (aCGH). *Human Mutation, 29*(4), 555–564. https://doi.org/10.1002/humu.20678.

Taylor, J. H., & Hozier, J. C. (1976). Evidence for a four micron replication unit in CHO cells. *Chromosoma, 57*(4), 341–350. https://doi.org/10.1007/BF00332159.

Técher, H., Koundrioukoff, S., Azar, D., Wilhelm, T., Carignon, S., Brison, O., … Le Tallec, B. (2013). Replication dynamics: Biases and robustness of DNA fiber analysis. *Journal of Molecular Biology, 425*(23), 4845–4855. https://doi.org/10.1016/j.jmb.2013.03.040.

Thakar, T., Leung, W., Nicolae, C. M., Clements, K. E., Shen, B., Bielinsky, A. K., & Moldovan, G. L. (2020). Ubiquitinated-PCNA protects replication forks from DNA2-mediated degradation by regulating Okazaki fragment maturation and chromatin assembly. *Nature Communications, 11*(1), 2147. https://doi.org/10.1038/s41467-020-16096-w.

Ungelenk, M. (2021). Sequencing approaches. In T. Liehr (Ed.), *Cytogenomics* (pp. 87–122). Academic Press. Chapter 7 (in this book).

van den Ende, T., Sharifi, S., van der Salm, S. M. A., & van Rootselaar, A. F. (2018). Familial cortical myoclonic tremor and epilepsy, an enigmatic disorder: From phenotypes to pathophysiology and genetics. A systematic review. *Tremor and Other Hyperkinetic Movements (New York, NY), 8*, 503. https://doi.org/10.7916/D85155WJ.

van Deutekom, J. C., Wijmenga, C., van Tienhoven, E. A., Gruter, A. M., Hewitt, J. E., Padberg, G. W., … Frants, R.R. (1993). FSHD associated DNA rearrangements are due to deletions of integral copies of a 3.2 kb tandemly repeated unit. *Human Molecular Genetics, 2*(12), 2037–2042. https://doi.org/10.1093/hmg/2.12.2037.

van Geel, M., Dickson, M. C., Beck, A. F., Bolland, D. J., Frants, R. R., van der Maarel, S. M., … Hewitt, J.E. (2002). Genomic analysis of human chromosome 10q and 4q telomeres suggests a common origin. *Genomics, 79*(2), 210–217. https://doi.org/10.1006/geno.2002.6690.

Venter, J. C., Adams, M. D., Myers, E. W., Li, P. W., Mural, R. J., Sutton, G. G., … Zhu, X. (2001). The sequence of the human genome. *Science, 291*(5507), 1304–1351. https://doi.org/10.1126/science.1058040.

Wakida, T., Ikura, M., Kuriya, K., Ito, S., Shiroiwa, Y., Habu, T., … Furuya, K. (2017). The CDK-PLK1 axis targets the DNA damage checkpoint sensor protein RAD9 to promote cell proliferation and tolerance to genotoxic stress. *eLife, 6*, e29953. https://doi.org/10.7554/eLife.29953.

Weise, A., & Liehr, T. (2021a). Cytogenetics. In T. Liehr (Ed.), *Cytogenomics* (pp. 25–34). Academic Press. Chapter 3 (in this book).

Weise, A., & Liehr, T. (2021b). Interchromosomal interactions with meaning for disease. In T. Liehr (Ed.), *Cytogenomics* (pp. 349–356). Academic Press. Chapter 17 (in this book).

Weise, A., & Liehr, T. (2021c). Molecular karyotyping. In T. Liehr (Ed.), *Cytogenomics* (pp. 73–85). Academic Press. Chapter 6 (in this book).

Yadav, H., & Sharma, P. (2019). A simple and novel DNA combing methodology for Fiber-FISH and optical mapping. *Genomics, 111*(4), 567–578. https://doi.org/10.1016/j.ygeno.2018.03.012.

Zeng, S., Zhang, M. Y., Wang, X. J., Hu, Z. M., Li, J. C., Li, N., … Tang, B.S. (2019). Long-read sequencing identified intronic repeat expansions in SAMD12 from Chinese pedigrees affected with familial cortical myoclonic tremor with epilepsy. *Journal of Medical Genetics, 56*(4), 265–270. https://doi.org/10.1136/jmedgenet-2018-105484.

Molecular karyotyping

Anja Weise and Thomas Liehr

Jena University Hospital, Friedrich Schiller University, Institute of Human Genetics,
Jena, Germany

Chapter outline

Background

Molecular karyotyping, also called "chromosomal microarray" (CMA) or "array-comparative genomic hybridization" (aCGH), is a cytogenomic approach (Liehr, 2021a, 2021b), which historically developed from molecular cytogenetics (Liehr, 2021c). In 1992 it was published that it is possible and informative if one applies whole genomic DNA (WG-DNA) in fluorescence in-situ hybridization (FISH) as probe (Kallioniemi et al., 1992). The trick is to hybridize simultaneously two WG-DNA-samples, labeled in two different colors, in the same ratio, to normal metaphase spreads. The reason this can be informative is best exemplified as follows: if the test or patient WG-DNA is derived from a male with Down syndrome (and labeled in green) and the control WG-DNA from a normal female (labeled in red) and is hybridized to normal male metaphase spreads, this leads to a result as shown in Fig. 6.1. Overall, all chromosomes are stained in a green-red mixed color, here resulting in yellow, except for chromosomes X, Y, 21, and noncovered heterochromatic regions (gray). As a competing hybridization took place along each chromosome, and in the male-derived probe there was only one copy for the X-chromosomal DNA compared to two copies

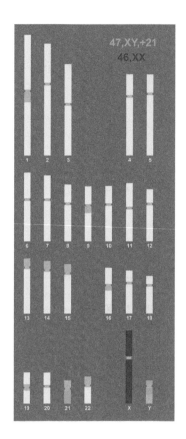

FIG. 6.1

Principle of CGH/aCGH.

DNA derived from a male with trisomy 21 (47,XY,+21) labeled in *green* is co-hybridized with DNA from a normal female (46,XX) labeled in *red* against a normal chromosome set or a micro-array, the results of which are projected on a human genome of 46,XY. At each target sequence without copy number variants, a balanced, *yellow color* is the result. At chromosome 21 there is more "green DNA" and at X-chromosome more "red DNA". Thus, gain of chromosome 21 material and presence of only one X- and one Y-chromosome (= male gender) can be visualized after this experiment.

in the female-derived probe and no copy for the Y-chromosomal DNA in the female control DNA, X-chromosomes on the hybridized metaphases will be stained preferentially in red and the Y-chromosome turns green. For chromosome 21, it is the other way around: there are three copies in the male and two copies in the female-derived probe, leading to a more greenish staining of chromosome 21. This method is called "comparative genomic hybridization" (CGH) (Kallioniemi et al., 1992, 1993).

This chromosome based CGH approach was applied in human genetics, specifically to study solid tumors, mainly between 1992 and 2000 (Gebhart & Liehr, 2000).

In addition, it was used exceptionally for clinical genetic questions (du Manoir et al., 1993). A rarely applied but smart technical amendment was to use a third control probe with a chromosomal trisomy in an additional color (Karhu et al., 1999). Nowadays, CGH is still successfully applied for comparison of closely related species in molecular cytogenetic studies to delineate further evolutionarily caused differences, preferentially visualized in rapidly evolving genomic regions (Spangenberg et al., 2020). In addition, CGH is helpful to identify cryptic sex chromosomes within a species (Deon et al., 2020).

From ~2000 onward, miniaturization approaches, together with cytogenomic knowledge on human DNA-sequence, and the ability to synthetize any desired DNA stretch were advanced enough to apply the principle of CGH not on chromosomes as hybridization targets, but to use instead glass slides with thousands of tiny DNA dots distributed on their surface. In other words, the whole human genome (also commercially available for the murine genome) can nowadays be split into thousands of exactly defined regions and "printed" on glass slides; accordingly, the latter are also called microchips/or microarrays (for a review, see Heller, 2002). Before we look at the different available variants of chromosomal microarrays used for molecular karyotyping/CMA/aCGH, the restrictions of CGH and aCGH need to be discussed.

Advantages and restrictions

The CGH and aCGH techniques have largely identical benefits and shortcuts. CGH gave a significant boost to tumor cytogenetics, as this approach is not in need of metaphases, which are hard to prepare from solid cancer tissues (Gebhart & Liehr, 2000). Thus, CGH delivered for many tumor-types the first ever available "cytogenetic information." Accordingly, it was of great importance that CGH does not need any living or dividing cells, but DNA (also cDNA from RNA) is sufficient for an analysis. The latter is also valid for aCGH (Liehr & Claussen, 2002).

As in aCGH, size and density of the used probes distributed in the microarrays can be more or less freely chosen, and the resolution can be much higher than achievable in banding cytogenetics (Weise & Liehr, 2021) or FISH (Liehr, 2021c). At the same time, a whole genomic view for all euchromatic regions of genomes is possible. For diagnostics, this means that an aCGH analysis can be indicated when gain or loss of genetic material is suggested for a patient (see below).

In addition to these indisputable advantages, the following restrictions must be kept in mind:

- It is exclusively possible to detect imbalanced chromosomal aberrations; a patient with chronic myelogenous leukemia (CML) and the typical acquired translocation t(9;22)(q34;q11.2) would present with an unaltered aCGH profile (Liehr & Claussen, 2002).
- The same is valid in the case of a prenatal case with a karyotype 92,XXYY; polyploidy is not depictable in CGH or regular (non-SNP-based, see below) aCGH (Heller et al., 2000).

- Cells with a chromosomal imbalance need to present the corresponding aberration in at least 20% of the studied tissue - otherwise the aberrant cell clone may be missed - for example, a woman with Turner syndrome with karyotype 45,X[80]/46,XX[20] may just appear to have loss of X-chromosome in aCGH, and mosaic would be missed. There are publications reporting 8% mosaic detection rates (Valli et al., 2011); however, this is normally not achieved in routine settings.
- Depending on the platform resolution used, the detection of imbalances is ~ 100 times higher than in cytogenetics but still not in the range to resolve small indels (insertion or deletion of bases), e.g., within an exon. There are huge efforts to overcome this limitation by NGS-based CNV analysis (Ungelenk, 2021) that might outperform aCGH and MLPA (multiplex ligation-dependent probe amplification) in the next few years (Roca et al., 2019).
- On top of that, it is important to consider that in CGH/aCGH there is no information available about centromeric-regions and heterochromatic regions, i.e., for ~ 10% of the human genome (Liehr, 2021d) this approach is blind. This is especially problematic, as this method-intrinsic lack of data may lead to several possibilities for how an aCGH result may be interpreted, and which chromosomal aberration may be the reason for a specific recognized imbalance (for examples, see Fig. 6.2) (Hochstenbach et al., 2021).
- Finally, there is the problem of submicroscopic copy number variations (CNVs), a problem only being valid for aCGH interpretation. Interestingly, the aCGH technique gave rise to detection of this previously unrecognized level of genetic heteromorphy within the human genome. These CNVs "include hundreds of previously undetected structural variants in the human genome such as deletions, gains and inversions. Surprisingly, CNVs can have sizes of ten to several hundred thousand base pairs and are located in euchromatic regions all over the genome" (Weise et al., 2008). Corresponding data are now collected in the "database of genomic variants" (Database of Genomic Variants, 2020) (see also Iafrate et al., 2004; Sebat et al., 2004). Some of these CNVs may even be large enough to visualize on a cytogenetic level; these are called "euchromatic variants" (Manvelyan et al., 2011). It must be considered that CNVs are not that much expressed in the preferred animal model *Mus musculus* as in human, this is due to the fact that it is not a natural model but based on inbred strains. Thus, genetic variability on a CNV-level has been depleted largely before it could be analyzed - still, actual studies show that "since inbred strains, in large part, reflect the genomic diversity of the individual mice that served as founders for the strains, their genomes reflect a snapshot in time of genome evolution. This developing picture of the mouse CNV landscape provides insight into the forces that shape genomes in both mice and humans" (Cutler & Kassner, 2009).

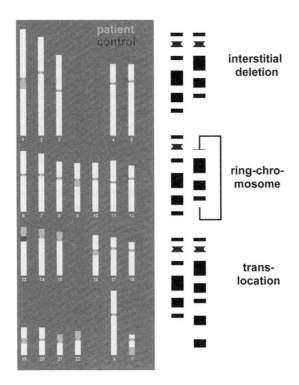

FIG. 6.2

Example of an aCGH result that needed to be checked by a second method.
aCGH revealed a loss of copy numbers near the centromere of chromosome 13 *(red)* -
left part of the figure. Without doing a karyotype or a fluorescence in-situ hybridization, it
remains questionable which chromosomal aberration led to this deletion. Three possible
underlying rearrangements are depicted on the *right part of the figure.* Most likely an
interstitial deletion was detected. However, as aCGH delivers no information on human
heterochromatin, loss of the *red* labeled euchromatic part near the centromere of
chromosome 13 could be accompanied by loss of the short arm of chromosome 13 and its
centromere, and the formation of a neocentric ring chromosome 13. The third possibility
shown here, but still not the last, is an unbalanced translocation of chromosome 13 and
20; here the break and loss of material at the tip of chromosome 20 would be so distal that
it was not detectable by aCGH. The patient would have 45 instead of 46 chromosomes and
a parental balanced translocation would be likely. Without knowing the chromosomes, the
identification of a genetic reason for corresponding clinical problems of the index patient
is possible - still, no well-founded genetic counselling to the family concerning risk of
repetition could be done without additional studies.

Roger Bumgarner, in his excellent review on human microarrays from 2013, also mentioned the following restrictions of the approach to be kept in mind:

"At their core, microarrays are simply devices to simultaneously measure the relative concentrations of many different DNA or RNA sequences. Arrays provide an indirect measure of relative concentration. That is the signal measured at a given position on a microarray is typically assumed to be proportional to the concentration of a presumed single species in solution that can hybridize to that location. However, due to the kinetics of hybridization, the signal level at a given location on the array is not linearly proportional to concentration of the species hybridizing to the array. At high concentrations the array will become saturated and at low concentrations, equilibrium favors no binding. Hence, the signal is linear only over a limited range of concentrations in solution. Also, especially for complex mammalian genomes, it is often difficult (if not impossible) to design arrays in which multiple related DNA/RNA sequences will not bind to the same probe on the array. A sequence on an array that was designed to detect 'gene A', may also detect 'genes B, C and D' if those genes have significant sequence homology to gene A. This can particularly problematic for gene families and for genes with multiple splice variants. Finally, a DNA array can only detect sequences that the array was designed to detect. That is, if the solution being hybridized to the array contains RNA or DNA species for which there is no complimentary sequence on the array, those species will not be detected"

(Bumgarner, 2013)

Material applied in molecular karyotyping

As already mentioned, in molecular karyotyping, the probe to be labeled, applied on microchips, and analyzed consists normally of DNA. According to the microchip used, different amounts of probe DNA must be available; this ranges from a few 50 ng to 500–1000 ng being necessary for a reliable result (Heller, 2002). Best suited is DNA extracted from living or cryofixed cells. Additionally, DNA from formalin fixed paraffin embedded (FFPE) tissue may be used; however, here the fixation must have been done using buffered formalin for a few hours only. Otherwise, FFPE-derived DNA may be degraded enough that it is not suitable for labeling and hybridization. In the case of very limited amounts of investigation material, i.e., a few pico- to a nanogram of DNA, this DNA may be amplified by a polymerase chain reaction using degenerated primers (DOP-PCR) (Heller et al., 2000) and applied in aCGH. Finally, RNA may be transcribed into cDNA and applied for expression array analysis in the field of transcriptomics (Heller, 2002). The latter research area also applies microarrays directed against proteins (Crecelius et al., 2015; Von Eggeling et al., 2007); however, this is not a topic of this chapter.

Microarrays in molecular karyotyping

Each traditional microarray is based on the same basic principle: it is a glass slide on which thousands of defined DNA-probes are spotted (spot size 10–500 µm). All the below-mentioned different kinds of microarrays are here just distinguished by the kind of DNA of which the probe on the array is composed. Due to the high grade of commercialization in human cytogenomics, microarrays based on bacterial artificial chromosome (BAC) probes do not play a major role here, although they may be the first choice for molecular karyotyping in other species (Ríos et al., 2008; Thomas et al., 2005). No attention is given here to the many kinds of resolution available for microarrays; it is a method-intrinsic advantage of this approach that probes (including resolution) may be selected according to scientific question and application.

In Table 6.1, DNA-probes spotted on microarrays applied in human molecular karyotyping are summarized; others, such as for microbiology or other organisms (Heller, 2002; Miller & Tang, 2009; Naidu & Suneetha, 2012), are not discussed here. In addition, the companies offering different devices suitable for molecular karyotyping are intentionally not mentioned. Corresponding information is available in detail in other literature on this topic.

Fluorescence imaging using corresponding scanners is the most common method for the reading out of microarrays. The obtained data can normally be exclusively evaluated with the support of bioinformatics, as described elsewhere (Heller, 2002; Naidu & Suneetha, 2012). Just as a reminder, the "database of genomic variants" (Database of Genomic Variants, 2020) is extremely important here to exclude CNVs currently marked as nondeleterious from analysis in diagnostics of so-called microdeletion or microduplication syndromes (Weise et al., 2012).

Table 6.1 List of DNA-derived probes for microarrays. DNA-derived probes that can serve as bases for microarrays are listed and their approximate lengths.

Microarrays can be based on	Probe size (base pairs)
BACs	150,000–350,000
Shotgun library clones	500–2000
dsDNA from PCR	200–800
cDNA oligonucleotides	25–80
Oligonucleotides synthesized directly on the surface by photochemistry	25–100
Single nucleotide polymorphism (SNP) genotyping platforms	25–100
Next-generation sequencing-based aCGH-like evaluation	Not applicable

Applications

Molecular karyotyping can be applied in cytogenomic research and human genetics diagnostics. It is impossible to review here all the different possible applications. Thus, only a few examples are picked out and presented for both mentioned fields.

Molecular karyotyping approach in human genetic diagnostics

The major advantage of molecular karyotyping in human genetics is its ability to detect potentially disease-causing, submicroscopic CNVs in children or adults with normal GTG-banding based karyotype. Since single nucleotide polymorphism (SNP)-based molecular karyotyping can be done, epigenetic changes (uniparental disomy (UPD) or loss of heterozygosity (LOH)) (Eggermann, 2021; Harutyunyan & Hovhannisyan, 2021) can also be grasped by this approach (Harper & Harton, 2010). However, it must be kept in mind that SNP-based CMA can exclusively detect relatively long stretches of isodisomy, thus UPD based on heterodisomy is missed (Liehr, 2020). In different countries, there are different ideas, suggestions, or guidelines regarding what to use as a first-line test for diagnostics of such cases. While, for example, in the UK, molecular karyotyping is a first-line test, and GTG-karyotype is only done as a second step in rare exceptions (Ahn et al., 2013), in Germany, molecular karyotyping cannot be reimbursed from insurance companies in case no GTG-karyotype was done before. Without a doubt, molecular karyotyping is an important component of modern human cytogenomic diagnostics (Kharbanda et al., 2015). However, it is also well-known that in connection with different "genetic background" in a parent and its offspring, a microdeletion may lead, in the child, to a known associated phenotype (e.g., DiGeorge syndrome), while the parent with the same microdeletion is (virtually) unaffected. To solve this Gordian knot, a few years ago the so-called two-hit model was suggested, proposing that a second large CNV may uncover the clinical picture in the affected individual, while the "healthy person" was spared from such a second CNV (Girirajan et al., 2010). Overall, finding a CNV in the genome of a patient is just the beginning of a process of comparison with the literature and interpretation of the finding in the individual context, or, vice versa, going back to the parents to search for predisposing rearrangements leading to recurrent microdeletions or microduplications in the offspring (Koolen et al., 2006; Liehr et al., 2018).

Molecular karyotyping approach in cytogenomic research

As already mentioned, ideas of what to study by molecular karyotyping are more or less unrestricted:

- One cytogenomic question already mentioned to be answered by this approach is to obtain transcriptomic information about cells or tissues by analyzing RNA/cDNA instead of genomic DNA by a microarray-platform (Heller, 2002).

- During recent years, SNP-aCGH platforms have identified major influences of acquired LOH in tumor biology; this finding has just begun to alter our understanding of tumor development and progression, including the as yet not understood processes of (erroneous) monosomic or trisomic rescue events (Erola et al., 2019).
- A major restriction of aCGH is that only imbalanced and/or such aberrations being present in > 20% of the studied cells may be grasped. However, in such an instance, chromosome microdissection may help. Small supernumerary marker chromosomes (sSMCs) present in low-level mosaic may be microdissected, and the specific DNA amplified and used to map the size and origin of the sSMCs (Backx et al., 2007). The same trick may be applied to map breakpoints of balanced chromosomal translocations (Jancuskova et al., 2013; Zamariolli et al., 2020).
- Another smart idea to get insights into the three-dimensional architecture of a nucleus is the ChIP-on-chip approach (for ChIP, see Harutyunyan & Hovhannisyan, 2021), which "is a technology that combines chromatin immunoprecipitation ('ChIP') with DNA microarray ('chip'). Like regular ChIP, ChIP-on-chip is used to investigate interactions between proteins and DNA *in vivo*. Specifically, it allows the identification of the cistrome, the sum of binding sites, for DNA-binding proteins on a genome-wide basis. The goal of ChIP-on-chip is to locate protein binding sites that may help identify functional elements in the genome. For example, in the case of a transcription factor as a protein of interest, one can determine its transcription factor binding sites throughout the genome. Overall, ChIP-on-chip offers both potential to complement our knowledge about the orchestration of the genome on the nucleotide level and information on higher levels of information and regulation as it is propagated by research on epigenetics" (Wiki/ChIP-on-Chip, 2020).
- In addition, chromothripsis can be detected and characterized by CMA; for more details, see Pellestor et al. (2021).

Conclusion

As summarized by Michael J. Heller (Heller, 2002), microarray technology is nowadays much more than application of glass slides as microarrays. Using this as a starting point, molecular karyotyping and many other approaches based on microarray technologies have been developed. Techniques using BeadArray (Sosnowski et al., 1997), sequencing by hybridization (SBH) (Heller, 1996), or LabCard device platform, the latter using electric fields to move fluids through capillaries on chips (Fodor et al., 1991), and many other sophisticated attempts have been established to achieve the same goal - to compare and/or determine copy numbers of specific DNA regions of a given genome at a given time point. The many different technical amendments can, at the same time, be more accurate and/or open to new possibilities to obtain additional information about the studied nucleic acids. Accordingly, molecular karyotyping/CMA/aCGH is and will remain a major player in the concert of cytogenomic approaches.

References

Ahn, J. W., Bint, S., Bergbaum, A., Mann, K., Hall, R. P., & Ogilvie, C. M. K. (2013). Array CGH as a first line diagnostic test in place of karyotyping for postnatal referrals—Results from four years' clinical application for over 8,700 patients. *Molecular Cytogenetics, 6*(1), 16. https://doi.org/10.1186/1755-8166-6-16.

Backx, L., Van Esch, H., Melotte, C., Kosyakova, N., Starke, H., Frijns, J. P., … Vermeesch, J.R. (2007). Array painting using microdissected chromosomes to map chromosomal breakpoints. *Cytogenetic and Genome Research, 116*(3), 158–166. https://doi.org/10.1159/000098181.

Bumgarner, R. (2013). Overview of dna microarrays: Types, applications, and their future. *Current Protocols in Molecular Biology, 101*, 22.1. https://doi.org/10.1002/0471142727.mb2201s101.

Crecelius, A. C., Schubert, U. S., & Von Eggeling, F. (2015). MALDI mass spectrometric imaging meets "omics": Recent advances in the fruitful marriage. *Analyst, 140*(17), 5806–5820. https://doi.org/10.1039/c5an00990a.

Cutler, G., & Kassner, P. D. (2009). Copy number variation in the mouse genome: Implications for the mouse as a model organism for human disease. *Cytogenetic and Genome Research, 123*(1–4), 297–306. https://doi.org/10.1159/000184721.

Database of Genomic Variants. (2020). http://dgv.tcag.ca/dgv/app/home.

Deon, A. G., Glugoski, G., Vicari, R. M., Nogaroto, V., Sassi, C. F. D. M., Cioffi, M. D. B., … Moreira-Filho, O. (2020). Highly rearranged karyotypes and multiple sex chromosome systems in armored catfishes from the genus Harttia (Teleostei, Siluriformes). *Genes, 11*(11), 1366. https://doi.org/10.3390/genes11111366.

du Manoir, S., Speicher, M. R., Joos, S., Schröck, E., Popp, S., Döhner, H., … Cremer, T. (1993). Detection of complete and partial chromosome gains and losses by comparative genomic in situ hybridization. *Human Genetics, 90*(6), 590–610. https://doi.org/10.1007/BF00202476.

Eggermann, T. (2021). Epigenetics. In T. Liehr (Ed.), *Cytogenomics* (pp. 389–401). Academic Press. Chapter 20 (in this book).

Erola, P., Torabi, K., Miró, R., & Camps, J. (2019). The non-random landscape of somatically-acquired uniparental disomy in cancer. *Oncotarget, 10*(40), 3982–3984. https://doi.org/10.18632/oncotarget.26987.

Fodor, S. P. A., Read, J. L., Pirrung, M. C., Stryer, L., Lu, A. T., & Solas, D. (1991). Light-directed, spatially addressable parallel chemical synthesis. *Science, 251*(4995), 767–773. https://doi.org/10.1126/science.1990438.

Gebhart, E., & Liehr, T. (2000). Patterns of genomic imbalances in human solid tumors (review). *International Journal of Oncology, 16*(2), 383–399. https://doi.org/10.3892/ijo.16.2.383.

Girirajan, S., Rosenfeld, J. A., Cooper, G. M., Antonacci, F., Siswara, P., Itsara, A., … Eichler, E.E. (2010). A recurrent 16p12.1 microdeletion supports a two-hit model for severe developmental delay. *Nature Genetics*, 203–209. https://doi.org/10.1038/ng.534.

Harper, J. C., & Harton, G. (2010). The use of arrays in preimplantation genetic diagnosis and screening. *Fertility and Sterility*, *94*(4), 1173–1177. https://doi.org/10.1016/j.fertnstert.2010.04.064.

Harutyunyan, T., & Hovhannisyan, G. (2021). Approaches for studying epigenetic aspects of the human genome. In T. Liehr (Ed.), *Cytogenomics* (pp. 155–209). Academic Press. Chapter 10 (in this book).

Heller, A., Chudoba, I., Bleck, C., Senger, G., Claussen, U., & Liehr, T. (2000). Microdissection based comparative genomic hybridization analysis (micro-CGH) of secondary acute myelogenous leukemias. *International Journal of Oncology*, *16*(3), 461–468. https://doi.org/10.3892/ijo.16.3.461.

Heller, M. J. (1996). An active microelectronics device for multiplex DNA analysis. *IEEE Engineering in Medicine and Biology Magazine*, *15*(2), 100–103. https://doi.org/10.1109/51.486725.

Heller, M. J. (2002). DNA microarray technology: Devices, systems, and applications. *Annual Review of Biomedical Engineering*, *4*, 129–153. https://doi.org/10.1146/annurev.bioeng.4.020702.153438.

Hochstenbach, R., Liehr, T., & Hastings, R. (2021). Chromosomes in the genomic age. Preserving cytogenomic competence of diagnostic genome laboratories. *European Journal of Human Genetics*. https://doi.org/10.1038/s41431-020-00780-y.

Iafrate, A. J., Feuk, L., Rivera, M. N., Listewnik, M. L., Donahoe, P. K., Qi, Y., … Lee, C. (2004). Detection of large-scale variation in the human genome. *Nature*, *36*(9), 949–951. https://doi.org/10.1038/ng1416.

Jancuskova, T., Plachy, R., Stika, J., Zemankova, L., Hardekopf, D., Liehr, T., … Pekova, S. (2013). A method to identify new molecular markers for assessing minimal residual disease in acute leukemia patients. *Leukemia Research*, *37*(10), 1363–1373. https://doi.org/10.1016/j.leukres.2013.06.009.

Kallioniemi, A., Kallioniemi, O. P., Sudar, D., Rutovitz, D., Gray, J. W., Waldman, F., & Pinkel, D. (1992). Comparative genomic hybridization for molecular cytogenetic analysis of solid tumors. *Science*, *258*(5083), 818–821. https://doi.org/10.1126/science.1359641.

Kallioniemi, O. P., Kallioniemi, A., Sudar, D., Rutovitz, D., Gray, J. W., Waldman, F., & Pinkel, D. (1993). Comparative genomic hybridization: A rapid new method for detecting and mapping DNA amplification in tumors. *Seminars in Cancer Biology*, *4*(1), 41–46.

Karhu, R., Rummukainen, J., Lörch, T., & Isola. (1999). Four-color CGH: A new method for quality control of comparative genomic hybridization. *Genes, Chromosomes & Cancer*, *24*(2), 112–118.

Kharbanda, M., Tolmie, J., & Joss, S. (2015). How to use… microarray comparative genomic hybridisation to investigate developmental disorders. *Archives of Disease in Childhood. Education and Practice Edition*, *100*(1), 24–29. https://doi.org/10.1136/archdischild-2014-306022.

Koolen, D. A., Vissers, L. E. L. M., Pfundt, R., De Leeuw, N., Knight, S. J. L., Regan, R., … De Vries, B.B.A. (2006). A new chromosome 17q21.31 microdeletion syndrome associated with a common inversion polymorphism. *Nature Genetics*, *38*(9), 999–1001. https://doi.org/10.1038/ng1853.

Liehr, T. (2020). *Cases with uniparental disomy*. http://cs-tl.de/DB/CA/UPD/0-Start.html. (Accessed 12 December 2020).

Liehr, T. (2021a). A definition for cytogenomics - Which also may be called chromosomics. In T. Liehr (Ed.), *Cytogenomics* (pp. 1–7). Academic Press. Chapter 1 (in this book).

Liehr, T. (2021b). Overview of currently available approaches used in cytogenomics. In T. Liehr (Ed.), *Cytogenomics* (pp. 11–24). Academic Press. Chapter 2 (in this book).

Liehr, T. (2021c). Molecular cytogenetics. In T. Liehr (Ed.), *Cytogenomics* (pp. 35–45). Academic Press. Chapter 4 (in this book).

Liehr, T. (2021d). Repetitive elements, heteromorphisms, and copy number variants. In T. Liehr (Ed.), *Cytogenomics* (pp. 373–388). Academic Press. Chapter 19 (in this book).

Liehr, T., & Claussen, U. (2002). Multicolor-FISH approaches for the characterization of human chromosomes in clinical genetics and tumor cytogenetics. *Current Genomics*, *3*(3), 213–235. https://doi.org/10.2174/1389202023350525.

Liehr, T., Schreyer, I., Kuechler, A., Manolakos, E., Singer, S., Dufke, A., … Weise, A. (2018). Parental origin of deletions and duplications—About the necessity to check for cryptic inversions. *Molecular Cytogenetics*, *11*(1), 20. https://doi.org/10.1186/s13039-018-0369-1.

Manvelyan, M., Cremer, F. W., Lancé, J., Kläs, R., Kelbova, C., Ramel, C., … Liehr, T. (2011). New cytogenetically visible copy number variant in region 8q21.2. *Molecular Cytogenetics*, *4*(1), 1.

Miller, M. B., & Tang, Y. W. (2009). Basic concepts of microarrays and potential applications in clinical microbiology. *Clinical Microbiology Reviews*, *22*(4), 611–633. https://doi.org/10.1128/CMR.00019-09.

Naidu, C. K., & Suneetha, Y. (2012). Current knowledge on microarray technology—An overview. *Tropical Journal of Pharmaceutical Research*, *11*(1), 153–164. https://doi.org/10.4314/tjpr.v11i1.20.

Pellestor, F., Gaillard, J.-B., Schneider, A., Puechberty, J., & Gatinois, V. (2021). Chromoanagenesis phenomena and their formation mechanisms. In T. Liehr (Ed.), *Cytogenomics* (pp. 213–245). Academic Press. Chapter 11 (in this book).

Ríos, G., Naranjo, M. A., Iglesias, D. J., Ruiz-Rivero, O., Geraud, M., Usach, A., & Talón, M. (2008). Characterization of hemizygous deletions in Citrus using array—Comparative genomic hybridization and microsynteny comparisons with the poplar genome. *BMC Genomics*, *9*(1), 381. https://doi.org/10.1186/1471-2164-9-381.

Roca, I., González-Castro, L., Fernández, H., Couce, M. L., & Fernández-Marmiesse, A. (2019). Free-access copy-number variant detection tools for targeted next-generation sequencing data. *Mutation Research, Reviews in Mutation Research*, *779*, 114–125. https://doi.org/10.1016/j.mrrev.2019.02.005.

Sebat, J., Lakshmi, B., Troge, J., Alexander, J., Young, J., Lundin, P., … Wigler, M. (2004). Large-scale copy number polymorphism in the human genome. *Science*, *305*(5683), 525–528. https://doi.org/10.1126/science.1098918.

Sosnowski, R. G., Tu, E., Butler, W. F., O'Connell, J. P., & Heller, M. J. (1997). Rapid determination of single base mismatch mutations in DNA hybrids by direct electric field control. *Proceedings of the National Academy of Sciences of the United States of America*, *94*(4), 1119–1123. https://doi.org/10.1073/pnas.94.4.1119.

Spangenberg, V., Kolomiets, O., Stepanyan, I., Galoyan, E., de Bello Cioffi, M., Martynova, E., … Arakelyan, M. (2020). Evolution of the parthenogenetic rock lizard hybrid karyotype: Robertsonian translocation between two maternal chromosomes in Darevskia rostombekowi. *Chromosoma*, *129*(3–4), 275–283. https://doi.org/10.1007/s00412-020-00744-7.

Thomas, R., Scott, A., Langford, C. F., Fosmire, S. P., Jubala, C. M., Lorentzen, T. D., … Breen, M. (2005). Construction of a 2-Mb resolution BAC microarray for CGH analysis of canine tumors. *Genome Research*, *15*(12), 1831–1837. https://doi.org/10.1101/gr.3825705.

Ungelenk, M. (2021). Sequencing approaches. In T. Liehr (Ed.), *Cytogenomics* (pp. 87–122). Academic Press. Chapter 7 (in this book).

Valli, R., Marletta, C., Pressato, B., Montalbano, G., Curto, L., Pasquali, F., & Maserati, F. (2011). Comparative genomic hybridization on microarray (a-CGH) in constitutional and acquired mosaicism may detect as low as 8% abnormal cells. *Molecular Cytogenetics*, *4*(1), 13. https://doi.org/10.1186/1755-8166-4-13.

Von Eggeling, F., Melle, C., & Ernst, G. (2007). Microdissecting the proteome. *Proteomics*, *7*(16), 2729–2737. https://doi.org/10.1002/pmic.200700079.

Weise, A., Gross, M., Mrasek, K., Mkrtchyan, H., Horsthemke, B., Jonsrud, C., … Liehr, T. (2008). Parental-origin-determination fluorescence in situ hybridization distiguishes homologous human chromosomes on a single-cell level. *International Journal of Molecular Medicine*, *21*(2), 189–200.

Weise, A., & Liehr, T. (2021). Cytogenetics. In T. Liehr (Ed.), *Cytogenomics* (pp. 25–34). Academic Press. Chapter 3 (in this book).

Weise, A., Mrasek, K., Klein, E., Mulatinho, M., Llerena, J. C., Hardekopf, D., … Liehr, T. (2012). Microdeletion and microduplication syndromes. *Journal of Histochemistry and Cytochemistry*, *60*(5), 346–358. https://doi.org/10.1369/0022155412440001.

Wiki/ChIP-on-Chip. (2020). https://en.wikipedia.org/wiki/ChIP-on-chip.

Zamariolli, M., Di-Battista, A., Moyses-Oliveira, M., De Mello, C. B., De Paula Ramos, M. A., Liehr, T., & Melaragno, M. I. (2020). Disruption of PCDH10 and TNRC18 genes due to a balanced translocation. *Cytogenetic and Genome Research*, *160*(6), 321–328. https://doi.org/10.1159/000508820.

Sequencing approaches

7

Martin Ungelenk

Jena University Hospital, Friedrich Schiller University, Institute of Human Genetics,
Jena, Germany

Chapter outline

Introduction

The primary structure of proteins is determined by their underlying sequence of DNA nucleotides - the genetic code. By transcription and translation, the distinct order of DNA nucleotides defines the corresponding order of amino acids. This forms

Cytogenomics. https://doi.org/10.1016/B978-0-12-823579-9.00021-7

the primary structure of every protein in living cells. Clearly an alteration within the genetic code can lead to dis- or nonfunctional proteins (Crick, 1962). That in turn can lead to various disease phenotypes. Therefore, understanding the genetic background of inherited or acquired diseases at base pair (bp) level is vital in a diagnostic setting, but also for cytogenomic research (Liehr, 2021a, 2021b).

Several sequencing approaches emerged in the last decades that increased in accuracy and foremost in throughput yet dramatically decreased in cost. The evolution of these sequencing techniques was termed "generations" and will be illustrated within this chapter.

Next to DNA, also RNA plays a very important role in molecular life science. Transcriptional states of individual cells are snap shots of the cells' activity at that time point. Entire transcriptomes of either whole tissues or single cells can be investigated by sequencing. In most cases, enzymatic reverse transcription is carried out to transcribe RNA into cDNA molecules. In this chapter, sequencing approaches are described that could also be used to sequence RNA molecules. For simplicity reasons, sequencing of DNA is mentioned most often but can be interchanged with sequencing of RNA.

The first generation

In 1953 the double-helix model of the DNA molecules was proposed by James Watson and Francis Crick based upon X-ray crystallography experiments (Franklin & Gosling, 1953; Wilkins et al., 1953). The four complementary nucleotides were described and how each nucleotide would be paired with its counterpart forming two separate strands running into opposite directions (Watson & Crick, 1953). As the structure was apparent, researchers would try to determine the sequence of these nucleotides. Early methods involved location-specific primer extensions using DNA-polymerase catalysis and distinct nucleotide labeling.

Maxam and Gilbert's chemical cleavage

In 1976, Maxam and Gilbert developed a technique that allowed sequencing of DNA fragments up to 100 bp. There, cleaving the DNA at chemically modified sites enabled the determination of the sequence by size selection using gel electrophoresis (Maxam & Gilbert, 1977).

Two distinct reactions enable the differentiation between the purines (adenine and guanine; A and G) and the pyrimidines (cytosine and thymine; C and T) (Fig. 7.1A). Methylation of the purines by dimethyl sulfate creates an unstable glycosidic bond that breaks easily on heating with 0.1 M alkali at 90°C. The pyrimidines are yet not affected by methylation. Instead, breakage of the pyrimidines can be induced by hydrozylation using hydrazine, followed by cleavage with piperidine. Subsequently, there is the need to separate the two bases within each subtype. Generally, guanines methylate fivefold faster than adenines upon treatment. Thus, using the methylation

Maxim and Gilbert's chemical cleaving

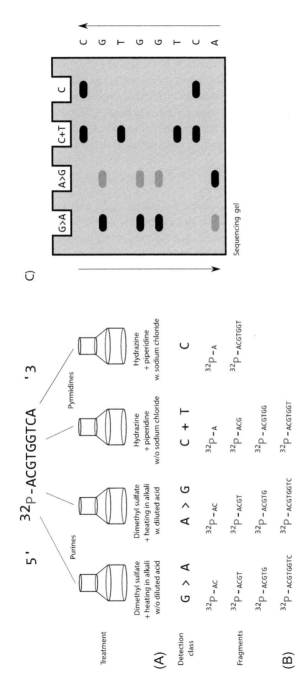

FIG. 7.1

Maxim and Gilbert's chemical cleaving. DNA molecules are cleaved at chemically modified sites. The sequence of the resulting fragments is then determined using gel electrophoresis and size selection. (A) Scheme of the four different chemical treatments. Dimethyl sulfate methylates purines and induces breaks at adenines (A) or guanines (G). Diluted acid suppresses methylation reaction of guanines. Hydrazine induces hydrozylation of pyrimidines followed by cleavage with piperidine at cytosines (C) or thymines (T). Sodium chloride represses the hydrozylation of thymines (T). (B) Resulting fragments of the example DNA sequence belonging to four distinct classes (G > A, A > G, C + T, C). Radioactive labeled phosphate (32P) is used for detection. (C) Gel electrophoresis of the fragments showing a distinct band pattern. From the presence or absence of specific bands, the actual sequence can be determined.

approach will result in an increased breakage at guanines (class G > A). To receive an adenine-enhanced cleavage, Maxam and Gilbert suppressed the methylation of guanines in a 3rd reaction using diluted acid (class A > G). In contrast, the hydrozylation of the pyrimidines does not favor one base over the other (class C + T). Therefore, sodium chloride was used in a fourth reaction to repress the reaction of thymines with hydrazine resulting in cytosine-only cleavage (class C) (Fig. 7.1B). The concentration of the chemical treatment is carefully controlled to enable only one modification event per DNA molecule. Side-by-side gel electrophoresis of the four reactions results in a specific band pattern. From the presence or absence of specific bands, the actual sequence can be determined (Fig. 7.1C; Maxam & Gilbert, 1977).

Sanger sequencing

Maxam and Gilbert's complex method using hazardous chemicals was difficult to upscale, as only short sequences could be determined. It was quickly replaced by Sanger's chain termination method, which was the method of choice for obtaining a DNA sequence for three decades (Kumar et al., 2019). The key idea of Sanger's method was to add chain termination nucleotides to an in vitro DNA replication reaction using a DNA polymerase (Sanger et al., 1977; Sanger & Coulson, 1975).

In general, DNA polymerase transcription incorporates complementary free deoxynucleotidetriphosphates (dNTPs) into a newly synthesized strand. Under controlled conditions in in vitro experiments, this growing oligonucleotide chain could be prematurely stopped by the introduced chain-terminating nucleotides that lack one 3′-hydroxyl group necessary for further elongation. A carefully controlled mixture of standard dNTPs and chain-elongation-inhibiting ddNTPs is applied, usually in a 100:1 ratio. Four different reactions, each containing only one of the ddNTPs (ddATPs, ddCTPs, ddGTPs, or ddTTPs), were carried out. When incubating with primer and template DNA, this resulted in an assortment of fragments of different length, originally radioactive labeled at the 5′-end, ending with one of the ddNTPs at the 3′-end (Fig. 7.2A). All reactions were then run in parallel on a gel leading to a specific band pattern. The order of the different bands can be read off from the bottom of the gel to the top to determine the actual template sequence (Fig. 7.2B; Sanger et al., 1977).

Further advances in Sanger sequencing

Over that period, great advances were made in the Sanger technique, such as prior PCR amplification, fluorescent labeling, capillary electrophoresis, and general automation. PCR amplification of the template DNA ensures enough DNA available for Sanger sequencing, and virtually every sequence can be determined as long primer pairs are available (Heather & Chain, 2016). Fluorescence labeling with different fluorophores enabled Sanger sequencing to be carried out within a single reaction including all four ddNTPs (Fig. 7.2C) (Smith et al., 1985). Each of these fluorophores emits light at a different spectrum and thus a separation is feasible in a single lane (Fig. 7.2D; Prober et al., 1987).

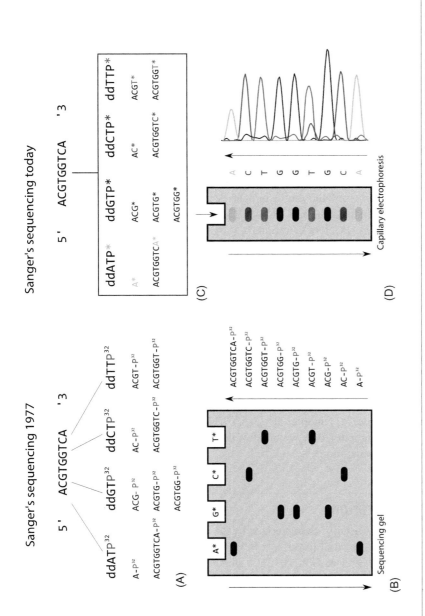

FIG. 7.2

Sanger sequencing. Irreversible chain termination nucleotides (ddNTPs) are used together with unmodified nucleotides (dNTPs) in an in vitro DNA replication reaction conducted by a DNA polymerase. Originally, four different reactions were carried out containing each a single type of ddNTPs and all dNTPs. The ddNTPs had been radioactively labeled (P32). Nowadays, fluorescently labeled ddNTPs (*) are used. All four ddNTP*s can be used within the same reaction. Automated capillary electrophoresis can process the size selection within one lane. (A) Example of DNA sequence fragments resulting of Sanger sequencing in four separate reactions with radioactively labeled ddNTPs. (B) Sequencing gel of the four separate reactions leading to a specific band pattern. From the presence or absence of specific bands, the actual sequence can be determined in the correct order. (C) Example of DNA sequence fragments resulting of Sanger sequencing in a single reaction with fluorescently labeled ddNTP*s. (D) Single lane of capillary electrophoresis. The emitted light of the fluorescent dyes (*) is detected and is used to determine the specific sequence.

Using fluorescence and the application of capillary electrophoresis led to automation of sequencing. Beck and Pohl (1984) described a new technique of direct blotting onto an immobilizing matrix during electrophoresis that marked the first step toward commercialization and automation of Sanger sequencing. This was followed by Applied Biosystems' marketing of the first fully automated sequencing machine, the ABI 370, in 1987 (Smith et al., 1986) and by Dupont's Genesis 2000 (Prober et al., 1987). The capillary sequencers of today have more than one capillary ranging from 4 to 24 capillaries and creating 600–1000 bases of accurate sequences (Slatko et al., 2018). Due to the nature of size selection, the first bases, roughly 40 bp, are usually of low quality, as they cannot be separated distinguishable within the capillary by size.

Yet only a small amount of DNA fragments could be analyzed at a time and sequencing entire genes was both expensive and labor-intensive (Heather & Chain, 2016). Nevertheless, a milestone in DNA sequencing was the isochronal publishing of a draft sequence of the entire human genome in Nature and Science in February 2001 by the International Human Genome Sequencing Consortium and by Craig Venter and his Celera Corporation (Lander et al., 2001; Venter et al., 2001). Both investigations were carried out using a vast amount of automated capillary sequencers based on Sanger technology among several facilities around the world, devouring billions of dollars. Furthermore, it took twelve years to complete 99% of the first draft (Human Genome Project Completion: Frequently Asked Questions, n.d.). This illustrated the need for faster and more cost-effective sequencing strategies.

The release of the human genome reference laid the foundation for new genetic research and development and new avenues for advances in cytogenomics, medicine, and biotechnology. With a reference at hand, researchers were able to identify differences between newly sequenced data and the said reference. As a result, genetic variants could be identified that increase the risk for genetic disorders like cancer or neurodegenerative disorders (Naidoo et al., 2011). The desperate need for cheaper methods with high throughput led to evolution of the "next-generation" of sequencing - the actual second generation.

The second generation

The second generation is usually characterized by the ability to sequence the entire genome or enriched parts of it at lower cost in vastly decreased time compared to Sanger sequencing (McGinn & Gut, 2013). A genome-wide shotgun fragmentation approach is used, followed by amplification and massive parallelization of automated sequencing of the smaller fragments, and subsequent reassembly of the determined sequences. The shotgun fragmentation splits the genomic DNA randomly into smaller pieces either using mechanical shearing (e.g., ultrasound) or enzymatic reactions (e.g., endonucleases, transposons) (Fig. 7.3A; Stranneheim & Lundeberg, 2012).

Second generation of sequencing

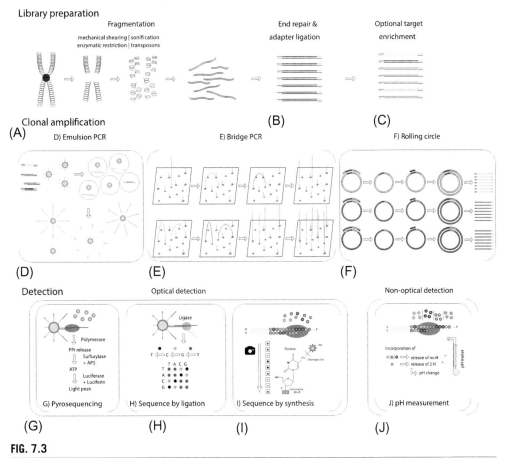

FIG. 7.3

Next-(second)-generation sequencing. In general, the library preparation of second-generation sequencing involves shotgun fragmentation of genomic DNA, end repair, and adapter ligation. Optionally, target enrichment can be conducted. Clonal amplification is another key feature of the second-generation sequencing. Either emulsion PCR, bridge PCR, or rolling circle amplification is commonly used. Most second-generation methods rely on optical detection, whereas also nonoptical pH measurements are being used. Library preparation (A, B, C)/Clonal amplification (D, E, F)/detection (G, H, I, J): (A) Shotgun fragmentation of genomic DNA can be achieved either by mechanical shearing, sonification, enzymatic restriction, or use of transposons. (B) Sticky ends are converted into blunt ends by end repair followed by specific adaptor ligation. (C) Target enrichment can be achieved via hybridization probes or specific primers (e.g., for exomes or smaller gene panels). (D) Emulsion PCR (ePCR) features small reaction volumes (droplets) where individual PCR can take place. To reach clonal amplification via ePCR, beads with fitting attached adapters and single library fragments become isolated in emulsion droplets followed by repeated PCR. Each bead carries then multiple copies of the original single library fragment. (E) Bridge PCR realizes clonal

(Continued)

Several different sequencing approaches were commercialized in the new millennium replacing Sanger as main protagonist in sequencing. The actual commercial breakthrough came with 454 Life Sciences' marketed massive parallelized version of pyrosequencing in 2004 (Voelkerding et al., 2009). The first version of their machine already reduced the costs sixfold in comparison to automated Sanger sequencing (Schuster, 2008; Wetterstrand, 2013). Subsequently, the Solexa 1G Genetic Analyzer from Illumina was commercialized in 2006, and the SOLiD (Supported Oligonucleotide Ligation and Detection) system from Applied Biosystems, based on polony sequencing, was launched in 2007 (Applied Biosystems, 2008). In 2010, Ion Torrent Systems started ion semiconductor sequencing, a new technology without optical detection of fluorescence signals, but relying on detection of ion flow changes (Rusk, 2011). In 2013, Beijing Genomics Institute (BGI) acquired Complete Genomics and took their combinatorial probe anchor synthesis sequencing approach

FIG. 7.3, CONT'D

amplification through cluster generation of individual molecules attached to the surface of a flow cell. Spatially distributed single-stranded library fragments hybridize to their counterpart oligonucleotide imprinted on the flow cell surface. The fragment is then bridging toward another fitting anchor probe on the flow cell surface followed by PCR creating two copies of covalently bound library fragments. This process is repeated 25 times leading to roughly 1000 clonal copies. (F) Rolling circle amplification is used when the library fragments can be made circularized. Then continued DNA synthesis can produce multiple single-stranded linear copies of the original DNA in a continuous head-to-tail series which can be cut into individual pieces. (G) Pyrosequencing makes use of the released pyrophosphate (PPi) each time a nucleotide is incorporated by a polymerase into a newly synthesized strand. The enzyme sulfurylase converts PPi together with the metabolite adenosine 5'-phosphosulfate (APS) into ATP. ATP can be used by the enzyme luciferase to metabolize luciferin, thereby creating a light peak which is detected by an optical system. Only one specific type of nucleotide can be used within one reaction as always, the same light signal is released upon nucleotide incorporation. (H) The detection method sequence-by-ligation uses specific oligonucleotide-probes. These probes hybridize to the target library sequence and are ligated to it and subsequently elongate the strand. The attached fluorescent signal is detected by an optical system. A two base encoding system of 16 specific oligonucleotide-probes is used - one for each possible combination of the four nucleotides. (I) The sequence-by-synthesis method features reversible chain terminators which are used in subsequent PCR cycles. The attached fluorescent dye and the terminator can be removed. All four nucleotides can be present during the reaction. Only one will be incorporated by the polymerase at the time. (J) Nonoptical detection using pH measurement makes use of the released H^+ ion when a nucleotide is incorporated by the polymerase. A semiconductor device acts as pH meter and can determine how many nucleotides have been introduced. Only one specific type of nucleotide can be used within one reaction.

a step further. These different techniques as part of the second generation of sequencing will be treated in the following sections and summarized in Fig. 7.3.

Pyrosequencing

One of the first "next-generation" of sequencing approaches started with the development of pyrosequencing in 1993 (Nyrén et al., 1993). Here, the real-time chemiluminescence of pyrophosphate release is recorded when a DNA polymerase incorporates a nucleotide during sequencing.

In principle the single stranded fragments of a precedent shotgun fragmentation become immobilized on a solid phase, usually on magnetic beads coated with streptavidin. Then, a mixture of polymerase, luciferase, ATP-sulfurylase, and one single type of dNTP is added to the beads. The polymerase incorporates the dNTP and the released phosphate is transformed into pyrophosphate by the sulfurylase, which in turn can be used by the luciferase to emit light (Fig. 7.3G) (Nyrén et al., 1993). Depending on the intensity of the emitted light, it can be determined how many dNTPs became incorporated by the polymerase. Subsequently the enzymatic mixture and nucleotides need to be removed and the procedure is repeated in several steps, each using another distinct type of dNTPs (Nyrén et al., 1993).

In 1996, the method was modified to add an additional enzyme (apyrase) that could remove nonincorporated dNTPs from reaction. Therefore, the enzymatic mixture did not have to be removed, speeding up the entire process (Ronaghi et al., 1996). The actual commercial breakthrough happened in 2004 with 454 Life Sciences' marketed massive parallelized version of pyrosequencing (Margulies et al., 2005). There, the solid phase was a microfabricated microarray, where either one or no shotgun fragment became immobilized in a single well of the microarray. This enables for automation and massive parallel sequencing (Voelkerding et al., 2009).

Polony sequencing

Polony sequencing was first described in 2005 by Shendure et al. (2005). Polymerase colony (in short polony) technologies perform multiplex amplification of individual molecules while maintaining spatial clustering of identical amplicons. It greatly reduced the necessary reaction volume while dramatically extending the number of sequencing reactions compared to Sanger sequencing (Schuster, 2008).

Shendure et al. used an emulsion PCR (ePCR) for separation. A paired end-tag library needs to be constructed. A-tailing is applied to fragmented genomic DNA (1 kb of size) that creates blunt ends with an adenine nucleotide. Then the fragments become circularized with the help of T-tailed 30 bp long synthetic oligonucleotides (T30) carrying restriction sites. Following rolling circle replication, which creates multiple copies of the fragment (Fig. 7.3F), enzymatic restriction is carried out releasing even smaller fragments with paired-end tags flanking the remnants of the T30. Within an ePCR, further specific adapters/primers are ligated resulting in an amplified library of roughly 135 bp fragments. These adapters attach to magnetic beads, under conditions that favor one DNA molecule per bead, and the fragments are again amplified using an ePCR following clean-up (Metzker, 2010). In the following step, the beads become immobilized within a polyacrylamide gel between a microscope coverslip/slide stack and form a monolayer. They are then sequenced by ligation with annealing fluorescent labeled, degenerated nonamers (Fig. 7.3H). The necessary four-color imaging could be done with a commonly available, inexpensive epifluorescence microscope and corresponding software. After imaging, the nonamers are stripped off again and a new cycle is initiated with new nonamers shifting one position further into the genomic tag.

A total of 26 bases could be determined (13 bp per genomic tag) in this way at low costs (Shendure et al., 2005). Polony sequencing was commercialized in an open source system called the Polonator that was sold by Dover Systems. Yet, the biggest drawback was the very small amount of usable information and the reliance on uniform amplification for good efficiency (Kircher & Kelso, 2010).

The Applied Biosystem SOLiD platform

The sequencing by oligonucleotide ligation and detection (SOLiD) system is the successor of the polony sequencing approach and was launched by Applied Biosystems in 2006. After shotgun fragmentation, again clonal amplified beads carrying the target sequences are generated. Known universal adaptors are ligated to the 5′ and 3′ end of the fragments within an ePCR (Fig. 7.3D). These adaptors enable the covalent binding of the fragments upon release of the emulsion on glass slides. Sequencing is performed by five rounds of cycles of oligonucleotide ligation and detection (Applied Biosystems, 2008).

A two base encoding system of 16 specific octamer probes is used for SOLiD, one for each possible combination of the four nucleotides (Fig. 7.3H). In particular, the first two bases resemble the possible nucleotide combinations, followed by six degenerated bases, which can pair with any other nucleotide. There is a cleavage site between the fifth and sixth base and labeled on the 5′ end with one of four specific fluorescent dyes (Dimalanta et al., 2009).

During sequencing the octamers compete for hybridization to the target sequence and remaining octamers are washed away. A DNA ligase is guided by a distinct primer complementary to the 5′ adapter of the target sequence. There, it ligates only the very first octamer in close proximity to the 5′ adapter. Other hybridized octamers are subsequently washed away. The fluorescent dye is imaged and then cleaved off together with the last three bases of the hybridized and ligated octamer.

In the next cycle, another distinct octamer can hybridize to the positions 6–10 of the target sequence. Thus, during the first round of ligation-dependent sequencing, the positions n + m*5 and n + m*5 + 1 are correctly paired, where n equals the starting position and m is the number of performed cycles; m also determines the read length. Usually, ten cycles are performed leading to read length of 50 bp (Kircher & Kelso, 2010). Intermediate positions are determined by using a slightly different primer complementary to the end of the 5′ adapter but the very last position. Hence, the entire target sequence is shifted one base in the next round of sequencing and new combinations of octamers can hybridize within each cycle. This is repeated for a total of five rounds of sequencing and every time shifting one base into the 5′ adapter. By knowing the exact sequence of the 5′ adapter, all incorporated dinucleotides can be determined. Parallelization can be achieved by multiple dispersions of covalently bound beads to a glass slide surface.

There are several drawbacks of this method. The preamplification step can introduce PCR-artifacts. If no reference genome is available for error correction and no assembly and consensus calling is performed, then the average error rate can be relatively high. Also, if dye removal is incomplete, then the subsequent read out of the sequence-by-ligation process is erred. Beads carrying a mixture of sequences and beads in close proximity to one another can create false reads and low-quality bases (Dimalanta et al., 2009). Palindromic sequences prove also to be a hindrance (Huang et al., 2012).

Applied BioScience merged with Invitrogen in 2008 creating a new company named Life Technologies, that in turn was acquired by Thermo Fisher Scientific in 2014 (Thermo Fisher Scientific Completes Acquisition of Life Technologies Corporation, 2014). As of May 1, 2016, Thermo Fisher Scientific discontinued their SOLiD sequencers after acquiring Ion Torrent and promoting these sequencers (5500 Discontinuance Letter). Due to the very short reads created, SOLiD sequencing remains as a niche in the field.

Ion torrents pH measurements with semiconductors

Another very innovative sequencing approach is pH measurements with semiconductors released by Ion Torrent in 2010 (Rusk, 2011). The key idea is to measure the change of pH induced by incorporation of a specific type of nucleotide into a newly synthesized strand. Thus, no optical imaging is necessary.

Very similar to 454 pyrosequencing, shotgun fragments (Fig. 7.3A) of the target sequence become immobilized in wells of a microplate, in this case, a semiconductor chip. The microwells are then flooded with a single nucleotide type at the time. A DNA polymerase subsequently incorporates one or many nucleotides if they are complementary to the template. Yet, the big difference to pyrosequencing is that no modified nucleotides with fluorophores are in use, and, also no optical imaging at all takes place. Instead, every time a nucleotide covalently included in a growing synthesized strand releases pyrophosphate and a single positively charged hydrogen ion.

The pyrophosphate is made of use in pyrosequencing, yet the free hydrogen is changing slightly the pH. This pH change can be recorded with the help of a very sensitive ion-sensitive field-effect transistor. The unattached dNTP molecules are washed out before the next cycle when a different dNTP species is introduced (Pennisi, 2010). Incorporation of several nucleotides of the same type leads to greater release of H^+ within one cycle and therefore to a greater change in pH (Fig. 7.3J). The series of electronic signals is directly translated into sequence by the analysis computer. As result, the entire measurement is really fast compared to the other sequencing approaches and could be almost carried out in real time. Yet in reality, sequence determination is gated by the speed of nucleotide exchange during and in between the cycles (Eid et al., 2009).

The biggest drawback of the Ion torrent sequencing approach is the increased inaccuracy, proportional to the number of nucleotides build in by the polymerase in the same cycle. The system can distinguish very accurate if none, one or two nucleotides of the same type are incorporated. On the other hand, it is very difficult to decide whether seven or eight nucleotides have been introduced (Slatko et al., 2018). The current version of ion semiconductor sequencers uses a range of chips producing read lengths of 200 to 600 bp with more than 99% accuracy and outputs up to 50 Gb (Kumar et al., 2019).

Illumina's sequence-by-synthesis approach

As of 2020, Illumina is estimated to hold a market share of ~75% of genetic sequencing industry driving off many other competitors (https://www.bizjournals.com/sanfrancisco/news/2020/01/03/sequencing-giant-illumina-scraps-1-2-billion.html). Their sequence-by-synthesis with reversible dye terminators and bridge amplification approach yields very good quality data. However, most shining are their sequencer systems which yield by far the highest output of all sequence technologies.

The basic principle was introduced by Solexa, a company that was acquired by Illumina in 2007. The strategy is analog to Sanger's chain termination sequencing but with engineered polymerases and reversible dye terminators. Massive parallel sequencing is achieved by creating clonal clusters of template sequences on a glass slide (flow cell) followed by sequencing (Bentley et al., 2008). During library preparation, genomic DNA becomes randomly shotgun fragmented by tagmentation using specific transposons (transposases) that have a rather short and arbitrary target sequence (Fig. 7.3A). At the same time, these transposases ligate Illumina-specific adapters onto the 5'- and 3'-end of the fragment (so called P7 and P5). In an intermediate step, further index adapters, also including P7 and P5, can be ligated that enable a postsequencing identification of different samples within one library (Fig. 7.3B). The P7 and P5 adapters are used as primer sites as well as binding sites for complementary oligonucleotide which are printed on the flow cell (www.illumina.com).

At the beginning of the cluster generation, the library is washed over the flow cell and hybridizes to the complementary adapter binding sites. The flow cell is loaded with the library at a concentration in the pM range. This ensures that statistically hybridizing fragments are not in close proximity to each other. Once the fragments hybridized to the flow cell, several cycles of cluster generation are initiated (in general about 25). There, bridge amplification is used, where each bound fragment bends over and hybridizes to a neighboring adapter sequence, hence forming a "bridge" (Fig. 7.3E). The P7 and P5 adapters are then used as primers by a polymerase to complete the complementary strand. The double-stranded bridge is denatured resulting again in single strand covalently bound fragments but doubled in number. This ensures a clonal amplification of each of the original library fragment as well as covalently binding to the flow cell. Roughly 1000 copies are generated which greatly enhances the fluorescent signal in later stages (www. illumina.com).

Usually paired end sequencing is carried out. All fragments become straightened out in the same direction as fragments bound to a P5 complementary sequence on the flow cell are cut and washed away. After sequencing of the first read, another round of five cycles of cluster generation is initiated. This time, all fragments that are attached to the P7 sequences are cut and washed away ensuring sequencing of only the second read.

Each cluster should emit a distinct fluorescent signal during the cycles of sequencing. However, clusters can overlap, if the original hybridized fragments were too close to each other. Then, these clusters form a hypercluster emitting different fluorescent signals at the same at the imaging stage. These clusters are then removed from analysis to ensure high sequence quality. Overloading of a flow cell will result in lots of clusters not passing this filter and overall reduce the gained yield. With the introduction of the NovaSeq system in 2018, a new patterned flow cell was presented. The wells can be more densely packed on the flow cell leading to even higher cluster density. Yet, with patterned flow cells, it is unlikely that more than one molecule hybridizes to each well before cluster generation if the appropriate concentration is loaded (www.illumina.com).

The actual sequence-by-synthesis is very similar to Sanger sequencing with fluorescent dyes, yet reversible terminators are used (Fig. 7.3I). In each cycle, only one specific nucleotide is incorporated, and an image is taken of the flow cell. After imaging the reversible block as well as the dye is removed, and another cycle of sequencing starts. The MiSeq and HiSeq sequencer distributed by Illumina use a four-color system, one for each type of nucleotide. That requires two different cameras and four images total for detection of all four different wavelengths. NextSeq, MiniSeq, and NovaSeq sequencers use a two-color system where the green channel encodes for thymine, the red channel marks cytosine, a combination of both encodes for adenine and, innovatively, the absence of a signal stands for guanines. Sequential images are then piled up to determine the sequence of each individual cluster resulting in millions of individual reads.

In fact, the NovaSeq, released in 2018, can generate up to 3 Tb of data using one S4 flow cell with an error rate of roughly 0.1%, which is unbeaten by any other competitor (Karst et al., 2021).

Complete genomics/BGI combinatorial probe anchor synthesis

The one competitor that really tries to take on Illumina at the moment is the company BGI (Shendure et al., 2017). Their sequencing approach is based upon a combinatorial probe anchor synthesis developed by Complete Genomics in 2010, which BGI acquired in 2013 (https://www.completegenomics.com/). In principle, accumulations of clonal amplified DNA in form of DNA-nanoballs become sequenced by hybridization of specific anchor probes acting as primers. At present, BGI offers the only sequencing by ligation method at the market (Kumar et al., 2019).

Like all the second-generation approaches, first genomic DNA needs to become shotgun-fragmented (Fig. 7.3A). Then, specific adapter sequences are ligated to the smaller fragments (Fig. 7.3B). These two adapter sequences are then ligated with the help of a splint oligonucleotide, thus forming a circularized DNA template. Rolling circle replication is carried out with a 29-DNA polymerase (Blanco et al., 1989) that creates a long newly synthesized single-stranded DNA including several copies attached head to tail, usually 300–500 copies long (Fig. 7.3F; Huang et al., 2017).

This long strand coils itself into a DNA-nanoball of ~ 250 nm diameter. Because of their negative charge, the DNA nanoballs repel each other (Drmanac et al., 2010). These nanoballs are then attached to a solid surface of a patterned array flow cell via binding to positively charged aminosilane printed on the flow cell. The intriguing idea is here to avoid the cluster generation on the flow cell that could introduce PCR bias, duplicate reads, and adapter dimers (Drmanac et al., 2010). For imaging high-accuracy combinatorial probe anchor ligation (cPAL) sequencing chemistry comes in use.

Originally, Complete Genomics applied an approach very similar to that described by Shendure et al. in 2005 and used in polony sequencing. A predefined sequence probe (anchor) hybridized as primer to the DNA-nanoball adapter sites followed by hybridization of fluorescently labeled but degenerated octamers (Fig. 7.3H). In subsequent cycles, different octamers and anchors would be used to determine about 10 bp of the template. Nowadays, also a specific anchor hybridizes to the adapter sites and acts as primer, but the sequence-by-synthesis approach from Illumina is adapted (Fig. 7.3I). Also, four nucleotides with different labeled fluorescent dyes and a reversible block become incorporated by a polymerase into new synthesized strand, yet at each copy within the DNA-nanoball at the same time. This leads to an increased signal upon laser agitation and camera recording. In the next cycle, dyes and blocks are removed and another round of sequencing can begin.

In this way, up to 400 bases can be sequenced of a template DNA. With highly dense packed flow cells up to six terabases of data can be generated (see for DNBSEQ-G400RS and MGI DNBSEQ-T7 on https://en.mgi-tech.com).

Third generation

The second generation of "next-generation" sequencing relies upon fragmentation, amplification, sequencing, and reassembly of the small pieces. The short-read length made it difficult to deal with longer repetitive sequences, full-length transcripts, or native base modifications (Kraft & Kurth, 2020). Also, clonal amplification can always lead to PCR duplicates and PCR-introduced artifacts. On top, although sequencing costs decreased dramatically within the last years, acquisition cost of new sequencers is still very high (e.g., NovaSeq costs € 1.2 million). As result, the third generation of "next-generation" sequencing features long read sequencing of single molecules and smaller sequencers, which are designed at lower cost, as less sample preprocessing is necessary. Processing of single molecules also significantly increased the speed of sequencing (Shendure et al., 2017). Two notable companies are selling third-generation sequencing technologies now: PacBio and Oxford Nanopore Technologies. Their two sequencing approaches are presented in the following two sections.

Single-molecule real-time sequencing

Pacific BioScience, often shortened to PacBio, offers one example of sequencing approaches of the third generation. PacBio facilitates the so called "Single-Molecule Real-Time" (SMRT) sequencing established in 2011 (Karow, 2011). The main idea was to generate very long reads that better address de novo genome assembly, HLA genotyping, structural variations or full-length transcriptomes, then with short reads. Yet, with the newest generation of sequencers, chemistry and SMARTcells, PacBio is not only complementary to second-generation sequencing but also can be used for the same applications (Kumar et al., 2019).

The key feature of SMRT sequencing is the zero-mode waveguides (ZMWs) in use (Fig. 7.4A). The ZMWs are wells on a chip consisting of a circular hole in an aluminum cladding film deposited on a clear silica substrate (Korlach et al., 2008). Amazingly the size of the hole is less than half the wavelength of light, and thus limiting the fluorescent excitation to the tiny volume at the very bottom (Shendure et al., 2017). The optical detector is located underneath the well essentially taking a real-time movie. In these ZMWs, a single active polymerase sits at the bottom of the ZMW and the single-stranded template DNA is immobilized next to it. Then in real time, the activity of the polymerase processing the single molecule of DNA template is monitored when incorporating free nucleotides into a new synthesized strand (Eid et al., 2009). A specific fluorescent dye is linked to the phosphate chain

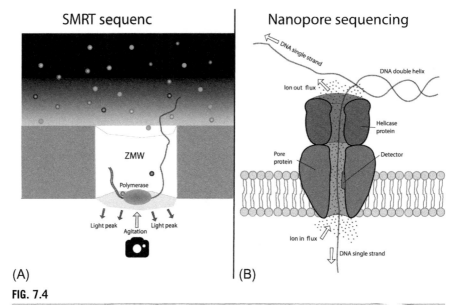

FIG. 7.4

Third generation of sequencing. Long read sequencing of single molecules is the key idea of the third generation of sequencing. No fragmentation and clonal amplification are needed speeding up the sequencing process tremendously. (A) Single molecule real-time (SMRT) sequencing. A single polymerase molecule is attached to the bottom of a very tiny zero-mode waveguide (ZMW) well on a chip. Individual single-stranded DNA molecules act as template. Fluorescently labeled nucleotides are incorporated in real time by the polymerase into a newly synthesized strand. Each incorporated nucleotide is emitting a small light peak which is recorded by an optical detector underneath the ZMW. (B) Nanopore sequencing features single molecules traversing nanopores within a membrane. The helicase protein of the pore is unraveling the DNA double helix creating single strands. The actual pore protein guides the single-stranded DNA to the other side of the membrane. The DNA is interfering with the ion flux going the other direction. Each individual nucleotide of the DNA strand is changing the ion flux in a specific manner which can be recognized by a detector.

of each nucleotide type (four color system). Only when a nucleotide is processed by the single polymerase molecule then its attached dye is in close enough proximity to be excited by laser light. When the polymerase incorporates the complementary nucleotide, the dye is cleaved off together with the phosphate group and diffuses out of the ZMW such that the fluorescent signal cannot be detected anymore. Hence, background noise is not a problem at all unlike in other sequence-by-synthesis approaches (Foquet et al., 2008).

The average read length is of 10–30 kilobases and a very fast turnaround time of hours (3–10) rather than days can be achieved (Kumar et al., 2019). Yet, the main drawbacks of the PacBio sequencer RS and RS II were the lower throughput, higher error rate (13%), and greater cost per base compared to second-generation sequencing approaches (Rhoads & Au, 2015). The latest PacBio sequencer Sequel, however, increased its performance tremendously from 150,000 ZMWs up to 1 million, and the error rates dropped as low as 3% (Nakano et al., 2017). With new chemistry in 2018, the average read lengths went up to 100 kb for shorter insert libraries and 30 kb for longer insert libraries. SMRTcell yield increased up to 50 billion bases for shorter insert libraries (PacBio Website). The Sequel II system released in April 2019 features a SMRT cell with incredible eight million ZMWs. Furthermore, the new single-molecule high-fidelity system (HiFi) was introduced combining both short reads' and long reads' advantages with a high accuracy of 99.99%. HiFi reads are produced using circular consensus sequencing (CCS). There, SMRTbell adapters are ligated to double-stranded template DNA which basically link the 5'- and 3'-end of each end of the double-strand together, thus forming a gigantic circle. Primers then anneal to the adapters and the entire circularized DNA molecule can be sequenced in a ZMW multiple times. This results in a consensus sequence of ~50x coverage yielding the high accuracy (Wenger et al., 2019).

Nanopore sequencing

The approach of nanopore sequencing is based on the idea that when single-stranded DNA molecules pass through a very narrow channel, then also the flow of ions is distinctly changed (Church et al., 1998; Deamer et al., 2016).

Mimicking bacterial proteins, Oxford Nanopore Technologies (ONT) created a protein-based nanopore embedded in electrical-resistant polymer membrane (Fig. 7.4B). This ensures that ion current changes are only measured inside the pores. The pores themselves are orderly dispersed using a micro scaffold structure (Kraft & Kurth, 2020). The membrane also separates two ionic solutions that create an electric current flowing through the nanopores. Usually, the translocation of the DNA molecules mediated by this electric current is ultra-fast. The number of ions changed per nucleotide is not enough to create a stable signal (Shendure et al., 2017). Therefore, an ATP-dependent motor protein is used to slow down the translocation. That way, the DNA can be ratchet through the nanopore one base at the time, thereby ensuring enough quality of the signal at high resolution. The different nucleotide bases induce characteristic changes in the electrical current running through the nanopore that are translated to base calls (Lu et al., 2016).

In 2014, ONT released nanopore sequencing in the form of the MinION, a handheld sequencer that uses a grid of membrane-embedded biological nanopores and is just as big as a larger USB stick which results in extreme portability. The small size can be achieved, as no optical detection device is necessary and only few depletable reagents are used (Shendure et al., 2017). As no polymerase is necessary, ONT sequencers are not limited to only sequence DNA molecules but can also process RNA molecules. Furthermore, modified nucleotides like methylation can also be assessed, as these particular modified nucleotides leave a distinct signal while passing through the nanopore (Giesselmann et al., 2019; Liu et al., 2019).

Size selection is very important when creating a library for nanopore sequencing. Very large molecules tend to block the nanopores, whereas too short molecules hamper the output of sequencing. Additionally, another drawback of nanopore sequencing is the inherent high error rate of 5%–25% (Wick et al., 2018). Yet, the latest pores and callers make us of a consensus sequence. At least five bases of the DNA molecule passing through are present at the same time in the nanopore. Therefore, the signal always belongs to a pentamer. This has the advantage that every base is measured five times while being pulled through the pore which improves the consensus accuracy to 96% with the most advanced pore (R10.3) and base caller (guppy version 3.6) (Kraft & Kurth, 2020). To further increase sequencing accuracy, the usage of unique molecular identifiers is proposed (Karst et al., 2021).

Currently, the SmidgION is being developed. This mini sequencer will attach to a smartphone for analysis and will be complemented by rapid library preparation (VolTRAX within 10 min) and simplified analysis pipelines. Potential applications may include remote monitoring of pathogens in a breakout or infectious disease like we are encountering in the moment with the Covid19 breakout. Also the on-site analysis of environmental samples such as water/metagenomics samples, real-time species ID for analysis of food, timber, wildlife, or even unknown samples and field-based analysis of agricultural environments could be further applications (https://nanoporetech.com/products/smidgion).

Recent advances in nanopore technology might even further increase the resolution of the nanopore sequencing. Engineered solid-state pores made of synthetic materials like carbon or graphene in combination with build in detectors could further enhance the speed of detection DNA molecules (Luan & Kuroda, 2020).

Fourth generation

Massive parallel sequencing marked the entry into the second-generation and single molecules sequencing into the third generation of sequencing. The fourth generation of sequencing takes it one step further by sequencing in situ instead in vitro, thereby taking the cellular context into account.

Already in 2010, a concept and according experiments were presented by the group of Mats Nilsson. They analyzed single-cell RNASeq data of neighboring tumor cells in parallel in their cellular context (Larsson et al., 2010). This was the first proof of principle of a fourth generation of sequencing approach (McGinn & Gut, 2013). In 2013, the same group demonstrated in situ sequencing of mRNA preserving histological context and morphologically conserved cells of breast cancer tissue (Ke et al., 2013). They then used Complete Genomics' sequencing by ligation to read the target region (Mignardi & Nilsson, 2014).

3D intact-tissue RNA sequencing was announced in 2018 by Wang et al. They resolved over 1000 genes simultaneously with spatially resolved transcript amplicon readout mapping. So instead of using a tissue section, they could perform sequencing on an entire thick tissue block revealing a gradient of excitatory neurons on a cubic millimeter scale (Wang et al., 2018).

A very interesting idea was proposed in 2017, when Kühnemund et al. (2017) carried out in situ sequencing analysis using a mobile phone camera paired with a self-developed 3D printed small portable fluorescent and bright-field microscope device (Kumar et al., 2019).

Another approach was carried out in 2020 to investigate the spatial transcriptome profiling of tissue domains around amyloid plaques in a study of Alzheimer's disease. Here, barcoded arrays allow unbiased transcriptome profiling in tissue, maintaining the spatial localization of the sequenced molecules (Chen et al., 2020). A novel technique was used by Qian et al. (2020) utilizing a probabilistic cell typing by in situ sequencing approach.

Complementary methods

There are several methods that are complementary to high-throughput sequencing. These techniques are usually used to confirm various DNA sequence findings and to support pathological or biological/cytogenomic relevance of them (Kumar et al., 2019; Slatko et al., 2018). Briefly, single-cell sequencing, optical mapping, and DNA microarrays as examples are illustrated in the following sections.

Single-cell sequencing

The advantages of sequencing single cells are at hand. When using single cells, subpopulations of cells within a tissue can be analyzed. For example, differently evolved tumor cells can be assessed and compared from a single tumor tissue (Nawy, 2014) or low abundant circulating tumor cells (Huang et al., 2015). Also, specific cell types of neurons can be investigated like the Purkinje cells when doing research on the cerebellum (Varga et al., 2015). Furthermore, cellular functions can be investigated as each cell will show their unique expression profile during RNASeq. Furthermore, analyses of microbes that cannot be cultured can be conducted (Alneberg et al., 2018).

In general, about 10 ng of total DNA is necessary to generate a library for sequencing. Yet, a single cell only contains about 6–7 pg of total DNA (Chen et al., 2018). Thus, profound methods to isolate single cells, amplify the genomic content of each single cell individually, and then generate distinct libraries are in use.

So far, there are more than 100 approaches to perform single-cell sequencing (#SingleCell #Omics v2.6.20 bit.ly/scellmarket @albertvilella). Next to simply diluting a cell suspension to very low concentrations, mainly manual micromanipulation (usually under a microscope) (Han et al., 2014; Konry et al., 2016), laser capture microdissection (LCM) (Foley et al., 2019; Frumkin et al., 2008; Kamme et al., 2004), fluorescence-activated cell sorting (FACS) (Muraro et al., 2016; Perkel, 2004; Villani & Shekhar, 2017), and droplet microfluidics (Pensold & Zimmer-Bensch, 2020) are used to isolate single cells. For micromanipulation, there are also automated machines like the CellenONE that aspirate single cells into a micropipette under visual control of a camera and subsequently transfer the cell to a new reaction compartment (Vallone et al., 2020). In 2015, Macosco et al. could demonstrate how to barcode the RNA of thousands of individual cells with the help of a droplet-based microfluidic system and then sequence them all together on an Illumina NextSeq, thereby conserving each cell individual transcriptome. Each cell becomes encapsulated in a tiny droplet formed by combining oil and aqueous flows creating nanoliter-scale aqueous compartments. Also, magnetic beads carrying unique molecular identifiers are incorporated into the droplets. At best, one cell and one bead are located within one droplet. Then cellular lysis is facilitated, and RNA molecules can bind to the bead and are processed in downstream sequencing-by-synthesis (Macosko et al., 2015).

Optical mapping

Optical mapping allows for the investigation of high-molecular weight DNA molecules (that is in the Mb range) that are beyond the detection range of single-base resolution NGS. Structural variation and genome assembly are main focus targets (Slatko et al., 2018). During synthesis, fluorochrome-labeled nucleotides are incorporated through DNA polymerases and tracked by fluorescence microscopy. The fluorescent signal provides a genome-wide, sequence-specific pattern. By combining images from numerous DNA molecules, an overlapping large consensus optical map of the physical localization is created. This can be directly visualized and be used to place other physical markers, sequences, or contigs in relation to the map (Ramanathan et al., 2005).

The commercial leader in this field is Bionano Genomics offering the instruments Irys and Saphyr that carry out optical mapping, alignment to reference genome, identification of structural variants and visualization (https://bionanogenomics.com/products/). A major advantage of this method is that it avoids cloning or PCR artifacts and analyzes a single molecule at a time (Delpu et al., 2021; Slatko et al., 2018).

Sequencing by hybridization

DNA microarrays are used in the sequencing-by-hybridization approach. On DNA microarrays, a large amount of specific and mapped oligonucleotides is immobilized on a standard glass slide. In general, DNA is fluorescently labeled and then hybridized to microarray (Weise & Liehr, 2021a). By knowing the sequence of each single probe on the array, fluorescent signals can be correctly annotated and hence be "sequenced." Microarrays are used for investigation of larger copy number variations (Shinawi et al., 2008) and gene expression profiling (Pensold & Zimmer-Bensch, 2020) but also for SNP detection (Hacia et al., 1999) and alternative splicing analysis (Pan et al., 2004).

In microarray comparative genomic hybridization (aCGH), the oligonucleotides on the glass slide act as hybridization matrix for DNA from a test and a reference sample. Like in normal CGH, both samples are differently labeled with fluorophores. Then both samples are mixed and hybridized onto the microarray. An image is then captured via a dual color laser and the relative fluorescence rate of each mapped oligonucleotide is determined (Solinas-Toldo et al., 1997). Misbalance of both signals indicates structural variation in the test sample. If the signal of the reference is stronger, then a deletion is apparent in the test sample. Accordingly, a stronger test sample signal indicates an amplification event in the test sample (Weise et al., 2015). The resolution is restricted only by the clone size and by the density of the clones on the microarray. However, balanced aberrations like translocations or inversions cannot be detected; and there is a problem with detection of low-level mosaics (Weise & Liehr, 2021a).

With SNP arrays, it is possible to identify loss of heterozygosity at individual nucleotides next to copy number alterations (Zhao et al., 2004). This also opens possibilities to detect epigenetic alterations (Eggermann, 2021; Harutyunyan & Hovhannisyan, 2021). SNP arrays are high-density oligonucleotide-based arrays with 100,000 or more SNPs, each a 25-mer oligonucleotide and represented by both sense and antisense strand (Matsuzaki et al., 2004). The three expected genotypes (for example, AA, BB, or AB) are presented by the probe intensities. Copy number alterations are identified using independent control hybridizations in comparison (Speicher & Carter, 2005).

The method linked to NanoString is a variation of DNA micro arrays. It uses molecular barcodes and microscopic imaging to detect and count up to several hundred unique transcripts in one hybridization reaction (Geiss et al., 2008). Sequencing with NanoString was used to investigate the driver of migrating preoptic-area-derived interneurons in mice, revealing a single enzyme (DNMT1) as key mediator (Pensold et al., 2017).

Summary and conclusion

During the last four decades, DNA sequencing underwent huge evolution from the first few determined nucleotides to entire genomes today undergoing several generations (Fig. 7.5). As they are at least four generations of sequencing, one should not speak any more about NGS but rather of high-throughput sequencing (HTS)

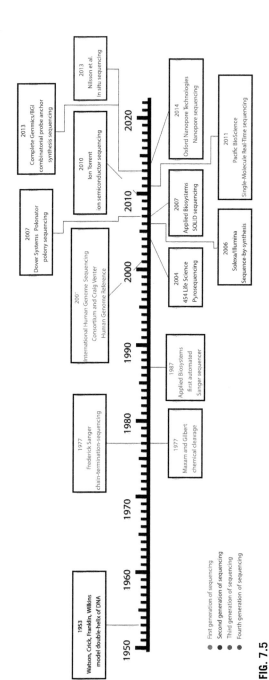

FIG. 7.5

Timeline of key events of the different generations of sequencing.

when referring to massively parallel sequencing. Such developments enabled new insights into the three dimensional genome, just to mention as an exemplary keyword topologically associating domains (TADs) (Daban, 2021; Weise & Liehr, 2021b; Yumiceba et al., 2021).

The biggest innovation in the history of sequencing with lots of emerging different technologies happened after the succession of the Human Genome Project in the new millennium. However, the nanopore technology was the last successfully commercialized sequencing approach in the recent past. There are a few promising approaches with the recent advances in nanopore fabrication and in using electron microscopy in sequencing which might lead to new companies emerging in the field. Yet, now Illumina (sequence-by-synthesis) keeps the biggest market share but is challenged by BGI (DNA nanoballs), Oxford Nanopore Technolgies (nanopores) and PacBio (Single-Molecule Real-Time Sequencing). Nevertheless, also Sanger sequencing is still widely used in clinical diagnostics either as verification tool of HTS findings or for small targeted applications (Alekseyev et al., 2018). See Table 7.1 for an overview of all presented technologies.

In general, targeted gene panels designed to address investigation of specific phenotype-related genes are most commonly in use in diagnostic laboratories (Caspar et al., 2018). As these lists are very restrictive and very fast outdated, a more comprehensive approach is to sequence entire exomes (WES) or even genomes (WGS). Sequencers of the newest generation can yield enormous output of sequencing data, thus making WES and WGS feasible and inexpensive. WES is restricted to the coding regions of the genome comprising roughly 2% of it. Yet, most disease-causing mutations (~85%) are estimated to be located in the coding regions of the genome (Mattick et al., 2018).

Nevertheless, intronic but also intergenic regions of the genome have been proven to play a critical role in some clinical phenotypes (Chen & Tian, 2016; Concolino et al., 2017; Nassisi et al., 2019; Zernant et al., 2018). Usually, WES does not investigate promotor regions or regulatory elements. WGS closes this gap but is more costly as WES. Another big advantage of WGS libraries is their uniformity of sequencing, as no amplification is required in PCR-free library preparations, leading to cleaner and more bias-free data (Meienberg et al., 2016). With the long reads of single molecule sequencing, also structural variations like imbalances, translocations, large insertions and deletions can be determined. Thus, WGS will likely replace WES as method of choice in diagnostics in the near future (Mattick et al., 2018). Another challenge in HTS is the huge data created proving to be a significant problem to deal with in terms of data analysis, data management, and storage. Yet, this is another story to tell.

Table 7.1 Overview of sequencing technologies and their respective classification. The table shows information about the amplification mode, the properties of the sequencing template, the actual sequencing chemistry used, the detection method, the output in bases, the read-length in basepairs, applications of the respective technology and pros and cons.

	Year of introduction	Amplification mode	Sequencing template	Chemistry	
Sanger sequencing 1st generation	1977	PCR	Pool of PCR-amplicons	Primer extension with chain termination	
Maxam-Gilbert 1st generation	1977	Sufficient template required	Pool	Chemical base-specific cleaving	
454 Pyrosequencing 2nd generation	2004	Emulsion PCR	Amplified clone	Primer extension one base at a time	
Solexa-Illumina 2nd generation	2006	Bridge PCR	Amplified clone	Single-base primer extension with reversible blocks	

Detection	Output	Read length	Application	Pros	Cons
Radioactivity, fluorescence 4 color	Up to 24× in kb range	40–900 bp	Single exon or gene sequencing, conformation of HTS results	High accuracy, widely used for conformation, easy to use	Expensive per base, no upscaling, limited by capilar count
Radioactivity	Very small	100 bp max	Bacterial DNA sequencing	First sequencing approach	Hazardous chemicals, difficult to upscale, only short sequences, radioactive labeling
Real-time chemiluminescence	25 Mb	600–800 bp	Bacterial and insect genome de novo assemblies; medium scale (<3 Mb) exome capture; 16S in metagenomics	Longer reads improve mapping in repetitive regions; fast run times	High reagent cost; high error rates in homopolymer repeats, intrinsically higher error rate relative to Sanger sequencing, and rely on high sequence coverage, out of market
Fluorescence imaging - 1|2|4 color	6 Tb max	50–300 bp	Variant discovery by whole-genome resequencing or whole-exome capture; gene discovery inmetagenomics	Market leader, widest profile of sequencers, high-quality reads, highest output	Intrinsically higher error rate relative to Sanger sequencing, and rely on high sequence coverage, high cost in chemicals and sequencers, cluster passing filter issues

Continued

Table 7.1 Overview of sequencing technologies and their respective classification. The table shows information about the amplification mode, the properties of the sequencing template, the actual sequencing chemistry used, the detection method, the output in bases, the read-length in basepairs, applications of the respective technology and pros and cons—cont'd

	Year of introduction	Amplification mode	Sequencing template	Chemistry	
Applied Biosystem SOLiD 2nd generation	2007	Emulsion PCR	Amplified clone	Oligonucleotide ligation	
Complete genomics 2nd generation	2010	Rolling circle replication	Amplified clone	Oligonucleotide ligation	
BGI 2nd generation	2013	Rolling circle replication	Amplified clone	Primer extension one base at a time	
IonTorrent 2nd generation	2010	Emulsion PCR	Amplified clone	Primer extension one base at a time at beads	

Detection	Output	Read length	Application	Pros	Cons
Fluorescence imaging - two-base encoding	9 Gb max	50 bp	Variant discovery by whole-genome resequencing or whole-exome capture; gene discovery in metagenomics	Two-base encoding provides inherent error correction	Long run times, intrinsically higher error rate relative to Sanger sequencing, and rely on high sequence coverage, out of market
Fluorescence imaging 4 color	1 Gb	200 bp	Variant discovery by whole-genome resequencing or whole-exome capture; gene discovery in metagenomics	Lower material costs	Out of market
Fluorescence imaging 4 color	6 Tb max	400 bp	Variant discovery by whole-genome resequencing or whole-exome capture; gene discovery in metagenomics	Significantly cheaper as Illumina at the same output	Legal issues and patent problems
pH measurement in semiconductor device	50 Gb	200–600 bp	De novo DNA and RNA sequencing, transcriptome sequencing, microbial sequencing, copy number variation detection, small RNA and miRNA sequencing and CHIP-seq	No optical device necessary, faster than other 2nd approaches	Homopolymer issues

Continued

Table 7.1 Overview of sequencing technologies and their respective classification. The table shows information about the amplification mode, the properties of the sequencing template, the actual sequencing chemistry used, the detection method, the output in bases, the read-length in basepairs, applications of the respective technology and pros and cons—cont'd

	Year of introduction	Amplification mode	Sequencing template	Chemistry	
Pacific BioScience 3rd generation	2011	None	Individual molecule	Primer extension in zero-mode wave guides	
Oxford nanopore Technologies 3rd generation	2013	None	Individual molecule	Direct measurement within nanopore	
In situ 4th generation	2013	In situ rolling circle	Rolling circle amplicon	Primer extension	

Detection	Output	Read length	Application	Pros	Cons
Real-time fluorescence imaging 4 color	50 Mb	> 10 kb	Full-length transcriptome sequencing; complements other resequencing efforts in discovering large structural variants and haplotype blocks, de novo assembly, methylation/ other modification assays	Very long reads improve assembly quality tremendously, HiFi reads increase accuracy > 99.99%, no optical device necessary	Originally with highest error rate, low output
Nanopore ion current measurement	2.1 Tb max	> 10 kb	Full-length transcriptome sequencing; whole-genome sequencing, de novo assembly, methylation/ other modification assays	Very long reads improve assembly quality tremendously, no optical device necessary, very low costs	Higher error ratio
Fluorescence imaging			Sequencing of tissue section	Preserving information of spatial surrounding	Difficult to conduct

References

5500 Discontinuance Letter. (2015). *Thermo Fischer Scientific.* https://www.thermofisher.com/content/dam/LifeTech/Documents/PDFs/5500_DiscontinuanceLetter_November2015.pdf (Accessed 02 November 2020).

Alekseyev, Y. O., Fazeli, R., Yang, S., Basran, R., Maher, T., Miller, N. S., & Remick, D. (2018). A next-generation sequencing primer—How does it work and what can it do? *Academic Pathology, 5.* https://doi.org/10.1177/2374289518766521. 2374289518766521.

Alneberg, J., Karlsson, C. M. G., Divne, A. M., Bergin, C., Homa, F., Lindh, M. V., … Pinhassi, J. (2018). Genomes from uncultivated prokaryotes: A comparison of metagenome-assembled and single-amplified genomes. *Microbiome, 6*(1), 173. https://doi.org/10.1186/s40168-018-0550-0.

Applied Biosystems. (2008). A theoretical understanding of 2 base color codes and its application to annotation, error detection, and error correction. In *Methods for annotating 2 base color encoded reads in the SOLiD™ system.* White Paper SOLiD™ System.

Beck, S., & Pohl, F. M. (1984). DNA sequencing with direct blotting electrophoresis. *The EMBO Journal, 3*(12), 2905–2909.

Bentley, D. R., Balasubramanian, S., Swerdlow, H. P., Smith, G. P., Milton, J., Brown, C. G., … Smith, A.J. (2008). Accurate whole human genome sequencing using reversible terminator chemistry. *Nature, 456*(7218), 53–59. https://doi.org/10.1038/nature07517.

Blanco, L., Bernad, A., Lázaro, J. M., Martín, G., Garmendia, C., & Salas, M. (1989). Highly efficient DNA synthesis by the phage phi 29 DNA polymerase. Symmetrical mode of DNA replication. *The Journal of Biological Chemistry, 264*(15), 8935–8940. 2498321.

Caspar, S. M., Dubacher, N., Kopps, A. M., Meienberg, J., Henggeler, C., & Matyas, G. (2018). Clinical sequencing: From raw data to diagnosis with lifetime value. *Clinical Genetics, 93*(3), 508–519. https://doi.org/10.1111/cge.13190.

Chen, W. T., Lu, A., Craessaerts, K., Pavie, B., Sala Frigerio, C., Corthout, N., … De Strooper, B. (2020). Spatial transcriptomics and in situ sequencing to study Alzheimer's disease. *Cell, 182*(4), 976–991.e19. https://doi.org/10.1016/j.cell.2020.06.038.

Chen, J., & Tian, W. (2016). Explaining the disease phenotype of intergenic SNP through predicted long range regulation. *Nucleic Acids Research, 44*(18), 8641–8654. https://doi.org/10.1093/nar/gkw519.

Chen, D., Zhen, H., Qiu, Y., Liu, P., Zeng, P., Xia, J., … Chen, F. (2018). Comparison of single cell sequencing data between two whole genome amplification methods on two sequencing platforms. *Scientific Reports, 8*(1), 4963. https://doi.org/10.1038/s41598-018-23325-2.

Church, G., Deamer, D. W., Branton, D., Baldarelli, R., & Kasianowicz, J. (1998). *Characterization of individual polymer molecules based on monomer-interface interactions.* Patent US5795782 https://patents.google.com/patent/US20120160687A1/en.

Concolino, P., Rizza, R., Costella, A., Carrozza, C., Zuppi, C., & Capoluongo, E. (2017). CYP21A2 intronic variants causing 21-hydroxylase deficiency. *Metabolism, 71,* 46–51. https://doi.org/10.1016/j.metabol.2017.03.003.

Crick, F. H. (1962). The genetic code. *Scientific American, 207,* 66–74. https://doi.org/10.1038/scientificamerican1062-66.

Daban, J.-R. (2021). Multilayer organization of chromosomes. In T. Liehr (Ed.), *Cytogenomics* (pp. 267–296). Academic Press. Chapter 13 (in this book).

Deamer, D., Akeson, M., & Branton, D. (2016). Three decades of nanopore sequencing. *Nature Biotechnology, 34*(5), 518–524. https://doi.org/10.1038/nbt.3423.

Delpu, Y., Barseghyan, H., Bocklandt, S., Hastie, A., & Chaubey, A. (2021). Next-generation cytogenomics: High-resolution structural variation detection by optical genome mapping. In T. Liehr (Ed.), *Cytogenomics* (pp. 123–146). Academic Press. Chapter 8 (in this book).

Dimalanta, E. T., Zhang, L., Hendrickson, C. L., Sokolsky, T. D., Sannicandro, A. E., Manning, J. M., ... Blanchard, A.P. (2009). *Increased read length on the SOLiD™ sequencing platform*. Poster SOLiD™ System.

Drmanac, R., Sparks, A. B., Callow, M. J., Halpern, A. L., Burns, N. L., Kermani, B. G., ... Reid, C.A. (2010). Human genome sequencing using unchained base reads on self-assembling DNA nanoarrays. *Science*, *327*(5961), 78–81. https://doi.org/10.1126/science.1181498.

Eggermann, T. (2021). Epigenetics. In T. Liehr (Ed.), *Cytogenomics* (pp. 389–401). Academic Press. Chapter 20 (in this book).

Eid, J., Fehr, A., Gray, J., Luong, K., Lyle, J., Otto, G., ... Turner, S. (2009). Real-time DNA sequencing from single polymerase molecules. *Science*, *323*(5910), 133–138. https://doi.org/10.1126/science.1162986.

Foley, J. W., Zhu, C., Jolivet, P., Zhu, S. X., Lu, P., Meaney, M. J., & West, R. B. (2019). Gene expression profiling of single cells from archival tissue with laser-capture microdissection and Smart-3SEQ. *Genome Research*, *29*(11), 1816–1825. https://doi.org/10.1101/gr.234807.118.

Foquet, M., Samiee, K. T., Kong, X., Chauduri, B. P., Lundquist, P. M., Turner, S. W., ... Roitman, D.B. (2008). Improved fabrication of zero-mode waveguides for single-molecule detection. *Journal of Applied Physics*, *103*(3), 034301. https://doi.org/10.1063/1.2831366.

Franklin, R. E., & Gosling, R. G. (1953). Molecular configuration in sodium thymonucleate. *Nature*, *421*(6921), 400–401. https://doi.org/10.1038/171740a0.

Frumkin, D., Wasserstrom, A., Itzkovitz, S., Harmelin, A., Rechavi, G., & Shapiro, E. (2008). Amplification of multiple genomic loci from single cells isolated by laser micro-dissection of tissues. *BMC Biotechnology*, *8*, 17. https://doi.org/10.1186/1472-6750-8-17.

Geiss, G. K., Bumgarner, R. E., Birditt, B., Dahl, T., Dowidar, N., Dunaway, D. L., ... Dimitrov, K. (2008). Direct multiplexed measurement of gene expression with color-coded probe pairs. *Nature Biotechnology*, *26*(3), 317–325. https://doi.org/10.1038/nbt1385.

Giesselmann, P., Brändl, B., Raimondeau, E., Bowen, R., Rohrandt, C., Tandon, R., ... Müller, F.J. (2019). Analysis of short tandem repeat expansions and their methylation state with nanopore sequencing. *Nature Biotechnology*, *37*(12), 1478–1481. https://doi.org/10.1038/s41587-019-0293-x.

Hacia, J. G., Fan, J. B., Ryder, O., Jin, L., Edgemon, K., Ghandour, G., ... Collins, F.S. (1999). Determination of ancestral alleles for human single-nucleotide polymorphisms using high-density oligonucleotide arrays. *Nature Genetics*, *22*(2), 164–167. https://doi.org/10.1038/9674.

Han, S. W., Nakamura, C., Miyake, J., Chang, S. M., & Adachi, T. (2014). Single-cell manipulation and DNA delivery technology using atomic force microscopy and nanoneedle. *Journal of Nanoscience and Nanotechnology*, *14*(1), 57–70. https://doi.org/10.1166/jnn.2014.9115.

Harutyunyan, T., & Hovhannisyan, G. (2021). Approaches for studying epigenetic aspects of the human genome. In T. Liehr (Ed.), *Cytogenomics* (pp. 155–209). Academic Press. Chapter 10 (in this book).

Heather, J. M., & Chain, B. (2016). The sequence of sequencers: The history of sequencing DNA. *Genomics*, *107*(1), 1–8. https://doi.org/10.1016/j.ygeno.2015.11.003.

Huang, Y. F., Chen, S. C., Chiang, Y. S., Chen, T. H., & Chiu, K. P. (2012). Palindromic sequence impedes sequencing-by-ligation mechanism. *BMC Systems Biology*, S10. https://doi.org/10.1186/1752-0509-6-S2-S10. 6 Suppl. 2(Suppl. 2).

Huang, J., Liang, X., Xuan, Y., Geng, C., Li, Y., Lu, H., … Gao, S. (2017). A reference human genome dataset of the BGISEQ-500 sequencer. *Gigascience, 6*(5), 1–9. https://doi.org/10.1093/gigascience/gix024.

Huang, L., Ma, F., Chapman, A., Lu, S., & Xie, X. S. (2015). Single-cell whole-genome amplification and sequencing: Methodology and applications. *Annual Review of Genomics and Human Genetics, 16*, 79–102. https://doi.org/10.1146/annurev-genom-090413-025352.

Human genome project completion: Frequently asked questions. National Human Genome Research Institute (NHGRI). https://www.genome.gov/11006943/human-genome-project-completion-frequently-asked-questions/.

Kamme, F., Zhu, J., Luo, L., Yu, J., Tran, D. T., Meurers, B., … Wan, J. (2004). Single-cell laser-capture microdissection and RNA amplification. *Methods in Molecular Medicine, 99*, 215–223. https://doi.org/10.1385/1-59259-770-X:215.

Karow, J. (2011). PacBio ships first two commercial systems; Order backlog grows to 44. *GenomeWeb.* 3 May 2011 https://www.genomeweb.com/sequencing/pacbio-ships-first-two-commercial-systems-order-backlog-grows-44#.X_8NghZ76Ul.

Karst, S. M., Ziels, R. M., Kirkegaard, R. H., Sørensen, E., McDonald, D., Zhu, Q., … Albertsen, M. (2021). High-accuracy long-read amplicon sequences using unique molecular identifiers with Nanopore or PacBio sequencing. *Nature Methods, 18*(2), 165–169. https://doi.org/10.1038/s41592-020-01041-y.

Ke, R., Mignardi, M., Pacureanu, A., Svedlund, J., Botling, J., Wählby, C., & Nilsson, M. (2013). In situ sequencing for RNA analysis in preserved tissue and cells. *Nature Methods, 10*(9), 857–860. https://doi.org/10.1038/nmeth.2563.

Kircher, M., & Kelso, J. (2010). High-throughput DNA sequencing—Concepts and limitations. *BioEssays, 32*(6), 524–536. https://doi.org/10.1002/bies.200900181.

Konry, T., Sarkar, S., Sabhachandani, P., & Cohen, N. (2016). Innovative tools and technology for analysis of single cells and cell-cell interaction. *Annual Review of Biomedical Engineering, 18*, 259–284. https://doi.org/10.1146/annurev-bioeng-090215-112735.

Korlach, J., Marks, P. J., Cicero, R. L., Gray, J. J., Murphy, D. L., Roitman, D. B., … Turner, S.W. (2008). Selective aluminum passivation for targeted immobilization of single DNA polymerase molecules in zero-mode waveguide nanostructures. *Proceedings of the National Academy of Sciences of the United States of America, 105*(4), 1176–1181. https://doi.org/10.1073/pnas.0710982105.

Kraft, F., & Kurth, I. (2020). Long-read sequencing to understand genome biology and cell function. *The International Journal of Biochemistry & Cell Biology, 126*, 105799. https://doi.org/10.1016/j.biocel.2020.105799.

Kühnemund, M., Wei, Q., Darai, E., Wang, Y., Hernandez-Neuta, I., Yang, Z., … Nilsson, M. (2017). Targeted DNA sequencing and in situ mutation analysis using mobile phone microscopy. *Nature Communications, 8*, 13913. https://doi.org/10.1038/ncomms13913.

Kumar, K. R., Cowley, M. J., & Davis, R. L. (2019). Next-generation sequencing and emerging technologies. *Seminars in Thrombosis and Hemostasis, 45*(7), 661–673. https://doi.org/10.1055/s-0039-1688446.

Lander, E. S., Linton, L. M., Birren, B., Nusbaum, C., Zody, M. C., Baldwin, J., … International Human Genome Sequencing Consortium. (2001). Initial sequencing and analysis of the human genome. *Nature, 409*(6822), 860–921. https://doi.org/10.1038/35057062.

Larsson, C., Grundberg, I., Söderberg, O., & Nilsson, M. (2010). In situ detection and genotyping of individual mRNA molecules. *Nature Methods, 7*(5), 395–397. https://doi.org/10.1038/nmeth.1448.

Liehr, T. (2021a). A definition for cytogenomics - Which also may be called chromosomics. In T. Liehr (Ed.), *Cytogenomics* (pp. 1–7). Academic Press. Chapter 1 (in this book).

Liehr, T. (2021b). Overview of currently available approaches used in cytogenomics. In T. Liehr (Ed.), *Cytogenomics* (pp. 11–24). Academic Press. Chapter 2 (in this book).

Liu, Q., Fang, L., Yu, G., Wang, D., Xiao, C. L., & Wang, K. (2019). Detection of DNA base modifications by deep recurrent neural network on Oxford Nanopore sequencing data. *Nature Communications, 10*(1), 2449. https://doi.org/10.1038/s41467-019-10168-2.

Lu, H., Giordano, F., & Ning, Z. (2016). Oxford nanopore MinION sequencing and genome assembly. *Genomics, Proteomics & Bioinformatics, 14*(5), 265–279. https://doi.org/10.1016/j.gpb.2016.05.004.

Luan, B., & Kuroda, M. A. (2020). Electrophoretic transport of single-stranded DNA through a two dimensional nanopore patterned on an in-plane heterostructure. *ACS Nano, 14*(10), 13137–13145. https://doi.org/10.1021/acsnano.0c04743.

Macosko, E. Z., Basu, A., Satija, R., Nemesh, J., Shekhar, K., Goldman, M., … McCarroll, S.A. (2015). Highly parallel genome-wide expression profiling of individual cells using nanoliter droplets. *Cell, 161*(5), 1202–1214. https://doi.org/10.1016/j.cell.2015.05.002.

Margulies, M., Egholm, M., Altman, W. E., Attiya, S., Bader, J. S., Bemben, L. A., … Rothberg, J.M. (2005). Genome sequencing in microfabricated high-density picolitre reactors. *Nature, 437*(7057), 376–380. https://doi.org/10.1038/nature03959.

Matsuzaki, H., Dong, S., Loi, H., Di, X., Liu, G., Hubbell, E., … Mei, R. (2004). Genotyping over 100,000 SNPs on a pair of oligonucleotide arrays. *Nature Methods, 1*(2), 109–111. https://doi.org/10.1038/nmeth718.

Mattick, J. S., Dinger, M., Schonrock, N., & Cowley, M. (2018). Whole genome sequencing provides better diagnostic yield and future value than whole exome sequencing. *The Medical Journal of Australia, 209*(5), 197–199. https://doi.org/10.5694/mja17.01176.

Maxam, A. M., & Gilbert, W. (1977). A new method for sequencing DNA. *Proceedings of the National Academy of Sciences of the United States of America, 74*(2), 560–564. https://doi.org/10.1073/pnas.74.2.560.

McGinn, S., & Gut, I. G. (2013). DNA sequencing—Spanning the generations. *New Biotechnology, 30*(4), 366–372. https://doi.org/10.1016/j.nbt.2012.11.012.

Meienberg, J., Bruggmann, R., Oexle, K., & Matyas, G. (2016). Clinical sequencing: Is WGS the better WES? *Human Genetics, 135*(3), 359–362. https://doi.org/10.1007/s00439-015-1631-9.

Metzker, M. L. (2010). Sequencing technologies—The next generation. *Nature Reviews. Genetics, 11*(1), 31–46. https://doi.org/10.1038/nrg2626.

Mignardi, M., & Nilsson, M. (2014). Fourth-generation sequencing in the cell and the clinic. *Genome Medicine, 6*(4), 31. https://doi.org/10.1186/gm548.

Muraro, M. J., Dharmadhikari, G., Grün, D., Groen, N., Dielen, T., Jansen, E., … van Oudenaarden, A. (2016). A single-cell transcriptome atlas of the human pancreas. *Cell Systems, 3*(4), 385–394.e3. https://doi.org/10.1016/j.cels.2016.09.002.

Naidoo, N., Pawitan, Y., Soong, R., Cooper, D. N., & Ku, C. S. (2011). Human genetics and genomics a decade after the release of the draft sequence of the human genome. *Human Genomics, 5*(6), 577–622. https://doi.org/10.1186/1479-7364-5-6-577.

Nakano, K., Shiroma, A., Shimoji, M., Tamotsu, H., Ashimine, N., Ohki, S., … Hirano, T. (2017). Advantages of genome sequencing by long-read sequencer using SMRT technology in medical area. *Human Cell, 30*(3), 149–161. https://doi.org/10.1007/s13577-017-0168-8.

Nassisi, M., Mohand-Saïd, S., Andrieu, C., Antonio, A., Condroyer, C., Méjécase, C., … Audo, I. (2019). Prevalence of ABCA4 deep-intronic variants and related phenotype in an unsolved "One-Hit" cohort with Stargardt disease. *International Journal of Molecular Sciences*, *20*(20), 5053. https://doi.org/10.3390/ijms20205053.

Nawy, T. (2014). Single-cell sequencing. *Nature Methods*, *11*(1), 18. https://doi.org/10.1038/nmeth.2771.

Nyrén, P., Pettersson, B., & Uhlén, M. (1993). Solid phase DNA minisequencing by an enzymatic luminometric inorganic pyrophosphate detection assay. *Analytical Biochemistry*, *208*(1), 171–175. https://doi.org/10.1006/abio.1993.1024.

Pan, Q., Shai, O., Misquitta, C., Zhang, W., Saltzman, A. L., Mohammad, N., … Blencowe, B.J. (2004). Revealing global regulatory features of mammalian alternative splicing using a quantitative microarray platform. *Molecular Cell*, *16*(6), 929–941. https://doi.org/10.1016/j.molcel.2004.12.004.

Pennisi, E. (2010). Genomics. Semiconductors inspire new sequencing technologies. *Science*, *327*(5970), 1190. https://doi.org/10.1126/science.327.5970.1190.

Pensold, D., Symmank, J., Hahn, A., Lingner, T., Salinas-Riester, G., Downie, B. R., … Zimmer, G. (2017). The DNA methyltransferase 1 (DNMT1) controls the shape and dynamics of migrating POA-derived interneurons fated for the murine cerebral cortex. *Cerebral Cortex*, *27*(12), 5696–5714. https://doi.org/10.1093/cercor/bhw341.

Pensold, D., & Zimmer-Bensch, G. (2020). Methods for single-cell isolation and preparation. *Advances in Experimental Medicine and Biology*, *1255*, 7–27. https://doi.org/10.1007/978-981-15-4494-1_2.

Perkel, J. (2004). Fluorescence-activated cell sorter. *The Scientist*. https://www.the-scientist.com/how-it-works/fluorescence-activated-cell-sorter-49796.

Prober, J. M., Trainor, G. L., Dam, R. J., Hobbs, F. W., Robertson, C. W., Zagursky, R. J., … Baumeister, K. (1987). A system for rapid DNA sequencing with fluorescent chain-terminating dideoxynucleotides. *Science*, *238*(4825), 336–341. https://doi.org/10.1126/science.2443975.

Qian, X., Harris, K. D., Hauling, T., Nicoloutsopoulos, D., Muñoz-Manchado, A. B., Skene, N., … Nilsson, M. (2020). Probabilistic cell typing enables fine mapping of closely related cell types in situ. *Nature Methods*, *17*(1), 101–106. https://doi.org/10.1038/s41592-019-0631-4.

Ramanathan, A., Pape, L., & Schwartz, D. C. (2005). High-density polymerase-mediated incorporation of fluorochrome-labeled nucleotides. *Analytical Biochemistry*, *337*(1), 1–11. https://doi.org/10.1016/j.ab.2004.09.043.

Rhoads, A., & Au, K. F. (2015). PacBio sequencing and its applications. *Genomics, Proteomics & Bioinformatics*, *13*(5), 278–289. https://doi.org/10.1016/j.gpb.2015.08.002.

Ronaghi, M., Karamohamed, S., Pettersson, B., Uhlén, M., & Nyrén, P. (1996). Real-time DNA sequencing using detection of pyrophosphate release. *Analytical Biochemistry*, *242*(1), 84–89. https://doi.org/10.1006/abio.1996.0432.

Rusk, N. (2011). Torrents of sequence. *Nature Methods*, *8*(1), 44. https://doi.org/10.1038/nmeth.f.330.

Sanger, F., & Coulson, A. R. (1975). A rapid method for determining sequences in DNA by primed synthesis with DNA polymerase. *Journal of Molecular Biology*, *94*(3), 441–448. https://doi.org/10.1016/0022-2836(75)90213-2.

Sanger, F., Nicklen, S., & Coulson, A. R. (1977). DNA sequencing with chain-terminating inhibitors. *Proceedings of the National Academy of Sciences of the United States of America*, *74*(12), 5463–5467. https://doi.org/10.1073/pnas.74.12.5463.

Schuster, S. C. (2008). Next-generation sequencing transforms today's biology. *Nature Methods*, *5*(1), 16–18. https://doi.org/10.1038/nmeth1156.

Shendure, J., Balasubramanian, S., Church, G. M., Gilbert, W., Rogers, J., Schloss, J. A., & Waterston, R. H. (2017). DNA sequencing at 40: Past, present and future. *Nature*, *550*(7676), 345–353. https://doi.org/10.1038/nature24286.

Shendure, J., Porreca, G. J., Reppas, N. B., Lin, X., McCutcheon, J. P., Rosenbaum, A. M., … Church, G.M. (2005). Accurate multiplex polony sequencing of an evolved bacterial genome. *Science, 309*(5741), 1728–1732. https://doi.org/10.1126/science.1117389.

Shinawi, M., Shao, L., Jeng, L. J., Shaw, C. A., Patel, A., Bacino, C., … Cheung, S.W. (2008). Low-level mosaicism of trisomy 14: Phenotypic and molecular characterization. *American Journal of Medical Genetics. Part A, 146A*(11), 1395–1405. https://doi.org/10.1002/ajmg.a.32287.

Slatko, B. E., Gardner, A. F., & Ausubel, F. M. (2018). Overview of next-generation sequencing technologies. *Current Protocols in Molecular Biology, 122*(1), e59. https://doi.org/10.1002/cpmb.59.

Smith, L. M., Fung, S., Hunkapiller, M. W., Hunkapiller, T. J., & Hood, L. E. (1985). The synthesis of oligonucleotides containing an aliphatic amino group at the 5' terminus: Synthesis of fluorescent DNA primers for use in DNA sequence analysis. *Nucleic Acids Research, 13*(7), 2399–2412. https://doi.org/10.1093/nar/13.7.2399.

Smith, L. M., Sanders, J. Z., Kaiser, R. J., Hughes, P., Dodd, C., Connell, C. R., … Hood, L.E. (1986). Fluorescence detection in automated DNA sequence analysis. *Nature, 321*(6071), 674–679. https://doi.org/10.1038/321674a0.

Solinas-Toldo, S., Lampel, S., Stilgenbauer, S., Nickolenko, J., Benner, A., Döhner, H., … Lichter, P. (1997). Matrix-based comparative genomic hybridization: Biochips to screen for genomic imbalances. *Genes, Chromosomes & Cancer, 20*(4), 399–407.

Speicher, M. R., & Carter, N. P. (2005). The new cytogenetics: Blurring the boundaries with molecular biology. *Nature Reviews Genetics, 6*(10), 782–792. https://doi.org/10.1038/nrg1692.

Stranneheim, H., & Lundeberg, J. (2012). Stepping stones in DNA sequencing. *Biotechnology Journal, 7*(9), 1063–1073. https://doi.org/10.1002/biot.201200153.

Thermo Fisher Scientific Completes Acquisition of Life Technologies Corporation. (2014). *The Wall Street Journal*, (3 February). Market Watch. http://www.marketwatch.com/story/thermo-fisher-scientific-completes-acquisition-of-life-technologies-corporation-2014-02-03.

Vallone, V. F., Telugu, N. S., Fischer, I., Miller, D., Schommer, S., Diecke, S., & Stachelscheid, H. (2020). Methods for automated single cell isolation and sub-cloning of human pluripotent stem cells. *Current Protocols in Stem Cell Biology, 55*(1), e123. https://doi.org/10.1002/cpsc.123.

Varga, R. E., Khundadze, M., Damme, M., Nietzsche, S., Hoffmann, B., Stauber, T., … Hübner, C.A. (2015). In vivo evidence for lysosome depletion and impaired autophagic clearance in hereditary spastic paraplegia type SPG11. *PLoS Genetics, 11*(8), e1005454. https://doi.org/10.1371/journal.pgen.1005454.

Venter, J. C., Adams, M. D., Myers, E. W., Li, P. W., Mural, R. J., Sutton, G. G., … Zhu, X. (2001). The sequence of the human genome. *Science, 291*(5507), 1304–1351. https://doi.org/10.1126/science.1058040.

Villani, A. C., & Shekhar, K. (2017). Single-cell RNA sequencing of human T cells. *Methods in Molecular Biology, 1514*, 203–239. https://doi.org/10.1007/978-1-4939-6548-9_16.

Voelkerding, K. V., Dames, S. A., & Durtschi, J. D. (2009). Next-generation sequencing: From basic research to diagnostics. *Clinical Chemistry, 55*(4), 641–658. https://doi.org/10.1373/clinchem.2008.112789.

Wang, X., Allen, W. E., Wright, M. A., Sylwestrak, E. L., Samusik, N., Vesuna, S., … Deisseroth, K. (2018). Three-dimensional intact-tissue sequencing of single-cell transcriptional states. *Science, 361*(6400), eaat5691. https://doi.org/10.1126/science.aat5691.

Watson, J. D., & Crick, F. H. (1953). The structure of DNA. *Cold Spring Harbor Symposia on Quantitative Biology, 18*, 123–131. https://doi.org/10.1101/sqb.1953.018.01.020.

Weise, A., Kosyakova, N., Voigt, M., Aust, N., Mrasek, K., Löhmer, S., … Fan, X. (2015). Comprehensive analyses of white-handed gibbon chromosomes enables access to 92 evolutionary conserved breakpoints compared to the human genome. *Cytogenetic and Genome Research, 145*(1), 42–49. https://doi.org/10.1159/000381764.

Weise, A., & Liehr, T. (2021a). Molecular karyotyping. In T. Liehr (Ed.), *Cytogenomics* (pp. 73–85). Academic Press. Chapter 6 (in this book).

Weise, A., & Liehr, T. (2021b). Interchromosomal interactions with meaning for disease. In T. Liehr (Ed.), *Cytogenomics* (pp. 349–356). Academic Press. Chapter 17 (in this book).

Wenger, A. M., Peluso, P., Rowell, W. J., Chang, P. C., Hall, R. J., Concepcion, G. T., … Hunkapiller, M.W. (2019). Accurate circular consensus long-read sequencing improves variant detection and assembly of a human genome. *Nature Biotechnology, 37*(10), 1155–1162. https://doi.org/10.1038/s41587-019-0217-9.

Wetterstrand, K. (2013). *DNA sequencing costs: Data from the NHGRI Genome Sequencing Program (GSP)*. National Human Genome Research Institute. https://www.genome.gov/about-genomics/fact-sheets/Sequencing-Human-Genome-cost.

Wick, R. R., Judd, L. M., & Holt, K. E. (2018). Deepbinner: Demultiplexing barcoded Oxford Nanopore reads with deep convolutional neural networks. *PLoS Computational Biology, 14*(11), e1006583. https://doi.org/10.1371/journal.pcbi.1006583.

Wilkins, M. H., Strokes, A. R., & Wilson, H. R. (1953). Molecular structure of deoxypentose nucleic acids. *Nature, 421*(6921), 398–400. https://doi.org/10.1038/171738a0.

Yumiceba, V., Souto Melo, U., & Spielmann, M. (2021). 3D cytogenomics: Structural variation in the three-dimensional genome. In T. Liehr (Ed.), *Cytogenomics* (pp. 247–266). Academic Press. Chapter 12 (in this book).

Zernant, J., Lee, W., Nagasaki, T., Collison, F. T., Fishman, G. A., Bertelsen, M., … Allikmets, R. (2018). Extremely hypomorphic and severe deep intronic variants in the ABCA4 locus result in varying Stargardt disease phenotypes. *Cold Spring Harbor Molecular Case Studies, 4*(4), a002733. https://doi.org/10.1101/mcs.a002733.

Zhao, X., Li, C., Paez, J. G., Chin, K., Jänne, P. A., Chen, T. H., … Meyerson, M. (2004). An integrated view of copy number and allelic alterations in the cancer genome using single nucleotide polymorphism arrays. *Cancer Research, 64*(9), 3060–3071. https://doi.org/10.1158/0008-5472.can-03-3308.

Next-generation cytogenomics: High-resolution structural variation detection by optical genome mapping

8

Yannick Delpu[a], Hayk Barseghyan[a,b], Sven Bocklandt[a], Alex Hastie[a], and Alka Chaubey[a]

[a]*Bionano Genomics, San Diego, CA, United States*
[b]*Children's National Research Institute, Washington, DC, United States*

Chapter outline

Introduction

As discussed in previous chapters, cytogenetic techniques including karyotyping (Weise & Liehr, 2021a) and fluorescence in situ hybridization (FISH) (Daban, 2021; Iourov et al., 2021; Ishii et al., 2021; Liehr, 2021a, 2021b; Weise & Liehr, 2021b), despite being several decades old, have retained their prominent role in cytogenetics/molecular cytogenetics due to their unique ability to detect all classes of chromosomal aberrations and structural variations (SVs). Despite being a low-resolution and labor-intensive technique, karyotyping is utilized worldwide in routine diagnostic

Cytogenomics. https://doi.org/10.1016/B978-0-12-823579-9.00022-9

testing as a whole-genome diagnostic tool. This genome wide approach has the ability to detect SVs such as inversions and balanced translocations not found by other untargeted molecular cytogenomic techniques (Liehr, 2021c, 2021d). FISH is more sensitive than karyotyping but is only useful for targeted loci and requires an assay for each targeted abnormality. Chromosomal microarray (CMA) (Weise & Liehr, 2021c) and next-generation sequencing (NGS) techniques (Ungelenk, 2021) have yielded significant improvements in genetic diagnostic power by increased sensitivity; however, they are primarily focused on copy number variations (CNV) and single-nucleotide variation (SNV) detection, respectively. With a priori information of translocation/fusion partners, it is possible to detect some chromosomal fusions using NGS-based methods; however, limitations render a very high fraction of the genome inaccessible to SV detection.

Two thirds of the human genome consist of repetitive sequences, making unambiguous mapping of short reads to those sequences and variant detection impossible (de Koning et al., 2011; Liehr, 2021e). The most common repetitive sequences in the genome are satellite DNA such as rDNA repeats and centromeric repeats, long interspersed nuclear elements (LINEs), short interspersed nuclear elements (SINEs), retrotransposons, and segmental duplications (de Koning et al., 2011; Liehr, 2021e). NGS provides short-read sequences that map with poor accuracy to these repeats because of ambiguity between different copies (Alkan et al., 2011). Alignment algorithms typically fail to identify the exact/unique genomic location for these short reads and are further confused by reference errors such as missing copies of nonallelic repeats (segmental duplications). Incorrect short-read alignments lead to calling of false-positive SVs and incorrect breakpoint localization. For these reasons, SV detection by NGS has mostly focused on CNVs. NGS tools do not reliably detect balanced SVs such as inversions and translocations; they have an unacceptably high level of both false positives and false negatives (Cameron et al., 2019; Silva et al., 2019). A plethora of algorithms have been developed for the detection of SVs, but because of the tremendous challenges discussed earlier, many groups have acquiesced to running multiple algorithms and using heuristics to merge call sets for a confident call set, albeit with high false negatives and complicated implementation (Chaisson et al., 2019; English et al., 2015).

While long-read sequencing has improved significantly over the years and occasionally reads reach hundreds of kilobase pairs in size, the median-read lengths are typically in the 5–30 kbp range and reads remain plagued with errors (Mantere et al., 2019; Ungelenk, 2021). This profile is not sufficient to span longer repetitive areas of the genome or elucidate many large, complex events, in fact, in a recent preprint, PacBio sequencing only detected 73% of insertions and deletions > 5 kbp (Ebert et al., 2020), the relatively easier SV to detect. Sequence-based approaches also lack the sensitivity to reliably detect heterozygous SVs, making the detection of rare variants in subclonal populations less likely (Hastie et al., 2017). Although PacBio sequencing has been used for somatic variation analysis, studies are mostly based on cell lines and complete validation has not been performed (Nattestad et al., 2018). The remaining limitations regarding cost, time, and coverage depth for long-read sequencing prevent them from being suitable for clinical diagnostics, cancer research, or large-scale population genetics analysis where whole-genome SV detection is needed.

The utilization of any technology for clinical use requires the ability to detect all SV classes at very high sensitivity, precision, and positive predictive value. Current workflows involve multiple low-resolution technologies to interrogate the genome, but the process is cumbersome and expensive and diagnostic yield remains in the 20%–50% range, depending on the indication (Deelen et al., 2019; Lee et al., 2014; Miller et al., 2010; Wright et al., 2018; Yang et al., 2013). Optical genome mapping (OGM) is able to assay all SV classes, genome wide in a single assay, making it the only contender to replace classic cytogenetic methods, on the horizon.

Introduction to Optical Genome Mapping
General principle

OGM leverages direct visualization of extremely long DNA molecules by the Saphyr System. The core of the technology is a nanochannel array where DNA molecules can be stretched very uniformly for imaging. The DNA is labeled at a specific 6 bp sequence motif such that imaged linearized molecules have a specific pattern that can be used to identify a chromosomal region, similar to a high-resolution karyotype. These patterns can be used to identify all classes of structural and copy number variants at very high resolution (Figs. 8.1 and 8.4).

The procedure utilizes ultra-high-molecular-weight (UHMW) DNA obtained by using commercial kits where cells are lysed in suspension, and the DNA binds to a paramagnetic disk, trapping the DNA and protecting it from any mechanical stress during the downstream wash and elution process. Next, fluorescent labels are attached to a specific 6-bp sequence motif, occurring approximately every 5 kbp in the human genome. The pattern is unique throughout the genome, and the specific set of labels bore by each molecule allows mapping to a reference genome (with the exception of the centromeres and some very large low complexity regions that are devoid of such motifs). Fluorescence labels are transferred to a specific sequence motif (CTTAAG) on the DNA using the Direct Label Enzyme (DLE1), a nondamaging process, that results in ultra-long DNA molecules bearing this uniquely identifiable label pattern that is loaded on a Saphyr chip and imaged on the Saphyr system. The resulting purified and labeled DNA has significant numbers of molecules of megabase length and typically presents a median size > 350 kbp and is sufficiently long to span essentially all repeat elements in the genome to analyze SV. The Saphyr Chip's nanochannels allow only a single linearized DNA molecule to travel through while controlling the molecule from tangling or folding back on itself. This nanofluidic environment allows molecules to move swiftly through hundreds of thousands of parallel nanochannels simultaneously, enabling high-throughput processing to build an accurate Bionano genome map. As the process is dynamic and controlled by cycles of electrophoresis and imaging, user can decide on the appropriate volume of

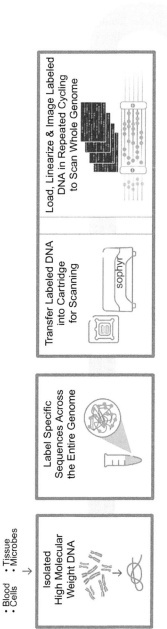

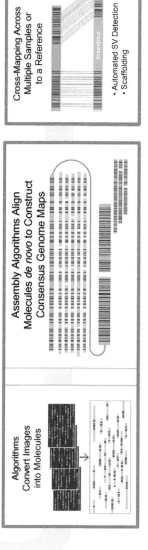

High-throughput, High-resolution Imaging of Megabase Length Molecules

Customer Sample
- Blood • Tissue
- Cells • Microbes

Isolated High Molecular Weight DNA

Label Specific Sequences Across the Entire Genome

Transfer Labeled DNA into Cartridge for Scanning

sophyr

Load, Linearize & Image Labeled DNA in Repeated Cycling to Scan Whole Genome

Algorithms Convert Images into Molecules

Assembly Algorithms Align Molecules *de novo* to Construct Consensus Genome Maps

Cross-Mapping Across Multiple Samples or to a Reference

Insertion

- Automated SV Detection
- Scaffolding

FIG. 8.1

Experimental workflow of optical genome mapping. From top left to bottom right: Ultra-high-molecular-weight DNA is extracted and subsequently labeled at a 6-bp motif by the Direct Label and Stain (DLS) labeling technology. The label pattern is unique throughout the genome. Labeled DNA is then loaded on a Syphyr Chip, where DNA molecules are linearized and imaged in repeated cycles. Images are digitalized into molecules that contain the linear positions of sequence motif labels. Multiple molecules are used to create consensus genome maps representing different alleles from the sample. The sample's optical genome map is aligned to the reference genome, and differences are automatically called allowing for detection of SVs in a genome-wide fashion.

data, and each chip allows the imaging of three samples in parallel and yields up to 5 Tbp per sample (equivalent to ~ 1600 × coverage for a human sample). The ability to reach such high coverage utilizing molecules strictly larger than 150 kbp (and up to 2.5 Mbp) makes this technology particularly attractive for the study of complex, heterogenous samples as in hematological malignancies and solid tumors.

Structural and copy number variation detection

Following acquisition, raw data are directed toward one of the two different and complementary analysis pipelines (Fig. 8.2):

- A de novo assembly pipeline, generally used for germline variation analysis, was designed to build consensus maps up to the size of a chromosome arms and

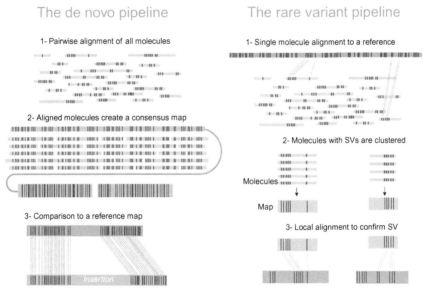

FIG. 8.2

Two complementary pipelines for automatic SV calling by map alignment. *The de novo pipeline*: (1) Based on the labeling patterns within imaged molecules *(thin blue lines)*, (2) they are aligned to one another in a pairwise fashion to produce a whole-genome de novo assembly of consensus genome maps. (3) Assembled maps *(thick blue lines)* are aligned against the reference genome *(thick green lines)*, where differences between the genome map and the reference are used to identify SVs. *The rare variant pipeline*: (1) Based on the labeling patterns, imaged molecules are directly aligned to a reference genome. Alignment differences in the labeling positions indicate the presence of SVs. (2) Molecules bearing the same SV are clustered and are used to assemble local consensus maps spanning the variation event. (3) The resultant consensus maps are re-aligned against the reference genome to make the final SV calls.

able to automatically detect homozygous or heterozygous variants with high sensitivity staring at 500 bp, genome wide.

- A rare variant pipeline, generally used for detection of somatic variation, was designed to specifically address heterogenous or mosaic samples and automatically detect SVs from single molecules, genome wide, starting at 5 kbp and as low as 1% allele fraction.

The automatic, genome-wide detection of SV by both pipelines relies on the comparison of labeling patterns between the sample's consensus maps and a reference map (Fig. 8.3). The reference map can be generated from parents, matched tissue, or in silico, from reference assemblies of the human genome such as HG19 or HG38 avoiding the need to run a control.

Moreover, both pipelines also run an independent copy number analysis algorithm that detects copy number variants (CNVs), generally larger than ~ 500 kbp, by identifying regions of the genome with significant coverage elevation or depression, providing information on the aneuploidy status of the sample.

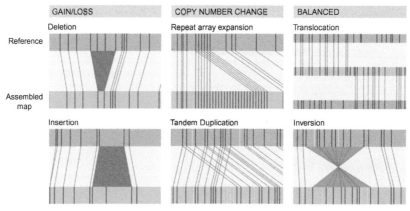

FIG. 8.3

Identification of distinct variant classes by optical genome map alignment. Distinct classes of SVs are detected by specific features following alignment of a sample's assembled maps *(blue lines)* against reference maps *(green lines)*. Deletions and insertions are identified by the decrease or increase in the distance between labels as measured in the corresponding region in the reference map, and they may or may not include gain or loss of labels. Repeat array expansions/contraction are identified following the same approach regardless of the presence or absence of labels within the tandem repeat. Tandem duplications are identified by a local repetition of a pattern of labels in the optical genome map matching a single labeling pattern in the reference map, and the extra copy can be direct or inverted. In the case of translocations, the assembled map aligns to two distinct genomic loci on different chromosomes. Inversions are characterized by the inversion of the label pattern in the assembled map in comparison with the reference map.

In addition, specialized pipelines have been created for analysis of complex disorders that require interpretation of SVs in the context of multiple chromosomes and alleles like facioscapulohumeral muscular dystrophy (FSHD) (Preston et al., 1993).

While the vast majority of SVs can be equally detected by the de novo assembly pipeline or the rare variant pipeline, their sensitivity differs for SVs smaller than 5 kbp, which is detected preferentially by the de novo approach (such as trinucleotides repeat expansions), and events affecting less than ~ 30% allele fraction will be detected by the rare variant pipeline more efficiently (Table 8.1).

Data visualization, filtering, and reviewing

Common challenges of genomics implementation in cytogenetic labs beyond variant detection limitations discussed earlier include the complexity of the bioinformatics pipelines and the lack of user-friendly tools to annotate and classify the large number of variants that are generally detected by NGS or other genomics methodologies. To account for these needs, the Bionano Access software has been designed as a user-friendly solution providing filtering options to allow one operator without extensive training to manage data and effectively filter and curate thousands of variants quickly. Key to this process is an embedded multiethnic control population database, which permits the user to eliminate the majority of the polymorphic, likely nonpathogenic variants encountered in a healthy population. The remaining SVs can be filtered to identify the ones overlapping with any known gene from the human reference genome, or more interestingly from a list of genes of interest or a panel of other chromosomal abnormalities such as a disease-specific FISH panel. The latter option using Browser Extensible Data (BED) files, containing a list of genomic coordinates of interest, is of particular value in routine practice as it allows to easily highlight classical rearrangements, well described for a given pathology while reserving access to the genome wide results, if needed. General feature format (GFF) files, containing more detailed DNA, RNA, and protein feature information, are also available for filtering against to help in further refining the understanding of discovered SVs.

Results are presented in the form of an automatically generated Circos plot reporting whole-genome information and consisting of 4 concentric circles (Fig. 8.4):

- Cytobands of corresponding chromosomes.
- Interstitial SVs—insertions/deletions/duplications/inversions identified from map alignments.
- Copy number variations.
- Translocations.

Each variant can be visualized in higher resolution in the genome browser application as which shows optical maps aligned to the reference and gene locations. A third application included for data visualization is a whole-genome CNV profile, similar to various microarray software.

Table 8.1 Specification of optical genome mapping for the detection of different variant classes.

Variant classes	Variant types	Variant classes	Bionano OGM
Aneuploidy	Monosomy	Monosomy	√
	Trisomy	Trisomy	√
	Triploidy	Triploidy (whole genome)	Not currently
	Tetraploidy	Tetraploidy (whole genome)	Not currently
Copy number variants	Deletions	Interstitial deletion	√ 500 bp or larger
		Terminal deletion	√ 500 kbp
	Duplications	Tandem duplications	√ 30 kb or larger
		Inverted duplications	√ 30 kb or larger
		Insertion duplications	√ 50 kb or larger
Structural variants	Insertions	Insertion of unknown sequence	√ 500 bp or larger
		Insertion (of known sequence, copy neutral)	√ 50 kbp or larger
	Translocations	Balanced translocations	√
		Unbalanced translocations	√
		Cryptic translocations	√
	Inversions	Inversion—pericentric fusion point	√
		Inversion—paracentric fusion point	√ 30 kbp or larger
	Ring chromosome	CNV and fusion	√ CNV > 500 kbp + fusion
Homozygosity mapping	LOH	AOH/ROH/LOH	Not currently
Macrosatellite/ microsatellite repeat contraction/ expansion	Repeat contraction	Repeat contraction (e.g., FSHD)	√ > 500 bp
	Repeat expansions	Repeat expansions (e.g., fragile X syndrome)	√ > 500 bp
Single-nucleotide variants	Small variants	SNVs, INDELs, frameshifts, stop-gains	No

Abbreviations: AOH, absence of heterozygosity; CNV, copy number variation; FSHD, facioscapulohumeral muscular dystrophy; INDEL, insertion and deletion; LOH, loss of heterozygosity; ROH, reduction of heterozygosity; SNV, single-nucleotide variation.

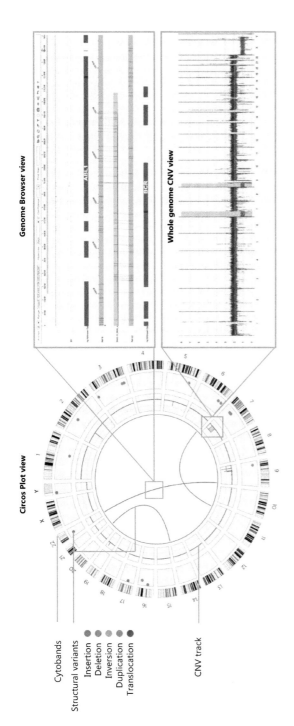

FIG. 8.4

Data visualization in the Bionano Access software graphical user interface. *Left panel:* Circos plots are automatically generated and results displayed in three concentric circles. The outer circle displays cytobands allowing genomic localization of the structural variants. The second circle reports interstitial SVs as dots following a color code to distinguish between insertions, deletions, inversions, and duplications. Copy number increases and decreases are plotted in the inner most circle. Translocations are reported as lines in the center connecting the genomic loci involved. Clicking on any SV switches to the genome browser view for a higher resolution view. *Top right panel:* The genome browser view details the alignment of the sample's consensus map *(light blue bar)* with the reference consensus maps *(light green bars)* and provides the detail of the SV and the breakpoint locations *(pink vertical line)* in the context of gene and/or exon location. Here, the sample's map alignment to the reference maps of chromosomes 9 and 22 illustrates a translocation. Genes in those regions on each chromosome are reported as purple bars. *Bottom right panel:* The whole-genome CNV view plots the measured copy number state at every label position as measured by the independent copy number tool. Segmental changes are reported for variations larger than 500 kbp (smaller CNVs are reported with the SV algorithm). Gains are highlighted in *blue,* while losses are highlighted in *red* (here, two losses larger than 500 kbp are identified on chromosomes 7 and 9).

Applications in constitutional cytogenetics

Current recommendation in cytogenetics relies on the combination of several approaches such as karyotyping, FISH, and chromosome microarray (CMA) for the detection of SVs or copy number changes in prenatal or postnatal samples (Baranov & Kuznetzova, 2021; Silva et al., 2019).

Each approach taken individually presents its own limitation either in term of the size cutoff, detection of balanced events, or restriction to predefined loci of clinical interest.

Moreover, while these methods are reasonably cost effective, individually, in many cases, multiple tests must be conducted on a single patient. These workflows can be labor intensive with important time-to-result implications and require extensive training for the certification and maintenance of the competency in the laboratory.

Some studies have been conducted to assess optical genome mapping for the modernization of current clinical cytogenetic methods.

Prenatal cytogenetics

In prenatal cytogenetics, noninvasive prenatal screening (NIPT) has become the standard of care for screening all pregnancies. However, every screen positive case and those that do not get an answer by a screening test are recommended to undergo invasive diagnostic testing to rule out a genetic abnormality (see as well Liehr et al., 2017).

In a milestone multicentric study, an international consortium benchmarked optical genome mapping technology against a combination of karyotyping, FISH, and array-based technology to evaluate its utility and analytical sensitivity on a cohort patients presenting of a wide range of simple and challenging chromosomal aberrations, representative of a broad range of clinical indications encountered routinely in the clinical practice (Mantere et al., 2020). The cohort composed of 85 patients selected to address a range of abnormalities and samples encountered in cytogenetic services for developmental or reproductive diseases. Among sample types which were included are amniotic fluid cells, chorionic villi cells, and lymphoblastoid cell cultures all generated from primary cultures according to standard diagnostic procedures. Optical genome mapping showed a 100% concordance compared with the combination of standard cytogenetic techniques, for all aberrations with noncentromeric breakpoints (Table 8.2). More interestingly, challenging chromosomal structures were perfectly identified by optical genome mapping illustrating the possibility to deduct the true structure of an abnormal or rearranged chromosome without visualizing cumbersome, condensed metaphase chromatids. In another study, studies were done on amniotic fluid as well as products of conception for the analysis of above-mentioned disorder FSHD1, caused by variation of tandem satellite repeats, and which will be covered in more detail in the following (Zheng et al., 2020).

Table 8.2 Performance of OGM in clinical benchmarking studies.

References	Cohort size	Clinical referral	Sample type	Variant types	Number of abnormalities included	Analysis cutoffs	Concordance with cytogenetic results	Average SVs/CNVs per sample identified by OGM (after filtration)	OGM additional findings
Mantere et al. (2020)	85	Autism spectrum disorders, intellectual disability associated or not with congenital malformations, reproductive disorders, familial history of chromosomal aberration, abnormal prenatal test results	Blood, amniotic fluid cell lines, chorionic villi cell lines, and lymphoblastoid cell lines	7 aneuploidies, 35 translocations, 6 inversions, 2 insertions, 39 copy number variations (20 deletions and 19 duplications), 6 isochromosomes, 1 ring chromosome, and 4 complex rearrangements	100	No size cutoff, homozygous/heterozygous SVs, CNVs down to 10% allele fraction	100%	80 (± 65) rare SVs, 11 CNVs per sample	NA
Neveling et al. (2020)	48	Myeloid and lymphoid neoplasms (AML, MDS, CML, CLL, ALL, MM, MPN, T-PLL, LYBM)	Peripheral blood or bone marrow	Copy number gains/losses, inversions, translocations, aneuploidies, LOH	112	No size cutoff, SVs affecting down to 10% allele fraction, CNVs affecting down to 10% allele fraction	100%	44 rare SVs, 62 CNV per sample	18 potential gene fusions absent from COSMIC database; 26 insertions/deletions overlapping with well-established cancer genes
Levy et al. (2020)	100	Acute myeloid leukemia	Peripheral blood or bone marrow	Copy number gains/losses, inversions, translocations, aneuploidies	NA	No size cutoff, SVs affecting down to 5% allele fraction, CNVs affecting down to 10% allele fraction	100%	NA	3 translocations, 1 inversion, 2 deletions, and 1 derivative chromosome
Lestringant et al. (2021)	10	B and T acute lymphoblastic leukemia	Peripheral blood or bone marrow	Copy number gains/losses, inversions, translocations, aneuploidies, LOH	78	20 kbp for insertions, deletions, and duplications, 1 Mbp for CNVs. SVs down to 10% allele fraction, CNVs down to 10% allele fraction	72 out of 78[a]	NA	4 fusions, 6 deletions, 2 gains, 1 duplication, 3 complex chromosomal rearrangements

[a] Missed calls correspond to 6 anomalies that were considered missed by OGM: four were either not automatically called but were visible within the Bionano Access visualization software or erroneously discarded by the operator. There remain only two false-negative calls corresponding to gains affecting the PAR region on X-chromosome.

Postnatal cytogenetics

Postnatal genetic diagnostics employs many of the same tools as prenatal testing but may see more diverse diseases and types of abnormalities. Several publications or ongoing works are illustrating OGM's potential for detection of currently diagnostic variants as well as improving the patient and family's experience in the, so-called, diagnostic odyssey of patients with suspected genetic disorders. The diagnostic odyssey is the term used to describe the journey of trying to provide a genetic diagnosis for a disorder though multiple rounds of genetic testing on different platforms often occurring over many years. The true hope and goal is to provide testing that is more streamlined and that can provide a higher diagnostic yield in a shorter time.

Benchmarking against clinical testing

The dystrophin gene on X-chromosome is one of the largest in the genome, and SVs within this gene can cause either Duchenne muscular dystrophy (DMD) or the milder form of DMD called Becker muscular dystrophy (BMD). The genomic abnormalities can be of any different variant class: SNV, deletion, insertion, duplication, inversion, and translocations. In a publication in Genome Medicine, Barseghyan et al. (2017) used optical genome mapping in patients with DMD. The team successfully mapped deletions/duplications and an inversion affecting the dystrophin gene, identifying deletions 45–250 kbp in size and an insertion of 13 kbp. The optical genome maps refined the location of deletion break points within the introns compared with current PCR-based clinical techniques. They detected heterozygous SVs in carrier mothers of DMD patients as well, demonstrating the ability of optical genome mapping to ascertain carrier status for large SVs. Of particular interest, they identified a copy neutral, 5.1 Mbp inversion disrupting the DMD gene. This inversion could previously only be identified by RNA sequencing from a muscle biopsy sample, from this, they inferred that there was a chromosomal inversion. Because of the difficulty to detect balanced SVs using genomic methods, it was missed by standard clinical methods including short-read-based genome sequencing (performed on a research basis). This clearly illustrates the increased value of optical genome mapping to improve patient management by improving the diagnostic workflow and eliminating some more invasive tests.

In addition to the prenatal specimens studied by Mantere et al. (2020), postnatal cases were included for which results from optical genome mapping a combination of karyotyping, FISH, and CMA technologies. As noted earlier, 100% concordance was achieved. In total, observed aberrations amounted to 100 balanced and unbalanced chromosomal aberrations. Among the 100 chromosomal aberrations included were 7 aneuploidies, 35 translocations, 6 inversions, 2 insertions, 39 copy number variations (20 deletions and 19 duplications), 6 isochromosomes, 1 ring chromosome, and 4 complex rearrangements. Among these, four patients harboring complex chromosomal rearrangements, involving three or more chromosomes or when at least four SVs are detected on the same chromosome, were successfully detected and interpreted by optical genome mapping (Mantere et al., 2020). Besides the clear potential to modernize classical cytogenetic workflows, OGM has the potential to answer unsolved cases and to provide higher resolution to identify SV events responsible for inherited disorders.

Resolving unsolved cases

As mentioned previously, the current diagnostic workflow can be a long and difficult process which leaves many patients with no genetic diagnosis. Because of the ability for OGM to detect clinical abnormalities of all classes but with much higher resolution and sensitivity compared with traditional cytogenetic methods, OGM can provide diagnosis in unsolved cases.

The underlying genetic cause of familial cancers remains undetermined in a subset of cases, leading to challenges for genetic counseling and preventing predictive testing in these families. Sabatella et al. (2021) recently described a family with two siblings born from healthy parents who were neonatally diagnosed with atypical teratoid rhabdoid tumor (ATRT). This rare and aggressive brain tumor is commonly associated with alterations at the tumor suppressor gene, *SMARCB1* (Eaton et al., 2011). Immunohistochemistry on both tumors showed loss of expression of SMARCB1, and multiple molecular techniques including whole-exome and whole-genome sequencing failed to identify germline mutation in the *SMARCB1* locus. OGM was used to detect variants missed by other technologies including whole-genome sequencing, and it was able to identify an insertion of ~ 2.8 kbp within intron 2 of the *SMARCB1* gene. The initial attempts to confirm the insertion by PCR and NGS failed. The authors finally confirmed the insertion SV using the PacBio HiFi genome sequencing and identified the insertion to be an SVA retrotransposon element (consisting of a CCCTC repeat, followed by an Alu-like repeat, a VNTR and SINE-R region, and a poly-A tail) affecting correct splicing of the gene and resulting in the loss of a functional allele. The initial failure to confirm by PCR and NGS is likely due to GC-rich nature of the region as well as the presence of a repeat involved.

In a study from the Institute for Human Genetics and the Benioff Children's Hospital at the University of California, San Francisco (UCSF), Shieh et al. (2020) evaluated the ability of Bionano's optical genome mapping combined with linked-read sequencing to diagnose children with genetic conditions, who were previously undiagnosed by standard of care methods alone (Shieh et al., 2020). The team performed whole-genome analysis by combining OGM with linked-read sequencing on 50 undiagnosed patients with a variety of rare genetic disorders and their parents to determine if this comprehensive genome analysis method could help improve diagnostic yield. Of the 50 children in the study, OGM results were sufficient to definitively diagnose 6 patients (or 12%), and for another 10 patients (or 20%), the OGM data revealed candidate pathogenic variants. Upon further analysis of the candidate SVs, additional 3 patients could be diagnosed with the optical genome mapping data, bringing the total of diagnosed patients to 9 (or 18%). In this study, OGM identified a number of pathogenic variants missed by CMA and undetectable by whole-exome sequencing, including duplications and deletions that were too small to be identified by CMA, or occurring in regions of the genome not typically covered by CMA or whole-exome sequencing. Of the additional 7 patients with variations considered to be candidates for pathogenic variants, the findings included deletions, duplications, and inversions. Before concluding that these variants are sufficient to diagnose the patients, further analysis is required since these variants had not previously been reported in patients with similar disease.

Research studies on microdeletion and microduplication syndromes

Low copy repeats (LCRs) are recognized as a significant source of genomic rearrangements, which are associated with most microdeletion and microduplication syndromes. These LCRs are extremely long and tremendously difficult to study; also, the size, complexity, and diversity of these elements are not well represented in the human genome reference and hence genotyping is impossible. Among microdeletion/duplication syndromes, 22q11 deletion syndrome (22q11DS) is one of the most prevalent disorders. In two recent studies, optical genome mapping has been used to elucidate LCR22 structure and variation and uncover potential risk factors for germline rearrangements leading to 22q11.2DS.

In a study by Demaerel et al. (2019), optical genome mapping was combined with a fiber FISH technique to assemble LCR22 alleles in a collection cell lines (Demaerel et al., 2019). The group successfully mapped these repeats and elucidated the LCR22 structures and their variability and further refined the 22q11DS rearrangement breakpoint regions. They observed an astonishing level of interindividual variability of LCR22A and LCR22D repeats and revealed at least 25 different alleles of LCR22A and six variants of LCR22D with size ranging from ∼ 250 kb to ∼ 2 Mbp. In a second study by Pastor et al. (2020), optical genome mapping was used to elucidate LCR22 structures and variation in 88 individuals from 30 families consisting of healthy parents and 22q11.2DS-affected probands. This study represents the most complete elucidation of 22q11.2 LCR22A and LCR22D haplotypes and nonallelic homologous recombination (NAHR) events associated with the deletions. The work may shed light on the mechanisms leading to 22q11 rearrangements and the different frequencies of the variation among populations, potentially paving the way for future prenatal counseling and testing for 22q11-related disorders.

Repeat expansion/contraction disorders

The determination of repeat expansion/contraction disorders is known to be challenging and often requires specialized and cumbersome tests (such as Southern blotting) which are often outsourced to specialized reference centers. Using extremely long molecules, OGM can span all of these repeat expansion/contraction loci and accurately measure repeat expansion/contractions from 500 bp and up to 10s or 100s of kbp. Using a single platform opens new avenues for the systematic detection of repeat expansion disorders such as Fragile X syndrome, FSHD, or Friedrich ataxia. Since the technology works on the same data for repeat analysis as it does for whole-genome SV analysis, a complete analysis of the patient's genomic SVs could be done alongside or as a follow-up after the analysis of repeat regions.

FSHD is the third most common form of muscular dystrophy with an incidence of 1:25,000 worldwide (Preston et al., 1993). The molecular mechanism underlying

FSHD is complex and mainly centered around the abnormal expression of DUX4 in muscle cells. The expression is impacted by the D4Z4 repeat array on chromosome 4 that is directly followed by a 10-kbp polymorphic region determining the haplotype A or B (depending, respectively, on the presence or absence of a particular regulatory sequence). The presence of a partially homologous D4Z4 array present on chromosome 10 makes the distinction of the possible haplotypes extremely difficult (Preston et al., 1993). There are two distinct forms of FSHD, both associated to the same specific set of symptoms namely scapular winging, facial and lower legs weaknesses, atrophy of the upper arms, retinopathy, abdominal weakness, and hearing loss.

Type 1 FSHD (FSHD1) represents 95% of the cases. It is autosomal dominant and typically caused by a contraction of the D4Z4 repeat array on chromosome 4 below 10 repeat units in the presence of a permissive allele (type A allele on chromosome 4) (Lemmers et al., 2010). Type 2 FSHD represents 5% of the cases and is associated with hypomethylation of the D4Z4 repeat array on chromosome 4. Accurately measuring the length of the repeat array is critical for the diagnosis of FSHD1. The traditional testing options available mainly rely on the labor-intensive Southern blots using radioactive labeling (Lemmers et al., 2012). Hence, new methods for a safe, rapid, and automatic characterization of the D4Z4 locus are needed to accurately diagnose these patients with FSHD1, thereby improving patient management. Several studies published in 2019 and 2020 used optical genome mapping for the accurate sizing of D4Z4 repeats (Dai et al., 2020; Zhang et al., 2019; Zheng et al., 2020).

Among them and of particular interest, Zhang et al. (2019) reported results of the application of OGM for the characterization of a five-generation FSHD1 pedigree. Optical genome mapping correctly diagnosed the pathogenic and normal haplotypes, identifying the founder 4qA disease allele as having 4 D4Z4 repeat units. Southern blot and molecular combing analysis confirmed OGM results for the 4qA disease and 4qB nondisease alleles (Zhang et al., 2019).

In an effort to streamline the analysis of FSHD cases, Bionano developed a focused analysis pipeline: EnFocus FSHD. This analytical tool allows a targeted measurement and analysis of the D4Z4 regions at chromosomes 4 and 10 and SMCHD1 as well as providing pass/fail criteria for data quality and an internal control measurement of control regions of the genome. This solution is currently implemented at FSHD diagnostic reference centers in replacement of conventional Southern blotting. The EnFocus analysis paradigm is a platform for clinical focused reporting of other abnormalities such as repeat expansion disorders like Fragile X syndrome or Friedrich ataxia.

Applications in hematological malignancies

One key hallmark of cancer is genomic instability, usually marked by structural rearrangements which accumulate in somatic tissues, and these rearrangements can disrupt genes, increase gene copy numbers, change gene functions, or change expression (Pellestor et al., 2021). As a result of the significant somatic structural changes in cancer genomes, genomic analysis becomes highly challenging. Within the milieu of somatic variation, particular mutations have been identified that can be used as biomarkers for diagnosis, prognosis, and therapy guidance. In hematologic malignancies, genetic testing is generally accomplished by karyotyping, FISH, and CMA, as well as gene panel sequencing. Different types of leukemias and lymphomas in different jurisdictions will have unique genetic testing workflows. For example, for acute myeloid leukemia (AML), the World Health Organization (WHO) and National Comprehensive Cancer Network guidelines recommend chromosomal analysis (karyotyping) plus FISH testing *CSF1R*, centromere 7, *MLL/KMT2A*, *RUNX1/RUNX1T1*, *CBF/MYH11*, among others, totaling 5 + FISH tests along with karyotyping (National Comprehensive Cancer Network, 2019). For acute lymphocytic leukemia, *BCR/ABL1*, trisomy 8, *ETV6/RUNX1*, *MLL* rearrangements, *IGH* rearrangements, numerical aberrations of chromosomes 4, 10, or 17, *cMYC*, and *P16* may be tested by FISH plus karyotyping (see as well Liehr et al., 2015). Differential diagnosis is often based on canonical structures but does not always account for the full complexity of the tumor, which may harbor rare somatic driver mutations in some clones composing the tumor tissue. A more complete characterization of structural aberrations of the tumor is paramount to the development of precision medicine.

Clinical benchmark studies

Several preclinical studies on hematological cancer cohorts are illustrating the capability of OGM to answer these needs. The first published work benchmarking Bionano's OGM against classical cytogenetics aimed at assessing performance of OGM to detect known chromosomal aberrations in a cohort of 48 patients with leukemia, on a large panel of hematological malignancies representative of the diversity of cases encountered in clinical practice (Neveling et al., 2020). Patients were referred for AML, myelodysplastic syndrome (MDS), chronic myelogeneous leukemia (CML), chronic lymphatic leukemia (CLL), acute lymphatic leukemia (ALL), multiple myeloma (MM), myeloproliferative neoplasm (MPN), T-cell prolymphocytic leukemia (T-PLL), and lymphoma of bone marrow (LYBM) with an abnormal cytogenetic report to represent a broad set of clinically relevant SVs and CNVs.

The cohort was comprised of 37 simple cases harboring less than 5 alterations and 11 complex case harboring more than 5 chromosomal alterations. An average of 44 chromosomal alterations were reported per patient after applying a filter against the embedded normal population database, eliminating the vast majority of variants corresponding to polymorphic, nonpathogenic variants. The study reports 100% concordance between optical genome mapping and the gold standard cytogenetic methods in the reporting of clinically relevant variants with variant allele frequencies higher than 10% (Table 8.2). For simple cases, the use of a custom filter setting led to 100% true positive rate for known aberrations. For complex cases, 10 of the 11 complex cases including a chromothripsis event showed full concordance with the previous findings, and only one case presenting very complex aberrations was not fully resolved, while the authors still identified the majority of the events involved. The authors pointed out that for several of the complex cases, the detected rearrangements showed a much higher level of complexity than previously seen by standard testing, with examples of additional translocations or marker chromosomes of unknown origin identified (Levy et al., 2020; Neveling et al., 2020).

Hematological neoplasms, such as acute lymphoblastic leukemia (ALL), are of notorious genetic diversity and require a combination of conventional and molecular cytogenetic techniques to efficiently profile the tumor samples. In recent work by Lestringant et al., results from a retrospective study on a cohort of 10 patients previously genetically typed B- or T-ALL by a combination of karyotype, FISH, SNP-array, and RT-MLPA (reverse transcriptase mixed ligation-dependent probe amplification) were presented (Lestringant et al., 2021). The cohort is representative of a panel of major prognostic impact anomalies and their large structural diversity. A total of 80 variants defined by the combined resolution of karyotype, FISH, SNP-array, and RT-MLPA were included in the scope of the study. Upon review, 2 variants were outside of the stated performance specifications for OGM, and should therefore not be included in concordance metrics. The analysis was conducted in a blinded fashion by two independent operators, one with training in hematological cytogenetics but without experience in either bioinformatics or optical genome mapping and the other trained in the use of the OGM tool only but without knowledge in the genetics of hematological malignancies. Interestingly, their reports deviated for only one variant (88 versus 89) illustrating the simplicity of the technology, not requiring specific bioinformatics training or complex data curation. All but 6 of the 78 variants were identified by both operators (Table 8.2). Among the 6 variants missed, 4 were detected by optical genome mapping but escaped the initial filtering or were overlooked by the operator during analysis and were further recovered by manual inspection. Only two copy number variations in the pseudoautosomal region (PAR) in Xp22.33 were truly missed by optical genome mapping due to the highly repeated sequences and interindividual variability of that regions frequently associated with false negative with SNP arrays and RT-MLPA. Based on these results, Bionano is making needed modifications to its software to enable detection of these X-chromosome CNVs. Overall, these results are in agreement with those from Neveling et al. (2020), who mentioned that fine tuning the filter settings on the Bionano Access software led to the recovery of variants detected but not reported, and considering manual classification of variants is common practice in clinical labs, the performance seems quite impressive (Neveling et al., 2020).

Detection of hallmark genomic aberrations in AML is essential for prognosis and patient management. A multicenter evaluation of AML cases by a consortium of opinion leaders in the United States on the clinical utility of OGM for assessment of genomic aberrations in 100 AML cases resulted in 100% concordance of OGM findings with diagnosis achieved with traditional karyotype analysis (Levy et al., 2020; Table 8.2). Moreover, OGM identified clinically relevant SVs in 11%

of cases that had been missed by the routine testing. In 24% of cases, OGM refined the underlying genomic structure reported by traditional cytogenetic methods (13%), identified additional clinically relevant variants (7%), or both (4%). Three of 48 (6.25%) cases reported with normal karyotypes were shown to have cryptic translocations involving gene fusions. Two of these cases included fusion between NSD1-NUP98. Based on the results of this comprehensive genomic profiling of the AML patients in this multiinstitutional study, authors recommend OGM to be considered the first-line test for detection and identification of clinically relevant SVs in AML.

Novel SV identification

While benchmarking studies comparing Bionano optical genome mapping with conventional cytogenetic techniques aimed at determining the ability of the former to reach sensitivities of the latter, it is interesting to observe that many novel findings were reported. This is anticipated because of OGM can detect all classes of SV and is unbiased, and there is a much higher resolution compared with karyotyping, CMA, and FISH. In the work by Neveling et al. (2020), 17 interchromosomal translocations were unique calls leading to potential gene fusions and none were reported previously in the cosmic database. Of particular interest were putative fusions of *RUNX1/AGBL4* and *BCR/EXOC2*, where the fusion involved one gene classically described as altered in leukemia fused to a partner never reported in hematological malignancies but previously reported in solid tumors. Similar findings were observed in the work by Lestringant et al. (2021) (Table 8.2) where 12 new anomalies missed by conventional cytogenetics were identified by optical genome mapping. In the majority of cases, gains and deletions were confirmed by reanalysis of the SNP-array raw data and translocations could be confirmed retrospectively by FISH. Of significant interest, new translocations identified included chromosomal rearrangements of *LMO2/TCRA, TCRB/MYC,* and *IGH/CEBPB* fusions and another translocation involving *DNMT3B,* all confirmed by orthogonal technologies.

In line with the previous findings, Levy et al. (2020) described that OGM brought additional information allowing to refine or resolve cytogenetics results in 11% of the cohort. Moreover, in 7 of these patients (7% of the cohort), OGM identified new clinically significant variants such as cryptic translocations and CNV changes completely missed by routine testing and results were confirmed by PCR or CMA. As an example and of particular interest, a cryptic unbalanced translocation leading to a derivative chromosome 5 was newly identified and could possibly lead to changes in patient management and prognosis. These novel findings nicely illustrate some of the benefits of optical genome mapping in precision medicine and more specifically in the better characterization of tumors and identification of novel actionable SVs. Validation and characterization of these events, especially the ones leading to novel fusions, are mandatory step toward measuring the clinical utility of Bionano in the clinical space.

Application in solid tumors

The complexity of solid tumor genomes presents a daunting challenge to clinical diagnostics as well as research studies. Complexity, like hematologic malignancies, results from genome instability and cellular heterogeneity; however, solid tumors are commonly understood to be more complex than hematologic tumors. Current guidelines in tumor histology for pathologic grading of tissues from biopsies rely on preservation using formalin-fixed paraffin-embedded (FFPE) samples restrict genetic testing to FISH, CMA, and short-read gene panels sequencing (Litton et al., 2019). Although FFPE tissue is not compatible with OGM at this time, fresh or fresh-frozen tissue biopsies or other specimens are completely compatible with OGM directly enabling whole-genome comprehensive SV analysis which has to date been elusive.

Recent work is highlighting the value proposition of OGM for the genome-wide identification of SVs in solid tumors. Jaratlerdsiri et al. (2017) aimed to compare OGM and high-coverage short-read NGS for SV identification in prostate cancer (Jaratlerdsiri et al., 2017). They identified 85 large somatic structural rearrangements by optical genome mapping; the vast majority of the large SVs (89%) observed by OGM were not detected by NGS; however, guided manual inspection of NGS reads at candidate regions allowed for confirmation of 94% of these large SVs. Strikingly, over a third of the large SVs impacted genes with oncogenic potential. Of particular interest, a 14.3-kbp somatic deletion was identified, leading to a *DUSP11/C2orf78* gene fusion event that was missed by NGS. This study provided the first evidence that OGM could uncover a broader spectrum of potentially oncogenic driver events that has been underdetected by molecular techniques, with significant potential for further cancer subclassification.

In preliminary work on head and neck cancer by Labarge et al. (internal communication, unpublished), OGM helped to elucidate events discriminating tongue cancer in nonsmoker nondrinker young female patients versus elderly smoker drinker patients (representing a higher risk group). While the literature conflicted on mutational profile specific to each group, the pilot study using OGM highlighted evidence of clear somatic genetic differences between age groups. Optical genome mapping further helped to better characterize the genetic landscape in highly lethal thyroid tumors. Labarge et al. identified SVs previously missed corresponding to actionable mutations involving kinases. Finally, human papilloma virus (HPV)-positive head and neck cancer can display three HPV statuses: integrated, viral episomal, or viral/human hybrid episomal. The group used OGM in combination with short-read whole-genome sequencing to characterize the molecular state of HPV to investigate its correlation with patient outcome and potentially enable tailored treatment (de-escalation for low-risk patients versus more aggressive therapy for high risks). Interestingly, they observed that episomal HPV cancer shows less-to-no SV while integrated forms have a global genome instability signature, and integration correlates with aggressiveness of the disease.

Comprehensive tumor profiling is the cornerstone of pharmacogenomics and precision medicine, and a fast, robust, and automated detection of genomic CNVs and

SVs is essential. Better detection of scars, specific types of structural abnormalities resulting from homologous recombination defects (HRD), could allow for improved identification of candidates for PARP-inhibitor treatment to a wider range of cancer types (Watkins et al., 2014). Current NGS-based techniques for tumor HRD profiling remain challenging and rely on the combination of many complicated analytical pipelines and require advanced bioinformatics skills.

The group of Jessica Zucman Rossi conducted preliminary work on a hepatocellular carcinoma (HCC) subgroup exhibiting cyclin activation through various mechanisms, including HBV and AAV2 viral insertions, gene fusions, and enhancer hijacking (internal communication, unpublished). Optical genome mapping has been used for the identification of cancer samples harboring signatures specific of replicative stress, very similar to a BRCAness signature, without the need of a complex bioinformatics pipelines. As a result of the high sensitivity and positive predictive value of OGM, quantitation of HRD signature abnormalities in hepatocellular carcinoma samples was simple, and they were sufficient to distinguish patients with a viral insertion upstream of a cyclin gene causing a specific pattern of rearrangements, informative for the potential treatment with PARPi.

Concluding remarks

To date, no single technology offers a comprehensive analysis of all classes of chromosomal aberrations and different combinations of genomic approaches are needed to compensate for the limitations of each individual technology. Despite decades of improvements to sequencing technologies, clinically, CMA, karyotyping, and/or FISH remain the first-line tests for many disorders worldwide, illustrating the persistence of unmet needs in terms of performance, sensitivity, cost/efficiency, and practicality. Regardless of plethora of traditional and newer technologies, disappointingly, diagnostic yield for many disorders remains dismally low. Newer, cytogenomic technologies have potential to significantly improve the diagnostic yield in cancer and inherited disorders, but they need be able to replicate the functions of current technologies and fit into a clinical workflow from a cost, consistency, usability, and performance perspective. Here, we reviewed several studies that demonstrate that OGM is able to functionally replace classical cytogenetics as well as improve clinically relevant information and diagnostic yield.

OGM like karyotyping effectively detects balanced and unbalanced chromosomal abnormalities, such as duplications, insertions, deletions, aneuploidies, translocations, and inversions. It provides the advantage of the elimination of cell culturing, which improves workflows and removes biases selected for during culture. While FISH is a targeted, a priori approach, OGM allows a genome wide interrogation of the sample and indicates the presence of gene fusions, uncovering cryptic translocations and identifying CNVs below the resolution of standard G-banded karyotyping (< 10 Mbp).

The different examples reported here illustrate the clinical utility of OGM in the study of hematological malignancies and inherited disorders. OGM provided a more detailed characterization of the genome of patient samples while maintaining the performance the standard of care:

- OGM achieves 100% concordance compared with the standard of care methods (e.g., karyotype, FISH, CMA) in several studies.
- OGM is a simple workflow not requiring specially trained technicians for wet lab or analysis.
- Undiagnosed rare genetic cases analyzed by OGM were able to find pathogenic mutations in 12%–18% of cases with no previous genetic diagnosis through the standard of care.
- Many cryptic and otherwise missed anomalies by current technologies are detected in cancer samples using OGM, including new putative fusion events of paramount clinical significance can be detected.

Currently, many additional retrospective and prospective clinical validation studies are being conducted, aiming to further evaluate and quantify clinical utility of OGM in a routine clinical context. These cytogenomic studies are expected to lead to additional laboratory developed tests for routine diagnostics in patients with developmental delay, intellectual disability, leukemia, lymphoma, and solid tumors.

References

Alkan, C., Sajjadian, S., & Eichler, E. E. (2011). Limitations of next-generation genome sequence assembly. *Nature Methods*, *8*, 61–65. https://doi.org/10.1038/nmeth.1527.

Baranov, V. S., & Kuznetzova, T. V. (2021). Nuclear stability in early embryo. Chromosomal aberrations. In T. Liehr (Ed.), *Cytogenomics* (pp. 307–325). Academic Press. Chapter 15 (in this book).

Barseghyan, H., Tang, W., Wang, R. T., Almalvez, M., Segura, E., Bramble, M. S., ... Vilain, E. (2017). Next-generation mapping: A novel approach for detection of pathogenic structural variants with a potential utility in clinical diagnosis. *Genome Medicine*, *9*, 90. https://doi.org/10.1186/s13073-017-0479-0.

Cameron, D. L., Di Stefano, L., & Papenfuss, A. T. (2019). Comprehensive evaluation and characterisation of short read general-purpose structural variant calling software. *Nature Communications*, *10*, 3240. https://doi.org/10.1038/s41467-019-11146-4.

Chaisson, M. J. P., Sanders, A. D., Zhao, X., Malhotra, A., Porubsky, D., Rausch, T., ... Lee, C. (2019). Multi-platform discovery of haplotype-resolved structural variation in human genomes. *Nature Communications*, *10*, 1784. https://doi.org/10.1038/s41467-018-08148-z.

Daban, J.-R. (2021). Multilayer organization of chromosomes. In T. Liehr (Ed.), *Cytogenomics* (pp. 267–296). Academic Press. Chapter 13 (in this book).

Dai, Y., Li, P., Wang, Z., Liang, F., Yang, F., Fang, L., ... Wang, K. (2020). Single-molecule optical mapping enables quantitative measurement of D4Z4 repeats in facioscapulohumeral muscular dystrophy (FSHD). *Journal of Medical Genetics*, *57*, 109–120. https://doi.org/10.1136/jmedgenet-2019-106078.

de Koning, A. P. J., Gu, W., Castoe, T. A., Batzer, M. A., & Pollock, D. D. (2011). Repetitive elements may comprise over two-thirds of the human genome. *PLoS Genetics, 7*, e1002384. https://doi.org/10.1371/journal.pgen.1002384.

Deelen, P., van Dam, S., Herkert, J. C., Karjalainen, J. M., Brugge, H., Abbott, K. M., ... Franke, L. (2019). Improving the diagnostic yield of exome-sequencing by predicting gene-phenotype associations using large-scale gene expression analysis. *Nature Communications, 10*, 2837. https://doi.org/10.1038/s41467-019-10649-4.

Demaerel, W., Mostovoy, Y., Yilmaz, F., Vervoort, L., Pastor, S., Hestand, M. S., ... Vermeesch, J.R. (2019). The 22q11 low copy repeats are characterized by unprecedented size and structural variability. *Genome Research, 29*, 1389–1401. https://doi.org/10.1101/gr.248682.119.

Eaton, K. W., Tooke, L. S., Wainwright, L. M., Judkins, A. R., & Biegel, J. A. (2011). Spectrum of SMARCB1/INI1 mutations in familial and sporadic rhabdoid tumors. *Pediatric Blood & Cancer, 56*, 7–15. https://doi.org/10.1002/pbc.22831.

Ebert, P., Audano, P. A., Zhu, Q., Rodriguez-Martin, B., Porubsky, D., Bonder, M. J., ... Eichler, E.E. (2020). De novo assembly of 64 haplotype-resolved human genomes of diverse ancestry and integrated analysis of structural variation. *bioRxiv*. https://doi.org/10.1101/2020.12.16.423102. 2020.12.16.423102.

English, A. C., Salerno, W. J., Hampton, O. A., Gonzaga-Jauregui, C., Ambreth, S., Ritter, D. I., ... Gibbs, R.A. (2015). Assessing structural variation in a personal genome—Towards a human reference diploid genome. *BMC Genomics, 16*, 286. https://doi.org/10.1186/s12864-015-1479-3.

Hastie, A. R., Lam, E. T., Pang, A. W. C., Zhang, L. X., Andrews, W., Lee, J., ... Cao, H. (2017). Rapid automated large structural variation detection in a diploid genome by nanochannel based next-generation mapping. *bioRxiv*. https://doi.org/10.1101/102764. 102764.

Iourov, I. Y., Vorsanova, S. G., & Yurov, Y. B. (2021). Cytogenomic landscape of the human brain. In T. Liehr (Ed.), *Cytogenomics* (pp. 327–348). Academic Press. Chapter 16 (in this book).

Ishii, T., Nagaki, K., & Houben, A. (2021). Application of CRISPR/Cas9 to visualize defined genomic sequences in fixed chromosomes and nuclei. In T. Liehr (Ed.), *Cytogenomics* (pp. 147–153). Academic Press. Chapter 9 (in this book).

Jaratlerdsiri, W., Chan, E. K. F., Petersen, D. C., Yang, C., Croucher, P. I., Bornman, M. S. R., ... Hayes, V.M. (2017). Next generation mapping reveals novel large genomic rearrangements in prostate cancer. *Oncotarget, 8*, 23588–23602. https://doi.org/10.18632/oncotarget.15802.

Lee, H., Deignan, J. L., Dorrani, N., Strom, S. P., Kantarci, S., Quintero-Rivera, F., ... Nelson, S.F. (2014). Clinical exome sequencing for genetic identification of rare Mendelian disorders. *JAMA, 312*, 1880–1887. https://doi.org/10.1001/jama.2014.14604.

Levy, B., Baughn, L. B., Chartrand, S., LaBarge, B., Claxton, D., Lennon, A., ... Broach, J. (2020). A national multicenter evaluation of the clinical utility of optical genome mapping for assessment of genomic aberrations in acute myeloid leukemia. *medRxiv*. https://doi.org/10.1101/2020.11.07.20227728. 2020.11.07.20227728.

Liehr, T. (2021a). Molecular cytogenetics. In T. Liehr (Ed.), *Cytogenomics* (pp. 35–45). Academic Press. Chapter 4 (in this book).

Liehr, T. (2021b). Nuclear architecture. In T. Liehr (Ed.), *Cytogenomics* (pp. 297–305). Academic Press. Chapter 14 (in this book).

Liehr, T. (2021c). A definition for cytogenomics – Which also may be called chromosomics. In T. Liehr (Ed.), *Cytogenomics* (pp. 1–7). Academic Press. Chapter 1 (in this book).

Liehr, T. (2021d). Overview of currently available approaches used in cytogenomics. In T. Liehr (Ed.), *Cytogenomics* (pp. 11–24). Academic Press. Chapter 2 (in this book).

Liehr, T. (2021e). Repetitive elements, heteromorphisms, and copy number variants. In T. Liehr (Ed.), *Cytogenomics* (pp. 373–388). Academic Press. Chapter 19 (in this book).

Liehr, T., Lauten, A., Schneider, U., Schleussner, E., & Weise, A. (2017). Noninvasive prenatal testing—When is it advantageous to apply. *Biomedicine Hub*, *2*(1), 1–11. https://doi.org/10.1159/000458432.

Liehr, T., Othman, M. A., Rittscher, K., & Alhourani, E. (2015). The current state of molecular cytogenetics in cancer diagnosis. *Expert Review of Molecular Diagnostics*, *15*(4), 517–526. https://doi.org/10.1586/14737159.2015.1013032.

Litton, J. K., Burstein, H. J., & Turner, N. C. (2019). Molecular testing in breast cancer. *American Society of Clinical Oncology Educational Book*, e1–e7. https://doi.org/10.1200/EDBK_237715.

Lestringant, V., Duployez, N., Penther, D., Luquet, I., Grardel, N., Lutun, A., et al. (2021). Optical genome mapping, a promising alternative to gold standard cytogenetic approaches in a series of acute lymphoblastic leukemias. *Genes, Chromosomes and Cancer*. Accepted.

Lemmers, R. J. L. F., O'Shea, S., Padberg, G. W., Lunt, P. W., & van der Maarel, S. M. (2012). Best practice guidelines on genetic diagnostics of facioscapulohumeral muscular dystrophy: Workshop 9th June 2010, LUMC, Leiden, The Netherlands. *Neuromuscular Disorders*, *22*, 463–470. https://doi.org/10.1016/j.nmd.2011.09.004.

Lemmers, R. J. L. F., van der Vliet, P. J., Klooster, R., Sacconi, S., Camaño, P., Dauwerse, J. G., … van der Maarel, S.M. (2010). A unifying genetic model for facioscapulohumeral muscular dystrophy. *Science*, *329*, 1650–1653. https://doi.org/10.1126/science.1189044.

Mantere, T., Kersten, S., & Hoischen, A. (2019). Long-read sequencing emerging in medical genetics. *Frontiers in Genetics*, *10*. https://doi.org/10.3389/fgene.2019.00426.

Mantere, T., Neveling, K., Pebrel-Richard, C., Benoist, M., van der Zande, G., Kater-Baats, E., … Khattabi, L.E. (2020). Next generation cytogenetics: Genome-imaging enables comprehensive structural variant detection for 100 constitutional chromosomal aberrations in 85 samples. *bioRxiv*. https://doi.org/10.1101/2020.07.15.205245.

Miller, D. T., Adam, M. P., Aradhya, S., Biesecker, L. G., Brothman, A. R., Carter, N. P., … Ledbetter, D.H. (2010). Consensus statement: Chromosomal microarray is a first-tier clinical diagnostic test for individuals with developmental disabilities or congenital anomalies. *American Journal of Human Genetics*, *86*, 749–764. https://doi.org/10.1016/j.ajhg.2010.04.006.

National Comprehensive Cancer Network. (2019). *NCCN clinical practice guidelines in oncology: Acute myeloid leukemia*. National Comprehensive Cancer Network.

Nattestad, M., Goodwin, S., Ng, K., Baslan, T., Sedlazeck, F., Rescheneder, P., … Schatz, M.C. (2018). Complex rearrangements and oncogene amplifications revealed by long-read DNA and RNA sequencing of a breast cancer cell line. *Genome Research*. https://doi.org/10.1101/gr.231100.117. gr.231100.117.

Neveling, K., Mantere, T., Vermeulen, S., Oorsprong, M., van Beek, R., Kater-Baats, E., … Hoischen, A. (2020). Next generation cytogenetics: Comprehensive assessment of 48 leukemia genomes by genome imaging. *bioRxiv*. https://doi.org/10.1101/2020.02.06.935742. 2020.02.06.935742.

Pastor, S., Tran, O., Jin, A., Carrado, D., Silva, B. A., Uppuluri, L., … Emanuel, B.S. (2020). Optical mapping of the 22q11.2DS region reveals complex repeat structures and preferred locations for non-allelic homologous recombination (NAHR). *Scientific Reports*, *10*. https://doi.org/10.1038/s41598-020-69134-4.

Pellestor, F., Gaillard, J.-B., Schneider, A., Puechberty, J., & Gatinois, V. (2021). Chromoanagenesis phenomena and their formation mechanisms. In T. Liehr (Ed.), *Cytogenomics* (pp. 213–245). Academic Press. Chapter 11 (in this book).

Preston, M. K., Tawil, R., & Wang, L. H. (1993). Facioscapulohumeral muscular dystrophy. In M. P. Adam, H. H. Ardinger, R. A. Pagon, S. E. Wallace, L. J. Bean, K. Stephens, & A. Amemiya (Eds.), *GeneReviews®*. Seattle, WA: University of Washington.

Sabatella, M., Mantere, T., Waanders, E., Neveling, K., Mensenkamp, A.R., van Dijk, F., et al. (2021). Optical genome mapping identifies a germline retrotransposon insertion in SMARCB1 in two siblings with Atypical Teratoid Rhabdoid Tumors. *Journal of Pathology*. Submitted.

Shieh, J. T., Penon-Portmann, M., Wong, K. H. Y., Levy-Sakin, M., Verghese, M., Slavotinek, A., … Boffelli, D. (2020). Application of full genome analysis to diagnose rare monogenic disorders. *medRxiv*. https://doi.org/10.1101/2020.10.22.20216531.

Silva, M., de Leeuw, N., Mann, K., Schuring-Blom, H., Morgan, S., Giardino, D., … Hastings, R. (2019). European guidelines for constitutional cytogenomic analysis. *European Journal of Human Genetics*, *27*, 1–16. https://doi.org/10.1038/s41431-018-0244-x.

Ungelenk, M. (2021). Sequencing approaches. In T. Liehr (Ed.), *Cytogenomics* (pp. 87–122). Academic Press. Chapter 7 (in this book).

Watkins, J. A., Irshad, S., Grigoriadis, A., & Tutt, A. N. (2014). Genomic scars as biomarkers of homologous recombination deficiency and drug response in breast and ovarian cancers. *Breast Cancer Research*, *16*, 211. https://doi.org/10.1186/bcr3670.

Weise, A., & Liehr, T. (2021a). Cytogenetics. In T. Liehr (Ed.), *Cytogenomics* (pp. 25–34). Academic Press. Chapter 3 (in this book).

Weise, A., & Liehr, T. (2021b). Interchromosomal interactions with meaning for disease. In T. Liehr (Ed.), *Cytogenomics* (pp. 349–356). Academic Press. Chapter 17 (in this book).

Weise, A., & Liehr, T. (2021c). Molecular karyotyping. In T. Liehr (Ed.), *Cytogenomics* (pp. 73–85). Academic Press. Chapter 6 (in this book).

Wright, C. F., FitzPatrick, D. R., & Firth, H. V. (2018). Paediatric genomics: Diagnosing rare disease in children. *Nature Reviews. Genetics*, *19*, 253–268. https://doi.org/10.1038/nrg.2017.116.

Yang, Y., Muzny, D. M., Reid, J. G., Bainbridge, M. N., Willis, A., Ward, P. A., … Eng, C.M. (2013). Clinical whole-exome sequencing for the diagnosis of mendelian disorders. *The New England Journal of Medicine*, *369*, 1502–1511. https://doi.org/10.1056/NEJMoa1306555.

Zhang, Q., Xu, X., Ding, L., Li, H., Xu, C., Gong, Y., … Tang, S. (2019). Clinical application of single-molecule optical mapping to a multigeneration FSHD1 pedigree. *Molecular Genetics & Genomic Medicine*, *7*, e565. https://doi.org/10.1002/mgg3.565.

Zheng, Y., Kong, L., Xu, H., Lu, Y., Zhao, X., Yang, Y., … Kong, X. (2020). Rapid prenatal diagnosis of facioscapulohumeral muscular dystrophy 1 by combined bionano optical mapping and karyomapping. *Prenatal Diagnosis*, *40*, 317–323. https://doi.org/10.1002/pd.5607.

Application of CRISPR/Cas9 to visualize defined genomic sequences in fixed chromosomes and nuclei

9

Takayoshi Ishii[a], Kiyotaka Nagaki[b], and Andreas Houben[c]

[a]*Arid Land Research Center (ALRC), Tottori University, Tottori, Japan*
[b]*Institute of Plant Science and Resources, Okayama University, Kurashiki, Japan*
[c]*Leibniz Institute of Plant Genetics and Crop Plant Research (IPK), Seeland, Germany*

Chapter outline

Introduction

In situ hybridization is an established method for the visualization of specific DNA sequences in fixed nuclei and metaphase chromosomes. For fluorescence standard in situ hybridization (FISH), denaturation of the chromosomal DNA is necessary to allow probe hybridization (Liehr, 2021a). However, denaturation by heat, formamide, or sodium hydroxide may affect the native structure of chromatin (Boettiger et al., 2016; Kozubek et al., 2000). Nowadays, with improved microscopes like superresolution microscopy, and great interest in analyzing the real native chromatin structure in the frame of cytogenomic research (Liehr, 2021b, 2021c), more specimen-sensitive DNA labeling techniques are required. Recently described CRISPR/dCas9-based DNA labeling methods like Cas9-mediated FISH (CAS-FISH) (Deng et al., 2015) and RNA-guided endonuclease-in situ labeling (RGEN-ISL) (synonym CRISPR-FISH) (Ishii et al., 2019) allows the labeling of repetitive DNA elements in fixed mammalian and plant nuclei and chromosomes. Both CRISPR-based methods do not require DNA denaturation and therefore permit better structural chromatin preservation. Further, both cytogenomic methods do not need the transformation of the specimen with any kind of construct and an enzymatic in vitro RNA synthesis. Besides, another variant of CRISPR/Cas9 technology also allows for a FISH-like labeling of living cells - this is described elsewhere (Wang et al., 2018, 2019).

Cytogenomics. https://doi.org/10.1016/B978-0-12-823579-9.00007-2

147

Application of CRISPR-FISH

CAS-FISH employs a fluorophore-coupled single guide RNA (sgRNA) in combination with a fluorophore as well as Halo-tag coupled nuclease-deficient recombinant Cas9 (dCas9) protein for the detection of high-copy sequences (Deng et al., 2015). CRISPR-FISH (synonym of RGEN-ISL) represents a further development of this method, which does not require the application of a Halo-tag approach, and therefore simplifies the handling of the enzyme-RNA complex-based labeling of fixed mammalian and plant nuclei and chromosomes as described elsewhere (Ishii et al., 2019). The CRISPR-FISH method involves a ribonucleoprotein (RNP) that consists of a target-specific CRISPR RNA (crRNA), a transactivating crRNA (tracrRNA), and Cas9 endonuclease (Fig. 9.1). Besides the application of the nuclease-deficient

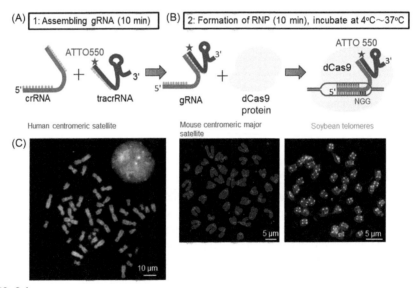

FIG. 9.1

Strategy of the CRISPR-FISH method. (A) Guide RNA (gRNA) complex formation after hybridization of CRISPR RNA (crRNA) and 5' ATTO 550 (star)-labeled trans-activating crRNA (tracrRNA). (B) Ribonucleoprotein (RNP) complex formation after the combination of the recombinant Cas9 protein with gRNA. Components of the CRISPR-FISH system to label genomic targets. The crRNA:tracrRNA complex uses optimized Alt-R crRNA and ATTO 550-labeled tracrRNA sequences that hybridize and then form a complex with Cas9 endonuclease to guide targeted binding to genomic DNA. The binding site is specified by the protospacer element of the crRNA *(light gray bar)*. The crRNA protospacer element recognizes 19 or 20 nucleotides on the opposite strand of the NGG PAM site. The PAM site must be present immediately downstream of the protospacer element that binding can occur. (C) Representative results of the CRISPR-FISH method showing human and mouse centromeres as well as soybean telomeres.

dCas9 protein, recombinant active Cas9, Cas9 with an MBP-tag, or Cas9 nickase were also successfully used for CRISPR-FISH. The addition of a fluorescent dye to the 5′ end of tracrRNA does not affect the functional performance of RNPs and allows the microscope-based detection of the interaction between RNPs and target DNA. The application of differentially labeled tracrRNAs (like Alexa488 and ATTO550) allows the multiplexing of CRISPR-FISH. Using maize as an example, a combination of CRISPR-FISH, immunostaining, and 5-ethynyl-2′-deoxyuridine (EdU) labeling to visualize in situ-specific repeats, histone marks, and DNA replication sites, respectively, was demonstrated (Němečková et al., 2019).

One limitation of the CRISPR-FISH is the requirement of the presence of an adjacent protospacer motif (PAM) at the target site. For the most used Cas9 from *S. pyogenes*, the required PAM sequence is NGG. To overcome this limitation, natural or engineered Cas9 variants with different PAM sequence requirements could be tested (Endo et al., 2019; Hu et al., 2018; Ma et al., 2015). The target-specific 20 base DNA sequence upstream of the PAM site could be selected with the help of the in silico tool CRISPRdirect (https://crispr.dbcls.jp/) (Naito et al., 2015) or others.

The RNP complex interrogates the target sites on the DNA via three-dimensional diffusion and recognizes the complementary target site with the NGG PAM diffusion (Shibata et al., 2017). To visualize the real-time dynamics of the RNP complex for the labeling of telomeres of fixed nuclei in action, time-lapse microscopy was employed (Ishii et al., 2019). Notably, already 20 seconds after applying the preassembled RNP complex, the first telomere signals became detectable. The intensity of fluorescence signals steadily increased during the next 390 seconds. The increase of the telomere signal intensity suggests, in line with previous biochemical experiments (Sternberg et al., 2014), that the dCas9-RNA complex remains tightly bound to the DNA after the target identification. However, to maintain the fluorescence signals after the CRISPR-FISH reaction, a short postfixation step in 4% formaldehyde on ice is required to prevent dissociation of the dCas9-RNA complex from the DNA. Incubation temperatures between 4°C and 37°C are suitable for this method (Ishii et al., 2019). For the detection of tandem repetitive sequences (see also Liehr, 2021d), a single guide RNA is sufficient for CRISPR-FISH imaging. However, the detection of nonrepetitive loci may require a simultaneous application of multiple guide RNAs as demonstrated for the CRISPR-based detection of a 5 kb region in living mammalian cells (Chen et al., 2013).

Further developments may improve the CRISPR-FISH detection method and potentially enable us to visualize even short single-copy sequences. Compared with plants, mammalian chromosomes displayed stronger CRISPR-FISH labeling without background noise. Maybe, this difference is due to the absence of a cell wall and less cytoplasm in mammalian cells. To evaluate the influence of denaturation on the morphology of chromatin, CRISPR-FISH was performed first, and superresolution microscopy images were acquired. Afterward, the same specimen was used for standard FISH, recorded again, and the images compared. Using standard microscopy, the overall morphology of chromosomes and nuclei was similar for both methods. However, the application of superresolution microscopy revealed subtle differences.

It seemed that standard FISH impaired and flattened the chromatin. In the case of CRISPR-FISH, the chromatin structure stays more compact. Hence, CRISPR-FISH is the method of choice for the visualization of repeats if the ultrastructure of chromatin is of interest.

CRISPR-FISH does not require denaturation, and therefore allows the visualization of target DNA sequences in tissues. However, CRISPR-FISH is sensitive to fixation, which creates a dilemma for tissue sectioning and detection of CRISPR-FISH signals; "To strongly fix, or not to strongly fix: that is the question," tissue sections which were fixed in a low-concentration of paraformaldehyde solution (1.5% (w/v)), result in intensive CRISPR-FISH signals. However, due to insufficient mechanical strength of the tissue, the tissue is fragile and may collapse. On the other hand, if the tissue is fixed strongly like in 3% paraformaldehyde, the tissue structure is maintained after slicing, but the CRISPR-FISH signals will be weak. This dilemma was recently solved by de-crosslinking of strongly fixed tissue after sectioning (Nagaki & Yamaji, 2020). First, the tissue was fixed in 3% paraformaldehyde and then sectioned using a vibratome. For decrosslinking, tissue slices were treated in 20 mM Tris-HCl (pH 9.0) for 2 h at 60°C. The decrosslinking treatment yielded CRISPR-FISH signals of sufficient intensity (Fig. 9.2A and B). This method is also compatible with immunohistochemical and tissue clearing methods (Nagaki et al., 2012, 2015, 2017), and allows simultaneous visualization of target DNA sequences and proteins while maintaining the tissue structure (Fig. 9.2C–F). As a result, it is possible to acquire information from tens of thousands of cells while retaining the positional information of cells in the tissue. This method not only allows a wide range of observations within the tissues but also enables high-resolution observation of individual cells within the tissue because it does not damage the cell structure (Fig. 9.2G). In other words, the CRISPR-FISH method with decrosslinking enables simultaneous acquisition of the bird's-eye view information of the entire tissue and high-resolution information of the part of interest. Recently, improvement of CRISPR-FISH signal intensity in alcohol:acetic acid fixed samples became possible after treatment of specimens before CRISPR-FISH with 40 mM Tris-HCl for 30 minutes at 37°C (Potlapalli et al., 2020). Since alcohol:acetic acid fixation is a widely used fixation method in cytogenetic studies, this decrosslinking step will dramatically extend the application of CRISPR-FISH. Table 9.1 summarizes the differences between standard FISH and CRISPR-FISH (Table 9.1).

As mentioned above, taking advantage of the intrinsic stability of CRISPR guide RNA, some groups (Wang et al., 2018, 2019) used fluorescent RNPs for live-cell imaging in transfected human T lymphocytes. Live-cell fluorescence in situ hybridization (LiveFISH) allowed tracking of multiple chromosomal loci in lymphocytes. Few hours after transfection of cells with labeled RNP complexes, robust chromosome labeling in living cells was visible. Using RNP complexes with chemically labeled fluorescent sgRNAs, they multiplexed the labeling of telomeres and major satellites in transfected mouse NIH 3T3 cells. Hence, the transient transformation of cells with fluorescent RNP complexes could become a cytogenomic option to label defined sequences in living cells of plants (reviewed in Khosravi et al., 2020) or other species.

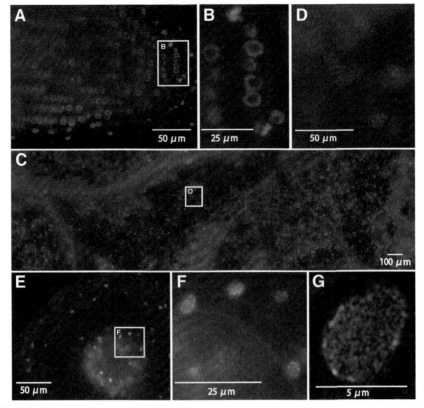

FIG. 9.2

Application of CRISPR-FISH to nuclear analysis in tissues. (A) CRISPR-FISH telomere signals detected in a tomato root section. Panel (B) is a magnified view of a boxed area in (A). (C) CRISPR-FISH telomere signals detected in a transparent tobacco leaf. Panel (D) shows a magnified image of an area indicated by a box in (C). (E) CRISPR-FISH and immunohistochemical signals in a rice root section. CRISPR-FISH signals of rice centromeric tandem repeats (*red*) and immunohistochemical signals of centromere-specific histone H3 variants (*yellow*) and K9 dimethylated histone H3 (*green*) were visualized on DAPI-stained nuclei (*gray*). Panel (F) shows a magnified image of a region indicated by a box in (E). (G) High-resolution confocal analysis of 3D CRISPR-FISH and immunohistochemical signals. The colors of the signals are the same as in panel (E).

Table 9.1 Standard FISH vs CRISPR-FISH.

	Standard FISH	CRISPR-FISH
Probe	Labeled DNA	In vitro assembled RNP
Global DNA denaturation	Yes	No
Temperature	37°C	Between 4°C and 42°C
Duration of experiment	Hours to days	Minutes to hours
Color	Multicolor	Multicolor
Preservation of chromatin structure	Altered structure	Preserved structure

Six major features of standard FISH vs. CRISPR-FISH are compared.

Funding information

Research of the authors have been supported by Deutsche Forschungsgemeinschaft ((DFG) HO1779/28-1), the JSPS KAKENHI (Grant Numbers JP 19K15817, 20K21317) and the Joint Research Program of Arid Land Research Center, Tottori University (No. 31C2002).

References

Boettiger, A. N., Bintu, B., Moffitt, J. R., Wang, S., Beliveau, B. J., Fudenberg, G., … Zhuang, X. (2016). Super-resolution imaging reveals distinct chromatin folding for different epigenetic states. *Nature*, *529*(7586), 418–422. https://doi.org/10.1038/nature16496.

Chen, B., Gilbert, L. A., Cimini, B. A., Schnitzbauer, J., Zhang, W., Li, G. W., … Huang, B. (2013). Dynamic imaging of genomic loci in living human cells by an optimized CRISPR/Cas system. *Cell*, *155*(7), 1479–1491. https://doi.org/10.1016/j.cell.2013.12.001.

Deng, W., Shi, X., Tjian, R., Lionnet, T., & Singer, R. H. (2015). CASFISH: CRISPR/Cas9-mediated in situ labeling of genomic loci in fixed cells. *Proceedings of the National Academy of Sciences of the United States of America*, *112*(38), 11870–11875. https://doi.org/10.1073/pnas.1515692112.

Endo, M., Mikami, M., Endo, A., Kaya, H., Itoh, T., Nishimasu, H., … Toki, S. (2019). Genome editing in plants by engineered CRISPR–Cas9 recognizing NG PAM. *Nature Plants*, *5*(1), 14–17. https://doi.org/10.1038/s41477-018-0321-8.

Hu, J. H., Miller, S. M., Geurts, M. H., Tang, W., Chen, L., Sun, N., … Liu, D.R. (2018). Evolved Cas9 variants with broad PAM compatibility and high DNA specificity. *Nature*, *556*(7699), 57–63. https://doi.org/10.1038/nature26155.

Ishii, T., Schubert, V., Khosravi, S., Dreissig, S., Metje-Sprink, J., Sprink, T., … Houben, A. (2019). RNA-guided endonuclease-in situ labelling (RGEN-ISL): A fast CRISPR/Cas9-based method to label genomic sequences in various species. *New Phytologist*, *222*(3), 1652–1661. https://doi.org/10.1111/nph.15720.

Khosravi, S., Ishii, T., Dreissig, S., & Houben, A. (2020). Application and prospects of CRISPR/Cas9-based methods to trace defined genomic sequences in living and fixed plant cells. *Chromosome Research*, *28*(1), 7–17. https://doi.org/10.1007/s10577-019-09622-0.

Kozubek, S., Lukášová, E., Amrichová, J., Kozubek, M., Lišková, A., & Šlotová, J. (2000). Influence of cell fixation on chromatin topography. *Analytical Biochemistry*, *282*(1), 29–38. https://doi.org/10.1006/abio.2000.4538.

Liehr, T. (2021a). Molecular cytogenetics. In T. Liehr (Ed.), *Cytogenomics* (pp. 35–45). Academic Press. Chapter 4 (in this book).

Liehr, T. (2021b). A definition for cytogenomics - Which also may be called chromosomics. In T. Liehr (Ed.), *Cytogenomics* (pp. 1–7). Academic Press. Chapter 1 (in this book).

Liehr, T. (2021c). Overview of currently available approaches used in cytogenomics. In T. Liehr (Ed.), *Cytogenomics* (pp. 11–24). Academic Press. Chapter 2 (in this book).

Liehr, T. (2021d). Repetitive elements, heteromorphisms, and copy number variants. In T. Liehr (Ed.), *Cytogenomics* (pp. 373–388). Academic Press. Chapter 19 (in this book).

Ma, H., Naseri, A., Reyes-Gutierrez, P., Wolfe, S. A., Zhang, S., & Pederson, T. (2015). Multicolor CRISPR labeling of chromosomal loci in human cells. *Proceedings of the National Academy of Sciences of the United States of America*, *112*(10), 3002–3007. https://doi.org/10.1073/pnas.1420024112.

Nagaki, K., Tanaka, K., Yamaji, N., Kobayashi, H., & Murata, M. (2015). Sunflower centromeres consist of a centromere-specific LINE and a chromosome-specific tandem repeat. *Frontiers in Plant Science*, *6*(1), 912. https://doi.org/10.3389/fpls.2015.00912.

Nagaki, K., & Yamaji, N. (2020). Decrosslinking enables visualization of RNA-guided endonuclease–in situ labeling signals for DNA sequences in plant tissues. *Journal of Experimental Botany*, *71*(6), 1792–1800. https://doi.org/10.1093/jxb/erz534.

Nagaki, K., Yamaji, N., & Murata, M. (2017). EPro-ClearSee: A simple immunohistochemical method that does not require sectioning of plant samples. *Scientific Reports*, *7*(1), 42203. https://doi.org/10.1038/srep42203.

Nagaki, K., Yamamoto, M., Yamaji, N., Mukai, Y., & Murata, M. (2012). Chromosome dynamics visualized with an anti-centromeric histone H3 antibody in allium. *PLoS One*, *7*(12), e51315. https://doi.org/10.1371/journal.pone.0051315.

Naito, Y., Hino, K., Bono, H., & Ui-Tei, K. (2015). CRISPRdirect: Software for designing CRISPR/Cas guide RNA with reduced off-target sites. *Bioinformatics*, *31*(7), 1120–1123. https://doi.org/10.1093/bioinformatics/btu743.

Němečková, A., Wäsch, C., Schubert, V., Ishii, T., Hřibová, E., & Houben, A. (2019). CRISPR/Cas9-based RGEN-ISL allows the simultaneous and specific visualization of proteins, DNA repeats, and sites of DNA replication. *Cytogenetic and Genome Research*, *159*(1), 48–53. https://doi.org/10.1159/000502600.

Potlapalli, B. P., Schubert, V., Metje-Sprink, J., Liehr, T., & Houben, A. (2020). Application of Tris-HCl allows the specific labeling of regularly prepared chromosomes by CRISPR-FISH. *Cytogenetic and Genome Research*, *160*(3), 156–165. https://doi.org/10.1159/000506720.

Shibata, M., Nishimasu, H., Kodera, N., Hirano, S., Ando, T., Uchihashi, T., & Nureki, O. (2017). Real-space and real-time dynamics of CRISPR-Cas9 visualized by high-speed atomic force microscopy. *Nature Communications*, *8*(1), 1430. https://doi.org/10.1038/s41467-017-01466-8.

Sternberg, S. H., Redding, S., Jinek, M., Greene, E. C., & Doudna, J. A. (2014). DNA interrogation by the CRISPR RNA-guided endonuclease Cas9. *Nature*, *507*(7490), 62–67. https://doi.org/10.1038/nature13011.

Wang, H., Nakamura, M., Abbott, T. R., Zhao, D., Luo, K., Yu, C., … Qi, L.S. (2019). CRISPR-mediated live imaging of genome editing and transcription. *Science*, *365*(6459), 1301–1305. https://doi.org/10.1126/science.aax7852.

Wang, H., Xu, X., Nguyen, C. M., Liu, Y., Gao, Y., Lin, X., … Qi, L.S. (2018). CRISPR-mediated programmable 3D genome positioning and nuclear organization. *Cell*, *175*(5), 1405–1417.e14. https://doi.org/10.1016/j.cell.2018.09.013.

Approaches for studying epigenetic aspects of the human genome

10

Tigran Harutyunyan and Galina Hovhannisyan

Department of Genetics and Cytology, Yerevan State University, Yerevan, Armenia

Chapter outline

Introduction

For many decades, the question how the tremendous diversity of specialized cells with the same genotype can develop from one cell, a zygote, was perplexing (Baranov & Kuznetzova, 2021; Pellestor et al., 2021). The branch of cytogenomic research (Liehr, 2021a, 2021b) dealing with this problem is "epigenomics." The term "epigenetics" was coined in the last century to categorize all the developmental steps leading from zygote to mature organism (Felsenfeld, 2014). Currently, a working definition of epigenetics is "mitotically and/or meiotically heritable changes in gene function that cannot be explained by changes in the DNA sequence" (Seo et al.,

Cytogenomics. https://doi.org/10.1016/B978-0-12-823579-9.00008-4

155

2015). Therefore, the epigenome consists of chemical compounds "above" the DNA level, and epigenetic changes can define "cell identity" among the diversity of tissues in the body (see also Eggermann, 2021).

Animal studies suggest that epigenetic effects may have a transgenerational impact, i.e., can be heritable along generations. Current data about the heritability of environmentally induced epigenetic changes in rodents are summarized elsewhere (Legoff et al., 2019). Nevertheless, there is a lack of evidence of transgenerational epigenetic inheritance in the human population (Nagy & Turecki, 2015; Van Otterdijk & Michels, 2016).

The main epigenetic factors affecting gene expression are DNA methylation, histone modifications, gene expression changes mediated by noncoding RNAs (Ichikawa & Saitoh, 2021), as well as higher-order chromatin structure (Daban, 2021; Liehr, 2021c; Weise & Liehr, 2021a; Yumiceba et al., 2021). Different epigenetic modifications interact with each other and modulate chromatin organization. The methyl-DNA binding protein MeCP2 provided the first and the most extensively studied example of a mediator between epigenetic modifications that bind to methylated DNA and then recruit histone deacetylases, as well as interact with different other proteins in the regulation of transcription and chromatin architecture (Schmidt et al., 2020). Accordingly, several layers of epigenetic modifications regulate gene expression and cell fate in combination with spatiotemporal factors. Currently, at least 5 types of DNA (Kumar et al., 2018) and 13 types of histone modifications are known (Audia & Campbell, 2016); 2300 human mature noncoding microRNAs were reported by 2019, 50% (1115) of which are yet annotated in miRBase V22 (Alles et al., 2019).

Epigenetic modifications can affect the strength of interactions between DNA and histone proteins leading to changes in chromatin structure and DNA packaging. Key players of DNA packaging are the basic histone proteins that form specific heterodimers, H2A-H2B and H3-H4 (Koyama & Kurumizaka, 2018; Pombo & Dillon, 2015). Two H2A-H2B and H3-H4 heterodimers associate via hydrophobic interactions to form the histone octamer, which wraps ~147 base pairs of DNA in the nucleosome, the basic repeating unit of chromatin (Kobayashi & Kurumizaka, 2019). Further packaging of DNA results in the formation of chromatosome consisting of the nucleosome bound to a linker histone H1 with ~10 bp of DNA at both the entry and the exit sites of the nucleosome core particle (Daban, 2021; Liehr, 2021d). There are two types of chromatin, which are spatially segregated within the nucleus, heterochromatin and euchromatin (Hildebrand & Dekker, 2020). Heterochromatin is mainly localized close to the nuclear periphery, while euchromatin is positioned in the interior of the nucleus (Bickmore & Van Steensel, 2013; Liehr, 2021c). Heterochromatin and euchromatin are associated with distinct DNA methylation and histone modification patterns that correlate with particular states of gene activity (Wang & Chang, 2018). In heterochromatin (see also Liehr, 2021d), the DNA is tightly bound to nucleosomal proteins and is generally transcriptionally inactive due to restricted access to DNA by transcriptional factors. By contrast in euchromatin, the DNA is loosely interacting with nucleosomal proteins and is freely accessible to the transcriptional machinery (Becker et al., 2016; Daban, 2021; Nakagawa & Okita, 2019). Finally, the organization of the nucleus into distinct compartments and chromosome territories (genome architecture) has been shown to contribute to the regulation of the cell cycle, cell differentiation, and senescence through the regulation of gene expression, which confirms the link between gene expression and genome architecture (Pombo & Dillon, 2015).

Several methods are used for the analysis of epigenetic changes in the human genome, enabling detection of epigenetic markings on the nucleotide and the whole-genome level (Eggermann, 2021). In this chapter, we will focus on widely used methods of epigenetics but did not cover all approaches due to their multiplicity and diversity. We have selected the most common cytogenomic methods, as well as the (subjectively) most interesting recent developments of epigenomics. A review of bioinformatic analysis tools for epigenomic data was not included. Various reviews on epigenetic methods are usually devoted to the analysis of one or several epigenetic mechanisms, but we tried to combine and compare all the main technologies. In addition, we discuss the implementation of advances in human epigenomics for the analysis of diseases and the epigenetic impact of environmental factors.

Analysis of DNA methylation

The covalent transfer of a methyl group from the cofactor *S*-adenosyl-L-methionine to the fifth carbon of cytosines (5mC) in DNA is the most prevalent epigenetic modification of DNA in mammalian genomes (Robertson, 2002). DNA methylation reaction is catalyzed by DNA methyltransferases (DNMTs); 5mC is established de novo by DNMT3A and DNMT3B and is maintained by DNMT1, which recognizes hemimethylated CpGs via interaction with UHRF1 (Angeloni & Bogdanovic, 2019). There are also a noncanonical DNMT2 and DNMT3L, which lack catalytic DNMT activity. The function of DNMT2 is poorly studied; however, it is mainly involved in the methylation of C38 of tRNA (Goll et al., 2006). Thus, the current official gene symbol is *TRDMT1* (tRNA aspartic acid methyltransferase 1) according to the HUGO Gene Nomenclature Committee (Gray et al., 2015). DNMT3L lacks the N-terminal part of the regulatory PWWP domain and the C-terminal part of the catalytic domain. It functions as a cofactor by interacting with DNMT3A, which is required for the establishment of distinct DNA methylation patterns found at imprinted genes (Barlow & Bartolomei, 2014).

DNA methylation predominantly (i.e., 70%–80% of it) occurs in CpG dinucleotides (CpGs; 28 million CpG sites have been mapped in the human genome) frequently found in CpG islands (CGIs) of gene promoters, which are mostly unmethylated (Angeloni & Bogdanovic, 2019). CGIs are defined as regions of > 200 bp with GC content of at least 50% and a ratio of observed to statistically expected CpG frequencies of at least 0.6 (Portela & Esteller, 2010); 5mC can also be identified at CGI shores, shelves, and open sea regions (Irizarry et al., 2009). Shores are defined as regions close to CGIs (0–2 kb from CGI), shelves are regions distributed

2–4 kb from CGIs, and open sea regions are isolated CpG sites in the genome that do not have a specific designation (Rechache et al., 2012). Methylation of DNA most frequently has a repressive effect on gene expression by preventing the binding of transcription factors to their motifs (Yin et al., 2017) and by contributing to heterochromatin formation (Du et al., 2015), although the opposite effect was also demonstrated in human cells (Rishi et al., 2010; Yin et al., 2017). Thus, CpG acts like a switch "ON → OFF" (unmethylated → methylated) for the gene expression.

DNA methylation is associated with numerous cellular processes, such as transcriptional repression, X-chromosome inactivation, stem cell differentiation, genomic imprinting, the alteration of chromatin structure, and transposon inactivation (Jones, 2012; Wilson et al., 2012). Nevertheless, 5mC is inherently mutagenic due to spontaneous deamination, leading to C → T transitions (Cooper et al., 2010; Greenberg & Bourc'his, 2019; Holliday & Grigg, 1993). Therefore, CpG sites tend to be evolutionarily depleted from vertebrate genomes (Angeloni & Bogdanovic, 2019).

An increasing number of cytogenomic studies demonstrate the implication of abnormal DNA methylation in the development of different pathological conditions, including cancer, neurological, immunological, and metabolic diseases, and aging (Jin & Liu, 2018). Thus, changes in DNA methylation are extensively studied for the identification of biomarkers and molecular targets for therapy (Locke et al., 2019). Recent reviews (Martin & Fry, 2018; Rider & Carlsten, 2019) summarize currently available literature that demonstrates a relationship between DNA methylation and environmental exposures resulting in altered global and gene-specific DNA methylation.

Although 5mC is the most common epigenetic modification in DNA, several other covalent modifications exist, such as 5-hydroxymethylcytosine (5hmC), 5-formylcytosine (5fC), and 5-carboxylcytosine (5caC) (Olkhov-Mitsel & Bapat, 2012). Since several technologies include or exclude these structures in their outputs indiscriminately, the target of analysis should be also considered.

The study of DNA methylation comprises (i) locus/gene-specific, (ii) genome-scale/wide, and (iii) "global" methylation approaches. The first one includes the examination of a small number of defined CpG sites in a limited genomic region. The second is based on building a DNA methylation profile by measuring many unique sites across the genome. The third referred to as global methylation is designed to assess the total 5-methylcytosine (5mC) content (but not 5hmC, 5fC, or 5caC) relative to total cytosine content (Vryer & Saffery, 2017).

Principles of DNA methylation analysis

DNA methylation can be analyzed based on several principles that differentially recognize 5-methylcytosine from cytosine, namely, (1) bisulfite-mediated DNA conversion, (2) restriction endonuclease-based analysis, and (3) DNA immunoprecipitation (Cheow et al., 2015; Feng & Lou, 2019; Harrison & Parle-McDermott, 2011).

Since during sequencing it is not possible to distinguish 5mC from cytosine, it was suggested to treat DNA with sodium bisulfite, which results in deamination of cytosine residues into uracil (Frommer et al., 1992). After PCR amplification and sequencing, converted cytosines will be read as thymine, while 5mC will stay intact and will be read as cytosine. Comparison of sequencing results of sodium bisulfite-treated DNA with untreated sequence will permit to identify and quantify the percentage of 5mC. Thus, the epigenetic modification can be identified in the same manner as a DNA base-pair change (Clark et al., 2006). This is one of the most preferred methods of global DNA methylation analysis due to its genome-wide coverage and high accuracy.

Although methods based on **methylation-specific restriction enzymes** are simple and cost-effective, they are hampered by availability of specific restriction sites, as only CpG sites found within the recognition sequences of restriction enzymes can be analyzed, which can lead to false-negative results (Tost, 2016).

Methylated DNA immunoprecipitation (MeDIP) consists of the selective immunoprecipitation of methylated DNA fragments using antibodies against 5-methylcytosine (Sørensen & Collas, 2009).

Methylated DNA recognition principles have been originally combined with PCR up to the late 1990s and later adapted to array-based or sequencing-based genome-scale DNA methylation analysis (Feng & Lou, 2019; Gupta et al., 2010; Harrison & Parle-McDermott, 2011).

Initially, **PCR technologies** were combined with methylation-sensitive restriction enzymes application (Singer-Sam et al., 1990). Follow-up methods are based on the use of bisulfite DNA conversion. PCR-based DNA methylation technologies are detailed elsewhere (Wani & Aldape, 2016). Methylation-specific PCR (MS-PCR or MSP) is one of the most commonly used methods for gene-specific detection of DNA methylation in which after bisulfite conversion, methylated sequences are selectively amplified with primers specific for methylation. MethyLight assay uses real-time methylation-specific PCR technology. It has been highlighted that PCR-based techniques permit a detailed analysis of specific regions of the genome (Hernández et al., 2013).

The underlying principle of **microarray technologies** (Weise & Liehr, 2021b) includes the separation of methylated and unmethylated fragments of the genome and their analysis through hybridization to a microarray of known probes (Harrison & Parle-McDermott, 2011). The main advantage of microarray is the ability to simultaneously analyze many methylated DNA regions at once. A limitation of microarray-based methods is that only those genomic regions that are probed by the microarrays will be interrogated (Ku et al., 2011). The microarray-based techniques are combined with enzyme digestion, immunoprecipitation with an antibody against 5mC, or bisulfite treatment (Harrison & Parle-McDermott, 2011). **Array-based analysis of restriction enzyme-digested DNA** is applied in the comprehensive high-throughput arrays for relative methylation (CHARM), a highly sensitive and specific approach to measure DNA methylation (DNAm) across the genome (Ladd-Acosta et al., 2010). This method makes no assumptions about where functionally important DNAm occurs, i.e., CpG island or promoter regions, and includes lower-CpG-density regions of the genome. Using CHARM, it was shown that epigenetic differences in paternal sperm may contribute to autism risk in offspring (Feinberg et al., 2015). **Methylated DNA immunoprecipitation and array-based hybridization (MeDIP-chip)** identify genes that are differentially methylated in acute myeloid leukemia cells in a genome-wide manner, thus are useful to identify new epigenetic targets for therapeutic or prognostic research (Yalcin et al., 2013). One of the most widely used **array platforms** for methylation profiling **on the base of bisulfite-converted DNA** is the Infinium Methylation Assay from Illumina, which made it possible to distinguish the methylation profiles of colorectal cancer and healthy tissue samples (Naumov et al., 2013).

Sequencing-based technologies, including next-generation sequencing (NGS) (Ungelenk, 2021), have been used to assess large fractions of the methylome. The 5mC sites after enzyme digestion, immunoprecipitation with an antibody against 5mC, or bisulfite treatment can then be identified by NGS. Compared to array-based technologies, NGS provides better coverage of all possible methylation sites in the human genome (Chatterjee et al., 2017; Ku et al., 2011). **Methyl-sensitive restriction enzyme sequencing (MRE-seq)** was applied to study the role of DNA methylation in cells of distinct brain regions (Maunakea et al., 2010). **Methylated DNA immunoprecipitation coupled with next-generation sequencing (MeDIP-seq)** is sensitive for analysis of archival dried blood spot samples routinely collected in perinatal screening programs, thereby holding great potential in epidemiological research (Staunstrup et al., 2016). **Whole-genome bisulfite sequencing (WGBS)** of bisulfite-converted DNA is considered as the "gold standard" for single-base resolution measurement of DNA methylation levels. WGBS application revealed reprogramming of DNA methylation in spermatozoa of type 2 diabetes mellitus patients (Chen et al., 2020) and differentially methylated regions in Down syndrome brain tissues (Laufer et al., 2019).

Global DNA methylation

Whole-genome techniques are based on methods that are routinely used for distinguishing 5-methylcytosine (5mC) from unmethylated cytosine (C) followed by a strategy of a global investigation of these modified sites.

Luminometric methylation assay

The Luminometric methylation assay (LUMA) is a tool for quantitative analysis of whole DNA methylation in any given genome. It requires digestion of DNA with *HpaII* and *MspI* restriction enzymes with different sensitivity to methylated and unmethylated CpG dinucleotides followed by pyrosequencing (Karimi et al., 2006). The ratio of *HpaII* and *MspI* digested DNA from parallel reactions provides global methylation values. Normalization between runs and for DNA input enables the addition of *EcoRI*, which recognizes 5′-GAATTC-3′ sequence; therefore, CpG methylation does not influence DNA digestion. During pyrosequencing, DNA polymerase incorporates the nucleotides (dNTP) in the restriction-enzyme-produced 3′-overhang. During the reaction, the released inorganic pyrophosphate converts to ATP, which is used to convert luciferin to oxyluciferen by luciferase. This reaction produces visible light, which is proportional to the amount of incorporated dNTP at restriction sites. The charge-coupled device camera detects the light and converts to peaks in the software (Karimi et al., 2011). Unincorporated nucleotides are degraded with apyrase before the next nucleotide is added. The global DNA methylation is calculated based on the following equation: $1 - [(HpaII(G)/EcoRI(T))/(MspI(G)/EcoRI(T))] \times 100$. The method has several advantages that make it an attractive tool for analysis of global DNA methylation:

- it requires a relatively small amount of DNA (250–500 ng);
- it demonstrates little variability and has the benefit of internal control;
- it has a short processing time (up to 48 samples can be run on the pyrosequencing platform in under 20 min);
- it can be used without a reference genome; and
- it is easily adapted to multiple species.

However, there are drawbacks to the method mentioned by several groups. It only measures CpG methylation within 5'-CCGG-3' sites, which are distributed unevenly in the genome and do not include all CpG dinucleotides (~8% of all CpG sites) (Delgado-Cruzata et al., 2015; Fazzari & Greally, 2004). In addition, high-quality DNA is required without spontaneous 5'-overhangs. Nevertheless, the method is widely used in different studies with the application of human and animal cells. LUMA was successfully used for the comparison of global DNA methylation between healthy and diseased individuals, for studies of psychotic disorders, as well as in environmental studies (Head et al., 2014; Kuchiba et al., 2014).

Applications of LUMA

- LUMA was successfully applied to identify epigenetic markers of diseases and the effects of environmental pollutants. Decrease of global DNA methylation, as well as the dependence of methylation levels on alcohol consumption, was revealed in peripheral blood leukocyte of Japanese breast cancer patients compared with the control group (Collin et al., 2019; Kuchiba et al., 2014). Opposite results were obtained in the US women population by several other studies (Delgado-Cruzata et al., 2012; McCullough et al., 2015; Xu et al., 2015), which showed that higher global methylation positively correlates with breast cancer risk. Promoter methylation of *APC* and *TWIST1*, as well as levels of global methylation assessed using LUMA, may modify the association between obesity and mortality following a breast cancer diagnosis (McCullough et al., 2017). Recently, the latter research group demonstrated a twofold increase in the LUMA methylation in breast cancer of women with first birth at age > 23 versus ≤ 23 years. The authors hypothesized that reproductive characteristics can modify the association of DNA methylation and breast cancer risk via hormonal changes, which contribute to epigenetic drift. Interestingly, no difference was reported in methylation levels by LUMA between affected and unaffected sisters from the United States, while [3H]-methyl acceptance assay revealed a 1.8-fold increased risk of breast cancer in women with lower global DNA methylation (Collin et al., 2019).

- Significant differences in global methylation levels between tumor and normal tissues of the thyroid in Swedish patients with follicular thyroid cancer were not identified. Nevertheless, hypermethylation of the promoter of a tumor suppressor gene *RASSF1A* was detected by pyrosequencing (Lee et al., 2008).
- Complete remission was achieved in 96% of Swedish acute myeloid leukemia (AML) patients with low global methylation level and 61% of patients with high global methylation levels (Deneberg et al., 2010). Also, an inverse correlation was detected between genome-wide promoter-associated methylation and global DNA methylation levels. The authors concluded that global and gene-specific methylation patterns are independently associated with the clinical outcome in AML patients. Infants from the Netherlands with mixed-lineage leukemia (MLL)-rearranged acute lymphoblastic leukemia (ALL) displayed increased levels of genome-wide methylation compared with normal bone marrow samples and promoter CpG hypermethylation. It was suggested that aberrant changes in gene promoter methylation are not always inversely correlating with genome-wide or nonpromoter CpG methylation levels. The authors concluded that the lack of global hypomethylation in MLL-rearranged infant ALL patients may reflect the unique development of this malignancy (Stumpel et al., 2013).
- Significant global DNA hypomethylation was detected in Swedish patients with schizophrenia compared to control (Melas et al., 2012).
- It was shown that indoor air pollution from solid fuels is associated with global DNA hypomethylation in peripheral blood in healthy Polish women (Tao et al., 2014). Although these findings are interesting, they should be interpreted carefully since several variables such as the quantity of fuels used in an enclosed area, the size of the area, and the frequency of ventilation in the exposure area were not analyzed.

Overall, it can be suggested that LUMA is an attractive tool for the analysis of global DNA methylation. Considering the discrepancies between the LUMA-obtained results in breast cancer studies, it can be speculated that ethnicity has a significant impact on global DNA methylation (Galanter et al., 2017; Zhang et al., 2011). In addition, the sample size and differences in LUMA protocols should be considered. In patients with AML and ALL, global DNA hypermethylation positively correlated with the cancer risk, while promoter-associated differences in methylation were observed only in AML (Deneberg et al., 2010; Stumpel et al., 2013). These findings indicate that different types of leukemias with characteristic genetic alterations can also have specific epigenetic changes. Global DNA hypomethylation was observed in patients with schizophrenia and healthy subjects exposed to air pollution (Melas et al., 2012; Tao et al., 2014). Metaanalysis between ambient air pollution and global DNA methylation revealed an inverse correlation with exposure to NO_2 (Plusquin et al., 2017). In recent studies, it was additionally shown that exposure to NO_2 increased odds of adolescent psychotic experiences (Newbury et al., 2019), and in schizophrenia being only partially explained by a polygenic risk score for schizophrenia (Horsdal et al., 2019). Is it possible that exposure to air pollutant NO_2

elevates the risk of psychotic disorders due to global DNA hypomethylation? Or is the low level of global DNA methylation a precondition for adverse effects of NO_2? Although some data on DNA methylation obtained with LUMA are contradictory and difficult to explain and many questions remain unanswered for now, the capability of the pathological changes in the organism, environmental, and lifestyle factors to modify the epigenome is clear and requires further elucidation.

Cytosine extension assay

Cytosine extension assay is another methylation-sensitive restriction enzyme-based method for the analysis of global methylation levels. Digestion of DNA with *HpaII* (or *AciI*, or *BssHII*) leaves a 5′-guanine overhang, which can be labeled with either a [³H]dCTP or a biotin-14-dCTP. The number of cleaved sites is directly proportional to the number of unmethylated sites, as unmethylated sites are end-labeled. A parallel reaction using *MspI* serves as a measurement for the total number of possible cut sites. Subsequently, labeled DNA is quantitatively analyzed by a scintillation counter (Bilichak & Kovalchuk, 2017; Zhou et al., 2017).

Applications of cytosine extension assay

- Global DNA methylation evaluation using the cytosine-extension method showed no difference between bipolar patients and healthy controls, which does not exclude the possibility that altered methylation of specific promoter regions is involved in the etiology of the disorder (Bromberg et al., 2009). Tritium-labeled cytosine extension assay ([³H]dCTP) was applied in colorectal mucosal biopsies of ulcerative colitis (UC) patients; however, no statistically significant difference was revealed in DNA methylation compared with age and sex-matched controls. However, the methylation levels of both *ESR-1* and *N-33* genes estimated by the COBRA technique (Combined Bisulfite Restriction Analysis) were significantly higher in UC subjects (Arasaradnam et al., 2010). Based on a literature review in PubMed, it should be noted that cytosine-extension assay does not belong to the widely used methods for the assessment of human global DNA methylation.

Bisulfite PCR of LINE-1 and Alu repetitive DNA elements

Among various methods of the DNA methylation analysis bisulfite sequencing (BS-seq) approach stands out as the gold standard due to its qualitative and quantitative ability to identify 5mC at a single base-pair resolution across an entire genome (Leti et al., 2018). Nevertheless, the BS-seq method can be challenging. Bisulfite conversion leads to DNA fragmentation, which complicates the amplification of long repetitive fragments and could potentially result in the generation of chimeric products (Kurdyukov & Bullock, 2016). The complete conversion of cytosines should be ensured since the final results of the analysis depend on it. Homogeneity of the cell population is of high importance since varying methylation profiles from the mixed population will hinder the accuracy of the global methylation analysis (see Krueger et al., 2012 for more details).

Recently, protocols of oxidative bisulfite sequencing (Kernaleguen et al., 2018) for discrimination of 5hmC and a new transposase-based library preparation assay for the Illumina HiSeq X platform (Suzuki et al., 2018) were developed, which significantly improved the accuracy and cost-effectiveness of whole-genome bisulfite sequencing. Thus, technological advances along with biological achievements permit to overcome the limitations of whole-genome methylation analysis approaches.

Analysis of 5mC in DNA repeats, such as *Alu* (*Arthrobacter luteus*) elements and long interspersed nucleotide elements (LINE) (Liehr, 2021d) using bisulfite treatment of DNA and simultaneous PCR was proposed by Yang et al. (2004) as a surrogate marker of global DNA methylation. This approach is less labor-intensive and requires less DNA than previous methods of assessing global DNA methylation.

LINE-1 is an autonomous retrotransposon that constitutes ∼17% of the human genome with around 500,000 copies (Lavasanifar et al., 2019). It was shown that any two human beings differ on average by ∼285 different LINE-1 insertions from which 124 LINE-1-mediated insertions are disease-causing (Ewing & Kazazian, 2010; Hancks & Kazazian, 2016). *Alu* elements of short interspersed elements family (SINEs) are primate-specific retrotransposons that are ∼300 bp in length accounting for at least 11% of the human genome (Hancks & Kazazian, 2016; Liehr, 2021d). Therefore, LINE-1 and *Alu* elements are mostly silenced by dense methylation at CGIs (∼25% and ∼10% of CpG sites, respectively). Considering that repetitive elements may constitute nearly two-thirds of the human genome quantitative assessment of DNA methylation in LINE-1 and *Alu* is of high importance (de Koning et al., 2011). A disadvantage of this assay is that 5mC of repetitive elements do not constitute total genomic 5mC, and therefore, this assay could not provide a perfect representation of total genomic methylation (Yang et al., 2004).

Applications of bisulfite PCR of LINE-1 and *Alu* repetitive DNA elements

- DNA hypomethylation of LINE-1 and *Alu* repeats is an early event in carcinogenesis that causes genome instability (Baba et al., 2018; Burns, 2017; Marinoni et al., 2017; Zeggar et al., 2020).
- Using the DNA bisulfite conversion followed by pyrosequencing-based methylation analysis, the association of hypermethylation of LINE-1 in white blood cells with high-grade cervical intraepithelial neoplasia was detected (Kosumi et al., 2019). It was assumed that higher LINE-1 methylation levels could mediate a positive effect on the immune response against human papillomaviruses infection (Piyathilake et al., 2011). This assumption is partially following results demonstrating that among esophageal cancer patients in the LINE-1 hypomethylation group, the level of peritumoral lymphocytic reaction is much lower compared to LINE-1 hypermethylation group (Barchitta et al., 2017).
- Bisulfite conversion of DNA and pyrosequencing analysis of LINE-1 revealed that global DNA methylation positively correlates with the consumption of dark green vegetables and inversely correlated with the consumption of saturated fat

(Pauwels et al., 2017). The authors concluded that dietary patterns characterized by a high intake of vegetables and fruits may protect against global DNA hypomethylation.

- Tissue-specific differences in methylation level of LINE-1 were discovered in blood and buccal cell samples from monozygotic (MZ) and dizygotic (DZ) Chinese twins using bisulfite pyrosequencing. The authors concluded that this method can be applied as a relatively cost- and time-effective forensic marker for discriminating MZ twins (Xu et al., 2015).

- It is known that environmental factors may influence the methylation pattern of DNA repetitive elements that in turn may become a risk factor for the development of adverse effects (Legoff et al., 2019).

- Accommodation of mothers from the US population near the major roadways in conditions of traffic-related pollution was shown to be associated with lower fetal growth and significant placental changes of LINE-1 methylation levels (Kingsley et al., 2016). Nevertheless, the authors precluded the impact of placental epigenetic changes in fetal growth since the association between residential proximity to major roadways and birth weight was nearly identical with or without adjustment for methylation levels of LINE-1, *AluYb8*, or both. In contrast, a significant association between maternal exposure to particulate matter with aerodynamic diameter $< 10\,\mu m$ (PM_{10}) during pregnancy with fetal growth restriction in the population of Chinese women was reported (Cai et al., 2017). Although the variations between these studies are unclear, it can be speculated that the differences between the study populations (the United States and China), the concentration, and composition of air pollutants are important variables that may impact on methylation levels of LINE-1. Thus, ethnicity and environmental factors should be considered in studies of global DNA methylation.

- Comparison of sperm DNA methylation between fertile men populations from Greenland, Poland, and Ukraine exposed to organic pollutants showed that LINE-1 and *Alu* methylation levels can vary in response to the same environmental pollutants and can be considered as unequal biomarkers of exposure. The biological significance and the mechanisms of these variations are presently unknown (Consales et al., 2016).

Enzyme-linked immunosorbent assay

Enzyme-linked immunosorbent assay (ELISA) is a widely used high-throughput method for the analysis of global DNA methylation. Currently, several companies have developed specific kits for detection of 5mC or 5hmC that require variable concentrations of input DNA (20–1000 ng). In general, the DNA is captured using a DNA binding buffer on an ELISA plate and treated with a primary antibody (capture) against 5mC. On the second step, labeled secondary antibodies against capture are used. After washing and blocking nonspecific signals, a colorimetric or fluorometric analysis is performed via a microplate spectrophotometer (Kurdyukov & Bullock, 2016). Positive (completely methylated DNA) and negative (unmethylated DNA) controls are crucial in these studies since the relative percentage of 5mC is quantified by normalizing to the standard curve generated from dilutions of these controls.

Currently, ELISA kit for LINE-1 methylation assay is commercially available (https://www.activemotif.com). In short, genomic DNA is fragmented by *MseI* (methylation nonsensitive) restriction enzyme and hybridized to a biotinylated human LINE-1 probe for 290 bp region of LINE-1. After immobilization onto a streptavidin-coated plate, the sample is washed and incubated with a secondary antibody conjugated to horseradish peroxidase (HRP). HRP catalyzes the reduction of hydrogen peroxide to water. HRP converts colorless hydrogen donor substrates into colored molecules, and the ELISA reader detects the optical density of the solution at 450 nm. Nevertheless, these methods are less accurate compared to sequencing or high-performance liquid chromatography (HPLC)-derived approaches, and adaptability to different species is questionable.

Applications of ELISA-based analysis

- Measurement of global DNA methylation using ELISA in patients with nonsmall-cell lung cancer (NSCLC) revealed hypomethylation of lung cancer tissues and showed the ability to predict the results of chemotherapy (Mo et al., 2014). The authors assumed that the efficiency of chemotherapy was higher in NSCLC patients with global DNA hypomethylation due to hypomethylation-caused genomic instability.
- Sperm global DNA methylation was shown to decrease after exercise training in the Australian population. The authors concluded that exercise training is capable of reprogramming the sperm methylome; nevertheless, it is not clear whether these changes can be heritable (Denham et al., 2015).
- Global DNA methylation levels were significantly higher in the blood DNA of Italian amyotrophic lateral sclerosis patients than in asymptomatic/paucisymptomatic carriers or noncarriers of *SOD1* mutations. Interestingly, a positive correlation between DNA methylation and disease duration was revealed (Coppedè et al., 2018).
- It was shown that the levels of 5mC and 5hmC in the blood DNA of healthy subjects from the US population positively correlated with a median percentage of dimethylarsinate and cadmium concentrations in urine. The authors suggested that further studies are required to develop epigenetic biomarkers of environmental exposure in the human population (Tellez-Plaza et al., 2014).

Overall, it can be suggested that ELISA-based analysis of global DNA methylation in the human population is a high-throughput approach and does not require expensive equipment. Nevertheless, it lacks accuracy and standardized analysis, which makes it less attractive compared to LUMA, LINE-1, or sequencing-based methods.

Liquid chromatography-tandem mass spectrometry

Mass spectrometry (MS) is based on the principle that a charged particle passing through a magnetic field is deflected along a circular path on a radius that is proportional to the mass-to-charge ratio. This approach has been adapted for DNA methylation analysis (Gupta et al., 2010). The abilities of MS have been significantly expanded by combining with other methods.

The most commonly used methods for profiling of DNA methylation, such as bisulfite sequencing and methylation-sensitive enzyme-based assays, are unable to distinguish between 5hmC and 5mC. Currently, liquid chromatography coupled with mass spectrometry (LC-MS/MS) is regarded as "the gold standard" for quantitative analysis of 5mC and discrimination of 5hmC, 5fC, and 5caC (Chowdhury et al., 2017).

LC-MS/MS can detect methylation levels ranging from 0.05% to 10%, and ~0.25% differences in the total cytosine residues between samples using 50–100 ng input DNA, although the much smaller amount of DNA is possible to use (Kurdyukov & Bullock, 2016). Another important advantage of the method is that the quality of DNA is not substantial for the analysis.

Briefly, the DNA is enzymatically degraded into nucleoside components that lack negatively charged phosphate. For calculating the actual percentage of 5mC and 5hmC in a test sample, the calibration curves should be constructed. Equal amounts of genomic DNA of the 897-bp DNA standard, cytosine, 5mC, and 5hmC should be hydrolyzed as sample DNA. In the next step, the calibration standards with a known degree of 5mC or 5hmC are prepared by mixing in separate tubes increasing amounts of hydrolyzed 5mC or 5hmC with the same amount of hydrolyzed cytosine. The results from the measurements of these samples are used for constructing calibration curves for 5mC and 5hmC, respectively. Subsequently, the sample and standards are analyzed by liquid chromatography-electrospray ionization tandem mass spectrometry with multiple reaction monitoring (Fernandez et al., 2018). For the simultaneous detection of four known cytosine modifications of DNA (i.e., 5mC, 5hmC, 5fC, and 5caC) in the human genome (Tang et al., 2015), a labeling technique with selective derivatization of cytosine moieties with 2-bromo-1-(4-dimethylamino-phenyl)-ethanone is developed. This approach significantly improved the liquid chromatography separation and increased detection sensitivities of cytosine modifications.

Applications of LC-MS/MS

- Based on the analysis of Italian patients with hepatocellular carcinoma, cholangiocarcinoma, and colon cancer with different one-carbon metabolism gene variants, it was shown that the carriers of the *MTHFD1* 1958GG genotype had lower global DNA methylation compared to carriers of the *A* allele (Moruzzi et al., 2017). The authors suggested that *A* allele may exert a protective effect on cancer risk by preserving from DNA hypomethylation.
- LC-MS/MS was implemented for the analysis of global DNA methylation in blood leukocytes of Taiwanese workers of nanomaterial manufacturing and/or handling factories. The hypomethylation was pronounced in the group of workers exposed to indium tin oxide. A negative correlation was detected between global DNA methylation and oxidative stress biomarker 8-hydroxydeoxyguanosine, which was significantly elevated in the blood of the exposed group (Liou et al., 2017).

- The mean global DNA methylation level in cord blood of Belgian infants was not influenced by the maternal dietary intake of methyl-group donors, while methylation changes were detected in *LEP* and *RXRA* genes (Pauwels et al., 2017). However, the biological importance of these findings is not clear yet.

Overall, it can be suggested that LC-MS/MS is a highly accurate method for the analysis of global DNA methylation changes in different tissues due to disease, lifestyle factors, or environmental pollution. LC-MS/MS is considered as a gold standard for the simultaneous analysis of cytosine derivatives. Nevertheless, complexities involved in data analysis and interpretation, as well as the technical expertise required to perform LC-MS/MS-based methods, make it costly and limit its applicability for the clinics (Chowdhury et al., 2017). It can be assumed that further technological advances in the field of LC-MS/MS will help to overcome current limitations.

Genome-wide DNA methylation

Whole methylome analyses covering each CpG in the genome remain technically challenging and costly. More than half of all DNA reads do not contain even a single CpG, and many CpGs will not show variable DNA methylation under any condition and therefore, several DNA methylation approaches have been developed to concentrate on the "potentially informative" fraction of the genome (Tost, 2016). One of the most common methods of genome-wide methylation analyses is based on bisulfite DNA conversion and subsequent hybridization to a chip with a specific set of probes that cover a wide range of CpG islands.

Bisulfite conversion-based microarray

Currently, cost-effective analysis of DNA methylome is an important limitation that drives high-throughput technologies to improve the coverage of the chips and reproducibility of the results. Microarray-based technologies enable the analysis of DNA methylation at thousands of loci simultaneously using a small amount of the sample (as low as 500 ng). HumanMethylation27K bead-chip introduced in 2008 by Infinium was an important achievement enabling analysis of the methylation of 14,475 genes with the implementation of over 27,000 probes for CpG dinucleotides (Bibikova et al., 2009). The development of Infinium HumanMethylation450 (450K) was a huge breakthrough enabling interrogation of the methylation status of more than 480,000 5mC distributed over the whole genome (Dedeurwaerder et al., 2011). It is a hybridization-based approach for quantitative genotyping of the C/T polymorphism generated by DNA bisulfite conversion (Dedeurwaerder et al., 2013). Recently, Infinium MethylationEPIC (850K) microarray was developed that contains the most modern available technology and provides coverage of $> 860,000$ CpG sites (Pidsley et al., 2016). It covers all known genes, intergenic regions, and regulatory sequences. DNA methylation (also called the methylation beta-value (β)) is calculated as the ratio of 5mC signal to unmethylated (T) signal by the formula: $\beta =$ intensity of the methylated signal/(intensity of the unmethylated signal + intensity of the methylated signal). A β-value ranges from 0 (completely unmethylated) to 1 (completely methylated) and can be expressed as a percentage by multiplying by 100 (Pidsley et al., 2016).

Briefly, after DNA isolation and sodium bisulfite treatment, the converted DNA is hybridized to Infinium bead-chip. Each bead contains 50 bp probes complementary to specific regions of bisulfite converted genomic DNA with a CpG site at the 3′ end of the probe. Two types of probes are used on Infinium bead-chips. Type I probe contains two separate probe sequences for methylated and unmethylated CpGs. Type II probes enable the measurement of unmethylated and methylated signals by the same bead. Thus, half of the physical space on the bead is used. The single-base extension of the probe incorporates a fluorochrome-conjugated ddNTP at the 3′ CpG site of the probe for detection of the C/T conversion that results from sodium bisulfite treatment. On the bead with type I probes methylation results in a mismatch at the 3′ end of the probe and inhibition of single-base extension while unmethylated CpG will permit incorporation of labeled ddNTP. Type II probes utilize differently labeled ddNTPs, and the color of the fluorescent signal depends on which nucleotide was incorporated. A scanner then measures the fluorescent signals on each bead (see Pidsley et al., 2016 for more details).

In a validation study of the MethylationEPIC BeadChip for fresh-frozen (FF) and formalin-fixed paraffin-embedded (FFPE) brain tumors from Swedish patients, methylation results obtained by 450K and 850K arrays were compared (Kling et al., 2017). Pearson correlations between β values of the 450K and 850K arrays for common probes of a FF (0.988–0.996) and FFPE (0.980–0.989) samples demonstrated that 850K array stably reproduces results from the 450K platform. Also, it was shown that a large portion of CpG sites measured as highly methylated in the FF samples have lost methylation in the FFPE sample. The authors concluded that they do not see any clear benefit for choosing one method over the other.

Applications of bisulfite conversion-based microarray analysis

- The Infinium MethylationEPIC microarray was used for the analysis of the methylation profile of serum circulating cell-free DNA (cfDNA) in patients with advanced adenomas and colorectal cancer in Spain. Global DNA hypomethylation was detected in patients with benign and malignant tumors compared to individuals without abnormalities in colorectal tissues. The authors concluded that MethylationEPIC microarray is a useful method for the early detection of colorectal cancer (Gallardo-Gómez et al., 2018).
- Technological advancements in the field of microarray methods lead to one of the most exciting discoveries, the DNA methylation age. Horvath (2013) and Hannum et al. (2013) discovered DNA methylation-based aging biomarkers and developed epigenetic clocks of tissues and populations. The epigenetic clock of Horvath relies on the measurement of percent methylation at 353 CpGs in a variety of tissues as measured on either the Illumina 27K or Illumina 450K arrays. The correlation between actual and predicted age was 0.96 and the error was 3.6 years. The Hannum DNA methylation age model uses the

percent methylation at 71 CpGs from the Illumina 450K array. The developed epigenetic clock demonstrated high accuracy with a correlation between actual and predicted age of 0.96 and an error of 3.9 years in blood cells. Strong predictive power for chronological age was detected for the breast, kidney, lung, and skin tissues (expected value $R=0.72$). Nearly all markers of DNA methylation were identified within or in proximity of genes associated with aging-related pathological conditions such as DNA damage, cancer, tissue degradation, Alzheimer's disease, as well as in genes regulating metabolic traits. It was shown that the aging velocity of the male methylome is approximately 4% faster than that of females. Although the models of Horvath and Hannum overlap by only six common CpGs, they both strongly correlate with chronological age. Nevertheless, the Horvath model is more frequently used in epidemiological studies (Dhingra et al., 2018).

- The discovery of DNA methylation age has led to discovering various factors that impact age according to the epigenetic clock. Nwanaji-Enwerem et al. (2016) revealed associations between DNA methylation age of older males residing in the Northeastern United States and particulate matter ($PM_{2.5}$) exposure using Illumina HumanMethylation450 BeadChip. It was shown that a $1 \mu g/m^3$ increase in $PM_{2.5}$ 1-year exposure was significantly associated with an approximately 0.52-year increase in DNA methylation age (Nwanaji-Enwerem et al., 2016).

- In patients diagnosed with human immunodeficiency virus (HIV) analysis of DNA methylation profiles from brain tissue and blood characterized a 7.4 and 5.2 years higher DNA methylation age, respectively, compared to controls. It was assumed that HIV infection leads to epigenomic instability as a result of genetic instability, and the observed methylation age acceleration may reflect the protective actions of the epigenomic maintenance system (Horvath & Levine, 2015).

- An association of DNA methylation age with low socioeconomic status (SES) was shown in over 5000 subjects using Horvath and Hannum models. It was uncovered that low SES was associated on average with 1-year higher methylation age acceleration and was only partially modulated by the unhealthy lifestyle (Fiorito et al., 2017; Miller et al., 2015).

- Interesting results were obtained in the study of African American teenagers from rural Georgia (United States) when analyzing the relationship between self-control and DNA methylation age in peripheral blood mononuclear cells. Among subjects with low-SES, it was revealed that self-control was associated with positive psychological outcomes but 1.46–2.27 year of acceleration of epigenetic aging (by Hannum and Horvath methods, respectively). In contrast among high-SES youth, better mid-adolescent self-control was associated with a 0.27 to 2.14 year deceleration (by Horvath and Hannum methods, respectively). The authors assumed that negative psychological experiences may cause a persistent release of stress-response hormonal products (e.g., glucocorticoids) that are elevated in youth who exhibit skin-deep resilience. Glucocorticoids can induce hypomethylation of stress-responsive genes such as *FKBP5*, which was observed in subjects exposed to childhood trauma (Klengel et al., 2013).

- A study of Danish twins has shown that DNA methylation age can discriminate between slow and fast agers, and this parameter can be a predictor of mortality (Christiansen et al., 2016).

Overall, it can be concluded that array approaches are high-throughput methods for individual and population-scale genome-wide methylation studies. The most popular systems for genome-wide methylation analysis are Illumina methylation arrays (over 850 publications on PubMed). Nevertheless, it is still expensive and not suitable for small projects; therefore, outsourcing to a big sequencing facility is the recommended option (Kurdyukov & Bullock, 2016). The Illumina bead-chip is more suitable for studies of human DNA, although can be adapted for the analysis of other species (Housman et al., 2020).

It is worth mentioning that there is an obvious difference in microarray-obtained data representation compared to global DNA methylation studies. While the latter is mainly directed to the analysis of the percentage of methylated cytosines in the genome (total or average), in microarray-based studies the methylation profiles within and between groups are compared. Currently, different tools are being developed to get a more comprehensive view of DNA methylation changes using Illumina methylation arrays (Chatterjee et al., 2017).

In addition to several advantages for implementation in population-based studies, microarray technology allowed the discovery of epigenetic clocks, which demonstrated remarkable results in predicting chronological age. Although for now microarray technologies still need standardization, further studies in this direction will certainly open new possibilities for the development of precise models with fewer errors.

Gene-specific methylation

While the analysis of global DNA methylation provides information about the distribution and dynamics of the 5mC on the whole-genome scale, gene-specific analysis of DNA methylation uncovers more specific and targeted effects of DNA methylation. Although methods of global methylation analysis have limited resolution for identification of the influence of environmental factors on specific genes, nevertheless the approaches overlap significantly. Bisulfite-specific PCR is one of the first (and still widely used) techniques for analyzing region-specific DNA methylation. This approach involves DNA bisulfite conversion followed by PCR of bisulfite conversion products (Chatterjee et al., 2017). The resulting PCR products can be analyzed for the DNA methylation status of specific genes or candidate regions using PCR, sequencing, or mass spectrometry approaches.

PCR-based approaches

Methylation-specific PCR (MSP) is performed after bisulfite conversion using two pairs of primers for methylated (M) and unmethylated (U) loci. The presence or absence of a PCR product can be analyzed by gel electrophoresis. The obvious limitation of this method is that it allows the detection of one or two CpGs. Currently, real-time MSP enables the interrogation of multiple promoters in a large number of samples without gel electrophoresis. MSP is very sensitive as it can detect <0.1% of methylated alleles in a specific locus. Also, the method requires a small amount of DNA and can be used for the analysis of DNA isolated from the plasma or paraffin-embedded tissues (Ramalho-Carvalho et al., 2018).

Methylation-sensitive high-resolution melting PCR (MS-HRM) allows the detection of methylation at specific loci by analysis of fluorescence depletion in qPCR reaction. The fluorescence intensity of intercalating dye such as SYBR or Eva green is higher when bound to double-stranded DNA. Due to an increase in the temperature during PCR, the DNA denatures, and the dye goes back to the solution resulting in a drop of the fluorescence. Since the methylated DNA retains cytosines after bisulfite conversion, it has a higher melting temperature compared to unmethylated loci. Thus, the level of methylation correlates with the melting profile of the sample (Ribeiro Ferreira et al., 2019). The method requires the DNA of high purity, which enables discrimination of methylation differences even between cell subpopulations in heterogeneous samples.

The implementation of the **TaqMan probe (MethyLight)** for specific methylated or unmethylated loci significantly improved the sensitivity of gene-specific DNA methylation analysis by PCR. A comparison of the amplification curves of a test sample with the standard samples containing known numbers of DNA molecules allows quantification of the absolute methylation level. MethyLight can be used in digital PCR, which makes it resistant to the background noise and PCR contaminants (Campan et al., 2018).

Another interesting approach is **COBRA**, which is a combination of PCR and enzyme digestion. COBRA assay consists of three major steps: bisulfite conversion, PCR amplification, and digestion with restriction enzymes. After bisulfite treatment, the recognition site of the restriction enzyme disappears in unmethylated loci, while in the case of methylated DNA, the enzyme is capable to digest it. PCR amplification (methylation-independent) results in the mix of fragments of the same length that have lost or retained CpG containing recognition sites. During the third step, PCR products are digested with a restriction enzyme (e.g., *Bst*UI, *Taq*I) and analyzed by gel electrophoresis. After digestion, the ratio of digested and undigested PCR products, i.e. the ratio of methylated and unmethylated DNA molecules can be quantified (Bilichak & Kovalchuk, 2017). COBRA assay is the method of choice when working with a small amount of biological material. Also, the rate of false positives is lower compared to MSP. The modified version Bio-COBRA enables electrophoresis of digested products in a microfluidics chip providing the opportunity of a quantitative assessment of DNA methylation patterns in large sample sets (Brena et al., 2006). Recently developed COBRA-seq assay for high-throughput genome sequencing platforms has higher sensitivity and accuracy. After digestion of the DNA with a restriction enzyme, streptavidin-coated magnetic beads are used to remove undigested (unmethylated) fragments (Varinli et al., 2015).

Applications of PCR-based approaches

- **MS-HRM** was successfully used to distinguish Prader-Willi and Angelman syndromes and healthy individuals on the base of methylation analysis at the *SNURF-SNRPN* locus (Ribeiro Ferreira et al., 2019).
- Detection of the methylation of the p16 gene is an important tool in epigenetic studies of various human cancers; therefore, different approaches are being developed to assess its epigenetic status. The 115-bp **MethyLight** assay was confirmed as a sensitive approach for the detection of the methylated-*p16* biomarker for clinical diagnosis (Zhou et al., 2011). Choudhury et al. (2018) presented the detection of promoter methylation of the p16 gene using MSP.
- Methylation analysis of frequently mutated protocadherins in pancreatic ductal adenocarcinomas patients using **COBRA** assay showed a negative correlation between high methylation levels of PCDH10 and progression-free survival rates (Curia et al., 2019).

Sequencing-based approaches

One of the most reliable and high-throughput methods of DNA methylation analysis is **BS-seq** that enables gene-specific detection of 5mC at the single-base resolution of PCR products (Wreczycka et al., 2017). Primers for PCR are designed based on the sequence around CGI for the amplification of bisulfite-converted DNA. Nevertheless, it cannot discriminate between 5mC and 5hmC since upon bisulfite treatment 5hmC converts to cytosine-5-methylenesulfonate, which then reads as a cytosine during sequencing. This limitation was overcome by oxidative bisulfite sequencing during which 5hmC oxidizes to 5fC and converts to uracil, which is read as a thymine during sequencing (Booth et al., 2013).

Pyrosequencing, which is designed to quantify single-nucleotide polymorphisms (SNPs), allows the detection of multiple CpGs (i.e., C/T SNPs) within amplicons produced during PCR of bisulfite-converted loci. The methylation within the amplicon is measured by comparing peak light emission of incorporation of cytosine or thymine at a CpG site (Delaney et al., 2015). Pyrosequencing analysis generates accurate and reproducible results of methylation level at several CpG sites, which allows early detection of small changes in methylation; therefore, it is an attractive method for clinical research (Poulin et al., 2018).

Applications of sequencing-based approaches

- Estival et al. (2019) have shown that MSP and pyrosequencing are suitable for detecting the methylation status of the O6-methylguanine-DNA methyltransferase (*MGMT*) gene in blood and tissue samples of glioblastoma patients.
- The link between air pollution exposure and methylation of genes related to coagulation and inflammation pathways in elderly men was demonstrated using the pyrosequencing approach (Bind et al., 2014). The authors concluded that observed changes in DNA methylation may reflect the biological impact of air pollution. Nevertheless, further studies are required to confirm the obtained results.

Mass spectrometry-based approaches

Mass spectrometry (MS) provides an attractive solution for DNA methylation analysis, as it enables direct, rapid, and quantitative detection of DNA products measuring the molecular weight. In MassArray (EpiTYPER assay), the bisulfite-converted DNA is amplified using forward primer and T7 promoter-tagged reverse primer. Thus, the RNA molecules are synthesized on the base of PCR products. After in vitro RNA transcription, base-specific RNA cleavage at uracil is performed, and the mass of cleavage products are analyzed by matrix-assisted laser desorption/ionization time-of-flight (MALDI-TOF) mass spectrometry. This approach has high reproducibility and is applicable for DNA methylation analysis at multiple CpG sites (Suchiman et al., 2015).

Applications of mass spectrometry-based approaches

- *BRCA1* promoter methylation status was analyzed in breast cancer patients using MassArray technology. It was shown that patients negative for estrogen and progesterone receptors, and excess HER2 protein (triple-negative) with tumors harboring *BRCA1* promoter hypermethylation derived the most benefit from adjuvant chemotherapy (Jacot et al., 2020).

Summary for methylated DNA detection

Methods for the identification of differentially methylated DNA regions are summarized in Table 10.1. On the base of the DNA methylation profiles of thousands of samples, their connection with pathological conditions and aging was discovered, besides epigenetic clocks were developed that correlate with chronological age. Several studies demonstrate the association between global and region-specific DNA methylation with environmental pollution. Thus, it was speculated that DNA methylation mediates the interplay between genetic and environmental components connecting the environment with the genome. Therefore, several methods have been developed with different levels of accuracy, reproducibility, sensitivity, and complexity that are suitable for clinical practice or scientific research.

Currently available methods of DNA methylation analysis allow detection of 5mC on the nucleotide and the whole-genome scale and open a huge avenue for a variety of studies. Global and genome-wide DNA methylation analysis is more suitable for population-scale studies, while gene-specific methylation analysis is more informative for interrogation of genes that regulate development or have implications in pathological conditions. Kurdyukov and Bullock (2016) recommended several key factors to be considered when choosing a method for the analysis of DNA methylation (Kurdyukov & Bullock, 2016): the…

- aims of the study (For analysis of methylation of the known genes, sodium-bisulfite conversion and digestion-based assay are recommended. If the candidate genes are unknown whole-genome methylation profiling and search of differentially methylated regions should be performed);
- amount and quality of the DNA sample;

Table 10.1 Methods for the analysis of DNA methylation.

Method	Principle of 5mC discrimination	Recognition of 5mC sites	CpG coverage	References
Luminometric methylation assay (LUMA)	Restriction enzyme-based	Pyrosequencing or light detection based on a chain reaction when pyrophosphate is released	~8%	Sant and Goodrich (2018)
Cytosine extension assay	Restriction enzyme-based	5'-Guanine overhang labeled with either a [³H]dCTP or a biotin-14-dCTP detected by a scintillation counter	~8%	Zhou et al. (2017)
Bisulfite PCR of LINE-1 and *Alu* repetitive DNA elements	Bisulfite conversion-based	Bisulfite PCR	~10%–15% each	Sant and Goodrich (2018)
Enzyme-linked immunosorbent assay (ELISA)	Antibodies bind methylated cytosines	Colorimetric or fluorometric analysis of antibodies	100%	Sant and Goodrich (2018)
Liquid chromatography-tandem mass spectrometry (LC-MS/MS)	DNA degradation into nucleoside components, liquid chromatography and mass spectrometry of products	Mass spectrometric analysis of DNA products	100%	Sant and Goodrich (2018)
Bisulfite conversion-based microarray	Bisulfite conversion-based	Array-based analysis	~1.6%–3%	Pidsley et al. (2016)
Methylation-specific PCR (MSP)	PCR of bisulfite-converted DNA using primers for methylated and unmethylated loci	Gel electrophoresis of PCR product	Region-specific	Ramalho-Carvalho et al. (2018)
Methylation-sensitive high-resolution melting PCR (MS-HRM)	Different melting profiles of methylation-specific PCR products of bisulfite converted DNA	Comparison of the melting profiles of PCR products	Region-specific	Ribeiro Ferreira et al. (2019)

Continued

Table 10.1 Methods for the analysis of DNA methylation—cont'd

Method	Principle of 5mC discrimination	Recognition of 5mC sites	CpG coverage	References
MethyLight	Amplification of bisulfite-converted genomic DNA by fluorescence-based, real-time quantitative PCR (Taq/Man) technology that requires no further manipulations after the PCR step	Real-time quantitative PCR	Region-specific	Campan et al. (2018)
Combined bisulfite restriction analysis (COBRA)	Bisulfite conversion, amplification of products by PCR, restriction by methylation-specific enzymes	Gel electrophoresis of restriction products	Region-specific	Bilichak and Kovalchuk (2017)
Bisulfite sequencing (BS-seq)	Bisulfite conversion-based	Any method that can analyze sequence	100%	Wreczycka et al. (2017)
Pyrosequencing	Bisulfite conversion-based	Comparison of peak light emission of incorporation of cytosine or thymine at a CpG site	Region-specific	Delaney et al. (2015)
EpiTYPER Assay	Mass spectrometry-based bisulfite sequencing method	Products analysis by matrix-assisted laser desorption/ionization time-of-flight (MALDI-TOF) mass spectrometry	Region-specific	Suchiman et al. (2015)

Thirteen different approaches for DNA methylation analyses are summarized and treated here.

- sensitivity and specificity requirements of the study;
- robustness and simplicity of the method;
- availability of bioinformatics software for the analysis and interpretation of the data;
- availability of specialized equipment and reagents; and
- cost per sample.

Analysis of histone modifications

Like DNA methylation, posttranslational modification of histones is another level of epigenetic regulation of gene expression. Histones affect both positively and negatively on gene expression, depending on posttranslational modifications on specific amino acid residues. Histone proteins have highly basic amino N-terminal tails densely populated with lysine and arginine residues. The histone tails protrude from the nucleosome and contact the neighboring nucleosome are the main targets for the histone posttranslational modifications (PTMs), which modulates the chromatin state (Audia & Campbell, 2016). The main factors that regulate chromatin state are histone-modifying and chromatin-remodeling enzymes. The best characterized PTMs of histones include acetylation, methylation, phosphorylation, and ubiquitination (Hyun et al., 2017). Also, there are several other modifications such as ADP-ribosylation, SUMOylation, citrullination, deamination, formylation, propionylation, butyrylation, crotonylation, and proline isomerization (Audia & Campbell, 2016). Most frequently PTMs involve amino acids such as lysine, arginine, serine, threonine, and tyrosine.

In the nucleosome, negatively charged DNA interacts with the positively charged histone octamer through electrostatic interactions at 14 different sites at intervals of approximately 10 bp. At the same time, genomic DNA should be accessible for proteins of gene transcription, DNA replication, and DNA repair. Therefore, the state of chromatin has to be dynamic switching between DNA-packaging and DNA-unpacking status, a process called nucleosome breathing (Zhang et al., 2016). PTMs influence the charge density of histones, which modify the interactions between histone proteins and the DNA.

The first reported histone PTM was **histone acetylation**, which can occur at all four core histones in the nucleosome octamer at various lysine residues (Zentner & Henikoff, 2013). Covalent binding of acetyl groups to the histone reduces the positive charge of the histone core and weakens the interaction of negatively charged DNA with histones. Therefore, histone acetylation is associated with gene transcription, DNA replication, and DNA double-strand breaks (DSB) repair. Histone acetyltransferases (HACs) catalyze the transfer of acetyl group from acetyl-CoA to the lysine side chain and form ε-*N*-acetyl-lysine inducing euchromatin state. In contrast, the removal of acetyl groups is catalyzed by histone deacetylases (HDACs) restoring the positive charge of histones (Guo et al., 2018). Thus, most HDACs function as gene transcription repressors. Currently, there are three FDA-approved HDAC inhibitors for cutaneous/peripheral T-cell lymphoma and many are in different stages of clinical development (Cappellacci et al., 2020).

Histone mono- (me1), di- (me2), or tri-methylation (me3) occurs at lysine (K) and as monomethylation or dimethylation at arginine residues. These types of modifications do not affect the electronic charge of the amino-acid side chain and serve as recognition sites for different chromatin-binding factors (chromatin readers) that modulate gene expression (Hyun et al., 2017). For example, methylations at H3K4me3, H3K36me3, and H3K79me3 mark active transcription sites, while methylations at H3K9me3, H3K27me3, and H4K20me3 are associated with silenced chromatin states (Black et al., 2012).

Histone methylation and demethylation are regulated by histone methyltransferases (HMTs) and histone demethylases (HDMs). The largest group of HMTs are the lysine methyltransferases (KMTs) with 51 members (McCabe et al., 2017). KMTs catalyze the transfer of methyl group from SAM to the lysine residues. Accumulating evidence demonstrates that alterations of these enzymes are frequent features for some diseases such as cancer and, therefore, are considered as potential targets for cancer therapy. In particular, inhibitors of EZH2 (KMT of H3K27) and Dot1L (KMT of H3K79) demonstrated promising results in several studies (Hoy, 2020; Sarno et al., 2020).

Histone phosphorylation mainly occurs on serine, threonine, and tyrosine residues. These reactions are catalyzed by kinases that attach a phosphorylate group to the hydroxyl group of the target residues while phosphatases remove the phosphorylate group from the target (Zhang et al., 2016). Phosphorylation reduces the histone's positive charge promoting transcription and DNA repair. For example, DSBs trigger H2AX phosphorylation at serine 139 (γ-H2AX) by ataxia telangiectasia mutated (ATM) kinase, which is recognized by DNA-repair machinery (Yang et al., 2019).

The ubiquitination of histones is the formation of an isopeptide bond between the terminal glycine of ubiquitin and the ε-amino group of a lysine residue from the target protein (Jeusset & McManus, 2019). The sequential reactions catalyzed by E1 activating, E2 conjugating, and E3 ligase enzymes target proteins for proteasome-mediated degradation. It was shown that all core and linker histones can be mono- or polyubiquitinated on multiple lysine residues (Tweedie-Cullen et al., 2009). Out of all histones, ubiquitination of H2A and H2B are best characterized, and targeting histone ubiquitination is now actively developing direction in cancer therapy (Jeusset & McManus, 2019).

Discrimination of the variety of histone modifications is one of the current challenges in chromatin studies. Thus, several methods were developed for low (e.g., Western blotting, immunofluorescence analysis, and chromatin immunoprecipitation (ChIP)) and high-throughput (e.g., mass spectrometry) studies (Kimura, 2013). Western blot, immunofluorescence analysis, and ChIP are based on the use of site-specific antibodies. The accurate choices of histone modification-specific antibodies from the Histone Antibody Specificity Database are discussed by Rothbart et al. (2015).

Western blot

Western blot allows the detection of specific proteins in the cell lysate of any culture or tissue. Cell lysis buffers should contain protease and phosphatase inhibitors to maintain the structure and phosphorylation status of the target protein. A commonly used radioimmunoprecipitation assay buffer contains a high concentration of salts and detergents facilitating nuclear disruption and increasing the yield of proteins (Mishra et al., 2017). After cell lysis, the concentration of proteins in the samples should be determined and normalized since unequal proteins per lane can skew the analysis. The obtained lysates are subjected to the heat for denaturation and sodium dodecyl sulfate-polyacrylamide gel electrophoresis (SDS-PAGE) is performed for the separation of proteins according to their weight. Electrophoresis is performed using a "sandwich" technique, which allows the transfer of the proteins from the gel to the nitrocellulose or polyvinylidene difluoride membrane. Following electrophoresis, the target proteins can be detected using specific primary and secondary antibodies (Rumbaugh & Miller, 2011). Blocking steps are required to reduce non-specific binding.

Major drawbacks of Western blot include its time-consuming nature, a relatively large amount of sample is required (usually 10–20 µg/assay), and usually, few proteins at a time can be detected. Current modifications of Western blot allow overcoming some of these limitations (Mishra et al., 2017). A capillary and microchip electrophoresis-based Western was developed. This approach has higher sensitivity and allows the measurement of multiple target proteins using a small amount of a single sample (Jin et al., 2016). Recently, an exciting approach of in situ single-cell Western blot of adherent cells was developed enabling measurement of proteins of single cells in the microwell (Zhang et al., 2019a).

Applications of Western blot
- Western blot analysis was implemented for validation of LC-MS/MS results using antibodies against H4K16ac, H3K9ac, H4K20me3, and H3K9me2-3 PTMs in human normal mammary epithelial cells (MCF10) and breast cancer cell lines (MCF7 and HCC1937). Low levels of H4K16ac and H4K20me3 and high levels of H3K9me2-3 were detected in cancer cells, which correlated with the results obtained by LC-MS/MS. Also, tyrosine phosphorylation on histone H1, a new PTM, was identified in cancer cells (Perri et al., 2019).

Immunofluorescence assay
Immunofluorescence utilizes fluorescent-labeled antibodies to detect specific target antigens. Using different antibody/dye combinations allows mapping the distribution of different structures simultaneously (Strickfaden & Hendzel, 2017).

Histone modifications can be detected using specific antibodies against particular histones or modification of them, and subsequently analyzed by **ELISA or fluorescent microscope**. For example, "Epigentek" offers an ELISA-based multiplex assay kit for the detection and quantification of 21 different histone H3 modification patterns of human, rat, mouse, and other species on the same plate. The method is based on capturing H3 with modification by specific antibodies immobilized on the strip wells of the 96-well plate. After histone extraction as low as 20–500 ng/well of input material is required. The captured histone can be detected by detection antibody and color development reagent. The ratio of modified histone is proportional to the intensity of absorbance measured by ELISA at 450 nm (https://www.epigentek.com/catalog/index.php).

Immunofluorescence is used to identify phosphorylated histone H2AX, which is essential for the recruitment of DSB repair proteins. Phosphorylation of H2AX at serine 139 generates γ-H2AX foci over mega base domains mediated by ATM kinase, DNA-dependent protein kinase (DNA-PK), and ATM-Rad3-related (ATR) kinase (Palla et al., 2017).

Fluorochrome-conjugated specific antibodies against γ-H2AX are developed enabling detection by ELISA, flow cytometry, and immunohistochemistry/immunofluorescence (single-cell analysis). γ-H2AX foci analysis can be performed in cell cultures or tissue sections, which makes it an attractive tool for basic research, animal, translational, and environmental studies. Moreover, it is widely used in clinical practice for the evaluation of the effectiveness of chemotherapy or radiation therapy (Redon et al., 2012).

Applications of ELISA/immunofluorescence assay

- An increase of H3K4me3, H3K27me2/3, H3K36me3, and H3K79me2/3 in microgravity conditions was observed in human blood-derived stem cells as a suitable osteogenic differentiation model using ELISA-like EpiQuik Histone H3 Modification Multiplex Assay Kit (Gambacurta et al., 2019). The authors concluded that microgravity may influence histone modifications and cell differentiation. Nevertheless, DNA sequences associated with histone modifications are not studied yet. The correlation between histone H3 modifications (H3K27me3 and H3K36me3) and traffic-derived particulate matter exposures were revealed using ELISA in Beijing truck drivers (Zheng et al., 2017). Although the biological impact of these findings is not clear, the data provide evidence for future air pollution studies in humans.

Applications of γ-H2AX immunofluorescence assay

Recently γ-H2AX analysis was implemented for the analysis of DSBs in postmortem brain tissues from patients with Alzheimer's disease and cognitively unimpaired controls. Increased proportions of γH2AX-labeled neurons and astrocytes in the hippocampus and frontal cortex of Alzheimer's disease patients were revealed compared to the control. The authors suggested that DSBs in neuronal and glial cells may result in neurodegeneration and cognitive impairment (Shanbhag et al., 2019).

Chromatin immunoprecipitation assay

Chromatin immunoprecipitation (ChIP) assay is implemented for the detection of specific PTMs in specific genomic loci. There are two main types of ChIP assay: native ChIP (N-ChIP) and crosslinked ChIP (X-ChIP). During X-ChIP cells are fixed with formaldehyde (crosslinking DNA and proteins) and lysed with the buffers containing protease inhibitors. During this step cell membrane and cytoplasmic proteins are washed out, the nuclear membrane is permeabilized, and noncrosslinked proteins are removed from the nucleus and chromatin. Afterward, the nuclei are separated from the cellular debris and subjected to a lysis buffer containing a high concentration of sodium dodecyl sulfate. The obtained sample is subjected to sonication or nuclease digestion for disruption of chromatin (0.3–1 kb fragments) followed by centrifugation to remove all insoluble debris. After immunoprecipitation with specific antibodies against a particular PTM or protein, the immunoprecipitated fraction is trapped using protein A, G, or L agarose, sepharose, or magnetic beads. The DNA is purified using phenol-chloroform-isoamyl alcohol (or DNA spin-column) and analyzed via qPCR (ChIP-qPCR), sequencing (ChIP-seq), or microarray (ChIP-chip) approaches (Wiehle & Breiling, 2016). The N-ChIP procedure precludes crosslinking agents, and native chromatin is digested by micrococcal nuclease to mononucleosome resolution. Therefore, N-ChIP can be used only for the analysis of histones, while the X-ChIP approach is suitable for nonhistone proteins as well.

ChIP-qPCR is recommended for studies of gene-specific loci while ChIP-seq and ChIP-chip would be useful for genome-wide studies. It is noteworthy that the lysate of the whole cell population is used in the ChIP assay while different cells in the tissue may have various histone modifications specific for the group of cells. Therefore, a new approach was developed based on proximity ligation assay in combination with immunofluorescence and in situ hybridization for the detection of histone modifications at specific gene loci in single cells in histological sections (Gomez et al., 2013).

Currently, several modifications of the ChIP assay are developed allowing to overcome some major limitations such as the amount of the initial sample, integrity of proteins during isolation, time of processing and analysis, and cost per sample. Recently, a tagmentation-assisted fragmentation of chromatin (TAF-ChIP) with hyperactive Tn5 transposase (cuts at linker DNA) was developed, which can generate high-quality data sets from as few as 100 human cells without prior isolation of nuclei and uses limited sonication power for nuclear lysis only (Akhtar et al., 2019). The method has several advantages such as avoiding chromatin fragmentation before immunoprecipitation, which prevents potential loss of DNA-protein interactions. Also, tagmentation reactions use the Tn5 transposomes (EZ-Tn5 transposase + transposon recognition sequence) preloaded with Illumina sequencing adaptors. This allows direct PCR-amplification of immunoprecipitated samples after the proteinase K inactivation step.

An ultralow-input N-ChIP-seq (ULI-N-ChIP-seq) method was developed for the analysis of genome-wide histone modifications in rare cells using as few as 10^3 cells. The authors demonstrated that ULI-N-ChIP-seq generated results of H3K27me3 profiles from E13.5 primordial germ cells isolated from a single male and female mouse embryos have high similarity to data sets generated using 50–180 times more material. Furthermore, sex-specific differences in expression and H3K27me3 enrichment in promoter regions of genes involved in meiosis and transforming growth factor-β receptor signaling were identified (Brind'Amour et al., 2015). Thus, the ULI-N-ChIP-seq method significantly minimizes the breeding colony size required for genome-wide analyses. This is of high importance since genetically manipulated animal embryos with the desired genotype may represent only a small fraction of each litter.

Applications of ChIP assay
- ChIP-seq analysis revealed increased levels of H3K27ac of genes involved in platelet activation, blood coagulation, and hemostasis in individuals subjected to different exposure levels of $PM_{2.5}$ (particles with diameters up to 2.5 μm) (Liu et al., 2015).

Mass spectrometry assay

The complexity of the histones caused by allelic variants and the diversity and multivalency of the PTMs are major challenges of histone analysis. Currently, more than 40 variants of human histones are described with the difference by few amino acid residues in the protein sequence. Therefore, very specific and sensitive analysis tools are required for their discrimination (Sidoli et al., 2012).

MS-based proteomics is a high-throughput strategy for accurate characterization of diverse histone PTMs in a single test. Moreover, it is a method of choice for the discovery of previously unknown modification patterns.

MS strategies can be divided into three major groups: **bottom-up, middle-down, and top-down**. The bottom-up approach requires proteolytic digestion of the target protein into short fragments (5–20 amino acids) before MS analysis. However, this approach makes it impossible to determine the global status of PTMs. Thus, the top-down approach overcomes this limitation and allows analysis of the intact protein along with combinations of PTMs. Although it is not suitable for the analysis of a large number of proteins with the same mass and nearly identical physicochemical properties (e.g., H3K27me1/K36me2 and H3K27me2/K36me1). The middle-down approach allows analysis of the histone N-terminal tails cleaved off by specific proteases, generating a polypeptide of 40–70 residues that contain a majority of the PTMs (Önder et al., 2015). Recently a new protocol for the analysis of 200 histone PTMs from tissue or cell lines in 7h was developed, which includes 4h of histone extraction, 3h of derivatization and digestion, and only 1min of MS analysis via direct injection. The authors claim that the novel workflow is suitable for high-throughput screening of > 1000 samples per day using a single mass spectrometer (Sidoli et al., 2019). Nevertheless, MS-based approaches have several major drawbacks, including the inability of analysis of genomic sequences associated with the detected histone PTMs. Also, it requires high expertise and specific equipment making it less attractive for clinical and environmental studies. Considering active developments in this field, it can be assumed that current limitations of MS-based approaches will not hinder the application of MS-based methods in the clinical practice in the future.

Applications of mass spectrometry assay
- Changes of histone acetylation were analyzed in the brain tissues of human fetuses with the failure in neural tube closure using Nano-HPLC-MS/MS. Comparison with the samples obtained from healthy fetuses revealed an aberrant increase of H4K5ac in tissue with neural tube defects. Additionally, 16 novel sites were discovered indicating the sensitivity of MS-based approaches for the identification of previously unknown PTMs (Li et al., 2019).

Summary for histone modifications

Methods for analyzing histone modifications are summarized in Table 10.2. Depending on the scope of the study, various approaches can be implemented for the identification of histone modifications alone (e.g., ELISA-based methods) or in combination with the analysis of genomic context (e.g., ChiP). Therefore, the amount of the sample, the integrity of proteins and DNA, sample processing time, and cost per sample should be justified.

Rothbart et al. (2015) noted that antibody-based techniques have well-known complications, including off-target recognition, strong influence by neighboring PTMs, and an inability to distinguish the modification state on a particular residue. Incorrect selection of antibodies can lead to wrong conclusions (Rothbart et al., 2015). In contrast to histone PTMs-specific antibodies, mass spectrometry represents a more unbiased technique and has emerged as a key platform for proteomic analyses (Simithy et al., 2018).

Extensive studies of chromatin modifications allowed to reveal specific biomarkers indicating alterations in "normal" histone code. These findings revealed associations of histone modifications and regulating systems with pathological conditions or environmental pollution. Technical advances in this field are helping to identify epigenetic targets of therapeutic interventions and facilitate the development of new drugs.

Higher-order chromatin structure analysis
Chromatin remodeling

Chromatin remodeling complexes (CRCs) alter nucleosome structure or conformation via the utilization of ATP-derived energy, regulating the interaction of transcription factors with their DNA binding sites. CRCs are generally made up of an ATPase subunit (SNF2 family of DNA helicases) and several associated subunits for modulation of catalytic activity and genomic binding specificity. Different combinations of ATPase and associated subunits determine cell- and tissue-specific functions of CRCs (Hota & Bruneau, 2016). CRCs modulate the binding of the nucleosomal DNA to histones by distorting or disrupting the histone-DNA contacts in the nucleosome. Based on ATPase domain sequences, CRCs can be divided into four major subfamilies: SWI/SNF (switch/sucrose nonfermentable), ISWI (imitation SWI), CHD (chromodomain helicase DNA-binding), and INO80 [SWI2/SNF2 related (SWR)] (Tyagi et al., 2016). Diseases associated with CRC are reviewed by Tyagi et al. (2016).

A major advance in assessing chromatin remodeling occurred with the development of the chromatin immunoprecipitation (ChIP) assay (Tollefsbol, 2004). This technique provides a powerful tool for monitoring and evaluating DNA-protein interactions and allows determination of the chromatin architecture of specific DNA sequences. Less strongly associated proteins are best assessed using XChIP. This approach provides a basic set of tools for assessing the chromatin state of DNA and its interactions with various proteins.

Table 10.2 Methods for the analysis of histone modifications.

Method	Principles of discrimination of histone modifications	Detection of histone modifications	Advantage	Disadvantage	References
Western blot	Use of site-specific antibodies	Sodium dodecyl sulfate-polyacrylamide gel electrophoresis (SDS-PAGE)	Detection of target proteins in the cell lysates of any culture or tissue	Laborious; a relatively large amount of sample is required; few proteins at a time can be detected; inability of analysis of genomic sequences associated with the detected histone PTMs	Mishra et al. (2017)
ELISA/ immunofluorescence assay	Use of site-specific fluorescent-labeled antibodies	Application of detection antibody and color development reagent	High-throughput assay; applicable for different species; can be used in single cells	Inability of analysis of genomic sequences associated with the detected histone PTMs	Strickfaden and Hendzel (2017)
γ-H2AX immunofluorescence assay	Use of fluorochrome-conjugated specific antibodies against γ-H2AX	ELISA or flow cytometry or fluorescent microscope	Can be used in single cells	Laborious; fluorescence can influence evaluation; inability of analysis of genomic sequences associated with the detected histone PTMs	Redon et al. (2012)
Chromatin immunoprecipitation (ChIP) assay	Use of site-specific antibodies, fixed with formaldehyde (crosslinking DNA and proteins) and sonication or nuclease digestion of samples	qPCR (ChIP-qPCR), sequencing (ChIP-seq), or microarray (ChIP-chip) approaches	PTMs in specific genomic loci can be detected; applicable for gene-specific and genome-wide studies	Laborious; high cost per sample	Wiehle and Breiling (2016)
Mass spectrometry assay	Proteolytic digestion of the target protein, ion generation	Mass spectrometric analysis	High-throughput assay; accurate characterization of diverse histone PTMs; discovery of previously unknown modification patterns	Laborious; inability of analysis of genomic sequences associated with the detected histone PTMs; requires high expertise and specific equipment	Sidoli et al. (2019)

Five different approaches for histone modification analysis are summarized and treated here.

3D nucleus organization and epigenetic regulation

Eukaryotic genomes are tightly folded and packaged in a highly organized manner to accommodate the spatial constraints of the nucleus while allowing regulatory factors access to the underlying sequences to affect transcriptional control. Understanding how chromosomes fold can provide insight into the complex relationships between chromatin structure, gene activity, and the functional state of the cell (Wang & Chang, 2018).

The existence of chromosome territories (CTs) in the interphase cell nucleus was proposed at the end of the 19th century by Carl Rabl, and the term CTs was coined by Theodor Boveri in 1909. Nearly 100 years later, Cremer et al. (1982) elegantly demonstrated the existence of CTs via ultraviolet microirradiation of the small region of the cell nucleus confirming the initial idea of Rabl. During DNA damage repair, the radioactively labeled nucleotides were incorporated in the DNA and repaired regions were detected by radiography in the chromosomes.

It was suggested that the location of a gene in the CT might influence its interaction with transcription and other factors indicating the potential epigenetic influence of CT on the regulation of gene expression (Cremer & Cremer, 2001; Daban, 2021; Liehr, 2021c). Over the last two decades, numerous studies have assessed the spatial proximity and nuclear organization of specific genomic loci, initially through microscopy techniques such as **fluorescence in situ hybridization (FISH)** (Liehr, 2021e) and more recently by **chromosome conformation capture (3C)** (Wang & Chang, 2018).

Chromosome conformation capture (3C) techniques estimate the contact frequencies between two chosen genomic sites in the mix of cell populations, making inferences on the 3D organization of the points of interest. There are several variations of the method, including chromosome conformation capture-on-chip (4C), chromosome conformation capture carbon copy (5C), and Hi-C with the application of high-throughput sequencing across an entire genome. After cell fixation with the crosslinking agent paraformaldehyde, the DNA is digested with four-base (*Dpn*II) or six-base (*Hin*dIII) recognizing restriction enzymes. Crosslinking of native chromatin captures 3D interacting fragments while digestion with restriction enzymes allows further amplification and analysis of these fragments. In the 3C method, digestion is followed by intramolecular ligation of restriction fragments. Thus, a one-dimensional 3C template (i.e., hybrid DNA molecule) representing the local 3D chromatin contacts at the point of crosslinking is formed. The relative abundance of a given 3C template is then analyzed by qPCR to determine an averaged contact frequency across the cell population (Sati & Cavalli, 2017). Although 3C-based methods can theoretically provide the information at a resolution of a few hundred base pairs, nevertheless, the results should be validated by FISH analysis. Therefore, the analysis of chromosome topology and nuclear architecture requires the implementation of both imaging and molecular-genetic methods. Improvements on Hi-C resolution (~20–40 kb) revealed a new level of chromatin folding denominated topologically associated domains (TADs) with an average size of around 880 kb (Mota-Gómez & Lupiáñez, 2019). TAD-adjacent regions are enriched in housekeeping genes and

CCCTC-binding factor loci. It is prominent that the disruption of TADs is associated with aberrant gene expression and is observed in different pathological conditions including cancer (Hnisz et al., 2016).

The major limitation of 3C-based or ligation-dependent methods is that the average contact maps are generated from the mixture of cells, which precludes the detection of cell-to-cell differences. **Single-cell combinatorial indexed Hi-C** (sciHi-C) overcomes this limitation by tagging the DNA within each nucleus with a specific combination of barcodes. Fixed cells are lysed and digested with a restriction enzyme followed by splitting whole nuclei in 96-well plates and indexed with individual barcodes. The high-throughput generation of sciHi-C libraries can be achieved by several rounds of indexing and in situ proximity ligation of pooled nuclei (see Kempfer & Pombo, 2020 for more details).

It was suggested that during ligation, the choice of only one or two other fragments dilutes the interactome of each DNA fragment. Thus, ligation-independent methods have been developed denoted as genome architecture mapping (GAM) (Beagrie et al., 2017), split-pool recognition of interactions by tag extension (SPRITE) (Quinodoz et al., 2018), and chromatin-interaction analysis via droplet-based and barcode-linked sequencing (ChIA-Drop) (Zheng et al., 2019). These methods allow genome-wide mapping of chromatin contacts that involve three or more DNA fragments and topological domains spanning tens of mega bases.

During **GAM analysis** cells are fixed in the 4% and 8% freshly depolymerized paraformaldehyde. Afterward, sucrose-embedded cells are frozen in liquid nitrogen on copper stubs and cryosectioned using ultracryomicrotome ($\sim$220 nm thickness). Each nuclear slice is then isolated directly from the cryosection by laser microdissection followed by a series of washing steps to remove sucrose solution. After extraction of the DNA from every slice, whole-genome amplification can be performed following attachment with indexed sequencing adapters and sequencing (Beagrie et al., 2017). Therefore, closely located loci in the nuclear space (but not necessarily on the linear genome) are detected in the same nuclear sections more often than distant loci.

SPRITE and ChIA-Drop allow genome-wide detection of several DNA interactions and interchromosomal spatial arrangement around nuclear bodies by tagging crosslinked chromatin complexes. During SPRITE and ChIA-Drop cells are fixed with paraformaldehyde and nuclei are disrupted using sonication. The chromatin is digested by DNAse (SPRITE) or *Hind*III (ChIA-Drop). For the SPRITE analysis purified DNA fragments are split in 96-well plates containing a unique barcode. After several rounds of pooling, splitting and barcoding the samples are sequenced. Thus, DNA fragments that were crosslinked with each other will display the same combinations of barcodes. Instead in ChIA-Drop analysis, DNA fragments are distributed into droplets containing reagents for barcoding and DNA amplification. Afterward, barcoded complexes are pooled and sequenced (Kempfer & Pombo, 2020).

A significant breakthrough in the understanding of how the individual territories are arranged within the 3D architecture of the nucleus has become possible due to improvements in microscopy and in situ hybridization techniques. Thus, **3D FISH** was developed, which enables direct visualization of the spatial arrangement of chromosomes and genes relative to the nuclear periphery and center in individual cells (Fritz et al., 2019; Iourov et al., 2021; Liehr, 2021c, 2021e; Manvelyan et al., 2008). In FISH analysis, fluorophore-conjugated DNA probes complementary to the genomic sequence of interest are used. For the hybridization of the probe and DNA in the nucleus, the permeabilization step is required using detergent or organic solvent, such as methanol followed by denaturation of DNA and overnight hybridization with the probe (Kempfer & Pombo, 2020). The hybridization can be visualized using a confocal laser scanning microscope to collect optical sections and then the position of a chromosome or gene is measured in these images from the geometric center of its signal to the nearest nuclear edge (Bridger et al., 2014).

For visualization of short fragments of genomic DNA, Oligopaint probes are used, which can target 5–15 kb sequences. Oligopaint probes are generated from synthetic libraries of ~60–100 bp oligonucleotides, which are produced by massively parallel synthesis. Also, molecular beacon FISH probes have been developed for targeting genomic regions as short as 2.5 kb. Due to the presence of the quencher in the unbound probe, the background signal is significantly reduced enabling the visualization of small genomic regions (Kempfer & Pombo, 2020). Furthermore, molecular combing (Bisht & Avarello, 2021) can be considered here.

CRISPR-Cas9 genome editing technology provided new opportunities for the live imaging of different genomic regions. The target sequence can be visualized using an endonuclease-deficient form of Cas9 (dCas9) fused with a fluorescent protein. dCas9 is interacting with target-specific gRNA, which can be differentially modified to act as a scaffold that brings fluorescent proteins to different genomic loci (Chen et al., 2013; Ishii et al., 2021).

Several studies demonstrate that FISH is a method of choice that allows discrimination of CTs and genes in individual nuclei enabling analysis of the distribution of cells according to their CTs (Liehr, 2021c). Nevertheless, the major drawback of the 3D FISH is its low throughput that allows only a small number of genomic loci to be analyzed at a time.

It was shown that chromosomes in somatic nuclei have nonrandom distribution: gene-rich chromosomes tend to locate nearer the nuclear center, while gene-poor ones are close to the periphery. Multicolor banding 3D FISH analysis in human sperm showed that large chromosomes are primarily located in the periphery while smaller and gene dense chromosomes can mainly be found in the central part of the nucleus. The only exceptions were chromosomes 18, 20, and 21 mainly found close to the periphery (Manvelyan et al., 2008). The authors suggested that in patients with idiopathic infertility, chromosomal localization in the nucleus should also be considered. This idea was substantiated by others indicating the association of chromosome localization in the sperm nucleus and fertility (Sarrate et al., 2018). Interestingly, nonrandom distribution of the chromosomes might have a favorable influence on sperm nuclear morphology, which is essential for the sperm movement within the female reproductive tract. Additionally, it was suggested that the preferential location of gene-rich chromosomes in the center of the nucleus could be attributed to the mechanisms of protection from DNA damage and the ordered reactivation of the paternal genome in the zygote (Sarrate et al., 2018).

Applications of FISH and chromosome conformation capture

- The repositioning of chromosomes in interphase nuclei is an energy-dependent process that can be observed in quiescent primary human fibroblast culture using 2D and 3D FISH. Moreover, the repositioning of chromosomes was inhibited by drugs affecting the polymerization of nuclear myosin and actin (Bridger et al., 2014). Thus, it was suggested that an active nuclear motor complex such as nuclear myosin 1β is required for the repositioning of chromosomes providing new insight into spatio-epigenetic mechanisms of gene regulation.
- FISH was used to map the chromatin interaction network bound by estrogen receptor alpha (ER-alpha) in the human genome (Fullwood et al., 2009). In later publications, the use of chromosome conformation captures techniques in the analysis of transcriptional regulation prevails. The Hi-C approach was applied to generate a comprehensive chromatin interaction map in human fibroblasts (Jin et al., 2013) and to identify interacting regions of 31,253 promoters in 17 human primary hematopoietic cell types (Javierre et al., 2016). Chromosome conformation capture sequencing (3C-Seq) permits to reveal more than a hundred loci in the rs6702619 locus interactome associating with colorectal cancer in the Polish population (Statkiewicz et al., 2019). Applying a modified version of Hi-C, capture Hi-C, allows identifying key long-range chromatin interactions involving 14 colorectal cancer risk loci (Jäger et al., 2015).

Summary for 3D nucleus organization and epigenetic regulation

Thus, it can be suggested that various methods for the analysis of 3D localization of chromatin in the nucleus have different limitations and advantages contributing to our understanding of the spatial organization of the genome. While some methods such as FISH, 3C-based, and ligation-independent methods can be used for the analysis of most human biopsies, others like CRISPR-based imaging approaches are more suitable for live-cell imaging studies. Overall, the spatial organization of the genome and its influence on gene expression is another level of epigenetics regulating cell fate via an orchestrated dynamic of intra- and interchromosomal interactions (Daban, 2021).

Analysis of noncoding RNAs

Noncoding RNAs (ncRNAs) do not encode functional proteins but regulate gene expression at the posttranscriptional level. NcRNAs include housekeeping (e.g., rRNA and tRNA) and regulatory ncRNAs. Based on their size, regulatory ncRNAs can be subdivided to long ncRNAs (> 200 nucleotides) and small ncRNAs (< 200 nucleotides). The main classes of small ncRNAs are microRNAs (miRNAs), small interfering RNAs (siRNAs), and piwi-interacting RNAs (piRNAs) (Zhang et al., 2019b). MiRNAs are the most abundant class of small ncRNAs in human tissues that are generated from transcribed hairpin loop structures (Zhang et al., 2019b). Endogenous siRNAs are mostly transcribed from transposon elements and similarly to miRNAs are double-stranded molecules that cleave target mRNAs via Argonaute 2 (Ago2) (Mahmoodi Chalbatani et al., 2019). SiRNA molecules were identified in human hepatocellular carcinoma cells; nevertheless, whether additional siRNAs expressed from human pseudogenes exist and contribute to carcinogenesis is not clear yet (Anastasiadou et al., 2018). In contrast to miRNA and siRNA, piRNAs are single-stranded molecules that interact with P-element-induced wimpy testis (PIWI) proteins and regulate the silencing of retrotransposons in germ cells (Romano et al., 2017).

At first, piRNAs, miRNAs, and siRNAs were thought to function independently, and this supposition was reinforced by the obvious differences between them. However, in recent years, some of these pathways appear to interact, thereby constituting a regulatory network (Wei et al., 2017).

Here, we will focus on studies of miRNA since it is the most abundant and widely analyzed class of ncRNAs detected practically in all human tissues and fluids.

MicroRNAs (miRNAs) are short molecules (22 nucleotides) that regulate gene expression. MiRNAs are involved in a variety of biological processes, including development, differentiation, apoptosis, and cell proliferation. It was shown that one miRNA can have several targets and one gene could be regulated by several miRNAs (Haas et al., 2012). Over 2500 mature miRNA encoding genes have been discovered in the human genome regulating ~30% of protein translation of coding genes (Hammond, 2015). MiRNA biogenesis is mediated by multiple steps starting with a transcription of pri-miRNA, processing into pre-miRNAs, and mature miRNAs (O'Brien et al., 2018). MiRNA genes can be located in the introns of known protein-coding genes or untranslated regions (UTRs) of coding sequences and repeat regions of the genome (Khan et al., 2019).

Alterations in miRNA expression have been identified in a wide range of human diseases. Additionally, the diagnostic value of miRNA was evaluated in many studies. Therefore, analysis of expression of miRNAs is of high importance (Szelenberger et al., 2019).

MiRNA expression can be detected in tissue samples, as well as in biological fluids such as serum or plasma. There are several methods for miRNA profiling: Real-time quantitative reverse transcription PCR (qRT-PCR), in situ hybridization, microarray profiling, and RNA-seq using next-generation sequencing (NGS).

Real-time quantitative reverse transcription PCR

Since the length of miRNA is typically short to design PCR primers is technically challenging to perform the conventional PCR. Thus, the obvious solution is to extend the length of miRNA using *Escherichia coli* poly(A) polymerase. The resulting polyadenylated RNA is then reverse transcribed using a universal oligo(dT) primer. Subsequently, specific PCR primers can be used for the amplification of cDNA (Pritchard et al., 2012).

In situ hybridization

For the detection of subcellular localization of miRNA, FISH is applied using locked nucleic acid (LNA) probes that have a higher binding affinity and mismatch discrimination. It was shown that LNA-FISH has a short hybridization time, high efficiency, and discriminatory power. Nevertheless, LNA probes are expensive and can generate strong background signals (Urbanek et al., 2015).

MiRNA microarray profiling

Arrays are typically used for parallel high-throughput analysis of large numbers of miRNAs. After isolation and reverse transcription of miRNAs, cDNA is tagged with fluorophore-labeled nucleotides. Subsequently, individual cDNAs are hybridized to complementary capture probes on the array. Following a series of washing steps to remove free DNA molecules, the fluorescence intensity of each well can be scanned and measured directly using specific scanners and software. As in the case of PCR and in situ hybridization, array methods also require sequence knowledge of the target. Although microarray-based methods have wide coverage, nevertheless high cost and low specificity for the discrimination of the miRNA with similar sequences are important limitations (Ye et al., 2019). Currently, Affymetrix GeneChip miRNA 4.0 arrays allow miRNA profiling without the cDNA synthesis step with direct labeling of isolated RNA (at least 130 ng).

MiRNA sequencing

NGS enables genome-wide analysis of miRNA expression and sequence in a given tissue or cells. It is the most suitable method for the identification and subsequent validation of novel diagnostic or prognostic miRNA signatures. NGS analysis of miRNAs includes RNA isolation, preparation of libraries, RNA sequencing, and sequence analysis (Kolanowska et al., 2018). Highly purified samples should be used for the analysis. The ligation of preadenylated 3′ adapter with RNA fragments is performed using T4 RNA ligase 2 while 5′ adapter is ligated using T4 RNA ligase 1. Following the ligation step, RNA can be converted to cDNA and amplified for the library preparation (Van Dijk et al., 2014). PCR amplification is performed using specific single barcoded primers complementary to the adapter sequences. During the second amplification step, indexed sequences (complementary to the adapter) are introduced into the cDNAs reaction mixture that tag the cDNA fragments and allow the parallel analysis of multiple samples in a single experiment. Afterward, cDNA libraries are applied to a flow cell where single cDNAs are captured and subsequently bridge-amplified, thus generating clonal spots of each cDNA in the flow cell. The flow cell with the respective spots is then sequenced (Kappel & Keller, 2017).

Summary for noncoding RNAs

Methods for detecting microRNAs are summarized in Table 10.3. MiRNAs have great diagnostic potential by being associated with many diseases (De Guire et al., 2013). One of the first diagnostic panels based on miRNA (miRview mets) was developed for the identification of cancers of unknown or uncertain primary origin. This panel measures the expression levels of 64 miRNAs using microarray technology and was capable to identify 90% of the 509 validation sample sets (Bonneau et al., 2019). Currently, many companies are offering miRNA-based diagnostic panels for different types of cancers (thyroid, pancreatic, nonsmall-cell lung carcinoma, melanoma, and breast cancer), neurodegenerative (multiple sclerosis, Alzheimer, Parkinson), cardiovascular (acute myocardial infarction and heart failure), and inflammatory bowel disease. Furthermore, various miRNAs as therapeutics are currently in different development phases demonstrating promising results in several trials (Bonneau et al., 2019). Nevertheless, this field can still be considered in its infancy and with no doubts, technical advances will drive it from the bench to the bedside.

Messenger RNA sequencing

Over the past decade, RNA sequencing (RNA-seq) has become an indispensable tool for transcriptome-wide analysis of differential gene expression and differential splicing of mRNAs (Stark et al., 2019). In the RNA-Seq method, cDNAs generated from the RNA of interest are directly sequenced using next-generation sequencing

Table 10.3 Methods for miRNA detection.

Method	Principles of microRNA detection	References
Real-time quantitative reverse transcription PCR (qRT-PCR)	Elongation of miRNA, conversion to cDNA, amplification of cDNA using specific PCR primers	Pritchard et al. (2012)
Locked nucleic acid (LNA)-FISH	Binding of LNA-FISH probes with miRNA, detection of subcellular localization of miRNA	Urbanek et al. (2015)
MiRNA microarray analysis	MiRNA isolation, conversion to cDNA, tagging of cDNA with fluorophore-labeled nucleotides, hybridization of individual cDNAs to complementary probes on the array	Ye et al. (2019)
MiRNA-seq using next-generation sequencing (NGS)	MiRNA isolation, conversion to cDNA, amplification for the library preparation, RNA sequencing and sequence analysis	Kolanowska et al. (2018)

Four different approaches for miRNA detection are summarized and treated here.

technologies (Nagalakshmi et al., 2010). The method includes RNA extraction, followed by mRNA enrichment or ribosomal RNA depletion, cDNA synthesis, and preparation of an adaptor-ligated sequencing library. The library is then sequenced with a read depth of 10–30 million reads per sample on a high-throughput platform (usually Illumina) (Stark et al., 2019). To date, almost 100 distinct methods have been derived from the standard RNA-seq protocol. Much of this method development has been achieved on Illumina short-read sequencing instruments, but recently long-read RNA-seq and direct RNA sequencing (dRNA-seq) methods have been developed (Stark et al., 2019). RNA-Seq has been successfully applied in ovarian cancer research for earlier detection, ascertaining pathological origin, and defining the aberrant genes and dysregulated molecular pathways across patient groups (Wang et al., 2019). Single-cell RNA-seq has enabled gene expression to be studied at an unprecedented resolution. Generating single-cell data requires single-cell dissociation, single-cell isolation, library construction, and sequencing (Luecken & Theis, 2019).

Using single-cell RNA sequencing based on the Illumina Hiseq2000 or Hiseq 25,000 with TruSeq SBS v3 chemistry, functionally distinct human dermal fibroblast subpopulations were identified (Philippeos et al., 2018).

Concluding remarks

Concluding the chapter one can say that epigenetics delivers new insight into the understanding of our genome and is a major part of cytogenomics (Liehr, 2021a, 2021b). Although each cell in the organism contains the same genome, the epigenome determines cellular identity within the diversity of cells; epigenetic alterations are very influential in both the normal and disease states of an organism (Rasool et al., 2015).

The complexity of epigenetic studies is that epigenetic alterations are far more frequent than genetic events, and in contrast to mutations, they are reversible (Machnik & Oleksiewicz, 2020). Although the techniques for profiling other genomic features (such as gene mutations and expression) are relatively standardized, there are diverse and evolving methods available in epigenetics (Chatterjee et al., 2017). Therefore, the selection of an appropriate approach can be challenging.

It should be mentioned that various methods of epigenetics have their advantages and limitations, which must be considered, based on the scope of the study. While some methods provide wider coverage of the human epigenome, there are no universal approaches suitable for every aspect of epigenetic studies. Thus, the combination of different methods is implemented to get a more comprehensive picture of epigenetic modifications, their interaction, and their impact on human health. It is anticipated that progress in epigenome mapping technologies will permit further development of biomarkers of human pathologies and environmental impacts.

Overall, epigenetics is like an "epistrand" connecting the strand of life and the environment. Certainly, future studies of mammalian epigenome will uncover exciting and impactful properties of cells. Just as the full sequence of a genome has greatly facilitated progress in genetics, a clearer understanding of epigenetics will likely earliest come when all the parts of the puzzle of the human genome will be known (Wang & Chang, 2018).

Acknowledgment

This work was supported by the Armenian-German joint laboratory project (AG-01/20).

References

Akhtar, J., More, P., Albrecht, S., Marini, F., Kaiser, W., Kulkarni, A., … Berger, C. (2019). TAF-ChIP: An ultra-low input approach for genome-wide chromatin immunoprecipitation assay. *Life Science Alliance*, 2(4), e201900318. https://doi.org/10.26508/lsa.201900318.

Alles, J., Fehlmann, T., Fischer, U., Backes, C., Galata, V., Minet, M., … Meese, E. (2019). An estimate of the total number of true human miRNAs. *Nucleic Acids Research*, 47(7), 3353–3364. https://doi.org/10.1093/nar/gkz097.

Anastasiadou, E., Jacob, L., & Frank, S. (2018). Non-coding RNA networks in cancer. *Nature Reviews Cancer*, 18(1), 5–18. https://doi.org/10.1038/nrc.2017.99.

Angeloni, A., & Bogdanovic, O. (2019). Enhancer DNA methylation: Implications for gene regulation. *Essays in Biochemistry*, 63(6), 707–715. https://doi.org/10.1042/EBC20190030.

Arasaradnam, R. P., Khoo, K. T. J., Bradburn, M., Mathers, J. C., & Kelly, S. B. (2010). DNA methylation of ESR-1 and N-33 in colorectal mucosa of patients with ulcerative colitis (UC). *Epigenetics*, 5(5), 422–426. https://doi.org/10.4161/epi.5.5.11959.

Audia, J. E., & Campbell, R. M. (2016). Histone modifications and cancer. *Cold Spring Harbor Perspectives in Biology*, 8(4), a019521. https://doi.org/10.1101/cshperspect.a019521.

Baba, Y., Yagi, T., Sawayama, H., Hiyoshi, Y., Ishimoto, T., Iwatsuki, M., … Baba, H. (2018). Long interspersed element-1 methylation level as a prognostic biomarker in gastrointestinal cancers. *Digestion*, 97(1), 26–30. https://doi.org/10.1159/000484104.

Baranov, V. S., & Kuznetzova, T. V. (2021). Nuclear stability in early embryo. Chromosomal aberrations. In T. Liehr (Ed.), *Cytogenomics* (pp. 307–325). Academic Press. Chapter 15 (in this book).

Barchitta, M., Quattrocchi, A., Maugeri, A., Canto, C., La Rosa, N., Cantarella, M. A., … Agodi, A. (2017). LINE-1 hypermethylation in white blood cell DNA is associated with high-grade cervical intraepithelial neoplasia. *BMC Cancer*, 17(1), 601. https://doi.org/10.1186/s12885-017-3582-0.

Barlow, D. P., & Bartolomei, M. S. (2014). Genomic imprinting in mammals. *Cold Spring Harbor Perspectives in Biology*, 6(2), a018382. https://doi.org/10.1101/cshperspect.a018382.

Beagrie, R. A., Scialdone, A., Schueler, M., Kraemer, D. C. A., Chotalia, M., Xie, S. Q., … Pombo, A. (2017). Complex multi-enhancer contacts captured by genome architecture mapping. *Nature*, 543(7646), 519–524. https://doi.org/10.1038/nature21411.

Becker, J. S., Nicetto, D., & Zaret, K. S. (2016). H3K9me3-dependent heterochromatin: Barrier to cell fate changes. *Trends in Genetics*, 32(1), 29–41. https://doi.org/10.1016/j.tig.2015.11.001.

Bibikova, M., Le, J., Barnes, B., Saedinia-Melnyk, S., Zhou, L., Shen, R., & Gunderson, K. L. (2009). Genome-wide DNA methylation profiling using Infinium® assay. *Epigenomics*, *1*(1), 177–200. https://doi.org/10.2217/epi.09.14.

Bickmore, W. A., & Van Steensel, B. (2013). Genome architecture: Domain organization of interphase chromosomes. *Cell*, *152*(6), 1270–1284. https://doi.org/10.1016/j.cell.2013.02.001.

Bilichak, A., & Kovalchuk, I. (2017). Analysis of global genome methylation using the cytosine-extension assay. *Methods in Molecular Biology*, *1456*, 73–79. https://doi.org/10.1007/978-1-4899-7708-3_6.

Bind, M. A., Lepeule, J., Zanobetti, A., Gasparrini, A., Baccarelli, A., Coull, B. A., … Schwartz, J. (2014). Air pollution and gene-specific methylation in the normative aging study: Association, effect modification, and mediation analysis. *Epigenetics*, *9*(3), 448–458. https://doi.org/10.4161/epi.27584.

Bisht, P., & Avarello, M. D. M. (2021). Molecular combing solutions to characterize replication kinetics and genome rearrangements. In T. Liehr (Ed.), *Cytogenomics* (pp. 47–71). Academic Press. Chapter 5 (in this book).

Black, J. C., Van Rechem, C., & Whetstine, J. R. (2012). Histone lysine methylation dynamics: Establishment, regulation, and biological impact. *Molecular Cell*, *48*(4), 491–507. https://doi.org/10.1016/j,molcel.2012.11.006.

Bonneau, E., Neveu, B., Kostantin, E., Tsongalis, G. J., & De Guire, V. (2019). How close are miRNAs from clinical practice? A perspective on the diagnostic and therapeutic market. *Electronic Journal of the International Federation of Clinical Chemistry and Laboratory Medicine*, *30*(2), 114–127. https://www.ifcc.org/media/477996/ejifcc2019vol-30no2pp114-27.pdf.

Booth, M. J., Ost, T. W. B., Beraldi, D., Bell, N. M., Branco, M. R., Reik, W., & Balasubramanian, S. (2013). Oxidative bisulfite sequencing of 5-methylcytosine and 5-hydroxymethylcytosine. *Nature Protocols*, *8*(10), 1841–1851. https://doi.org/10.1038/nprot.2013.115.

Brena, R. M., Auer, H., Kornacker, K., & Plass, C. (2006). Quantification of DNA methylation in electrofluidics chips (Bio-COBRA). *Nature Protocols*, *1*(1), 52–58. https://doi.org/10.1038/nprot.2006.8.

Bridger, J. M., Arican-Gotkas, H. D., Foster, H. A., Godwin, L. S., Harvey, A., Kill, I. R., … Ahmed, M.H. (2014). The non-random repositioning of whole chromosomes and individual gene loci in interphase nuclei and its relevance in disease, infection, aging, and cancer. *Advances in Experimental Medicine and Biology*, *773*, 263–279. https://doi.org/10.1007/978-1-4899-8032-8_12.

Brind'Amour, J., Liu, S., Hudson, M., Chen, C., Karimi, M. M., & Lorincz, M. C. (2015). An ultra-low-input native ChIP-seq protocol for genome-wide profiling of rare cell populations. *Nature Communications*, *6*(1), 6033. https://doi.org/10.1038/ncomms7033.

Bromberg, A., Bersudsky, Y., Levine, J., & Agam, G. (2009). Global leukocyte DNA methylation is not altered in euthymic bipolar patients. *Journal of Affective Disorders*, *118*(1–3), 234–239. https://doi.org/10.1016/j.jad.2009.01.031.

Burns, K. H. (2017). Transposable elements in cancer. *Nature Reviews Cancer*, *17*(7), 415–424. https://doi.org/10.1038/nrc.2017.35.

Cai, J., Zhao, Y., Liu, P., Xia, B., Zhu, Q., Wang, X., … Zhang, Y. (2017). Exposure to particulate air pollution during early pregnancy is associated with placental DNA methylation. *Science of the Total Environment*, *607–608*, 1103–1108. https://doi.org/10.1016/j.scitotenv.2017.07.029.

Campan, M., Weisenberger, D. J., Trinh, B., & Laird, P. W. (2018). MethyLight and digital methylight. *Methods in Molecular Biology*, *1708*, 497–513. https://doi.org/10.1007/978-1-4939-7481-8_25.

Cappellacci, L., Perinelli, D. R., Maggi, F., Grifantini, M., & Petrelli, R. (2020). Recent progress in histone deacetylase inhibitors as anticancer agents. *Current Medicinal Chemistry*, *27*(15), 2449–2493. https://doi.org/10.2174/0929867325666181016163110.

Chatterjee, A., Rodger, E. J., Morison, I. M., Eccles, M. R., & Stockwell, P. A. (2017). Tools and strategies for analysis of genome-wide and gene-specific DNA methylation patterns. *Methods in Molecular Biology*, *1537*, 249–277. https://doi.org/10.1007/978-1-4939-6685-1_15.

Chen, B., Gilbert, L. A., Cimini, B. A., Schnitzbauer, J., Zhang, W., Li, G. W., ... Huang, B. (2013). Dynamic imaging of genomic loci in living human cells by an optimized CRISPR/Cas system. *Cell*, *155*(7), 1479–1491. https://doi.org/10.1016/j.cell.2013.12.001.

Chen, X., Lin, Q., Wen, J., Lin, W., Liang, J., Huang, H., ... Chen, G. (2020). Whole genome bisulfite sequencing of human spermatozoa reveals differentially methylated patterns from type 2 diabetic patients. *Journal of Diabetes Investigation*, *11*(4), 856–864. https://doi.org/10.1111/jdi.13201.

Cheow, L. F., Quake, S. R., Burkholder, W. F., & Messerschmidt, D. M. (2015). Multiplexed locus-specific analysis of DNA methylation in single cells. *Nature Protocols*, *10*(4), 619–631. https://doi.org/10.1038/nprot.2015.041.

Choudhury, J. H., Das, R., Laskar, S., Kundu, S., Kumar, M., Das, P. P., ... Ghosh, S.K. (2018). Detection of p16 promoter hypermethylation by methylation-specific PCR. *Methods in Molecular Biology*, *1726*, 111–122. https://doi.org/10.1007/978-1-4939-7565-5_11.

Chowdhury, B., Cho, I. H., & Irudayaraj, J. (2017). Technical advances in global DNA methylation analysis in human cancers. *Journal of Biological Engineering*, *11*(1), 10. https://doi.org/10.1186/s13036-017-0052-9.

Christiansen, L., Lenart, A., Tan, Q., Vaupel, J. W., Aviv, A., Mcgue, M., & Christensen, K. (2016). DNA methylation age is associated with mortality in a longitudinal Danish twin study. *Aging Cell*, *15*(1), 149–154. https://doi.org/10.1111/acel.12421.

Clark, S. J., Statham, A., Stirzaker, C., Molloy, P. L., & Frommer, M. (2006). DNA methylation: Bisulphite modification and analysis. *Nature Protocols*, *1*(5), 2353–2364. https://doi.org/10.1038/nprot.2006.324.

Collin, L. J., McCullough, L. E., Conway, K., White, A. J., Xu, X., Cho, Y. H., ... Gammon, M.D. (2019). Reproductive characteristics modify the association between global DNA methylation and breast cancer risk in a population-based sample of women. *PLoS One*, *14*(2), e0210884. https://doi.org/10.1371/journal.pone.0210884.

Consales, C., Toft, G., Leter, G., Bonde, J. P. E., Uccelli, R., Pacchierotti, F., ... Spanò, M. (2016). Exposure to persistent organic pollutants and sperm DNA methylation changes in Arctic and European populations. *Environmental and Molecular Mutagenesis*, *57*(3), 200–209. https://doi.org/10.1002/em.21994.

Cooper, D. N., Mort, M., Stenson, P. D., Ball, E. V., & Chuzhanova, N. A. (2010). Methylation-mediated deamination of 5-methylcytosine appears to give rise to mutations causing human inherited disease in CpNpG trinucleotides, as well as in CpG dinucleotides. *Human Genomics*, *4*(6), 406–410. https://doi.org/10.1186/1479-7364-4-6-406.

Coppedè, F., Stoccoro, A., Mosca, L., Gallo, R., Tarlarini, C., Lunetta, C., ... Penco, S. (2018). Increase in DNA methylation in patients with amyotrophic lateral sclerosis carriers of not fully penetrant SOD1 mutations. *Amyotrophic Lateral Sclerosis and Frontotemporal Degeneration*, *19*(1–2), 93–101. https://doi.org/10.1080/21678421.2017.1367401.

Cremer, T., & Cremer, C. (2001). Chromosome territories, nuclear architecture and gene regulation in mammalian cells. *Nature Reviews Genetics, 2*(4), 292–301. https://doi.org/10.1038/35066075.

Cremer, T., Cremer, C., Baumann, H., Luedtke, E. K., Sperling, K., Teuber, V., & Zorn, C. (1982). Rabl's model of the interphase chromosome arrangement tested in Chinise hamster cells by premature chromosome condensation and laser-UV-microbeam experiments. *Human Genetics, 60*(1), 46–56. https://doi.org/10.1007/BF00281263.

Curia, M. C., Fantini, F., Lattanzio, R., Tavano, F., Di Mola, F., Piantelli, M., … Cama, A. (2019). High methylation levels of PCDH10 predict poor prognosis in patients with pancreatic ductal adenocarcinoma. *BMC Cancer, 19*(1), 452. https://doi.org/10.1186/s12885-019-5616-2.

Daban, J.-R. (2021). Multilayer organization of chromosomes. In T. Liehr (Ed.), *Cytogenomics* (pp. 267–296). Academic Press. Chapter 13 (in this book).

De Guire, V., Robitaille, R., Tétreault, N., Guérin, R., Ménard, C., Bambace, N., & Sapieha, P. (2013). Circulating miRNAs as sensitive and specific biomarkers for the diagnosis and monitoring of human diseases: Promises and challenges. *Clinical Biochemistry, 46*(10–11), 846–860. https://doi.org/10.1016/j.clinbiochem.2013.03.015.

de Koning, A. P. J., Gu, W., Castoe, T. A., Batzer, M. A., & Pollock, D. D. (2011). Repetitive elements may comprise over two-thirds of the human genome. *PLoS Genetics, 7*(12), e1002384. https://doi.org/10.1371/journal.pgen.1002384.

Dedeurwaerder, S., Defrance, M., Bizet, M., Calonne, E., Bontempi, G., & Fuks, F. (2013). A comprehensive overview of infinium human methylation450 data processing. *Briefings in Bioinformatics, 15*(6), 929–941. https://doi.org/10.1093/bib/bbt054.

Dedeurwaerder, S., Defrance, M., Calonne, E., Denis, H., Sotiriou, C., & Fuks, F. (2011). Evaluation of the infinium methylation 450K technology. *Epigenomics, 3*(6), 771–784. https://doi.org/10.2217/epi.11.105.

Delaney, C., Garg, S. K., & Yung, R. (2015). Analysis of DNA methylation by pyrosequencing. *Methods in Molecular Biology, 1343*, 249–264. https://doi.org/10.1007/978-1-4939-2963-4_19.

Delgado-Cruzata, L., Wu, H. C., Perrin, M., Liao, Y., Kappil, M. A., Ferris, J. S., … Terry, M.B. (2012). Global DNA methylation levels in white blood cell DNA from sisters discordant for breast cancer from the New York site of the breast cancer family registry. *Epigenetics, 7*(8), 868–874. https://doi.org/10.4161/epi.20830.

Delgado-Cruzata, L., Zhang, W., McDonald, J. A., Tsai, W. Y., Valdovinos, C., Falci, L., … Greenlee, H. (2015). Dietary modifications, weight loss, and changes in metabolic markers affect global DNA methylation in Hispanic, African American, and Afro-Caribbean breast cancer survivors. *The Journal of Nutrition, 145*(4), 783–790. https://doi.org/10.3945/jn.114.202853.

Deneberg, S., Grövdal, M., Karimi, M., Jansson, M., Nahi, H., Corbacioglu, A., … Lehmann, S. (2010). Gene-specific and global methylation patterns predict outcome in patients with acute myeloid leukemia. *Leukemia, 24*(5), 932–941. https://doi.org/10.1038/leu.2010.41.

Denham, J., O'Brien, B. J., Harvey, J. T., & Charchar, F. J. (2015). Genome-wide sperm DNA methylation changes after 3 months of exercise training in humans. *Epigenomics, 7*(5), 717–731. https://doi.org/10.2217/epi.15.29.

Dhingra, R., Nwanaji-Enwerem, J. C., Samet, M., & Ward-Caviness, C. K. (2018). DNA methylation age—Environmental influences, health impacts, and its role in environmental epidemiology. *Current Environmental Health Reports, 5*(3), 317–327. https://doi.org/10.1007/s40572-018-0203-2.

Du, J., Johnson, L. M., Jacobsen, S. E., & Patel, D. J. (2015). DNA methylation pathways and their crosstalk with histone methylation. *Nature Reviews Molecular Cell Biology*, *16*(9), 519–532. https://doi.org/10.1038/nrm4043.

Eggermann, T. (2021). Epigenetics. In T. Liehr (Ed.), *Cytogenomics* (pp. 389–401). Academic Press. Chapter 20 (in this book).

Estival, A., Sanz, C., Ramirez, J. L., Velarde, J. M., Domenech, M., Carrato, C., … Balana, C. (2019). Pyrosequencing versus methylation-specific PCR for assessment of MGMT methylation in tumor and blood samples of glioblastoma patients. *Scientific Reports*, *9*(1), 11125. https://doi.org/10.1038/s41598-019-47642-2.

Ewing, A. D., & Kazazian, H. H. (2010). High-throughput sequencing reveals extensive variation in human-specific L1 content in individual human genomes. *Genome Research*, *20*(9), 1262–1270. https://doi.org/10.1101/gr.106419.110.

Fazzari, M. J., & Greally, J. M. (2004). Epigenomics: Beyond CpG islands. *Nature Reviews Genetics*, *5*(6), 446–455. https://doi.org/10.1038/nrg1349.

Feinberg, J. I., Bakulski, K. M., Jaffe, A. E., Tryggvadottir, R., Brown, S. C., Goldman, L. R., … Feinberg, A.P. (2015). Paternal sperm DNA methylation associated with early signs of autism risk in an autism-enriched cohort. *International Journal of Epidemiology*, *44*(4), 1199–1210. https://doi.org/10.1093/ije/dyv028.

Felsenfeld, G. (2014). A brief history of epigenetics. *Cold Spring Harbor Perspectives in Biology*, *6*(1), a018200. https://doi.org/10.1101/cshperspect.a018200.

Feng, L., & Lou, J. (2019). DNA methylation analysis. *Methods in Molecular Biology*, *1894*, 181–227. https://doi.org/10.1007/978-1-4939-8916-4_12.

Fernandez, A. F., Valledor, L., Vallejo, F., Cañal, M. J., & Fraga, M. F. (2018). Quantification of global DNA methylation levels by mass spectrometry. *Methods in Molecular Biology*, *1708*, 49–58. https://doi.org/10.1007/978-1-4939-7481-8_3.

Fiorito, G., Polidoro, S., Dugué, P. A., Kivimaki, M., Ponzi, E., Matullo, G., … Vineis, P. (2017). Social adversity and epigenetic aging: A multi-cohort study on socioeconomic differences in peripheral blood DNA methylation. *Scientific Reports*, *7*(1), 16266. https://doi.org/10.1038/s41598-017-16391-5.

Fritz, A. J., Sehgal, N., Pliss, A., Xu, J., & Berezney, R. (2019). Chromosome territories and the global regulation of the genome. *Genes, Chromosomes and Cancer*, *58*(7), 407–426. https://doi.org/10.1002/gcc.22732.

Frommer, M., McDonald, L. E., Millar, D. S., Collis, C. M., Watt, F., Grigg, G. W., … Paul, C.L. (1992). A genomic sequencing protocol that yields a positive display of 5-methylcytosine residues in individual DNA strands. *Proceedings of the National Academy of Sciences of the United States of America*, *89*(5), 1827–1831. https://doi.org/10.1073/pnas.89.5.1827.

Fullwood, M. J., Liu, M. H., Pan, Y. F., Liu, J., Xu, H., Mohamed, Y. B., … Ruan, Y. (2009). An oestrogen-receptor-α-bound human chromatin interactome. *Nature*, *462*(7269), 58–64. https://doi.org/10.1038/nature08497.

Galanter, J. M., Gignoux, C. R., Oh, S. S., Torgerson, D., Pino-Yanes, M., Thakur, N., … Zaitlen, N. (2017). Differential methylation between ethnic sub-groups reflects the effect of genetic ancestry and environmental exposures. *eLife*, *6*(1), e20532. https://doi.org/10.7554/eLife.20532.

Gallardo-Gómez, M., Moran, S., Páez de la Cadena, M., Martínez-Zorzano, V. S., Rodríguez-Berrocal, F. J., Rodríguez-Girondo, M., … De Chiara, L. (2018). A new approach to epigenome-wide discovery of non-invasive methylation biomarkers for colorectal cancer screening in circulating cell-free DNA using pooled samples. *Clinical Epigenetics*, *10*(1), 53. https://doi.org/10.1186/s13148-018-0487-y.

Gambacurta, A., Merlini, G., Ruggiero, C., Diedenhofen, G., Battista, N., Bari, M., … Maccarrone, M. (2019). Human osteogenic differentiation in space: Proteomic and epigenetic clues to better understand osteoporosis. *Scientific Reports*, 9(1), 8343. https://doi.org/10.1038/s41598-019-44593-6.

Goll, M. G., Kirpekar, F., Maggert, K. A., Yoder, J. A., Hsieh, C. L., Zhang, X., … Bestor, T.H. (2006). Methylation of tRNAAsp by the DNA methyltransferase homolog Dnmt2. *Science*, 311(5759), 395–398. https://doi.org/10.1126/science.1120976.

Gomez, D., Shankman, L. S., Nguyen, A. T., & Owens, G. K. (2013). Detection of histone modifications at specific gene loci in single cells in histological sections. *Nature Methods*, 10(2), 171–177. https://doi.org/10.1038/nmeth.2332.

Gray, K. A., Yates, B., Seal, R. L., Wright, M. W., & Bruford, E. A. (2015). Genenames.org: The HGNC resources in 2015. *Nucleic Acids Research*, 43(1), D1079–D1085. https://doi.org/10.1093/nar/gku1071.

Greenberg, M. V. C., & Bourc'his, D. (2019). The diverse roles of DNA methylation in mammalian development and disease. *Nature Reviews Molecular Cell Biology*, 20(10), 590–607. https://doi.org/10.1038/s41580-019-0159-6.

Guo, P., Chen, W., Li, H., Li, M., & Li, L. (2018). The histone acetylation modifications of breast cancer and their therapeutic implications. *Pathology and Oncology Research*, 24(4), 807–813. https://doi.org/10.1007/s12253-018-0433-5.

Gupta, R., Nagarajan, A., & Wajapeyee, N. (2010). Advances in genome-wide DNA methylation analysis. *BioTechniques*, 49(4), iii–xi. https://doi.org/10.2144/000113493.

Haas, U., Sczakiel, G., & Laufer, S. D. (2012). MicroRNA-mediated regulation of gene expression is affected by disease-associated SNPs within the 3′-UTR via altered RNA structure. *RNA Biology*, 9(6), 924–937. https://doi.org/10.4161/rna.20497.

Hammond, S. M. (2015). An overview of microRNAs. *Advanced Drug Delivery Reviews*, 87, 3–14. https://doi.org/10.1016/j.addr.2015.05.001.

Hancks, D. C., & Kazazian, H. H. (2016). Roles for retrotransposon insertions in human disease. *Mobile DNA*, 7(1), 9. https://doi.org/10.1186/s13100-016-0065-9.

Hannum, G., Guinney, J., Zhao, L., Zhang, L., Hughes, G., Sadda, S. V., … Zhang, K. (2013). Genome-wide methylation profiles reveal quantitative views of human aging rates. *Molecular Cell*, 49(2), 359–367. https://doi.org/10.1016/j.molcel.2012.10.016.

Harrison, A., & Parle-McDermott, A. (2011). DNA methylation: A timeline of methods and applications. *Frontiers in Genetics*, 2(1), 74. https://doi.org/10.3389/fgene.2011.00074.

Head, J. A., Mittal, K., & Basu, N. (2014). Application of the LUminometric Methylation Assay to ecological species: Tissue quality requirements and a survey of DNA methylation levels in animals. *Molecular Ecology Resources*, 14(5), 943–952. https://doi.org/10.1111/1755-0998.12244.

Hernández, H. G., Tse, M. Y., Pang, S. C., Arboleda, H., & Forero, D. A. (2013). Optimizing methodologies for PCR-based DNA methylation analysis. *BioTechniques*, 55(4), 181–197. https://doi.org/10.2144/000114087.

Hildebrand, E. M., & Dekker, J. (2020). Mechanisms and functions of chromosome compartmentalization. *Trends in Biochemical Sciences*, 45(5), 385–396. https://doi.org/10.1016/j.tibs.2020.01.002.

Hnisz, D., Weintrau, A. S., Day, D. S., Valton, A. L., Bak, R. O., Li, C. H., … Young, R.A. (2016). Activation of proto-oncogenes by disruption of chromosome neighborhoods. *Science*, 351(6280), 1454–1458. https://doi.org/10.1126/science.aad9024.

Holliday, R., & Grigg, G. W. (1993). DNA methylation and mutation. *Mutation Research, Fundamental and Molecular Mechanisms of Mutagenesis*, 285(1), 61–67. https://doi.org/10.1016/0027-5107(93)90052-H.

Horsdal, H. T., Agerbo, E., McGrath, J. J., Vilhjálmsson, B. J., Antonsen, S., Closter, A. M., … Pedersen, C.B. (2019). Association of childhood exposure to nitrogen dioxide and polygenic risk score for schizophrenia with the risk of developing schizophrenia. *JAMA Network Open*, *2*(11), e1914401. https://doi.org/10.1001/jamanetworkopen.2019.14401.

Horvath, S. (2013). DNA methylation age of human tissues and cell types. *Genome Biology*, *14*(10), R115. https://doi.org/10.1186/gb-2013-14-10-r115.

Horvath, S., & Levine, A. J. (2015). HIV-1 infection accelerates age according to the epigenetic clock. *Journal of Infectious Diseases*, *212*(10), 1563–1573. https://doi.org/10.1093/infdis/jiv277.

Hota, S. K., & Bruneau, B. G. (2016). ATP-dependent chromatin remodeling during mammalian development. *Development (Cambridge, England)*, *143*(16), 2882–2897. https://doi.org/10.1242/dev.128892.

Housman, G., Quillen, E. E., & Stone, A. C. (2020). Intraspecific and interspecific investigations of skeletal DNA methylation and femur morphology in primates. *American Journal of Physical Anthropology*, *173*(1), 34–49. https://doi.org/10.1002/ajpa.24041.

Hoy, S. M. (2020). Tazemetostat: First approval. *Drugs*, *80*(5), 513–521. https://doi.org/10.1007/s40265-020-01288-x.

Hyun, K., Jeon, J., Park, K., & Kim, J. (2017). Writing, erasing and reading histone lysine methylations. *Experimental and Molecular Medicine*, *49*(4), E324. https://doi.org/10.1038/emm.2017.11.

Ichikawa, Y., & Saitoh, N. (2021). Shaping of genome by long noncoding RNAs. In T. Liehr (Ed.), *Cytogenomics* (pp. 357–372). Academic Press. Chapter 18 (in this book).

Iourov, I. Y., Vorsanova, S. G., & Yurov, Y. B. (2021). Cytogenomic landscape of the human brain. In T. Liehr (Ed.), *Cytogenomics* (pp. 327–348). Academic Press. Chapter 16 (in this book).

Irizarry, R. A., Ladd-Acosta, C., Wen, B., Wu, Z., Montano, C., Onyango, P., … Feinberg, A.P. (2009). The human colon cancer methylome shows similar hypo- and hypermethylation at conserved tissue-specific CpG island shores. *Nature Genetics*, *41*(2), 178–186. https://doi.org/10.1038/ng.298.

Ishii, T., Nagaki, K., & Houben, A. (2021). Application of CRISPR/Cas9 to visualize defined genomic sequences in fixed chromosomes and nuclei. In T. Liehr (Ed.), *Cytogenomics* (pp. 147–153). Academic Press. Chapter 9 (in this book).

Jacot, W., Lopez-Crapez, E., Mollevi, C., Boissière-Michot, F., Simony-Lafontaine, J., Ho-Pun-Cheung, A., … Guiu, S. (2020). BRCA1 promoter hypermethylation is associated with good prognosis and chemosensitivity in triple-negative breast cancer. *Cancers*, *12*(4), 828. https://doi.org/10.3390/cancers12040828.

Jäger, R., Migliorini, G., Henrion, M., Kandaswamy, R., Speedy, H. E., Heindl, A., … Houlston, R.S. (2015). Capture Hi-C identifies the chromatin interactome of colorectal cancer risk loci. *Nature Communications*, *6*(1), 6178. https://doi.org/10.1038/ncomms7178.

Javierre, B. M., Burren, O. S., Wilder, S. P., Kreuzhuber, R., Hill, S. M., Sewitz, S., … Fraser, P. (2016). Lineage-specific genome architecture links enhancers and non-coding disease variants to target gene promoters. *Cell*, *167*(5), 1369–1384.e19. https://doi.org/10.1016/j.cell.2016.09.037.

Jeusset, L., & McManus, K. (2019). Developing targeted therapies that exploit aberrant histone ubiquitination in cancer. *Cell*, *8*(2), 165. https://doi.org/10.3390/cells8020165.

Jin, S., Furtaw, M. D., Chen, H., Lamb, D. T., Ferguson, S. A., Arvin, N. E., … Kennedy, R.T. (2016). Multiplexed Western blotting using microchip electrophoresis. *Analytical Chemistry*, *88*(13), 6703–6710. https://doi.org/10.1021/acs.analchem.6b00705.

Jin, F., Li, Y., Dixon, J. R., Selvaraj, S., Ye, Z., Lee, A. Y., … Ren, B. (2013). A high-resolution map of the three-dimensional chromatin interactome in human cells. *Nature, 503*(7475), 290–294. https://doi.org/10.1038/nature12644.

Jin, Z., & Liu, Y. (2018). DNA methylation in human diseases. *Genes and Diseases, 5*(1), 1–8. https://doi.org/10.1016/j.gendis.2018.01.002.

Jones, P. A. (2012). Functions of DNA methylation: Islands, start sites, gene bodies and beyond. *Nature Reviews Genetics, 13*(7), 484–492. https://doi.org/10.1038/nrg3230.

Kappel, A., & Keller, A. (2017). MiRNA assays in the clinical laboratory: Workflow, detection technologies and automation aspects. *Clinical Chemistry and Laboratory Medicine, 55*(5), 636–647. https://doi.org/10.1515/cclm-2016-0467.

Karimi, M., Johansson, S., Stach, D., Corcoran, M., Grandér, D., Schalling, M., … Ekström, T. J. (2006). LUMA (LUminometric methylation assay)—A high throughput method to the analysis of genomic DNA methylation. *Experimental Cell Research, 312*(11), 1989–1995. https://doi.org/10.1016/j.yexcr.2006.03.006.

Karimi, M., Luttropp, K., & Ekström, T. J. (2011). Global DNA methylation analysis using the luminometric methylation assay. *Methods in Molecular Biology, 791*, 135–144. https://doi.org/10.1007/978-1-61779-316-5_11.

Kempfer, R., & Pombo, A. (2020). Methods for mapping 3D chromosome architecture. *Nature Reviews Genetics, 21*(4), 207–226. https://doi.org/10.1038/s41576-019-0195-2.

Kernaleguen, M., Daviaud, C., Shen, Y., Bonnet, E., Renault, V., Deleuze, J. F., … Tost, J. (2018). Whole-genome bisulfite sequencing for the analysis of genome-wide DNA methylation and hydroxymethylation patterns at single-nucleotide resolution. *Methods in Molecular Biology, 1767*, 311–349. https://doi.org/10.1007/978-1-4939-7774-1_18.

Khan, S., Ayub, H., Khan, T., & Wahid, F. (2019). MicroRNA biogenesis, gene silencing mechanisms and role in breast, ovarian and prostate cancer. *Biochimie, 167*, 12–24. https://doi.org/10.1016/j.biochi.2019.09.001.

Kimura, H. (2013). Histone modifications for human epigenome analysis. *Journal of Human Genetics, 58*(7), 439–445. https://doi.org/10.1038/jhg.2013.66.

Kingsley, S. L., Eliot, M. N., Whitsel, E. A., Huang, Y. T., Kelsey, K. T., Marsit, C. J., & Wellenius, G. A. (2016). Maternal residential proximity to major roadways, birth weight, and placental DNA methylation. *Environment International, 92–93*(1), 43–49. https://doi.org/10.1016/j.envint.2016.03.020.

Klengel, T., Mehta, D., Anacker, C., Rex-Haffner, M., Pruessner, J. C., Pariante, C. M., … Binder, E. B. (2013). Allele-specific FKBP5 DNA demethylation mediates gene-childhood trauma interactions. *Nature Neuroscience, 16*(1), 33–41. https://doi.org/10.1038/nn.3275.

Kling, T., Wenger, A., Beck, S., & Carén, H. (2017). Validation of the MethylationEPIC BeadChip for fresh-frozen and formalin-fixed paraffin-embedded tumours. *Clinical Epigenetics, 9*(1), 33. https://doi.org/10.1186/s13148-017-0333-7.

Kobayashi, W., & Kurumizaka, H. (2019). Structural transition of the nucleosome during chromatin remodeling and transcription. *Current Opinion in Structural Biology, 59*, 107–114. https://doi.org/10.1016/j.sbi.2019.07.011.

Kolanowska, M., Kubiak, A., Jażdżewski, K., & Wójcicka, A. (2018). MicroRNA analysis using next-generation sequencing. *Methods in Molecular Biology, 1823*, 87–101. https://doi.org/10.1007/978-1-4939-8624-8_8.

Kosumi, K., Yoshifumi, B., Kazuo, O., Taisuke, Y., Yuki, K., Naoya, Y., … Hideo, B. (2019). Tumor long-interspersed nucleotide element-1 methylation level and immune response to esophageal cancer. *Annals of Surgery*. https://doi.org/10.1097/sla.0000000000003264.

Koyama, M., & Kurumizaka, H. (2018). Structural diversity of the nucleosome. *Journal of Biochemistry*, *163*(2), 85–95. https://doi.org/10.1093/jb/mvx081.

Krueger, F., Kreck, B., Franke, A., & Andrews, S. R. (2012). DNA methylome analysis using short bisulfite sequencing data. *Nature Methods*, *9*(2), 145–151. https://doi.org/10.1038/nmeth.1828.

Ku, C. S., Naidoo, N., Wu, M., & Soong, R. (2011). Studying the epigenome using next generation sequencing. *Journal of Medical Genetics*, *48*(11), 721–730. https://doi.org/10.1136/jmedgenet-2011-100242.

Kuchiba, A., Iwasaki, M., Ono, H., Kasuga, Y., Yokoyama, S., Onuma, H., … Yoshida, T. (2014). Global methylation levels in peripheral blood leukocyte DNA by LUMA and breast cancer: A case-control study in Japanese women. *British Journal of Cancer*, *110*(11), 2765–2771. https://doi.org/10.1038/bjc.2014.223.

Kumar, S., Chinnusamy, V., & Mohapatra, T. (2018). Epigenetics of modified DNA bases: 5-Methylcytosine and beyond. *Frontiers in Genetics*, *9*(1), 640. https://doi.org/10.3389/fgene.2018.00640.

Kurdyukov, S., & Bullock, M. (2016). DNA methylation analysis: Choosing the right method. *Biology*, *5*(1), 3. https://doi.org/10.3390/biology5010003.

Ladd-Acosta, C., Aryee, J., Ordway, J. M., & Feinberg, A. P. (2010). Comprehensive high-throughput arrays for relative methylation (CHARM). *Current Protocols in Human Genetics*, *20*(65), 19. https://doi.org/10.1002/0471142905.hg2001s65.

Laufer, B. I., Hwang, H., Vogel Ciernia, A., Mordaunt, C. E., & LaSalle, J. M. (2019). Whole genome bisulfite sequencing of down syndrome brain reveals regional DNA hypermethylation and novel disorder insights. *Epigenetics*, *14*(7), 672–684. https://doi.org/10.1080/15592294.2019.1609867.

Lavasanifar, A., Sharp, C. N., Korte, E. A., Yin, T., Hosseinnejad, K., & Jortani, S. A. (2019). Long interspersed nuclear element-1 mobilization as a target in cancer diagnostics, prognostics and therapeutics. *Clinica Chimica Acta*, *493*, 52–62. https://doi.org/10.1016/j.cca.2019.02.015.

Lee, J. J., Geli, J., Larsson, C., Wallin, G., Karimi, M., Zedenius, J., … Foukakis, T. (2008). Gene-specific promoter hypermethylation without global hypomethylation in follicular thyroid cancer. *International Journal of Oncology*, *33*(4), 861–869. https://doi.org/10.3892/ijo_00000074.

Legoff, L., D'Cruz, S. C., Tevosian, S., Primig, M., & Smagulova, F. (2019). Transgenerational inheritance of environmentally induced epigenetic alterations during mammalian development. *Cell*, *8*(12), 1559. https://doi.org/10.3390/cells8121559.

Leti, F., Llaci, L., Malenica, I., & DiStefano, J. K. (2018). Methods for CPG methylation array profiling via bisulfite conversion. *Methods in Molecular Biology*, *1706*, 233–254. https://doi.org/10.1007/978-1-4939-7471-9_13.

Li, D., Wan, C., Bai, B., Cao, H., Liu, C., & Zhang, Q. (2019). Identification of histone acetylation markers in human fetal brains and increased H4K5ac expression in neural tube defects. *Molecular Genetics & Genomic Medicine*, *7*(12), e1002. https://doi.org/10.1002/mgg3.1002.

Liehr, T. (2021a). A definition for cytogenomics - Which also may be called chromosomics. In T. Liehr (Ed.), *Cytogenomics* (pp. 1–7). Academic Press. Chapter 1 (in this book).

Liehr, T. (2021b). Overview of currently available approaches used in cytogenomics. In T. Liehr (Ed.), *Cytogenomics* (pp. 11–24). Academic Press. Chapter 2 (in this book).

Liehr, T. (2021c). Nuclear architecture. In T. Liehr (Ed.), *Cytogenomics* (pp. 297–305). Academic Press. Chapter 14 (in this book).

Liehr, T. (2021d). Repetitive elements, heteromorphisms, and copy number variants. In T. Liehr (Ed.), *Cytogenomics* (pp. 373–388). Academic Press. Chapter 19 (in this book).

Liehr, T. (2021e). Molecular cytogenetics. In T. Liehr (Ed.), *Cytogenomics* (pp. 35–45). Academic Press. Chapter 4 (in this book).

Liou, S. H., Wu, W. T., Liao, H. Y., Chen, C. Y., Tsai, C. Y., Jung, W. T., & Lee, H. L. (2017). Global DNA methylation and oxidative stress biomarkers in workers exposed to metal oxide nanoparticles. *Journal of Hazardous Materials, 331*, 329–335. https://doi.org/10.1016/j.jhazmat.2017.02.042.

Liu, C., Xu, J., Chen, Y., Guo, X., Zheng, Y., Wang, Q., … Hou, L. (2015). Characterization of genome-wide H3K27ac profiles reveals a distinct PM2.5-associated histone modification signature. *Environmental Health, 14*(1), 65. https://doi.org/10.1186/s12940-015-0052-5.

Locke, W. J., Guanzon, D., Ma, C., Liew, Y. J., Duesing, K. R., Fung, K. Y. C., & Ross, J. P. (2019). DNA methylation cancer biomarkers: Translation to the clinic. *Frontiers in Genetics, 10*(1), 1150. https://doi.org/10.3389/fgene.2019.01150.

Luecken, M. D., & Theis, F. J. (2019). Current best practices in single-cell RNA-seq analysis: A tutorial. *Molecular Systems Biology, 15*(6), e8746. https://doi.org/10.15252/msb.20188746.

Machnik, M., & Oleksiewicz, U. (2020). Dynamic signatures of the epigenome: Friend or foe? *Cell, 9*(3), 653. https://doi.org/10.3390/cells9030653.

Mahmoodi Chalbatani, G., Dana, H., Gharagouzloo, E., Grijalvo, S., Eritja, R., Logsdon, C. D., … Marmari, V. (2019). Small interfering RNAs (siRNAs) in cancer therapy: A nano-based approach. *International Journal of Nanomedicine, 14*, 3111–3128. https://doi.org/10.2147/IJN.S200253.

Manvelyan, M., Friederike, H., Samarth, B., Kristin, M., Franck, P., Anja, W., … Thomas, L. (2008). Chromosome distribution in human sperm—A 3D multicolor banding-study. *Molecular Cytogenetics, 1*(1), 25. https://doi.org/10.1186/1755-8166-1-25.

Marinoni, I., Wiederkeher, A., Wiedmer, T., Pantasis, S., Di Domenico, A., Frank, R., … Perren, A. (2017). Hypo-methylation mediates chromosomal instability in pancreatic NET. *Endocrine-Related Cancer, 24*(3), 137–146. https://doi.org/10.1530/ERC-16-0554.

Martin, E. M., & Fry, R. C. (2018). Environmental influences on the epigenome: Exposure-associated DNA methylation in human populations. *Annual Review of Public Health, 39*, 309–333. https://doi.org/10.1146/annurev-publhealth-040617-014629.

Maunakea, A. K., Nagarajan, R. P., Bilenky, M., Ballinger, T. J., Dsouza, C., Fouse, S. D., … Costello, J.F. (2010). Conserved role of intragenic DNA methylation in regulating alternative promoters. *Nature, 466*(7303), 253–257. https://doi.org/10.1038/nature09165.

McCabe, M. T., Mohammad, H. P., Barbash, O., & Kruger, R. G. (2017). Targeting histone methylation in cancer. *Cancer Journal, 23*(5), 292–301. https://doi.org/10.1097/PPO.0000000000000283.

McCullough, L. E., Chen, J., Cho, Y. H., Khankari, N. K., Bradshaw, P. T., White, A. J., … Gammon, M.D. (2017). Modification of the association between recreational physical activity and survival after breast cancer by promoter methylation in breast cancer-related genes. *Breast Cancer Research, 19*(1), 19. https://doi.org/10.1186/s13058-017-0811-z.

McCullough, L. E., Chen, J., White, A. J., Xu, X., Cho, Y. H., Bradshaw, P. T., … Gammon, M.D. (2015). Global DNA methylation, measured by the luminometric methylation assay (LUMA), associates with postmenopausal breast cancer in non-obese and physically active women. *Journal of Cancer, 6*(6), 548–554. https://doi.org/10.7150/jca.11359.

Melas, P. A., Rogdaki, M., Ösby, U., Schalling, M., Lavebratt, C., & Ekström, T. J. (2012). Epigenetic aberrations in leukocytes of patients with schizophrenia: Association of global DNA methylation with antipsychotic drug treatment and disease onset. *FASEB Journal, 26*(6), 2712–2718. https://doi.org/10.1096/fj.11-202069.

Miller, G. E., Yu, T., Chen, E., & Brody, G. H. (2015). Self-control forecasts better psychosocial outcomes but faster epigenetic aging in low-SES youth. *Proceedings of the National Academy of Sciences of the United States of America*, *112*(33), 10325–10330. https://doi.org/10.1073/pnas.1505063112.

Mishra, M., Tiwari, S., & Gomes, A. V. (2017). Protein purification and analysis: Next generation western blotting techniques. *Expert Review of Proteomics*, *14*(11), 1037–1053. https://doi.org/10.1080/14789450.2017.1388167.

Mo, M. L., Ma, J., Chen, Z., Wei, B., Li, H., Zhou, Y., … Zhou, H.M. (2014). Measurement of genome-wide DNA methylation predicts survival benefits from chemotherapy in non-small cell lung cancer. *Journal of Cancer Research and Clinical Oncology*, *141*(5), 901–908. https://doi.org/10.1007/s00432-014-1860-7.

Moruzzi, S., Guarini, P., Udali, S., Ruzzenente, A., Guglielmi, A., Conci, S., … Friso, S. (2017). One-carbon genetic variants and the role of MTHFD1 1958G>A in liver and colon cancer risk according to global DNA methylation. *PLoS One*, *12*(10), e0185792. https://doi.org/10.1371/journal.pone.0185792.

Mota-Gómez, I., & Lupiáñez, D. G. (2019). A (3D-nuclear) space odyssey: Making sense of Hi-C maps. *Genes*, *10*(6), 415. https://doi.org/10.3390/genes10060415.

Nagalakshmi, U., Waern, K., & Snyder, M. (2010). RNA-Seq: A method for comprehensive transcriptome analysis. *Current Protocols in Molecular Biology*, *4*, 1–13. https://doi.org/10.1002/0471142727.mb0411s89.

Nagy, C., & Turecki, G. (2015). Transgenerational epigenetic inheritance: An open discussion. *Epigenomics*, *7*(5), 781–790. https://doi.org/10.2217/epi.15.46.

Nakagawa, T., & Okita, A. K. (2019). Transcriptional silencing of centromere repeats by heterochromatin safeguards chromosome integrity. *Current Genetics*, *65*(5), 1089–1098. https://doi.org/10.1007/s00294-019-00975-x.

Naumov, V. A., Generozov, E. V., Zaharjevskaya, N. B., Matushkina, D. S., Larin, A. K., Chernyshov, S. V., … Govorun, V.M. (2013). Genome-scale analysis of DNA methylation in colorectal cancer using Infinium HumanMethylation450 BeadChips. *Epigenetics*, *8*(9), 921–934. https://doi.org/10.4161/epi.25577.

Newbury, J. B., Arseneault, L., Beevers, S., Kitwiroon, N., Roberts, S., Pariante, C. M., … Fisher, H.L. (2019). Association of air pollution exposure with psychotic experiences during adolescence. *JAMA Psychiatry*, *76*(6), 614–623. https://doi.org/10.1001/jamapsychiatry.2019.0056.

Nwanaji-Enwerem, J. C., Elena, C., Trevisi, L., Kloog, I., Just, A. C., Shen, J., … Baccarelli, A.A. (2016). Long-term ambient particle exposures and blood DNA methylation age: Findings from the VA normative aging study. *Environmental Epigenetics*, *2*(2), dvw006. https://doi.org/10.1093/eep/dvw006.

O'Brien, J., Hayder, H., Zayed, Y., & Peng, C. (2018). Overview of microRNA biogenesis, mechanisms of actions, and circulation. *Frontiers in Endocrinology*, *9*(1), 402. https://doi.org/10.3389/fendo.2018.00402.

Olkhov-Mitsel, E., & Bapat, B. (2012). Strategies for discovery and validation of methylated and hydroxymethylated DNA biomarkers. *Cancer Medicine*, *1*(2), 237–260. https://doi.org/10.1002/cam4.22.

Önder, Ö., Sidoli, S., Carroll, M., & Garcia, B. A. (2015). Progress in epigenetic histone modification analysis by mass spectrometry for clinical investigations. *Expert Review of Proteomics*, *12*(5), 499–517. https://doi.org/10.1586/14789450.2015.1084231.

Palla, V. V., Karaolanis, G., Katafigiotis, I., Anastasiou, I., Patapis, P., Dimitroulis, D., & Perrea, D. (2017). gamma-H2AX: Can it be established as a classical cancer prognostic factor? *Tumor Biology*, *39*(3), 1–11. https://doi.org/10.1177/1010428317695931.

Pauwels, S., Ghosh, M., Duca, R. C., Bekaert, B., Freson, K., Huybrechts, I., ... Godderis, L. (2017). Dietary and supplemental maternal methyl-group donor intake and cord blood DNA methylation. *Epigenetics*, *12*(1), 1–10. https://doi.org/10.1080/15592294.2016. 1257450.

Pellestor, F., Gaillard, J.-B., Schneider, A., Puechberty, J., & Gatinois, V. (2021). Chromoanagenesis phenomena and their formation mechanisms. In T. Liehr (Ed.), *Cytogenomics* (pp. 213–245). Academic Press. Chapter 11 (in this book).

Perri, A. M., Agosti, V., Olivo, E., Concolino, A., Angelis, M. T. D., Tammè, L., ... Scumaci, D. (2019). Histone proteomics reveals novel post-translational modifications in breast cancer. *Aging*, *11*(23), 11722–11755. https://doi.org/10.18632/aging.102577.

Philippeos, C., Telerman, S. B., Oulès, B., Pisco, A. O., Shaw, T. J., Elgueta, R., ... Watt, F.M. (2018). Spatial and single-cell transcriptional profiling identifies functionally distinct human dermal fibroblast subpopulations. *Journal of Investigative Dermatology*, *138*(4), 811–825. https://doi.org/10.1016/j.jid.2018.01.016.

Pidsley, R., Zotenko, E., Peters, T. J., Lawrence, M. G., Risbridger, G. P., Molloy, P., ... Clark, S.J. (2016). Critical evaluation of the Illumina MethylationEPIC BeadChip microarray for whole-genome DNA methylation profiling. *Genome Biology*, *17*(1), 208. https://doi. org/10.1186/s13059-016-1066-1.

Piyathilake, C. J., Macaluso, M., Alvarez, R. D., Chen, M., Badiga, S., Siddiqui, N. R., ... Johanning, G.L. (2011). A higher degree of LINE-1 methylation in peripheral blood mononuclear cells, a one-carbon nutrient related epigenetic alteration, is associated with a lower risk of developing cervical intraepithelial neoplasia. *Nutrition*, *27*(5), 513–519. https://doi. org/10.1016/j.nut.2010.08.018.

Plusquin, M., Guida, F., Polidoro, S., Vermeulen, R., Raaschou-Nielsen, O., Campanella, G., ... Chadeau-Hyam, M. (2017). DNA methylation and exposure to ambient air pollution in two prospective cohorts. *Environment International*, *108*, 127–136. https://doi. org/10.1016/j.envint.2017.08.006.

Pombo, A., & Dillon, N. (2015). Three-dimensional genome architecture: Players and mechanisms. *Nature Reviews Molecular Cell Biology*, *16*(4), 245–257. https://doi.org/10.1038/ nrm3965.

Portela, A., & Esteller, M. (2010). Epigenetic modifications and human disease. *Nature Biotechnology*, *28*(10), 1057–1068. https://doi.org/10.1038/nbt.1685.

Poulin, M., Zhou, J.Y., Yan, L., & Shioda, T. (2018). Pyrosequencing methylation analysis. *Methods in Molecular Biology*, *1856*, 283–296. https://doi.org/10.1007/978-1-4939-8751-1_17.

Pritchard, C. C., Cheng, H. H., & Tewari, M. (2012). MicroRNA profiling: Approaches and considerations. *Nature Reviews Genetics*, *13*(5), 358–369. https://doi.org/10.1038/nrg3198.

Quinodoz, S. A., Ollikainen, N., Tabak, B., Palla, A., Schmidt, J. M., Detmar, E., ... Guttman, M. (2018). Higher-order inter-chromosomal hubs shape 3D genome organization in the nucleus. *Cell*, *174*(3), 744–757.e24. https://doi.org/10.1016/j.cell.2018.05.024.

Ramalho-Carvalho, J., Henrique, R., & Jerónimo, C. (2018). Methylation-specific PCR. *Methods in Molecular Biology*, *1708*, 447–472. https://doi.org/10.1007/978-1-4939-7481-8_23.

Rasool, M., Malik, A., Naseer, M. I., Manan, A., Ansari, S. A., Begum, I., ... Gan, S.H. (2015). The role of epigenetics in personalized medicine: Challenges and opportunities. *BMC Genomics*, *8*(1), S5. https://doi.org/10.1186/1755-8794-8-S1-S5.

Rechache, N. S., Wang, Y., Stevenson, H. S., Killian, J. K., Edelman, D. C., Merino, M., ... Kebebew, E. (2012). DNA methylation profiling identifies global methylation differences and markers of adrenocortical tumors. *Journal of Clinical Endocrinology and Metabolism*, *97*(6), E1004–E1013. https://doi.org/10.1210/jc.2011-3298.

Redon, C. E., Weyemi, U., Parekh, P. R., Huang, D., Burrell, A. S., & Bonner, W. M. (2012). γ-H2AX and other histone post-translational modifications in the clinic. *Biochimica et Biophysica Acta, Gene Regulatory Mechanisms*, 743–756. https://doi.org/10.1016/j.bbagrm.2012.02.021.

Ribeiro Ferreira, I., dos Santos Cunha, W. D., Henrique Ferreira Gomes, L., Azevedo Cintra, H., Cabral Guimarães Fonseca, L. L., Ferreira Bastos, E., … da Cunha Guida, L. (2019). A rapid and accurate methylation-sensitive high-resolution melting analysis assay for the diagnosis of Prader Willi and Angelman patients. *Molecular Genetics & Genomic Medicine*, *7*(6), e637. https://doi.org/10.1002/mgg3.637.

Rider, C. F., & Carlsten, C. (2019). Air pollution and DNA methylation: Effects of exposure in humans. *Clinical Epigenetics*, *11*(1), 131. https://doi.org/10.1186/s13148-019-0713-2.

Rishi, V., Bhattacharya, P., Chatterjee, R., Rozenberg, J., Zhao, J., Glass, K., … Vinson, C. (2010). CpG methylation of half-CRE sequences creates C/EBPα binding sites that activate some tissue-specific genes. *Proceedings of the National Academy of Sciences of the United States of America*, *107*(47), 20311–20316. https://doi.org/10.1073/pnas.1008688107.

Robertson, K. D. (2002). DNA methylation and chromatin—Unraveling the tangled web. *Oncogene*, *21*(35), 5361–5379. https://doi.org/10.1038/sj.onc.1205609.

Romano, G., Veneziano, D., Acunzo, M., & Croce, C. M. (2017). Small non-coding RNA and cancer. *Carcinogenesis*, *38*(5), 485–491. https://doi.org/10.1093/carcin/bgx026.

Rothbart, S. B., Dickson, B. M., Raab, J. R., Grzybowski, A. T., Krajewski, K., Guo, A. H., … Strahl, B.D. (2015). An interactive database for the assessment of histone antibody specificity. *Molecular Cell*, *59*(3), 502–511. https://doi.org/10.1016/j.molcel.2015.06.022.

Rumbaugh, G., & Miller, C. A. (2011). Epigenetic changes in the brain: Measuring global histone modifications. *Methods in Molecular Biology*, *670*, 263–274. https://doi.org/10.1007/978-1-60761-744-0_18.

Sant, K., & Goodrich, J. (2018). Methods for analysis of DNA methylation. *Toxicoepigenetics. Core principles and applications*. Elsevier.

Sarno, F., Nebbioso, A., & Altucci, L. (2020). DOT1L: A key target in normal chromatin remodelling and in mixed-lineage leukaemia treatment. *Epigenetics*, *15*(5), 439–453. https://doi.org/10.1080/15592294.2019.1699991.

Sarrate, Z., Solé, M., Vidal, F., Anton, E., & Blanco, J. (2018). Chromosome positioning and male infertility: It comes with the territory. *Journal of Assisted Reproduction and Genetics*, *35*(11), 1929–1938. https://doi.org/10.1007/s10815-018-1313-3.

Sati, S., & Cavalli, G. (2017). Chromosome conformation capture technologies and their impact in understanding genome function. *Chromosoma*, *126*(1), 33–44. https://doi.org/10.1007/s00412-016-0593-6.

Schmidt, A., Zhang, H., & Cardoso, M. C. (2020). MeCP2 and chromatin compartmentalization. *Cell*, *9*(4), 878. https://doi.org/10.3390/cells9040878.

Seo, J.-Y., Yoon-Jung, P., Young-Ah, Y., Ji-Yun, H., In-Bog, L., Byeong-Hoon, C., … Deog-Gyu, S. (2015). Epigenetics: General characteristics and implications for oral health. *Restorative Dentistry & Endodontics*, *40*(1), 14–22. https://doi.org/10.5395/rde.2015.40.1.14.

Shanbhag, N. M., Evans, M. D., Mao, W., Nana, A. L., Seeley, W. W., Adame, A., … Mucke, L. (2019). Early neuronal accumulation of DNA double strand breaks in Alzheimer's disease. *Acta Neuropathologica Communications*, *7*(1), 77. https://doi.org/10.1186/s40478-019-0723-5.

Sidoli, S., Cheng, L., & Jensen, O. N. (2012). Proteomics in chromatin biology and epigenetics: Elucidation of post-translational modifications of histone proteins by mass spectrometry. *Journal of Proteomics*, *75*(12), 3419–3433. https://doi.org/10.1016/j.jprot.2011.12.029.

Sidoli, S., Kori, Y., Lopes, M., Yuan, Z. F., Kim, H. J., Kulej, K., … Garcia, B.A. (2019). One minute analysis of 200 histone posttranslational modifications by direct injection mass spectrometry. *Genome Research*, *29*(6), 978–987. https://doi.org/10.1101/gr.247353.118.

Simithy, J., Sidoli, S., & Garcia, B. A. (2018). Integrating proteomics and targeted metabolomics to understand global changes in histone modifications. *Proteomics*, *18*(18), e1700309. https://doi.org/10.1002/pmic.201700309.

Singer-Sam, J., LeBon, J., Tanguay, R., & Riggs, A. (1990). A quantitative HpaII-PCR assay to measure methylation of DNA from a small number of cells. *Nucleic Acids Research*, *18*(3), 687. https://doi.org/10.1093/nar/18.3.687.

Sørensen, A. L., & Collas, P. (2009). Immunoprecipitation of methylated DNA. *Methods in Molecular Biology*, *567*, 249–262. https://doi.org/10.1007/978-1-60327-414-2_16.

Stark, R., Grzelak, M., & Hadfield, J. (2019). RNA sequencing: The teenage years. *Nature Reviews Genetics*, *20*(11), 631–656. https://doi.org/10.1038/s41576-019-0150-2.

Statkiewicz, M., Maryan, N., Kulecka, M., Kuklinska, U., Ostrowski, J., & Mikula, M. (2019). Functional analyses of a low-penetrance risk variant rs6702619/1p21.2 associating with colorectal cancer in polish population. *Acta Biochimica Polonica*, *66*(3), 305–313. https://doi.org/10.18388/abp.2019_2775.

Staunstrup, N. H., Starnawska, A., Nyegaard, M., Christiansen, L., Nielsen, A. L., Børglum, A., & Mors, O. (2016). Genome-wide DNA methylation profiling with MeDIP-seq using archived dried blood spots. *Clinical Epigenetics*, *8*(1), 81. https://doi.org/10.1186/s13148-016-0242-1.

Strickfaden, H., & Hendzel, M. J. (2017). Immunofluorescence of histone proteins. *Methods in Molecular Biology*, *1528*, 165–171. https://doi.org/10.1007/978-1-4939-6630-1_10.

Stumpel, D. J. P. M., Schneider, P., Van Roon, E. H. J., Pieters, R., & Stam, R. W. (2013). Absence of global hypomethylation in promoter hypermethylated mixed lineage leukaemia-rearranged infant acute lymphoblastic leukaemia. *European Journal of Cancer*, *49*(1), 175–184. https://doi.org/10.1016/j.ejca.2012.07.013.

Suchiman, H. E. D., Slieker, R. C., Kremer, D., Slagboom, P. E., Heijmans, B. T., & Tobi, E. W. (2015). Design, measurement and processing of region-specific DNA methylation assays: The mass spectrometry-based method EpiTYPER. *Frontiers in Genetics*, *6*(1), 287. https://doi.org/10.3389/fgene.2015.00287.

Suzuki, M., Liao, W., Wos, F., Johnston, A. D., DeGrazia, J., Ishii, J., … Greally, J.M. (2018). Whole-genome bisulfite sequencing with improved accuracy and cost. *Genome Research*, *28*(9), 1364–1371. https://doi.org/10.1101/gr.232587.117.

Szelenberger, R., Kacprzak, M., Saluk-Bijak, J., Zielinska, M., & Bijak, M. (2019). Plasma MicroRNA as a novel diagnostic. *Clinica Chimica Acta*, *499*, 98–107. https://doi.org/10.1016/j.cca.2019.09.005.

Tang, Y., Zheng, S. J., Qi, C. B., Feng, Y. Q., & Yuan, B. F. (2015). Sensitive and simultaneous determination of 5-methylcytosine and its oxidation products in genomic DNA by chemical derivatization coupled with liquid chromatography-tandem mass spectrometry analysis. *Analytical Chemistry*, *87*(6), 3445–3452. https://doi.org/10.1021/ac504786r.

Tao, M. H., Zhou, J., Rialdi, A. P., Martinez, R., Dabek, J., Scelo, G., … Boffetta, P. (2014). Indoor air pollution from solid fuels and peripheral blood DNA methylation: Findings from a population study in Warsaw, Poland. *Environmental Research*, *134*, 325–330. https://doi.org/10.1016/j.envres.2014.08.017.

Tellez-Plaza, M., Tang, W. Y., Shang, Y., Umans, J. G., Francesconi, K. A., Goessler, W., … Navas-Acien, A. (2014). Association of global DNA methylation and global DNA hydroxymethylation with metals and other exposures in human blood DNA samples. *Environmental Health Perspectives*, *122*(9), 946–954. https://doi.org/10.1289/ehp.1306674.

Tollefsbol, T. O. (2004). Methods of epigenetic analysis. *Methods in Molecular Biology, 287*(1), 1–8. https://doi.org/10.1385/1-59259-828-5:001.

Tost, J. (2016). Current and emerging technologies for the analysis of the genome-wide and locus-specific DNA methylation patterns. *Advances in Experimental Medicine and Biology, 945*, 343–430. https://doi.org/10.1007/978-3-319-43624-1_15.

Tweedie-Cullen, R. Y., Reck, J. M., & Mansuy, I. M. (2009). Comprehensive mapping of post-translational modifications on synaptic, nuclear, and histone proteins in the adult mouse brain. *Journal of Proteome Research, 8*(11), 4966–4982. https://doi.org/10.1021/pr9003739.

Tyagi, M., Imam, N., Verma, K., & Patel, A. K. (2016). Chromatin remodelers: We are the drivers!! *Nucleus, 7*(4), 388–404. https://doi.org/10.1080/19491034.2016.1211217.

Ungelenk, M. (2021). Sequencing approaches. In T. Liehr (Ed.), *Cytogenomics* (pp. 87–122). Academic Press. Chapter 7 (in this book).

Urbanek, M. O., Nawrocka, A. U., & Krzyzosiak, W. J. (2015). Small RNA detection by in situ hybridization methods. *International Journal of Molecular Sciences, 16*(6), 13259–13286. https://doi.org/10.3390/ijms160613259.

Van Dijk, E. L., Jaszczyszyn, Y., & Thermes, C. (2014). Library preparation methods for next-generation sequencing: Tone down the bias. *Experimental Cell Research, 322*(1), 12–20. https://doi.org/10.1016/j.yexcr.2014.01.008.

Van Otterdijk, S. D., & Michels, K. B. (2016). Transgenerational epigenetic inheritance in mammals: How good is the evidence? *FASEB Journal, 30*(7), 2457–2465. https://doi.org/10.1096/fj.201500083.

Varinli, H., Statham, A. L., Clark, S. J., Molloy, P. L., & Ross, J. P. (2015). COBRA-seq: Sensitive and quantitative methylome profiling. *Genes, 6*(4), 1140–1163. https://doi.org/10.3390/genes6041140.

Vryer, R., & Saffery, R. (2017). What's in a name? Context-dependent significance of 'global' methylation measures in human health and disease. *Clinical Epigenetics, 9*(1), 2. https://doi.org/10.1186/s13148-017-0311-0.

Wang, K. C., & Chang, H. Y. (2018). Epigenomics technologies and applications. *Circulation Research, 122*(9), 1191–1199. https://doi.org/10.1161/CIRCRESAHA.118.310998.

Wang, J., Dean, D. C., Hornicek, F. J., Shi, H., & Duan, Z. (2019). RNA sequencing (RNA-Seq) and its application in ovarian cancer. *Gynecologic Oncology, 152*(1), 194–201. https://doi.org/10.1016/j.ygyno.2018.10.002.

Wani, K., & Aldape, K. D. (2016). PCR techniques in characterizing DNA methylation. *Methods in Molecular Biology, 1392*, 177–186. https://doi.org/10.1007/978-1-4939-3360-0_16.

Wei, J. W., Huang, K., Yang, C., & Kang, C. S. (2017). Non-coding RNAs as regulators in epigenetics (review). *Oncology Reports, 37*(1), 3–9. https://doi.org/10.3892/or.2016.5236.

Weise, A., & Liehr, T. (2021a). Interchromosomal interactions with meaning for disease. In T. Liehr (Ed.), *Cytogenomics* (pp. 349–356). Academic Press. Chapter 17 (in this book).

Weise, A., & Liehr, T. (2021b). Molecular karyotyping. In T. Liehr (Ed.), *Cytogenomics* (pp. 73–85). Academic Press. Chapter 6 (in this book).

Wiehle, L., & Breiling, A. (2016). Chromatin immunoprecipitation. *Methods in Molecular Biology, 1480*(1), 7–21. https://doi.org/10.1007/978-1-4939-6380-5_2.

Wilson, G. A., Dhami, P., Feber, A., Cortázar, D., Suzuki, Y., Schulz, R., … Beck, S. (2012). Resources for methylome analysis suitable for gene knockout studies of potential epigenome modifiers. *GigaScience, 1*(1), 3. https://doi.org/10.1186/2047-217X-1-3.

Wreczycka, K., Gosdschan, A., Yusuf, D., Grüning, B., Assenov, Y., & Akalin, A. (2017). Strategies for analyzing bisulfite sequencing data. *Journal of Biotechnology, 261*, 105–115. https://doi.org/10.1016/j.jbiotec.2017.08.007.

Xu, J., Fu, G., Yan, L., Craig, J. M., Zhang, X., Fu, L., … Cong, B. (2015). LINE-1 DNA methylation: A potential forensic marker for discriminating monozygotic twins. *Forensic Science International: Genetics*, *19*, 136–145. https://doi.org/10.1016/j.fsigen.2015.07.014.

Yalcin, A., Kreutz, C., Pfeifer, D., Abdelkarim, M., Klaus, G., Timmer, J., … Hackanson, B. (2013). MeDIP coupled with a promoter tiling array as a platform to investigate global DNA methylation patterns in AML cells. *Leukemia Research*, *37*(1), 102–111. https://doi.org/10.1016/j.leukres.2012.09.014.

Yang, A. S., Estécio, M. R. H., Doshi, K., Kondo, Y., Tajara, E. H., & Issa, J. P. J. (2004). A simple method for estimating global DNA methylation using bisulfite PCR of repetitive DNA elements. *Nucleic Acids Research*, *32*(3), e38. https://doi.org/10.1093/nar/gnh032.

Yang, G., Komaki, Y., Yoshida, I., & Ibuki, Y. (2019). Formaldehyde inhibits UV-induced phosphorylation of histone H2AX. *Toxicology In Vitro*, *61*(1), 104687. https://doi.org/10.1016/j.tiv.2019.104687.

Ye, J., Xu, M., Tian, X., Cai, S., & Zeng, S. (2019). Research advances in the detection of miRNA. *Journal of Pharmaceutical Analysis*, *9*(4), 217–226. https://doi.org/10.1016/j.jpha.2019.05.004.

Yin, Y., Morgunova, E., Jolma, A., Kaasinen, E., Sahu, B., Khund-Sayeed, S., … Taipale, J. (2017). Impact of cytosine methylation on DNA binding specificities of human transcription factors. *Science*, *356*(6337), eaaj2239. https://doi.org/10.1126/science.aaj2239.

Yumiceba, V., Souto Melo, U., & Spielmann, M. (2021). 3D cytogenomics: Structural variation in the three-dimensional genome. In T. Liehr (Ed.), *Cytogenomics* (pp. 247–266). Academic Press. Chapter 12 (in this book).

Zeggar, H. R., How-Kit, A., Daunay, A., Bettaieb, I., Sahbatou, M., Rahal, K., … Kharrat, M. (2020). Tumor DNA hypomethylation of LINE-1 is associated with low tumor grade of breast cancer in Tunisian patients. *Oncology Letters*, *20*(2), 1999–2006. https://doi.org/10.3892/ol.2020.11745.

Zentner, G. E., & Henikoff, S. (2013). Regulation of nucleosome dynamics by histone modifications. *Nature Structural and Molecular Biology*, *20*(3), 259–266. https://doi.org/10.1038/nsmb.2470.

Zhang, F. F., Cardarelli, R., Carroll, J., Fulda, K. G., Kaur, M., Gonzalez, K., … Morabia, A. (2011). Significant differences in global genomic DNA methylation by gender and race/ethnicity in peripheral blood. *Epigenetics*, *6*(5), 623–629. https://doi.org/10.4161/epi.6.5.15335.

Zhang, Y., Naguro, I., & Herr, A. E. (2019a). In situ single-cell Western blot on adherent cell culture. *Angewandte Chemie International Edition*, *58*(39), 13929–13934. https://doi.org/10.1002/anie.201906920.

Zhang, P., Torres, K., Liu, X., Liu, C. G., & Pollock, R. E. (2016). An overview of chromatin-regulating proteins in cells. *Current Protein and Peptide Science*, *17*(5), 401–410. https://doi.org/10.2174/1389203717666160122120310.

Zhang, P., Wu, W., Chen, Q., & Chen, M. (2019b). Non-coding RNAs and their integrated networks. *Journal of Integrative Bioinformatics*, *16*(3), 20190027. https://doi.org/10.1515/jib-2019-0027.

Zheng, Y., Sanchez-Guerra, M., Zhang, Z., Joyce, B. T., Zhong, J., Kresovich, J. K., … Hou, L. (2017). Traffic-derived particulate matter exposure and histone H3 modification: A repeated measures study. *Environmental Research*, *153*, 112–119. https://doi.org/10.1016/j.envres.2016.11.015.

Zheng, M., Tian, S. Z., Capurso, D., Kim, M., Maurya, R., Lee, B., … Ruan, Y. (2019). Multiplex chromatin interactions with single-molecule precision. *Nature*, *566*(7745), 558–562. https://doi.org/10.1038/s41586-019-0949-1.

Zhou, J., Cao, J., Lu, Z., Liu, H., & Deng, D. (2011). A 115-bp MethyLight assay for detection of p16 (CDKN2A) methylation as a diagnostic biomarker in human tissues. *BMC Medical Genetics*, *12*(1), 67. https://doi.org/10.1186/1471-2350-12-67.

Zhou, G., Parfett, C., Cummings-Lorbetskie, C., Xiao, G. H., & Desaulniers, D. (2017). Two-color fluorescent cytosine extension assay for the determination of global DNA methylation. *BioTechniques*, *62*(4), 157–164. https://doi.org/10.2144/000114533.

Current cytogenomic research

Chromoanagenesis phenomena and their formation mechanisms

11

Franck Pellestor[a,b], Jean-Baptiste Gaillard[a], Anouck Schneider[a], Jacques Puechberty[a], and Vincent Gatinois[a,b]

[a]*Unit of Chromosomal Genetics and Research Platform Chromostem, Department of Medical Genetics, Arnaud de Villeneuve Hospital, Montpellier, France*
[b]*INSERM 1183 Unit "Genome and Stem Cell Plasticity in Development and Aging" Institute of Regenerative Medicine and Biotherapies, St Eloi Hospital, Montpellier, France*

Chapter outline

Introduction

While the identification and classification of complex chromosomal rearrangements, as well as their mechanisms of formation and transmission, seemed to be well analyzed and understood (Madan, 2012; Pellestor et al., 2011), the emergence of innovative genome sequencing technologies and the advances in computational biology have led to the identification of new types of chromosomal rearrangements being much more complex and massive than what had been imagined until then. These unanticipated complex chromosomal phenomena are termed chromothripsis, chromoanasynthesis, and chromoplexy. They have been grouped under the name of chromoanagenesis (Holland & Cleveland, 2012), meaning chromosome rebirth or creation of new chromosomes. These phenomena are new emerged fields of cytogenomic research (Liehr, 2021a, 2021b).

Cytogenomics. https://doi.org/10.1016/B978-0-12-823579-9.00009-6

Accumulation of sequencing data (Ungelenk, 2021) and information drawn from experimental models have gradually made it possible to better understand the underlying mechanisms of these phenomena and their effective impact both in cancer and congenital disorders (Kloosterman & Cuppen, 2013). Thus, in cancer, the paradigm that genome alteration occurs gradually through the progressive accumulation of mutational events has been challenged by the observations of chromoanagenesis events in a broad spectrum of tumors (Kloosterman et al., 2014). In congenital disorders, the notion of inheritance and viability of massive chromosomal rearrangements had to be reconsidered following the identification of chromoanagenesis-related rearrangements, not only in numerous patients with developmental disorders, but also in phenotypically normal subjects (de Pagter et al., 2015). In addition, the reanalysis of many chromosomal structural abnormalities considered as "simple" balanced rearrangements, revealed an unsuspected complexity in connection with chromoanagenesis (Chiang et al., 2012; Weckselblatt et al., 2015).

Both the complexity and the diversity of chromoanagenesis events raise important questions on the stability of cell genome, as well as on the capacity of cells to manage such crisis (Fukami & Kurahashi, 2018), and on a potential driving role of chromoanagenesis in species evolution (Pellestor & Gatinois, 2020). The discovery of these new forms of massive chromosomal rearrangements in human pathologies (Baranov & Kuznetzova, 2021) has generated a wave of interest for genomic plasticity and the role of genome maintenance pathways, especially because similar catastrophic phenomena have been described not only in other mammalian species like gibbons (Carbone et al., 2014; Meyer et al., 2016) or rodents (Romanenko et al., 2017) but also in plants (Carbonell-Bejerano et al., 2017; Henry et al., 2018; Tan et al., 2015), plankton (Blanc-Mathieu et al., 2017), nematode *Caenorhabditis elegans* (Itani et al., 2016), and *Saccharomyces cerevisiae* (Anand et al., 2014), indicating that the cellular pathways responsible for generating such massive patterns of chromosomal rearrangements are highly conserved.

An overview of chromoanagenesis phenomena: Definitions and hallmarks

Although the biological consequences of these three chaotic chromosome phenomena are close with the formation of derivative chromosomes highly remodeled, their molecular mechanisms differ. Specific features have been described, allowing each catastrophic process to be distinguished from each other and from other types of complex genomic alterations.

Chromothripsis

First described in cancers (Stephens et al., 2011) and then in congenital disorders (Kloosterman et al., 2011), chromothripsis (for chromosome breaking into small pieces) is defined as a relatively clustered chromosomal shattering followed by a random restitching of chromosomal fragments, resulting in the formation of complex genomic rearrangements. Since its discovery, chromothripsis-like events have been observed in a wide range of cancers (Cai et al., 2014; Forment et al., 2012; Hadi et al., 2020; Luijten et al., 2018) and in patients harboring congenital and developmental disorders (Anderson et al., 2016; Collins et al., 2017) or apparently balanced simple rearrangements (Kloosterman & Cuppen, 2013; Weckselblatt et al., 2015), as well as in asymptomatic subject (de Pagter et al., 2015). Familial studies showed that derivative chromothriptic chromosomes can be stably inherited (Bertelsen et al., 2016), and some reports documented the possible reversibility of chromothripsis (Bassaganyas et al., 2013) and its potential curative effect (McDermott et al., 2015).

Remarkably, the genomic chaotic alterations that characterize chromothripsis appear to arise in a single cellular event, and all reported chromothripsis-mediated rearrangements share similar patterns. Consequently, a set of common hallmarks has been described consistent with the proposition of a single catastrophic event scenario. These key features are (i) the generation of numerous genomic rearrangements clustered in one or a few chromosomal loci, (ii) the restitching of fragments without order and preferential orientation, (iii) the lack of any sequence homology at their breakpoints or only microhomology of a few nucleotides, (iv) the low DNA copy number change, and (v) the preservation of heterozygosity in the rearranged chromosomal segments. Altogether, these criteria allowed to establish a molecular signature of chromothripsis (Korbel & Campbell, 2013), and these have been used to develop bioinformatic tools for the identification and the annotation of chromothripsis-like pattern in large-scale genomic data. CTLPScanner (Yang et al., 2016) and ShatterProof (Govind et al., 2014) are examples of such software accessible through web browsers.

In cancer cells, chromothripsis events often involve tens to hundreds of inter- and intrachromosomal rearrangements. The phenomenon was observed in 2% to 3% of all human cancers, with a particularly high incidence in bone cancers and glioblastoma (up to 39%) (Cai et al., 2014). Recent estimates report frequencies in the order of 45% for all cancers, indicating how the prevalence of chromothripsis in cancer may have been underestimated (Cortés-Ciriano et al., 2020; Voronina et al., 2020). In all cases, chromothripsis is associated with aggressive cancer-forms and poor patient survival (Kloosterman et al., 2014; Lee et al., 2017). A feature of chromothripsis in tumors is the formation of neochromosomes, giant extrachromosomes found in 3% of cancers (Papenfuss & Thomas, 2015), or the generation of circular, extra double minute chromosome markers that often involve amplified oncogenes (Fontana et al., 2018; Ly et al., 2017). Constitutional chromothripsis usually results in less complicated alterations than in cancers. Accumulative data indicated that chromothripsis occurs approximately in 5% of case with karyotypically identified chromosomal rearrangements (Redin et al., 2017), and that most reported de novo chromothripsis events are of paternal origin, suggesting that chromothripsis takes place predominantly in male meiosis (Fukami et al., 2017; Pellestor, 2014).

Analyses of breakpoint junction sequences indicate that the shattered chromosomal fragments are reassembled primarily by classical nonhomologous end joining (c-NHEJ) or alternative form of end joining (alt-EJ), both error-prone repair system operating in all phases of the cell cycle (Iliakis et al., 2015; Willis et al., 2015). Some other processes of DNA repair may also operate in chromothripsis. Masset et al. (2016) described three cases of chromothripsis-like rearrangements with multiple focalized duplications and insertions mediated by a DNA polymerase Polθ-dependent pathway of alternative NHEJ, which may create aberrant end-to-end fusion of multiple chromosomes (Mateos-Gomez et al., 2015).

High-resolution investigations have revealed additional complexity in chromothripsis such as the complexification of an initial benign chromosomal alteration in chromothripsis through unequal crossing over during meiosis (Pettersson et al., 2018), or the extension of remodeling events to nonchomothriptic chromosomes, wherein submicroscopic insertions of shattered fragments occur (Kurtas et al., 2019). Also, it is possible that constitutional chromothripsis and multiple de novo CNVs (copy number variations) could occur in the genome as the result of the same multifocal crisis events (Brás et al., 2020), raising the possibility of "CNV mutator" phenotypes (Hattori et al., 2019; Liu et al., 2017).

In this context, Oesper et al. (2018) proposed a variation on the initial signature, termed the H/T alternating fraction, allowing to overcome some of the limitations of the initial signature and to provide a more precise method for identifying genome rearrangements as simultaneous or sequential.

Chromoplexy

Chromoplexy (for chromosome restructuring) constitutes another form of "all-at-one" mechanism of massive genome reshuffling, initially described in human prostate cancer (Baca et al., 2013) and subsequently in bone and soft tissue tumors (Anderson et al., 2018; Wang et al., 2013).

This phenomenon is characterized by the interdependent occurrence of multiple inter- and intratranslocations and deletions resulting from double-strand breaks (DSBs) with precise junctions. Chromoplexy events may involve up to eight chromosomes and this "close-chain" process leads to the generation of derivative chromosomes that present little or no CNVs. Chromoplexy breakpoints appear to cluster with active DNA replicator or transcription regions and open chromatin configurations. The involvement of the *TMPRSS2-ERG* gene fusion (EST+) at chromoplexy breakpoints suggests that chromoplexy events could occur from the same transcriptional mechanism driven by the androgen receptor (AR) that induce *TMPRSS2-ERG* fusion. The AR-mediated transcription has been implicated in the occurrence of DSBs through direct interaction with topoisomerase. By triggering the formation of clustered DSBs, AR transcription could promote the creation of chained rearrangements within a restricted nuclear domain. Chromoplexy could contribute to the aggressive evolution of high-grade prostate cancers. The description of similar chained rearrangements in melanomas, lung, and neck cancers suggests that chromoplexy can occur in a large spectrum of cancer types (Shen, 2013).

Both chromothripsis and chromoplexy processes may occur concurrently or asynchronously within the same cell, generating different patterns of chromosome complexity (Zepeda-Mendoza & Morton, 2019).

Chromoanasynthesis

The third type of one-time chaotic cellular event leading to the constitution of massive chromosomal rearrangement is termed chromoanasynthesis (for chromosome reconstitution).

First described by Liu et al. (2011), chromoanasynthesis is defined as a replication-based complex rearrangement process. The key features of chromoanasynthesis are the localized presence of copy number gains (duplications and triplications) in combination with deletions, and involve copy neutral chromosomal segments. Such multiple copy number changes cannot be explained by a process of chromosome shattering and NHEJ-mediated restitching of chromosomal fragments. The breakpoint junctions of these rearranged segments show microhomology and template insertions, consistent with defective DNA replications and suggesting the involvement of error-prone DNA replication pathways such as folk stalling and template switching (FoSTes) and microhomology-mediated break-induced replication (MMBIR) (Hastings et al., 2009; Lee et al., 2007). Both mechanisms are highly mutagenic (Ottaviani et al., 2014). In these processes, when a replication fork encounters an obstacle or a DNA lesion, the lagging strand can serially switch to another intra- or interchromosomal area with microhomology to establish a new active replication fork and restart DNA synthesis. The new template strands are not necessarily adjacent to the initial replication fork but in spatial proximity. Multiple fork disengaging and strand invasions can occur before the resumption of replication on the original template (Piazza et al., 2017). FoSTes and MMBIR pathways generate hybrid chromosomes with complex rearrangements involving duplications and triplications with short stretches of microhomology. Recent studies have also reported complex intra- and interchromosomal insertions as part of chromoanasynthesis events (Gu et al., 2016; Kato et al., 2017).

Numerous exogenous factors and a variety of cellular events can create conditions of replication stress and chromoanasynthesis occurrence by interfering with progression of the replication fork (Aguilera & Gómez-González, 2008; Venkatesan et al., 2015). In particular, interference between transcription and DNA replication represents an important source of genomic instability, because RNA and DNA polymerases operate on the same template (Lang et al., 2017). A codirectional orientation of the two processes can lead to the displacement of RNA polymerases and the eviction of the preformed transcriptional R loops (Hamperl et al., 2017). In contrast, when transcription is convergent to replication, the head-on collision of the two systems may increase the R-loop formation and promote fork collapse and chromoanasynthesis events (García-Muse & Aguilera, 2016; Lin & Pasero, 2012).

To date, the observation of chromoanasynthesis-linked rearrangements has been mostly limited to constitutional frame (Gudipati et al., 2019;

Zepeda-Mendoza & Morton, 2019). Chromoanasynthesis seems to operate both during gametogenesis and during the preimplantation period (Pellestor & Gatinois, 2018; Plaisancié et al., 2014). In contrast to chromothripsis, no preferential paternal origin of chromoanasynthesis events has been reported.

Chromoanasynthesis process appears to be nonexclusive. It can act in conjunction with other chromoanagenesis phenomena and various cellular mechanisms leading to generation of complex structural variants (SVs) pattern (Brás et al., 2020; Hattori et al., 2019; Masset et al., 2016; Piazza & Heyer, 2019).

Similar to chromothripsis, developmental delay, autism spectrum disorders, intellectual disability, and dysmorphic features are the mains clinical features described in patients with chromoanasynthesis (Fukami & Kurahashi, 2018; Grochowski et al., 2018; Zanardo et al., 2014), but phenotypically unaffected carriers have also been reported (Collins et al., 2017; Sabatini et al., 2018; Suzuki et al., 2016).

Mechanisms for chromoanagenesis occurrence

After the discovery and the characterization of these three unanticipated catastrophic phenomena, the mechanisms driving chromoanagenesis emergence have been actively investigated. To date, several nonexclusive causative mechanisms have been proposed that could trigger chaotic rearrangements of one or a few chromosomes within a unique cellular event, and thus explain the diversity and the complexity of chromoanagenesis phenomena.

Various exogenous factors can cause massive chromosomal rearrangements that may suggest chromoanagenesis events, such as free radicals, environmental toxins, or chemotherapeutic drugs. For instance, ionizing radiation has been shown to induce chromothripsis-like chromosomal rearrangements (Durante & Formenti, 2018; Morishita et al., 2016). Another possible cause of chromoanagenesis could be certain viral infections (Akagi et al., 2014; Schütze et al., 2016). Even cannabis exposure has been associated with chromothripsis occurrence (Reece & Hulse, 2016).

However, most of the research on causative factors has focused on endogenous mechanisms. Among the speculative factors inferring chromoanagenesis-linked rearrangements, abortive apoptosis has been regarded as a potential initiating event. In a cell-populations undergoing apoptosis, it is possible that a small portion of these cells engages in a restricted and incomplete form of apoptosis and thus survive (Tubio & Estivill, 2011). The subsequent DNA repair might be performed through a fast and incorrect process leading to the emergence of chromoanagenesis-mediated genomic alterations. Another plausible trigger mechanism is the premature chromosome condensation (PCC) (Stevens et al., 2010). Induced by the fusion of a mitotic cell with an interphasic cell, the PCC process causes the premature condensation of chromosomes still undergoing DNA replication and subsequently the partial pulverization of mitotic chromosomes and various degrees of chromosome breakage and rearrangements in a single cellular event (Pantelias et al., 2019; Terzoudi et al., 2015). An alternative scenario is that mitotic errors and replication stress could synergize to promote genome instability and chromoanagenesis occurrence (Jones & Jallepalli, 2012). In model cell lines, Mardin et al. (2015) showed that chromothripsis arise significantly more often in hyperploid than in diploid cells. Passerini et al. (2016) demonstrated that the presence of extrachromosomes triggers chromosomal instability and replication-related DNA damage and rearrangements. This link between numerical and structural chromosomal abnormalities could be explained by the downregulation of several replication factors, such as helicase MCM (minichromosome maintenance) or ORC (origin recognition complex) in aneuploid cells (Burrell et al., 2013; Dürrbaum et al., 2014). However, most of these exogenous and endogenous processes affect the whole genome and not just chromosomal regions confined to one or a few chromosomes, indicating that other cellular mechanisms should participate in the occurrence of chromoanagenesis.

Two alternative cellular mechanisms that can explain the restricted localization of breakpoints and complex alterations produced by chromoanagenesis phenomena are (i) the formation of micronuclei incorporating chromosomal material and (ii) the generation of chromatid bridges during cell division.

The micronucleus-mediated model

Micronuclei are frequently observed in tumor cells and they were considered for a long time as passive indicators of chromosomal instability (Luzhna et al., 2013). Accumulated data showed that the micronucleus is an important source of DNA alterations and genomic instability, and constitutes a key platform for chromoanagenesis emergence (Russo & Degrassi, 2018). Recently, it was reported that even the CRISPR-Cas9 genome editing could induce the formation of micronuclei and then the occurrence of chromoanagenesis phenomena (Leibowitz et al., 2020).

As starting point, there are defects in chromosomal segregation during cell division or presence of acentric chromosome fragments (Fig. 11.1). The underlying pathways that drive micronucleus formation are multiple and heterogeneous. Various defects in DNA replication and repair machinery, as well as clastogen agents producing DSBs may lead to anaphase lag of chromosomal material (Ly & Cleveland, 2017). On the other hand, numerous factors affecting chromosome segregation and cell cycle checkpoints can result in the spatial isolation of chromosomes or chromatids and the encapsulation of the lagging material into a micronucleus. These events include spindle disruption, centromere dysfunction, merotelic kinetochore microtubule attachment, centrosome dysfunction, spindle assembly checkpoint (SAC) dysfunction, and chromatid cohesion defects (for recent review, see Guo et al., 2019).

Following the anaphasic loss of a chromosome (or of chromosomal fragments), a micronucleus sequestering this chromosomal material is formed at the exit of mitosis. In these micronuclei involving a defective envelope, the chromosomal material will undergo premature condensation, and then rupture of the envelope allowing cytoplasmic exonucleases such as TREX1 to fragment DNA. The repair and inefficient replication of this DNA is accompanied by fragmentation of chromosomal material. The micronucleus can persist over several cell cycles, but frequently the

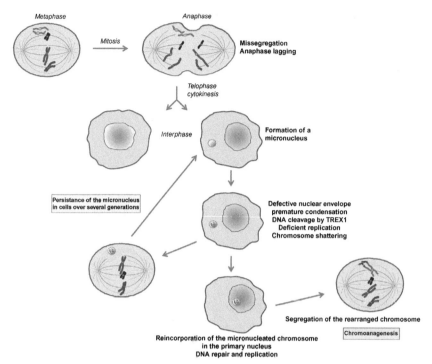

FIG. 11.1

Micronucleus-mediated model for the occurrence of chromoanagenesis.

micronucleus is reincorporated into the primary nucleus of a daughter cell where the repair, replication, and reassembly of chromosomal fragments results in the creation of a highly rearranged chromosome.

Quickly after the identification of chromothripsis and chromoanasynthesis phenomena, experimental studies have demonstrated that micronucleus formation offered an appealing mechanistic explanation for chromoanagenesis occurrence. By generating micronuclei in the human cell line HT 1080, Crasta et al. (2012) observed that lagging chromosome trapped in micronuclei underwent defective DNA replications, asynchronous with the primary nuclei, which led to extensive chromosomal damage and fragmentation in the micronuclei. Also, these experiments showed that micronuclei could persist in cells over several generations without degradation and that micronucleated chromosomal material could be reincorporated into daughter cell nuclei at a significant frequency (Crasta et al., 2012; Huang et al., 2012).

Micronuclei are structurally similar to regular nuclei, with a double membrane and nuclear pores (Liu & Pellman, 2020; Ungricht & Kutay, 2017), but ultrastructural studies evidenced important defects in the micronuclei envelope. The micronucleus envelope has a low density of nuclear pore complexes and defective nuclear lamina (Hatch et al., 2013; Sutyagina & Kisurina-Evgenieva, 2019; Terradas et al., 2010). The lack of lamin B impairs the structural integrity of the micronuclear lamina and the proper assembly of micronuclear envelope. The increased curvature of the micronuclear membrane could also negatively affect the assembly of the lamina (Kneissig et al., 2019). Liu et al. (2018) reported that only the "core" nuclear envelope proteins, and not the "noncore" nuclear proteins, such as nuclear pore complex proteins, assemble around lagging chromosomal material. The authors demonstrated that the high density of microtubules in the area of the main spindle block the recruitment of noncore proteins on lagging chromosomal material. This defective constitution leads to a significant defect in micronuclear recruitment and retention of essential components for DNA replication, DNA damage response, and the maintenance of micronucleus envelope integrity (Liu et al., 2018).

So, the damages to chromosomal material sequestered in the micronucleus could therefore be primarily mediated by defective DNA replication, potentially due to the lack of important factors such as origin recognition complexes (ORCs) and replication DNA helicases (Crasta et al., 2012). This could ultimately lead to chromoanasynthesis-like alterations. On the other hand, it was reported that a significant proportion of newly formed micronuclei underwent an irreversible disruption during interphase following their generation, and that almost 50% of disrupted micronuclear material rejoined the primary nucleus chromatin (Hatch et al., 2013). The disruption of micronuclei is caused by the micronuclear envelope collapse, triggered by lamin disorganization (Vargas et al., 2012). This loss of compartmentalization causes drastic reduction of transcription, replication, and proteasome functions inside the micronucleus (Hatch et al., 2013; Maass et al., 2018; Soto et al., 2018). The third possibility is that the micronucleus and its content can be degraded or extruded from the cell (Hintzsche et al., 2017; Reimann et al., 2020; Russo & Degrassi, 2018).

To investigate the timing of DNA damage in micronuclei, Zhang et al. (2015) performed in vitro experiments combining live cell imaging and single-cell sequencing technique (referred to as Look-Seq). They identified cells with disrupted micronuclei, tracked them through the next mitosis, and analyzed their DNA alterations. They showed that damage of sequestered material does not occur during G1 phase following micronucleus formation, but accumulate as cells progress through the next S and G2 phases, while the micronuclei have initiated DNA replication. In daughter cells, they subsequently evidenced the underreplication of DNA and the high degree of chromosomal rearrangements and fragmentation in micronuclei. Most of these rearrangements recapitulated the key features of not only chromothripsis but also chromoanagenesis, with typical short template insertions and microhomology. This experimental model provided the first direct evidence for a cellular mechanism that could explain the occurrence of chromoanagenesis, according to an "all-at-once" catastrophic process. It also clearly identified mitotic defects as an origin of chromoanasynthesis.

It was established that the disruption of micronuclear envelope initiated during the interphase induces the premature condensation of the encapsulated DNA and exposes it to cytoplasmic endo- and exonucleases, such as the exonuclease TREX1, which is activated upon the recognition of intermediate DNA structures, incompletely replicated, and collapsed replication forks (Piazza et al., 2017). This process induces double-stranded DNA breaks and the subsequent shattering of micronuclear chromosome material in acentric fragments, which will be randomly reassembled after their reintegration into newly formed daughter cell nuclei (Terradas et al., 2016).

Alternatively, the premature condensation of micronuclear DNA induced by mitotic entry can cause mechanical and replicative stress, delay active replication fork progression, and then initiate MMBIR or FoSTes pathways. Subsequently, this leads to the generation of a derivative chromosome, with localized chromoanagenesis-like rearrangements, and its potential reintegration into a daughter cell nucleus during the following mitosis (Russo & Degrassi, 2018; Terzoudi et al., 2015). The collapse or stall of replication forks initiating the chromoanasynthesis process could result from the premature condensation of the delayed replication regions into micronuclei. Indeed, such a hasty chromatin condensation generates a stress on replication forks still acting in the physiologically late replicating domains of entrapped chromatin. The observation of breakpoint enrichment in late replicating chromatin area is consistent with the role of PCC in the initiation of chromoanasynthesis (Chatron et al., 2020).

Once micronucleus-derived chromosomal material has been reincorporated into a primary nucleus and reassembled, one can speculate that the newly created derivative chromosome will be managed by appropriate replication machinery and thus become stabilized over subsequent cell generations.

In addition, the disruption of micronucleus envelope promotes immune preinflammatory signaling. Recent works evidenced that micronucleus envelope rupture initiates the recruitment of the cytosolic DNA recognition cyclic GMP-AMP synthase (cGAS) which detect foreign DNAs in the cytoplasm (Sun et al., 2013) to induce the innate immune response through the activation of the membrane adaptor *stimulator of interferon genes* (STING) and its interaction with many cytokines (Chen et al., 2016; Margolis et al., 2017). Such cGAS-STING cascade can act on the micronuclear DNA after envelope breakdown (Gekara, 2017; Mackenzie et al., 2017). This process not only activates the innate immune system, but also impacts DNA repair in micronuclei, because cGAS prevents DNA repair (Jiang et al., 2019).

In 2017, Ly and Cleveland described an alternative experimental model, based on the inducible centromere inactivation of the Y-chromosome and the specific sequestration into micronuclei of missegregated Y-chromosome, in order to analyze the fate of micronucleated chromosomes through several consecutive cell cycles (Ly and Cleveland, 2017). Following the micronuclei formation through the first cell cycle, it was observed that micronucleated Y-chromosome accumulated numerous DNA lesions and fragmented prior to or during the mitosis of the second cell cycle. Using calyculin A to drive premature chromosome condensation, it was established that premature chromosome condensation was the main cause of chromosome fragmentation and that this PCC-induced fragmentation was dependent on passage through S phase. These data confirmed that PCC-process constitutes a leading cause of shattering in micronuclei and an important mechanism underlying chromoanagenesis initiation (Pantelias et al., 2019). Most of chromosomal pieces are acentric fragments, which are lost or passively distributed into daughter cell nuclei. During the next interphase, the DNA fragments are rapidly repaired through c-NHEJ process and the resulting religated Y-chromosome displays typical hallmark of chromothripsis. Furthermore, by inhibiting the c-NHEJ machinery, the authors observed a twofold increase of Y-chromosome fragmentation in the second mitosis. These data clearly demonstrated that the c-NHEJ pathway is the major mechanism for rejoining micronuclei-derived chromosomal fragments in chromothripsis. The c-NHEJ pathway appears not to be inefficient in micronuclei. Therefore, c-NHEJ-dependent reassembly certainly occurs in the main nucleus, following fragment reintegration. This elegant model provided new and important insight into the mechanisms driving chromothripsis events and endorses the idea of a multicell cycle process for chromoanagenesis occurrence. This inducible chromosome missegregation strategy also offers the possibility of applying this approach to other chromosomes, which could be more amenable to genomic analysis (Ly & Cleveland, 2017). In 2019, Ly et al. improved their experimental system by combining the induced Y-chromosome missegregation and micronucleated fragmentation with the CRISPR-Cas9-mediated integration of a neomycin resistance gene (Neo^R), in order to identify cell carrying a reactivated Y-chromosome centromere (Ly et al., 2019). Coupling with appropriate cytogenetic (Liehr, 2021c; Weise & Liehr, 2021a) and whole-genome sequencing techniques, this strategy, termed CEN-SELECT, led to the highlighting of the high frequency and the diversity of chromosomal abnormalities induced by a single chromosome missegregation. Ly et al. (2019) reported not only a large spectrum of intrachromosomal abnormalities of the Y-chromosome but also complex patterns of interchromosomal rearrangements, most of them driven by c-NHEJ pathway and recapitulating the signature of chromothripsis.

Recently, in a model of trisomic and tetrasomic cell lines generated by micronucleus-mediated chromosome transfer, Kneissig et al. (2019) confirmed that transferred chromosomes often display complex pattern of localized chromosome rearrangements recapitulating the characteristics of chromoanagenesis.

Accumulating evidence supports that de novo alterations frequently occur in micronuclei. The incidence of DSBs in micronuclei is estimated to be up to 30 times higher than in primary nuclei (Umbreit et al., 2020). To date, single-cell sequencing approaches established that chromoanasynthesis may occur in more than 60% of micronuclei that are persisted for one cell cycle (Leibowitz et al., 2020), reinforcing the idea that micronucleus is the most credible hypothesis of chromoanagenesis origin. The observation of micronuclei in mammalian, plant, fish or reptile cells (Fenech, 2020), as well as in yeast and protozoa (Iwamoto et al., 2018) suggests considerable similarities in the way that cells can react to crisis, and confirms the connection between micronuclei, chromosomal abnormalities, and genomic instability (Comai & Tan, 2019).

Understanding the mechanisms of chromoanagenesis phenomena changes traditional thinking about micronuclei as a passive vector of damaged genetic material to become an active player in the genesis of DNA alterations and rearrangements. For Guo et al. (2020), the micronucleus can be compared to an isolated mutation factory in which single mutation patterns can be exacerbated into chromoanagenesis events. The generation of micronuclei can be considered as an effective driver for the cellular adaptation and the reorganization of the genome (Ye et al., 2019).

The chromatin bridge model

Micronucleation is not the only process driving chromoanagenesis phenomena as suggested by some cases of chromothripsis seen in different types of cancer (Hadi et al., 2020). Because chromothripsis-linked rearrangements are often localized within the telomeric regions, an alternative mechanism for chromoanagenesis occurrence has been proposed (Fig. 11.2), based on the telomere crisis (Ernst et al., 2016).

The formation of dicentric chromosomes due to end-to-end telomere fusion leads to the formation of chromatin bridges between daughter cells, at the exit of mitosis. In this chromatin bridge, surrounded by a defective envelope, the chromatin segment is stretched. The rupture of the envelope allows the resection of the DNA and its fragmentation by cytoplasmic exonucleases such as TREX1. The chromatin fragments are reintegrated into the primary nucleus of daughter cells where they can be repaired, replicated, and reassembled to generate a chromoanagenesis (± kataegis) event. New end-to-end telomere fusions can occur, giving rise to breakage-fusion-bridge (BFB) cycles. The rearranged chromosomal material can also be missegregated during a new mitosis and lead to the formation of a micronucleus, and then to chromoanagenesis.

Telomere crisis encompasses various alterations or dysfunctions of telomeres, such as replication-mediated shortening, defective telomere-associated proteins usually generating chromosomal instability (Dewhurst, 2020; Hackett et al., 2001; Mathieu et al., 2004). Due to such events, an end-to-end fusion between two uncapped or broken sister chromatids or two nonhomologous chromosomes may occur and give rise to the formation of a dicentric chromosome (Cleal & Baird, 2020; Poot, 2016). During mitosis, when the two centromeres of a dicentric chromosome are pulled to opposite poles, an anaphase bridge is created. At the end of mitosis, such a chromatid bridge may break and cause the formation of fragments detached from the spindle microtubule and eventually sequestered into a micronucleus, after the new nuclear envelope formation. After replication, these broken chromosome fragments can fuse again, giving rise to new dicentric chromosomes and thus undergo repeated cycles of breakage-fusion-bridge (BFB), leading to regional amplification or reorganization of small chromosome segments, potentially with chromoanagenesis-like rearrangements (Maciejowski & de Lange, 2017; Mardin et al., 2015; Sorzano et al., 2013). Also, in case of deficiencies in nuclear lamin proteins, some fragments may be involved in the formation of nuclear blebs, kinds of protrusion of the nuclear envelope (Broedersz & Brangwynne, 2013; Chen et al., 2018; Pampalona et al., 2010).

However, the chromatin bridge may also remain unbroken after cytokinesis, leading to the formation of a nucleocytoplasmic bridge connecting the two newly formed daughter cells. As in the case of the micronucleus, this chromatin bridge structure impairs the integrity of the nuclear envelope during interphase (Terradas et al., 2016).

In experimental conditions, after inducing telomere fusion and creation of anaphase chromatin bridges through the inactivation of the shelterin subunit TRF2, Maciejowski et al. (2015) observed a significant depletion of lamina components, such as lamin B1, and nuclear pore complexes (NPCs) as the bridge extended. Localized ruptures and loss of compartmentalization in the bridge allow access of

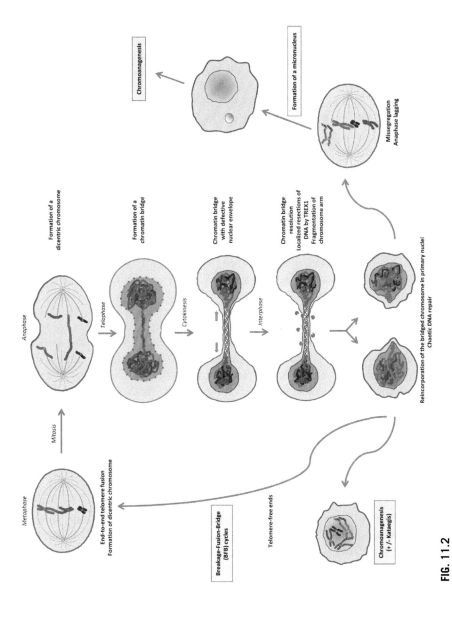

FIG. 11.2

Chromatin bridge-mediated model for the occurrence of chromoanagenesis.

cytoplasmic exonucleases to the bridging chromatin, such as the $3'$ exonuclease TREX1. The digestive action of these enzymes on the stretched chromatin will result in the generation of single-stranded DNA (ssDNA) bound by replication protein A (RPA). This also initiates the ultimate resolution of the bridge and the fragmentation of the chromatin it contains, before the primary nucleus entered S phase. In contrast with rearrangements resulting from micronuclei, the chromatin alteration generated chromatin bridge breakages do not require ongoing DNA synthesis. The damages are localized in the short area of the TREX1-mediated DNA fragmentation and bridge resolution. These sites are often associated with clusters of point mutation, termed kataegis, exhibiting cytosine substitution in TpC dinucleotides (Nik-Zainal et al., 2012) due to the activity of the apolipoprotein B mRNA-editing catalytic subunit (APOBEC) family of enzymes which target ssDNA and can function as a initiator of DNA fragmentation (Petljak & Maciejowski, 2020). These data support the hypothesis of a close mechanic relation between hypermutations and chromothripsis events (Luijten et al., 2018).

The multiple chromatin fragments can be involved in the formation of micronuclei, in one or both daughter cell at the end of mitosis, with the possibility of subsequent chromoanagenesis occurrence. The damaged bridging chromatin can also retract and join the daughter cell nuclei, then undergo illegitimate DNA repair and result in a reassembled chromosome potentially with chromothripsis-like rearrangements (Terradas et al., 2016).

By sequencing the posttelomere crisis cells, Maciejowski et al. (2015) established that 50% of them exhibit cluster of genome alterations with hallmarks of chromothripsis. Most of these chromothriptic alterations derived from two chromosomes and included terminal deletions, which is consistent with the hypothesis of an end-to-end fusion-mediated chromothripsis process. Such telomere crisis pathway can also generate reciprocal translocations, and even chromoplexy (Maciejowski & de Lange, 2017). To analyze these different hypotheses and to follow the fate of chromosome bridges and their DNA content, Umbreit et al. (2020) developed an experimental system of chromosome bridge induction and combined live cell imaging with single-cell whole genome sequencing (Look-Seq) technique. First, they observed that the bridge resolution required an actin/myosin-mediated mechanical stretching and that it produced a relatively simple pattern of chromothripsis-like rearrangements and local DNA fragmentation. Also, severe disturbances occur in the DNA replication process, leading to template-switching errors (Cleal et al., 2019) and to the generation, in a few daughter cells, of distinct chains of template short insertions, termed tandem short template (TST) jumps and originated from various distant breakpoint hotspots. Similar short insertion patterns were identified in cancer genomes (Li et al., 2020). This finding is to be compared to the observation of short sequence insertions throughout rearranged chromothriptic chromosomes (Collins et al., 2017; Slamova et al., 2018) and the description of short genomic area (rearrangement hubs) in different loops of chromothripsis-like rearrangements (Chatron et al., 2020). More complex rearrangements were then identified in fragmented chromosomes from broken bridges during the next mitosis due to aberrant mitotic DNA replication of this incompletely replicated chromatin. Consequently, more than 50% of divisions resulted in micronucleus formation in second-generation cells, promoting the occurrence of additional chromosome bridge formation or chromoanagenesis events in the following cell cycles, as described in the previous part. These data provide clear evidence that telomere dysfunction and BFB process can trigger chromoanagenesis events.

Strong analogies exist between micronucleus-mediated and chromatin bridge-mediated processes; in both cases defective nuclear envelope assembly around chromosomal material and severe alterations in DNA replication take place. These studies evidenced that the two processes are nonmutually exclusive but can coexist in a same context of cellular crisis. Altogether, these data provide a novel insight into the cellular mechanisms of chromoanagenesis formation.

Factors promoting the emergence of chromoanagenesis

Although the mechanisms of formation of micronuclei and chromatin bridges provide convincing models to explain the emergence of chromoanagenesis phenomena from a simple error in chromosomal segregation, it is also observed that other alternative pathways may lead to atypical forms of chromoanagenesis (Masset et al., 2016; Nazaryan-Petersen et al., 2016). This suggests that the phenomena of chromoanagenesis are perceptible reflections of the complex and multiple cellular processes of managing genomic instability. So, it is important to ask what are the cellular and genomic factors that promote genome instability and contribute to trigger the development of chromoanagenesis.

Defective mitosis progression is increasingly recognized as an essential cause of chromoanagenesis in mammals (Hattori & Fukami, 2020) and plants (Comai & Tan, 2019). Previous studies have highlighted the link between errors in chromosomal segregation and the formation of structural chromosomal abnormalities (Janssen et al., 2011). In vitro models like the one developed by Mardin et al. (2015) under the name of "complex alterations after selection and transformation (CAST)," as well as the experimental systems discussed in the previous part, have made it possible to define aneuploidy as an important risk factor, and to analyze the functional consequences of aneuploidy-linked chromoanagenesis in cell lines.

The correct assembly of the mitotic or meiotic spindle is crucial for the efficient segregation of chromosomes during cell divisions. The spindle is a dynamic self-organized bipolar array of microtubules whose assembly mobilizes several hundred proteins and several control systems such as the spindle assembly checkpoint (SAC) or the chromosome passenger complex (CPC) (Hutchins et al., 2010; Prosser & Pelletier, 2017). All alterations and disturbances in the pathways that contribute to spindle assembly and the timely completion of mitosis or in the gene controlling spindle formation and cell division may potentially give rise to instability and potentially to chromoanagenesis occurrence. Thus, abnormal centromere duplication, maturation, or separation (Pihan, 2013) as well as the presence of supernumerary centrosomes trigger chromothripsis by promoting kinetochore misattachment and anaphasic chromosome lagging (Ganem et al., 2009; Nam et al., 2015). The heterochromatin protein 1 (HP1) is an essential component of the CPC whose reduction was directly correlated with the improper kinetochore-microtubule attachments and the occurrence of segregation errors (Abe et al., 2016). The sister chromatin cohesion is also crucial for the biorientation of chromosomes on the mitotic and the meiotic spindle, and for the repair of damaged DNA (Litwin et al., 2018; Skibbens, 2019). Defects in chromosome cohesion and its control compromise the fidelity of chromosome segregation and trigger premature loss of cohesion, anaphase lagging, and aneuploidy (Barbero, 2011). The existence of a spindle matrix corresponding to a network of nuclear lamina components involved in the spindle assembly has been suggested (Schweizer et al., 2014; Zheng, 2010). The lamin B, whose role has been evoked both in the micronucleus formation and the chromatin bridge process, could be included in this spindle matrix since the depletion of lamin B was found to impede spindle assembly (Shi et al., 2014).

To date, numerous studies have shown that chromosome segregation defects and aneuploidy may induce replication damages and lead to the accumulation of chromosomal rearrangements in aneuploid cells (Storchova & Kuffer, 2008; Thompson & Compton, 2008). Basically, replication stress and perturbations of DNA repair process are mainly associated with chromosomal instability (Aguilera & Gómez-González, 2008; Arlt et al., 2012; Wilhelm et al., 2020). In classical conditions, the coordination between DNA replication and cell cycle progression is regulated by several checkpoints which play a crucial role of quality control processes for harmonizing replication with DNA repair, chromosome segregation, and cell division. Replication stress may alter the cell cycle process by slowing or blocking the progression of the "missegregating" cells through S or G2/M phase, thus leading to slowed DNA replication and potentially to the formation of chromatin bridges or micronucleation. Depending on the cell cycle position, DNA repair mechanisms will differ. Chromosomal instability is more likely to occur if the DNA repair efficiency is suboptimal (Streffer, 2010). The NHEJ pathway is operational throughout the cell cycle, whereas replicative repair mechanisms such as FoSTes and MMBIR can only work during S phase and eventually in G2 phase (Kass et al., 2016). Another determining factors can also be the number of DSBs to repair and the energy required by the cell to manage DNA break repair on a short timescale. A large number of breaks can quickly saturate the capacity of cellular DNA repair pathways. Gudjonsson et al. (2012) evidenced that more than 20 DSBs could alter standard error-free DNA repair mechanism, such as homologous recombination (HR) pathway, in favor of faster but error-prone repair processes, such as NHEJ and alt-EJ. The aneuploidy-linked genome instability could also be explained by a reduced expression of replication factors such as replicative helicase MCM and ORC (Passerini et al., 2016). Analysis of gene expression revealed downregulation of several replication factors in aneuploid cells with extrachromosome (Sheltzer et al., 2012).

Given the essential role of the genome organization and the local DNA sequence environment in driving and modulating genome functions (and reciprocally), various specific genomic features must be considered for a clear understanding of the sudden onset of chromoanagenesis phenomena. Repetitive sequences are known to facilitate genomic rearrangements (Harel & Lupski, 2018). Sequence analyses of chromoanagenesis-mediated rearrangements have evidenced the frequent presence of low copy repeats (LCRs) sequences or tandem repetitive sequences, such as minisatellite or ALU sequences, in the vicinity of breakpoint junctions (Gudipati et al., 2019; Kloosterman et al., 2011; Nazaryan-Petersen et al., 2018; Weckselblatt et al., 2015). These repeat and repetitive sequences create areas of genomic instability and potentially facilitate template switching (Zhang et al., 2009) and the subsequent occurrence of intra- and interchromosomal rearrangements. Fragile sites are other DNA sequences frequently associated with rearrangements and genetic diseases. They contribute to genome instability and replication impairment (Admire et al., 2006; Barlow et al., 2013; Wilhelm et al., 2018), and consequently a plausible link between chromoanagenesis events and fragile sites have been considered, since they can be the sites of multiple chromoanagenesis-mediated breakage events (Mackinnon & Campbell, 2013). Fragile sites, Alu sequences, microsatellites, and LCRs as well as other particular motifs like palindromic sequences can also promote instability by inducing the formation of unusual chromatin secondary structures such as hairpins, cruciforms, or DNA triplexes, which are able to disrupt replication and cause DSBs (Cooper et al., 2011; Kurahashi et al., 2009; Liu et al., 2011). Also, several recent studies have documented the contribution of transposable elements to the development of chromoanagenesis (Hancks, 2018). Nazaryan-Petersen et al. (2016) reported a retrotransposition-mediated chromothripsis displaying the insertion of an SVA (SINE_VNTE-Alu) element at one pair of breakpoints and several other DNA breaks produced by L1 endonuclease activity. The L1 endonuclease activity could be an important contributory factor to chromosome shattering. In the gibbon genome, the insertion of the retrotransposon LAVA is at the origin of a high rate of chromothripsis-mediated rearrangements leading to the accelerated evolution of the gibbon karyotype and the emergence of different gibbon lineages (Carbone et al., 2014; Meyer et al., 2016). Approximately 45% of the human genome derives from transposable elements (Pace & Feschotte, 2007). Only a small proportion of transposable elements, in particular the retrotransposons LINE and SINE (Liehr, 2021e), retain the capacity to change their position within the genome by transposition (Faulkner & Garcia-Perez, 2017; Kawakami et al., 2017). They can act as mutagen and consequently they are frequently associated with genetic disorders (Burns, 2017). They also can induce DSBs leading to genome instability (Gasior et al., 2006; Symer et al., 2002). The abundance of retrotransposons in the human genome provides numerous potential substrates for mitotic or meiotic recombination structural variations and rearrangements. Consequently, transposable elements are potential threats to genome stability because they can initiate ectopic recombination between nonhomologous regions or nonhomologous chromosomes, and thus lead to structural variations and large chromosomal rearrangements (Song et al., 2018; Underwood & Choi, 2019).

Beyond these characteristics linked to the DNA sequence, these data suggest that the conformation of the chromatin and the dynamic architecture of the nucleus are key factors in the emergence of chromoanasynthesis. The genome structure is clearly related to genome function (Agbleke et al., 2020). The modulatory role of chromatin is evident throughout the cell cycle for essential functions such as transcription, replication, and repair (Meschini et al., 2015; Misteli, 2020). It has been proposed that the chromatin looping can regulate DSBs repair through histone modifications and nucleosome remodeling within approximately 50 kb on each side of DSBs, in order to induce the formation of open and relaxed chromatin conformation and thus to facilitate loading of DNA repair proteins (Price & D'Andrea, 2013). In accordance with this chromatid regulation disturbances in epigenetic mechanisms could alter the dynamics of the organization of chromatin and the replication initiation, for example, in gametes or during early embryonic development, and thus promote the occurrence of more or less complex rearrangements (Åsenius et al., 2020; Franzago et al., 2019). The ultimate level of nuclear organization and regulation is the spatial partitioning of the genome into chromatin domains and chromosome territories, referred to as topologically associating domains (TADs), lamina-associated domains (LADs), or cis-regulatory domains (CRDs) (Daban, 2021; Delaneau et al., 2019; Dixon et al., 2012; Iourov et al., 2021; Liehr, 2021d; Luperchio et al., 2014; Lupiáñez et al., 2016; Weise & Liehr, 2021b; Yumiceba et al., 2021). This global hierarchical organization of chromatin is proposed to be lost during mitosis and reestablished in G1 phase (Bonev & Cavalli, 2016). It has been established that the occurrence of complex chromosomal rearrangements disrupts not only the conserved organization of these territories, transforming their architecture and their interaction but also the regulation of the genes (Yauy et al., 2018; Ye et al., 2019). Studies conducted on induced pluripotent cell lines (iPSCs) derived from a patient carrying a de novo germline chromothripsis have shown how chromothripsis can alter the functioning and regulation of genes, thus contributing to the patient's complex congenital phenotype (Middelkamp et al., 2017). In addition to the 3D genome architecture, it is important to consider the mobility of chromatin within the nucleus. An increase in chromatin mobility has been observed in response to DSBs (Dion & Gasser, 2013; Miné-Hattab & Rothstein, 2012). In the vicinity of DSBs, the damaged site is found to be more mobile and the chromatin becomes more rigid. The nuclear response triggered by a DSB is not limited to the damaged site and it is the dynamics of the whole genome that increases in proportion to the number of DSBs (Miné-Hattab et al., 2017). This chromatin mobility promotes the recombination of homologous chromosomal regions. The Rad5 protein, which plays an essential role in the recombination process, is directly involved in chromatin movements (Dion et al., 2012). It can therefore be speculated that increased chromatin mobility facilitates the search for homologous template sequences in the replication-based DNA repair processes such as FoSTes and MMBIR.

All these data reinforce the notion that chromoanagenesis events occur in cellular and genomic context which compromise genomic stability. Genome instability is a broad concept of which chromosome instability is the most prevalent form (Geigl et al., 2008). Growing literature indicates that this instability is linked to a large spectrum of genomic and nongenomic factors and mechanisms (Venkatesan et al., 2015).

Conclusion

To date, the mechanisms of chromoanagenesis are among the most fascinating cellular events evidenced over the past decade. It has been a controversial notion that these massive and complex rearrangement phenomena usually occur through a single cell cycle. However, the experimental data obtained have allowed to prove the existence of chromoanagenesis events and to shed light on their mechanisms of formation. In particular, the description of micronucleation and chromatid bridges formation highlighted the close links that may exist between apparently simple errors in chromosome segregation, the cellular stress and instability processes inducing chromoanagenesis. In light of these new cytogenomic data, it appears that the human genome can tolerate important modifications of its conformation, and that chromoanagenesis pathways are mechanisms of rapid and profound restructuring of the genome. Their discovery has renewed interest in questions of genome plasticity, the role of nuclear topology, and the capacity of cells to manage or not such crises and cellular chaos. Even if DNA remains the support of heredity, a new framework more focused on the concept of genome (Liehr, 2021a) must be taken into consideration to better understand the cellular, clinical, and evolutionary impact of chromoanagenesis phenomena.

References

Abe, Y., Sako, K., Takagaki, K., Hirayama, Y., Uchida, K. S. K., Herman, J. A., … Hirota, T. (2016). HP1-assisted Aurora B kinase activity prevents chromosome segregation errors. *Developmental Cell, 36*, 487–497. https://doi.org/10.1016/j.devcel.2016.02.008.

Admire, A., Shanks, L., Danzl, N., Wang, M., Weier, U., Stevens, W., … Weinert, T. (2006). Cycles of chromosome instability are associated with a fragile site and are increased by defects in DNA replication and checkpoint controls in yeast. *Genes & Development, 20*, 159–173. https://doi.org/10.1101/gad.1392506.

Agbleke, A. A., Amitai, A., Buenrostro, J. D., Chakrabarti, A., Chu, L., Hansen, A. S., … Wadduwage, D. (2020). Advances in chromatin and chromosome research: Perspectives from multiple fields. *Molecular Cell, 79*, 881–901. https://doi.org/10.1016/j.molcel.2020.07.003.

Aguilera, A., & Gómez-González, B. (2008). Genome instability: A mechanistic view of its causes and consequences. *Nature Reviews Genetics, 9*, 204–217. https://doi.org/10.1038/nrg2268.

Akagi, K., Li, J., Broutian, T. R., Padilla-Nash, H., Xiao, W., Jiang, B., … Gillison, M.L. (2014). Genome-wide analysis of HPV integration in human cancers reveals recurrent, focal genomic instability. *Genome Research, 24*, 185–199. https://doi.org/10.1101/gr.164806.113.

Anand, R. P., Tsaponina, O., Greenwell, P. W., Lee, C.-S., Du, W., Petes, T. D., & Haber, J. E. (2014). Chromosome rearrangements via template switching between diverged repeated sequences. *Genes & Development, 28*, 2394–2406. https://doi.org/10.1101/gad.250258.114.

Anderson, S. E., Kamath, A., Pilz, D. T., & Morgan, S. M. (2016). A rare example of germ-line chromothripsis resulting in large genomic imbalance. *Clinical Dysmorphology, 25*, 58–62. https://doi.org/10.1097/MCD.0000000000000113.

Anderson, N. D., de Borja, R., Young, M. D., Fuligni, F., Rosic, A., Roberts, N. D., … Shlien, A. (2018). Rearrangement bursts generate canonical gene fusions in bone and soft tissue tumors. *Science, 361*(6405), eaam8419. https://doi.org/10.1126/science.aam8419.

Arlt, M. F., Wilson, T. E., & Glover, T. W. (2012). Replication stress and mechanisms of CNV formation. *Current Opinion in Genetics & Development, 22,* 204–210. https://doi.org/10.1016/j.gde.2012.01.009.

Åsenius, F., Danson, A. F., & Marzi, S. J. (2020). DNA methylation in human sperm: A systematic review. *Human Reproduction Update, 26,* 841–873. https://doi.org/10.1093/humupd/dmaa025.

Baca, S. C., Prandi, D., Lawrence, M. S., Mosquera, J. M., Romanel, A., Drier, Y., … Garraway, L. A. (2013). Punctuated evolution of prostate cancer genomes. *Cell, 153,* 666–677. https://doi.org/10.1016/j.cell.2013.03.021.

Baranov, V. S., & Kuznetzova, T. V. (2021). Nuclear stability in early embryo. Chromosomal aberrations. In T. Liehr (Ed.), *Cytogenomics* (pp. 307–325). Academic Press. Chapter 15 (in this book).

Barbero, J. L. (2011). Sister chromatid cohesion control and aneuploidy. *Cytogenetic and Genome Research, 133,* 223–233. https://doi.org/10.1159/000323507.

Barlow, J. H., Faryabi, R. B., Callén, E., Wong, N., Malhowski, A., Chen, H. T., … Nussenzweig, A. (2013). Identification of early replicating fragile sites that contribute to genome instability. *Cell, 152,* 620–632. https://doi.org/10.1016/j.cell.2013.01.006.

Bassaganyas, L., Beà, S., Escaramís, G., Tornador, C., Salaverria, I., Zapata, L., … Estivill, X. (2013). Sporadic and reversible chromothripsis in chronic lymphocytic leukemia revealed by longitudinal genomic analysis. *Leukemia, 27,* 2376–2379. https://doi.org/10.1038/leu.2013.127.

Bertelsen, B., Nazaryan-Petersen, L., Sun, W., Mehrjouy, M. M., Xie, G., Chen, W., … Tümer, Z. (2016). A germline chromothripsis event stably segregating in 11 individuals through three generations. *Genetics in Medicine, 18,* 494–500. https://doi.org/10.1038/gim.2015.112.

Blanc-Mathieu, R., Krasovec, M., Hebrard, M., Yau, S., Desgranges, E., Martin, J., … Piganeau, G. (2017). Population genomics of picophytoplankton unveils novel chromosome hypervariability. *Science Advances, 3,* e1700239. https://doi.org/10.1126/sciadv.1700239.

Bonev, B., & Cavalli, G. (2016). Organization and function of the 3D genome. *Nature Reviews Genetics, 17,* 661–678. https://doi.org/10.1038/nrg.2016.112.

Broedersz, C. P., & Brangwynne, C. P. (2013). Nuclear mechanics: Lamin webs and pathological blebs. *Nucleus, 4,* 156–159. https://doi.org/10.4161/nucl.25019.

Brás, A., Rodrigues, A. S., & Rueff, J. (2020). Copy number variations and constitutional chromothripsis (review). *Biomedical Reports, 13,* 11. https://doi.org/10.3892/br.2020.1318.

Burns, K. H. (2017). Transposable elements in cancer. *Nature Reviews Cancer, 17,* 415–424. https://doi.org/10.1038/nrc.2017.35.

Burrell, R. A., McClelland, S. E., Endesfelder, D., Groth, P., Weller, M.-C., Shaikh, N., … Swanton, C. (2013). Replication stress links structural and numerical cancer chromosomal instability. *Nature, 494,* 492–496. https://doi.org/10.1038/nature11935.

Cai, H., Kumar, N., Bagheri, H. C., von Mering, C., Robinson, M. D., & Baudis, M. (2014). Chromothripsis-like patterns are recurring but heterogeneously distributed features in a survey of 22,347 cancer genome screens. *BMC Genomics, 15,* 82. https://doi.org/10.1186/1471-2164-15-82.

Carbone, L., Harris, R. A., Gnerre, S., Veeramah, K. R., Lorente-Galdos, B., Huddleston, J., … Gibbs, R. A. (2014). Gibbon genome and the fast karyotype evolution of small apes. *Nature, 513,* 195–201. https://doi.org/10.1038/nature13679.

Carbonell-Bejerano, P., Royo, C., Torres-Pérez, R., Grimplet, J., Fernandez, L., Franco-Zorrilla, J. M., … Martínez-Zapater, J.M. (2017). Catastrophic unbalanced genome rearrangements cause somatic loss of berry color in grapevine. *Plant Physiology*, *175*, 786–801. https://doi.org/10.1104/pp.17.00715.

Chatron, N., Giannuzzi, G., Rollat-Farnier, P.-A., Diguet, F., Porcu, E., Yammine, T., … Schluth-Bolard, C. (2020). *The enrichment of breakpoints in late-replicating chromatin provides novel insights into chromoanagenesis mechanisms*. bioRxiv, 2020.07.17.206771. https://doi.org/10.1101/2020.07.17.206771.

Chen, Q., Sun, L., & Chen, Z. J. (2016). Regulation and function of the cGAS-STING pathway of cytosolic DNA sensing. *Nature Immunology*, *17*, 1142–1149. https://doi.org/10.1038/ni.3558.

Chen, S., Luperchio, T. R., Wong, X., Doan, E. B., Byrd, A. T., Roy Choudhury, K., … Krangel, M.S. (2018). A lamina-associated domain border governs nuclear lamina interactions, transcription, and recombination of the Tcrb locus. *Cell Reports*, *25*, 1729–1740.e6. https://doi.org/10.1016/j.celrep.2018.10.052.

Chiang, C., Jacobsen, J. C., Ernst, C., Hanscom, C., Heilbut, A., Blumenthal, I., … Talkowski, M.E. (2012). Complex reorganization and predominant non-homologous repair following chromosomal breakage in karyotypically balanced germline rearrangements and transgenic integration. *Nature Genetics*, *44*, 390–397. S1 https://doi.org/10.1038/ng.2202.

Cleal, K., & Baird, D. M. (2020). Catastrophic endgames: Emerging mechanisms of telomere-driven genomic instability. *Trends in Genetics*, *36*, 347–359. https://doi.org/10.1016/j.tig.2020.02.001.

Cleal, K., Jones, R. E., Grimstead, J. W., Hendrickson, E. A., & Baird, D. M. (2019). Chromothripsis during telomere crisis is independent of NHEJ, and consistent with a replicative origin. *Genome Research*, *29*, 737–749. https://doi.org/10.1101/gr.240705.118.

Collins, R. L., Brand, H., Redin, C. E., Hanscom, C., Antolik, C., Stone, M. R., … Talkowski, M.E. (2017). Defining the diverse spectrum of inversions, complex structural variation, and chromothripsis in the morbid human genome. *Genome Biology*, *18*, 36. https://doi.org/10.1186/s13059-017-1158-6.

Comai, L., & Tan, E. H. (2019). Haploid induction and genome instability. *Trends in Genetics*, *35*, 791–803. https://doi.org/10.1016/j.tig.2019.07.005.

Cooper, D. N., Bacolla, A., Férec, C., Vasquez, K. M., Kehrer-Sawatzki, H., & Chen, J.-M. (2011). On the sequence-directed nature of human gene mutation: The role of genomic architecture and the local DNA sequence environment in mediating gene mutations underlying human inherited disease. *Human Mutation*, *32*, 1075–1099. https://doi.org/10.1002/humu.21557.

Cortés-Ciriano, I., Lee, J. J.-K., Xi, R., Jain, D., Jung, Y. L., Yang, L., … Park, P.J. (2020). Comprehensive analysis of chromothripsis in 2,658 human cancers using whole-genome sequencing. *Nature Genetics*, *52*, 331–341. https://doi.org/10.1038/s41588-019-0576-7.

Crasta, K., Ganem, N. J., Dagher, R., Lantermann, A. B., Ivanova, E. V., Pan, Y., … Pellman, D. (2012). DNA breaks and chromosome pulverization from errors in mitosis. *Nature*, *482*, 53–58. https://doi.org/10.1038/nature10802.

Daban, J.-R. (2021). Multilayer organization of chromosomes. In T. Liehr (Ed.), *Cytogenomics* (pp. 267–296). Academic Press. Chapter 13 (in this book).

de Pagter, M. S., van Roosmalen, M. J., Baas, A. F., Renkens, I., Duran, K. J., van Binsbergen, E., … Kloosterman, W.P. (2015). Chromothripsis in healthy individuals affects multiple protein-coding genes and can result in severe congenital abnormalities in offspring. *American Journal of Human Genetics*, *96*, 651–656. https://doi.org/10.1016/j.ajhg.2015.02.005.

Delaneau, O., Zazhytska, M., Borel, C., Giannuzzi, G., Rey, G., Howald, C., … Dermitzakis, E.T. (2019). Chromatin three-dimensional interactions mediate genetic effects on gene expression. *Science*, *364*. https://doi.org/10.1126/science.aat8266.

Dewhurst, S. M. (2020). Chromothripsis and telomere crisis: Engines of genome instability. *Current Opinion in Genetics & Development*, *60*, 41–47. https://doi.org/10.1016/j.gde.2020.02.009.

Dion, V., & Gasser, S. M. (2013). Chromatin movement in the maintenance of genome stability. *Cell*, *152*, 1355–1364. https://doi.org/10.1016/j.cell.2013.02.010.

Dion, V., Kalck, V., Horigome, C., Towbin, B. D., & Gasser, S. M. (2012). Increased mobility of double-strand breaks requires Mec1, Rad9 and the homologous recombination machinery. *Nature Cell Biology*, *14*, 502–509. https://doi.org/10.1038/ncb2465.

Dixon, J. R., Selvaraj, S., Yue, F., Kim, A., Li, Y., Shen, Y., … Ren, B. (2012). Topological domains in mammalian genomes identified by analysis of chromatin interactions. *Nature*, *485*, 376–380. https://doi.org/10.1038/nature11082.

Durante, M., & Formenti, S. C. (2018). Radiation-induced chromosomal aberrations and immunotherapy: Micronuclei, cytosolic DNA, and interferon-production pathway. *Frontiers in Oncology*, *8*, 192. https://doi.org/10.3389/fonc.2018.00192.

Dürrbaum, M., Kuznetsova, A. Y., Passerini, V., Stingele, S., Stoehr, G., & Storchová, Z. (2014). Unique features of the transcriptional response to model aneuploidy in human cells. *BMC Genomics*, *15*, 139. https://doi.org/10.1186/1471-2164-15-139.

Ernst, A., Jones, D. T. W., Maass, K. K., Rode, A., Deeg, K. I., Jebaraj, B. M. C., … Lichter, P. (2016). Telomere dysfunction and chromothripsis. *International Journal of Cancer*, *138*, 2905–2914. https://doi.org/10.1002/ijc.30033.

Faulkner, G. J., & Garcia-Perez, J. L. (2017). L1 mosaicism in mammals: Extent, effects, and evolution. *Trends in Genetics*, *33*, 802–816. https://doi.org/10.1016/j.tig.2017.07.004.

Fenech, M. (2020). Cytokines is-block micronucleus cytome assay evolution into a more comprehensive method to measure chromosomal instability. *Genes (Basel)*, *11*. https://doi.org/10.3390/genes11101203.

Fontana, M. C., Marconi, G., Feenstra, J. D. M., Fonzi, E., Papayannidis, C., Ghelli Luserna di Rorá, A., … Martinelli, G. (2018). Chromothripsis in acute myeloid leukemia: Biological features and impact on survival. *Leukemia*, *32*, 1609–1620. https://doi.org/10.1038/s41375-018-0035-y.

Forment, J. V., Kaidi, A., & Jackson, S. P. (2012). Chromothripsis and cancer: Causes and consequences of chromosome shattering. *Nature Reviews Cancer*, *12*, 663–670. https://doi.org/10.1038/nrc3352.

Franzago, M., La Rovere, M., Guanciali Franchi, P., Vitacolonna, E., & Stuppia, L. (2019). Epigenetics and human reproduction: The primary prevention of the noncommunicable diseases. *Epigenomics*, *11*, 1441–1460. https://doi.org/10.2217/epi-2019-0163.

Fukami, M., & Kurahashi, H. (2018). Clinical consequences of chromothripsis and other catastrophic cellular events. *Methods in Molecular Biology*, *1769*, 21–33. https://doi.org/10.1007/978-1-4939-7780-2_2.

Fukami, M., Shima, H., Suzuki, E., Ogata, T., Matsubara, K., & Kamimaki, T. (2017). Catastrophic cellular events leading to complex chromosomal rearrangements in the germline. *Clinical Genetics*, *91*, 653–660. https://doi.org/10.1111/cge.12928.

Ganem, N. J., Godinho, S. A., & Pellman, D. (2009). A mechanism linking extra centrosomes to chromosomal instability. *Nature*, *460*, 278–282. https://doi.org/10.1038/nature08136.

García-Muse, T., & Aguilera, A. (2016). Transcription-replication conflicts: How they occur and how they are resolved. *Nature Reviews Molecular Cell Biology*, *17*, 553–563. https://doi.org/10.1038/nrm.2016.88.

Gasior, S. L., Wakeman, T. P., Xu, B., & Deininger, P. L. (2006). The human LINE-1 retrotransposon creates DNA double-strand breaks. *Journal of Molecular Biology*, *357*, 1383–1393. https://doi.org/10.1016/j.jmb.2006.01.089.

Geigl, J. B., Obenauf, A. C., Schwarzbraun, T., & Speicher, M. R. (2008). Defining "chromosomal instability". *Trends in Genetics*, *24*, 64–69. https://doi.org/10.1016/j.tig.2007.11.006.

Gekara, N. O. (2017). DNA damage-induced immune response: Micronuclei provide key platform. *The Journal of Cell Biology*, *216*, 2999–3001. https://doi.org/10.1083/jcb.201708069.

Govind, S. K., Zia, A., Hennings-Yeomans, P. H., Watson, J. D., Fraser, M., Anghel, C., … Boutros, P.C. (2014). ShatterProof: Operational detection and quantification of chromothripsis. *BMC Bioinformatics*, *15*, 78. https://doi.org/10.1186/1471-2105-15-78.

Grochowski, C. M., Gu, S., Yuan, B., Tcw, J., Brennand, K. J., Sebat, J., … Carvalho, C.M.B. (2018). Marker chromosome genomic structure and temporal origin implicate a chromoanasynthesis event in a family with pleiotropic psychiatric phenotypes. *Human Mutation*, *39*, 939–946. https://doi.org/10.1002/humu.23537.

Gu, S., Szafranski, P., Akdemir, Z. C., Yuan, B., Cooper, M. L., Magriñá, M. A., … Lupski, J.R. (2016). Mechanisms for complex chromosomal insertions. *PLoS Genetics*, *12*, e1006446. https://doi.org/10.1371/journal.pgen.1006446.

Gudipati, M. A., Waters, E., Greene, C., Goel, N., Hoppman, N. L., Pitel, B. A., … Zou, Y. (2019). Stable transmission of complex chromosomal rearrangements involving chromosome 1q derived from constitutional chromoanagenesis. *Molecular Cytogenetics*, *12*, 43. https://doi.org/10.1186/s13039-019-0455-z.

Gudjonsson, T., Altmeyer, M., Savic, V., Toledo, L., Dinant, C., Grøfte, M., … Lukas, C. (2012). TRIP12 and UBR5 suppress spreading of chromatin ubiquitylation at damaged chromosomes. *Cell*, *150*, 697–709. https://doi.org/10.1016/j.cell.2012.06.039.

Guo, X., Ni, J., Liang, Z., Xue, J., Fenech, M. F., & Wang, X. (2019). The molecular origins and pathophysiological consequences of micronuclei: New insights into an age-old problem. *Mutation Research*, *779*, 1–35. https://doi.org/10.1016/j.mrrev.2018.11.001.

Guo, X., Dai, X., Wu, X., Cao, N., & Wang, X. (2020). Small but strong: Mutational and functional landscapes of micronuclei in cancer genomes. *International Journal of Cancer*. https://doi.org/10.1002/ijc.33300.

Hackett, J.A., Feldser, D. M., & Greider, C. W. (2001). Telomere dysfunction increases mutation rate and genomic instability. *Cell*, *106*, 275–286. https://doi.org/10.1016/s0092-8674(01)00457-3.

Hadi, K., Yao, X., Behr, J. M., Deshpande, A., Xanthopoulakis, C., Tian, H., … Imielinski, M. (2020). Distinct classes of complex structural variation uncovered across thousands of cancer genome graphs. *Cell*, *183*, 197–210.e32. https://doi.org/10.1016/j.cell.2020.08.006.

Hamperl, S., Bocek, M. J., Saldivar, J. C., Swigut, T., & Cimprich, K. A. (2017). Transcription-replication conflict orientation modulates R-loop levels and activates distinct DNA damage responses. *Cell*, *170*, 774–786.e19. https://doi.org/10.1016/j.cell.2017.07.043.

Hancks, D. C. (2018). A role for retrotransposons in chromothripsis. *Methods in Molecular Biology*, *1769*, 169–181. https://doi.org/10.1007/978-1-4939-7780-2_11.

Harel, T., & Lupski, J. R. (2018). Genomic disorders 20 years on-mechanisms for clinical manifestations. *Clinical Genetics*, *93*, 439–449. https://doi.org/10.1111/cge.13146.

Hastings, P. J., Ira, G., & Lupski, J. R. (2009). A microhomology-mediated break-induced replication model for the origin of human copy number variation. *PLoS Genetics*, *5*, e1000327. https://doi.org/10.1371/journal.pgen.1000327.

Hatch, E. M., Fischer, A. H., Deerinck, T. J., & Hetzer, M. W. (2013). Catastrophic nuclear envelope collapse in cancer cell micronuclei. *Cell*, *154*, 47–60. https://doi.org/10.1016/j.cell.2013.06.007.

Hattori, A., & Fukami, M. (2020). Established and novel mechanisms leading to de novo genomic rearrangements in the human germline. *Cytogenetic and Genome Research, 160*, 167–176. https://doi.org/10.1159/000507837.

Hattori, A., Okamura, K., Terada, Y., Tanaka, R., Katoh-Fukui, Y., Matsubara, Y., … Fukami, M. (2019). Transient multifocal genomic crisis creating chromothriptic and non-chromothriptic rearrangements in prezygotic testicular germ cells. *BMC Medical Genomics, 12*, 77. https://doi.org/10.1186/s12920-019-0526-3.

Henry, I. M., Comai, L., & Tan, E. H. (2018). Detection of chromothripsis in plants. *Methods in Molecular Biology, 1769*, 119–132. https://doi.org/10.1007/978-1-4939-7780-2_8.

Hintzsche, H., Hemmann, U., Poth, A., Utesch, D., Lott, J., Stopper, H., & Working Group "In vitro micronucleus test", Gesellschaft für Umwelt-Mutationsforschung (GUM, German-speaking section of the European Environmental Mutagenesis and Genomics Society EEMGS). (2017). Fate of micronuclei and micronucleated cells. *Mutation Research, 771*, 85–98. https://doi.org/10.1016/j.mrrev.2017.02.002.

Holland, A. J., & Cleveland, D. W. (2012). Chromoanagenesis and cancer: Mechanisms and consequences of localized, complex chromosomal rearrangements. *Nature Medicine, 18*, 1630–1638. https://doi.org/10.1038/nm.2988.

Huang, Y., Jiang, L., Yi, Q., Lv, L., Wang, Z., Zhao, X., … Shi, Q. (2012). Lagging chromosomes entrapped in micronuclei are not "lost" by cells. *Cell Research, 22*, 932–935. https://doi.org/10.1038/cr.2012.26.

Hutchins, J. R. A., Toyoda, Y., Hegemann, B., Poser, I., Hériché, J.-K., Sykora, M. M., … Peters, J.-M. (2010). Systematic analysis of human protein complexes identifies chromosome segregation proteins. *Science, 328*, 593–599. https://doi.org/10.1126/science.1181348.

Iliakis, G., Murmann, T., & Soni, A. (2015). Alternative end-joining repair pathways are the ultimate backup for abrogated classical non-homologous end-joining and homologous recombination repair: Implications for the formation of chromosome translocations. *Mutation Research, Genetic Toxicology and Environmental Mutagenesis, 793*, 166–175. https://doi.org/10.1016/j.mrgentox.2015.07.001.

Iourov, I. Y., Vorsanova, S. G., & Yurov, Y. B. (2021). Cytogenomic landscape of the human brain. In T. Liehr (Ed.), *Cytogenomics* (pp. 327–348). Academic Press. Chapter 16 (in this book).

Itani, O. A., Flibotte, S., Dumas, K. J., Guo, C., Blumenthal, T., & Hu, P. J. (2016). N-ethyl-N-nitrosourea (ENU) mutagenesis reveals an intronic residue critical for Caenorhabditis elegans 3' splice site function in vivo. *G3 (Bethesda), 6*, 1751–1756. https://doi.org/10.1534/g3.116.028662.

Iwamoto, M., Mori, C., Osakada, H., Koujin, T., Hiraoka, Y., & Haraguchi, T. (2018). Nuclear localization signal targeting to macronucleus and micronucleus in binucleated ciliate *Tetrahymena thermophila*. *Genes to Cells, 23*, 568–579. https://doi.org/10.1111/gtc.12602.

Janssen, A., van der Burg, M., Szuhai, K., Kops, G. J. P. L., & Medema, R. H. (2011). Chromosome segregation errors as a cause of DNA damage and structural chromosome aberrations. *Science, 333*, 1895–1898. https://doi.org/10.1126/science.1210214.

Jiang, H., Xue, X., Panda, S., Kawale, A., Hooy, R. M., Liang, F., … Gekara, N.O. (2019). Chromatin-bound cGAS is an inhibitor of DNA repair and hence accelerates genome destabilization and cell death. *The EMBO Journal, 38*, e102718. https://doi.org/10.15252/embj.2019102718.

Jones, M. J. K., & Jallepalli, P. V. (2012). Chromothripsis: Chromosomes in crisis. *Developmental Cell, 23*, 908–917. https://doi.org/10.1016/j.devcel.2012.10.010.

Kass, E. M., Moynahan, M. E., & Jasin, M. (2016). When genome maintenance goes badly awry. *Molecular Cell, 62*, 777–787. https://doi.org/10.1016/j.molcel.2016.05.021.

Kato, T., Ouchi, Y., Inagaki, H., Makita, Y., Mizuno, S., Kajita, M., … Kurahashi, H. (2017). Genomic characterization of chromosomal insertions: Insights into the mechanisms underlying chromothripsis. *Cytogenetic and Genome Research*, *153*, 1–9. https://doi.org/10.1159/000481586.

Kawakami, K., Largaespada, D. A., & Ivics, Z. (2017). Transposons as tools for functional genomics in vertebrate models. *Trends in Genetics*, *33*, 784–801. https://doi.org/10.1016/j.tig.2017.07.006.

Kloosterman, W. P., & Cuppen, E. (2013). Chromothripsis in congenital disorders and cancer: Similarities and differences. *Current Opinion in Cell Biology*, *25*, 341–348. https://doi.org/10.1016/j.ceb.2013.02.008.

Kloosterman, W. P., Guryev, V., van Roosmalen, M., Duran, K. J., de Bruijn, E., Bakker, S. C. M., … Cuppen, E. (2011). Chromothripsis as a mechanism driving complex de novo structural rearrangements in the germline. *Human Molecular Genetics*, *20*, 1916–1924. https://doi.org/10.1093/hmg/ddr073.

Kloosterman, W. P., Koster, J., & Molenaar, J. J. (2014). Prevalence and clinical implications of chromothripsis in cancer genomes. *Current Opinion in Oncology*, *26*, 64–72. https://doi.org/10.1097/CCO.0000000000000038.

Kneissig, M., Keuper, K., de Pagter, M. S., van Roosmalen, M. J., Martin, J., Otto, H., … Storchova, Z. (2019). Micronuclei-based model system reveals functional consequences of chromothripsis in human cells. *eLife*, *8*. https://doi.org/10.7554/eLife.50292.

Korbel, J. O., & Campbell, P. J. (2013). Criteria for inference of chromothripsis in cancer genomes. *Cell*, *152*, 1226–1236. https://doi.org/10.1016/j.cell.2013.02.023.

Kurahashi, H., Bolor, H., Kato, T., Kogo, H., Tsutsumi, M., Inagaki, H., & Ohye, T. (2009). Recent advance in our understanding of the molecular nature of chromosomal abnormalities. *Journal of Human Genetics*, *54*, 253–260. https://doi.org/10.1038/jhg.2009.35.

Kurtas, N. E., Xumerle, L., Giussani, U., Pansa, A., Cardarelli, L., Bertini, V., … Zuffardi, O. (2019). Insertional translocation involving an additional nonchromothriptic chromosome in constitutional chromothripsis: Rule or exception? *Molecular Genetics & Genomic Medicine*, *7*, e00496. https://doi.org/10.1002/mgg3.496.

Lang, K. S., Hall, A. N., Merrikh, C. N., Ragheb, M., Tabakh, H., Pollock, A. J., … Merrikh, H. (2017). Replication-transcription conflicts generate R-loops that orchestrate bacterial stress survival and pathogenesis. *Cell*, *170*, 787–799.e18. https://doi.org/10.1016/j.cell.2017.07.044.

Lee, J. A., Carvalho, C. M. B., & Lupski, J. R. (2007). A DNA replication mechanism for generating nonrecurrent rearrangements associated with genomic disorders. *Cell*, *131*, 1235–1247. https://doi.org/10.1016/j.cell.2007.11.037.

Lee, K. J., Lee, K. H., Yoon, K.-A., Sohn, J. Y., Lee, E., Lee, H., … Kong, S.-Y. (2017). Chromothripsis in treatment resistance in multiple myeloma. *Genome Informatics*, *15*, 87–97. https://doi.org/10.5808/GI.2017.15.3.87.

Leibowitz, M. L., Papathanasiou, S., Doerfler, P. A., Blaine, L. J., Yao, Y., Zhang, C.-Z., … Pellman, D. (2020). Chromothripsis as an on-target consequence of CRISPR-Cas9 genome editing (preprint). *Genetics*. https://doi.org/10.1101/2020.07.13.200998.

Li, Y., Roberts, N. D., Wala, J. A., Shapira, O., Schumacher, S. E., Kumar, K., … PCAWG Consortium. (2020). Patterns of somatic structural variation in human cancer genomes. *Nature*, *578*, 112–121. https://doi.org/10.1038/s41586-019-1913-9.

Liehr, T. (2021a). A definition for cytogenomics - Which also may be called chromosomics. In T. Liehr (Ed.), *Cytogenomics* (pp. 1–7). Academic Press. Chapter 1 (in this book).

Liehr, T. (2021b). Overview of currently available approaches used in cytogenomics. In T. Liehr (Ed.), *Cytogenomics* (pp. 11–24). Academic Press. Chapter 2 (in this book).

Liehr, T. (2021c). Molecular cytogenetics. In T. Liehr (Ed.), *Cytogenomics* (pp. 35–45). Academic Press. Chapter 4 (in this book).

Liehr, T. (2021d). Nuclear architecture. In T. Liehr (Ed.), *Cytogenomics* (pp. 297–305). Academic Press. Chapter 14 (in this book).

Liehr, T. (2021e). Repetitive elements, heteromorphisms, and copy number variants. In T. Liehr (Ed.), *Cytogenomics* (pp. 373–388). Academic Press. Chapter 19 (in this book).

Lin, Y.-L., & Pasero, P. (2012). Interference between DNA replication and transcription as a cause of genomic instability. *Current Genomics, 13*, 65–73. https://doi.org/10.2174/138920212799034767.

Litwin, I., Pilarczyk, E., & Wysocki, R. (2018). The emerging role of cohesin in the DNA damage response. *Genes (Basel), 9*(12), 581. https://doi.org/10.3390/genes9120581.

Liu, S., & Pellman, D. (2020). The coordination of nuclear envelope assembly and chromosome segregation in metazoans. *Nucleus, 11*, 35–52. https://doi.org/10.1080/19491034.2020.1742064.

Liu, P., Erez, A., Nagamani, S. C. S., Dhar, S. U., Kołodziejska, K. E., Dharmadhikari, A. V., … Bi, W. (2011). Chromosome catastrophes involve replication mechanisms generating complex genomic rearrangements. *Cell, 146*, 889–903. https://doi.org/10.1016/j.cell.2011.07.042.

Liu, P., Yuan, B., Carvalho, C. M. B., Wuster, A., Walter, K., Zhang, L., … Lupski, J.R. (2017). An organismal cnv mutator phenotype restricted to early human development. *Cell, 168*, 830–842.e7. https://doi.org/10.1016/j.cell.2017.01.037.

Liu, S., Kwon, M., Mannino, M., Yang, N., Renda, F., Khodjakov, A., & Pellman, D. (2018). Nuclear envelope assembly defects link mitotic errors to chromothripsis. *Nature, 561*, 551–555. https://doi.org/10.1038/s41586-018-0534-z.

Luijten, M. N. H., Lee, J. X. T., & Crasta, K. C. (2018). Mutational game changer: Chromothripsis and its emerging relevance to cancer. *Mutation Research, 777*, 29–51. https://doi.org/10.1016/j.mrrev.2018.06.004.

Luperchio, T. R., Wong, X., & Reddy, K. L. (2014). Genome regulation at the peripheral zone: Lamina associated domains in development and disease. *Current Opinion in Genetics & Development, 25*, 50–61. https://doi.org/10.1016/j.gde.2013.11.021.

Lupiáñez, D. G., Spielmann, M., & Mundlos, S. (2016). Breaking TADs: How alterations of chromatin domains result in disease. *Trends in Genetics, 32*, 225–237. https://doi.org/10.1016/j.tig.2016.01.003.

Luzhna, L., Kathiria, P., & Kovalchuk, O. (2013). Micronuclei in genotoxicity assessment: From genetics to epigenetics and beyond. *Frontiers in Genetics, 4*, 131. https://doi.org/10.3389/fgene.2013.00131.

Ly, P., & Cleveland, D. W. (2017). Rebuilding chromosomes after catastrophe: Emerging mechanisms of chromothripsis. *Trends in Cell Biology, 27*, 917–930. https://doi.org/10.1016/j.tcb.2017.08.005.

Ly, P., Teitz, L. S., Kim, D. H., Shoshani, O., Skaletsky, H., Fachinetti, D., … Cleveland, D.W. (2017). Selective Y centromere inactivation triggers chromosome shattering in micronuclei and repair by non-homologous end joining. *Nature Cell Biology, 19*, 68–75. https://doi.org/10.1038/ncb3450.

Ly, P., Brunner, S. F., Shoshani, O., Kim, D. H., Lan, W., Pyntikova, T., … Cleveland, D.W. (2019). Chromosome segregation errors generate a diverse spectrum of simple and complex genomic rearrangements. *Nature Genetics, 51*, 705–715. https://doi.org/10.1038/s41588-019-0360-8.

Maass, K. K., Rosing, F., Ronchi, P., Willmund, K. V., Devens, F., Hergt, M., … Ernst, A. (2018). Altered nuclear envelope structure and proteasome function of micronuclei. *Experimental Cell Research*, *371*, 353–363. https://doi.org/10.1016/j.yexcr.2018.08.029.

Maciejowski, J., & de Lange, T. (2017). Telomeres in cancer: Tumour suppression and genome instability. *Nature Reviews Molecular Cell Biology*, *18*, 175–186. https://doi.org/10.1038/nrm.2016.171.

Maciejowski, J., Li, Y., Bosco, N., Campbell, P. J., & de Lange, T. (2015). Chromothripsis and kataegis induced by telomere crisis. *Cell*, *163*, 1641–1654. https://doi.org/10.1016/j.cell.2015.11.054.

Mackenzie, K. J., Carroll, P., Martin, C.-A., Murina, O., Fluteau, A., Simpson, D. J., … Jackson, A.P. (2017). cGAS surveillance of micronuclei links genome instability to innate immunity. *Nature*, *548*, 461–465. https://doi.org/10.1038/nature23449.

Mackinnon, R. N., & Campbell, L. J. (2013). Chromothripsis under the microscope: A cytogenetic perspective of two cases of AML with catastrophic chromosome rearrangement. *Cancer Genetics*, *206*, 238–251. https://doi.org/10.1016/j.cancergen.2013.05.021.

Madan, K. (2012). Balanced complex chromosome rearrangements: Reproductive aspects. A review. *American Journal of Medical Genetics. Part A*, *158A*, 947–963. https://doi.org/10.1002/ajmg.a.35220.

Mardin, B. R., Drainas, A. P., Waszak, S. M., Weischenfeldt, J., Isokane, M., Stütz, A. M., … Korbel, J.O. (2015). A cell-based model system links chromothripsis with hyperploidy. *Molecular Systems Biology*, *11*, 828. https://doi.org/10.15252/msb.20156505.

Margolis, S. R., Wilson, S. C., & Vance, R. E. (2017). Evolutionary origins of cGAS-STING signaling. *Trends in Immunology*, *38*, 733–743. https://doi.org/10.1016/j.it.2017.03.004.

Masset, H., Hestand, M. S., Van Esch, H., Kleinfinger, P., Plaisancié, J., Afenjar, A., … Vermeesch, J.R. (2016). A distinct class of chromoanagenesis events characterized by focal copy number gains. *Human Mutation*, *37*, 661–668. https://doi.org/10.1002/humu.22984.

Mateos-Gomez, P. A., Gong, F., Nair, N., Miller, K. M., Lazzerini-Denchi, E., & Sfeir, A. (2015). Mammalian polymerase θ promotes alternative NHEJ and suppresses recombination. *Nature*, *518*, 254–257. https://doi.org/10.1038/nature14157.

Mathieu, N., Pirzio, L., Freulet-Marrière, M.-A., Desmaze, C., & Sabatier, L. (2004). Telomeres and chromosomal instability. *Cellular and Molecular Life Sciences*, *61*, 641–656. https://doi.org/10.1007/s00018-003-3296-0.

McDermott, D. H., Gao, J.-L., Liu, Q., Siwicki, M., Martens, C., Jacobs, P., … Murphy, P.M. (2015). Chromothriptic cure of WHIM syndrome. *Cell*, *160*, 686–699. https://doi.org/10.1016/j.cell.2015.01.014.

Meschini, R., Morucci, E., Berni, A., Lopez-Martinez, W., & Palitti, F. (2015). Role of chromatin structure modulation by the histone deacetylase inhibitor trichostatin A on the radio-sensitivity of ataxia telangiectasia. *Mutation Research*, *777*, 52–59. https://doi.org/10.1016/j.mrfmmm.2015.04.009.

Meyer, T. J., Held, U., Nevonen, K. A., Klawitter, S., Pirzer, T., Carbone, L., & Schumann, G. G. (2016). The flow of the gibbon LAVA element is facilitated by the LINE-1 retrotransposition machinery. *Genome Biology and Evolution*, *8*, 3209–3225. https://doi.org/10.1093/gbe/evw224.

Middelkamp, S., van Heesch, S., Braat, A. K., de Ligt, J., van Iterson, M., Simonis, M., … Cuppen, E. (2017). Molecular dissection of germline chromothripsis in a developmental context using patient-derived iPS cells. *Genome Medicine*, *9*, 9. https://doi.org/10.1186/s13073-017-0399-z.

Miné-Hattab, J., & Rothstein, R. (2012). Increased chromosome mobility facilitates homology search during recombination. *Nature Cell Biology*, *14*, 510–517. https://doi.org/10.1038/ncb2472.

Miné-Hattab, J., Recamier, V., Izeddin, I., Rothstein, R., & Darzacq, X. (2017). Multi-scale tracking reveals scale-dependent chromatin dynamics after DNA damage. *Molecular Biology of the Cell*. https://doi.org/10.1091/mbc.E17-05-0317.

Misteli, T. (2020). The self-organizing genome: Principles of genome architecture and function. *Cell*, *183*, 28–45. https://doi.org/10.1016/j.cell.2020.09.014.

Morishita, M., Muramatsu, T., Suto, Y., Hirai, M., Konishi, T., Hayashi, S., … Inazawa, J. (2016). Chromothripsis-like chromosomal rearrangements induced by ionizing radiation using proton microbeam irradiation system. *Oncotarget*, *7*, 10182–10192. https://doi.org/10.18632/oncotarget.7186.

Nam, H.-J., Naylor, R. M., & van Deursen, J. M. (2015). Centrosome dynamics as a source of chromosomal instability. *Trends in Cell Biology*, *25*, 65–73. https://doi.org/10.1016/j.tcb.2014.10.002.

Nazaryan-Petersen, L., Bertelsen, B., Bak, M., Jønson, L., Tommerup, N., Hancks, D. C., & Tümer, Z. (2016). Germline chromothripsis driven by L1-mediated retrotransposition and Alu/Alu homologous recombination. *Human Mutation*, *37*, 385–395. https://doi.org/10.1002/humu.22953.

Nazaryan-Petersen, L., Eisfeldt, J., Pettersson, M., Lundin, J., Nilsson, D., Wincent, J., … Lindstrand, A. (2018). Replicative and non-replicative mechanisms in the formation of clustered CNVs are indicated by whole genome characterization. *PLoS Genetics*, *14*, e1007780. https://doi.org/10.1371/journal.pgen.1007780.

Nik-Zainal, S., Alexandrov, L. B., Wedge, D. C., Van Loo, P., Greenman, C. D., Raine, K., … Breast Cancer Working Group of the International Cancer Genome Consortium. (2012). Mutational processes molding the genomes of 21 breast cancers. *Cell*, *149*, 979–993. https://doi.org/10.1016/j.cell.2012.04.024.

Oesper, L., Dantas, S., & Raphael, B. J. (2018). Identifying simultaneous rearrangements in cancer genomes. *Bioinformatics*, *34*, 346–352. https://doi.org/10.1093/bioinformatics/btx745.

Ottaviani, D., LeCain, M., & Sheer, D. (2014). The role of microhomology in genomic structural variation. *Trends in Genetics*, *30*, 85–94. https://doi.org/10.1016/j.tig.2014.01.001.

Pace, J. K., & Feschotte, C. (2007). The evolutionary history of human DNA transposons: Evidence for intense activity in the primate lineage. *Genome Research*, *17*, 422–432. https://doi.org/10.1101/gr.5826307.

Pampalona, J., Soler, D., Genescà, A., & Tusell, L. (2010). Telomere dysfunction and chromosome structure modulate the contribution of individual chromosomes in abnormal nuclear morphologies. *Mutation Research*, *683*, 16–22. https://doi.org/10.1016/j.mrfmmm.2009.10.001.

Pantelias, A., Karachristou, I., Georgakilas, A. G., & Terzoudi, G. I. (2019). Interphase cytogenetic analysis of micronucleated and multinucleated cells supports the premature chromosome condensation hypothesis as the mechanistic origin of chromothripsis. *Cancers (Basel)*, *11*. https://doi.org/10.3390/cancers11081123.

Papenfuss, A. T., & Thomas, D. M. (2015). The life history of neochromosomes revealed. *Molecular & Cellular Oncology*, *2*. https://doi.org/10.1080/23723556.2014.1000698.

Passerini, V., Ozeri-Galai, E., de Pagter, M. S., Donnelly, N., Schmalbrock, S., Kloosterman, W. P., … Storchová, Z. (2016). The presence of extra chromosomes leads to genomic instability. *Nature Communications*, *7*, 10754. https://doi.org/10.1038/ncomms10754.

Pellestor, F. (2014). Chromothripsis: How does such a catastrophic event impact human reproduction? *Human Reproduction*, *29*, 388–393. https://doi.org/10.1093/humrep/deu003.

Pellestor, F., & Gatinois, V. (2018). Chromoanasynthesis: Another way for the formation of complex chromosomal abnormalities in human reproduction. *Human Reproduction*, *33*, 1381–1387. https://doi.org/10.1093/humrep/dey231.

Pellestor, F., & Gatinois, V. (2020). Chromoanagenesis: A piece of the macroevolution scenario. *Molecular Cytogenetics*, *13*, 3. https://doi.org/10.1186/s13039-020-0470-0.

Pellestor, F., Anahory, T., Lefort, G., Puechberty, J., Liehr, T., Hédon, B., & Sarda, P. (2011). Complex chromosomal rearrangements: Origin and meiotic behavior. *Human Reproduction Update*, *17*, 476–494. https://doi.org/10.1093/humupd/dmr010.

Petljak, M., & Maciejowski, J. (2020). Molecular origins of APOBEC-associated mutations in cancer. *DNA Repair (Amst)*, *94*, 102905. https://doi.org/10.1016/j.dnarep.2020.102905.

Pettersson, M., Eisfeldt, J., Syk Lundberg, E., Lundin, J., & Lindstrand, A. (2018). Flanking complex copy number variants in the same family formed through unequal crossing-over during meiosis. *Mutation Research*, *812*, 1–4. https://doi.org/10.1016/j.mrfmmm.2018.10.001.

Piazza, A., & Heyer, W.-D. (2019). Homologous recombination and the formation of complex genomic rearrangements. *Trends in Cell Biology*, *29*, 135–149. https://doi.org/10.1016/j.tcb.2018.10.006.

Piazza, A., Wright, W. D., & Heyer, W.-D. (2017). Multi-invasions are recombination by-products that induce chromosomal rearrangements. *Cell*, *170*, 760–773.e15. https://doi.org/10.1016/j.cell.2017.06.052.

Pihan, G. A. (2013). Centrosome dysfunction contributes to chromosome instability, chromoanagenesis, and genome reprograming in cancer. *Frontiers in Oncology*, *3*, 277. https://doi.org/10.3389/fonc.2013.00277.

Plaisancié, J., Kleinfinger, P., Cances, C., Bazin, A., Julia, S., Trost, D., … Vigouroux, A. (2014). Constitutional chromoanasynthesis: Description of a rare chromosomal event in a patient. *European Journal of Medical Genetics*, *57*, 567–570. https://doi.org/10.1016/j.ejmg.2014.07.004.

Poot, M. (2016). From telomere crisis via dicentric chromosomes to kataegis and chromothripsis. *Molecular Syndromology*, *6*, 259–260. https://doi.org/10.1159/000443805.

Price, B. D., & D'Andrea, A. D. (2013). Chromatin remodeling at DNA double-strand breaks. *Cell*, *152*, 1344–1354. https://doi.org/10.1016/j.cell.2013.02.011.

Prosser, S. L., & Pelletier, L. (2017). Mitotic spindle assembly in animal cells: A fine balancing act. *Nature Reviews Molecular Cell Biology*, *18*, 187–201. https://doi.org/10.1038/nrm.2016.162.

Redin, C., Brand, H., Collins, R. L., Kammin, T., Mitchell, E., Hodge, J. C., … Talkowski, M.E. (2017). The genomic landscape of balanced cytogenetic abnormalities associated with human congenital anomalies. *Nature Genetics*, *49*, 36–45. https://doi.org/10.1038/ng.3720.

Reece, A. S., & Hulse, G. K. (2016). Chromothripsis and epigenomics complete causality criteria for cannabis- and addiction-connected carcinogenicity, congenital toxicity and heritable genotoxicity. *Mutation Research*, *789*, 15–25. https://doi.org/10.1016/j.mrfmmm.2016.05.002.

Reimann, H., Stopper, H., & Hintzsche, H. (2020). Long-term fate of etoposide-induced micronuclei and micronucleated cells in Hela-H2B-GFP cells. *Archives of Toxicology*, *94*, 3553–3561. https://doi.org/10.1007/s00204-020-02840-0.

Romanenko, S. A., Serdyukova, N. A., Perelman, P. L., Pavlova, S. V., Bulatova, N. S., Golenishchev, F. N., … Graphodatsky, A.S. (2017). Intrachromosomal rearrangements in rodents from the perspective of comparative region-specific painting. *Genes (Basel)*, *8*. https://doi.org/10.3390/genes8090215.

Russo, A., & Degrassi, F. (2018). Molecular cytogenetics of the micronucleus: Still surprising. *Mutation Research, Genetic Toxicology and Environmental Mutagenesis, 836*, 36–40. https://doi.org/10.1016/j.mrgentox.2018.05.011.

Sabatini, P. J. B., Ejaz, R., Stavropoulos, D. J., Mendoza-Londono, R., & Joseph-George, A. M. (2018). Stable transmission of an unbalanced chromosome 21 derived from chromoanasynthesis in a patient with a SYNGAP1 likely pathogenic variant. *Molecular Cytogenetics, 11*, 50. https://doi.org/10.1186/s13039-018-0394-0.

Schweizer, N., Weiss, M., & Maiato, H. (2014). The dynamic spindle matrix. *Current Opinion in Cell Biology, 28*, 1–7. https://doi.org/10.1016/j.ceb.2014.01.002.

Schütze, D. M., Krijgsman, O., Snijders, P. J. F., Ylstra, B., Weischenfeldt, J., Mardin, B. R., … Steenbergen, R.D.M. (2016). Immortalization capacity of HPV types is inversely related to chromosomal instability. *Oncotarget, 7*, 37608–37621. https://doi.org/10.18632/oncotarget.8058.

Sheltzer, J. M., Torres, E. M., Dunham, M. J., & Amon, A. (2012). Transcriptional consequences of aneuploidy. *Proceedings of the National Academy of Sciences of the United States of America, 109*, 12644–12649. https://doi.org/10.1073/pnas.1209227109.

Shen, M. M. (2013). Chromoplexy: A new category of complex rearrangements in the cancer genome. *Cancer Cell, 23*, 567–569. https://doi.org/10.1016/j.ccr.2013.04.025.

Shi, C., Channels, W. E., Zheng, Y., & Iglesias, P. A. (2014). A computational model for the formation of lamin-B mitotic spindle envelope and matrix. *Interface Focus, 4*, 20130063. https://doi.org/10.1098/rsfs.2013.0063.

Skibbens, R. V. (2019). Condensins and cohesins—One of these things is not like the other! *Journal of Cell Science, 132*. https://doi.org/10.1242/jcs.220491.

Slamova, Z., Nazaryan-Petersen, L., Mehrjouy, M. M., Drabova, J., Hancarova, M., Marikova, T., … Sedlacek, Z. (2018). Very short DNA segments can be detected and handled by the repair machinery during germline chromothriptic chromosome reassembly. *Human Mutation, 39*, 709–716. https://doi.org/10.1002/humu.23408.

Song, X., Beck, C. R., Du, R., Campbell, I. M., Coban-Akdemir, Z., Gu, S., … Lupski, J.R. (2018). Predicting human genes susceptible to genomic instability associated with Alu/Alu-mediated rearrangements. *Genome Research, 28*, 1228–1242. https://doi.org/10.1101/gr.229401.117.

Sorzano, C. O. S., Pascual-Montano, A., Sánchez de Diego, A., Martínez-A, C., & van Wely, K. H. M. (2013). Chromothripsis: Breakage-fusion-bridge over and over again. *Cell Cycle, 12*, 2016–2023. https://doi.org/10.4161/cc.25266.

Soto, M., García-Santisteban, I., Krenning, L., Medema, R. H., & Raaijmakers, J. A. (2018). Chromosomes trapped in micronuclei are liable to segregation errors. *Journal of Cell Science, 131*. https://doi.org/10.1242/jcs.214742.

Stephens, P. J., Greenman, C. D., Fu, B., Yang, F., Bignell, G. R., Mudie, L. J., … Campbell, P.J. (2011). Massive genomic rearrangement acquired in a single catastrophic event during cancer development. *Cell, 144*, 27–40. https://doi.org/10.1016/j.cell.2010.11.055.

Stevens, J. B., Abdallah, B. Y., Regan, S. M., Liu, G., Bremer, S. W., Ye, C. J., & Heng, H. H. (2010). Comparison of mitotic cell death by chromosome fragmentation to premature chromosome condensation. *Molecular Cytogenetics, 3*, 20. https://doi.org/10.1186/1755-8166-3-20.

Storchova, Z., & Kuffer, C. (2008). The consequences of tetraploidy and aneuploidy. *Journal of Cell Science, 121*, 3859–3866. https://doi.org/10.1242/jcs.039537.

Streffer, C. (2010). Strong association between cancer and genomic instability. *Radiation and Environmental Biophysics, 49*, 125–131. https://doi.org/10.1007/s00411-009-0258-4.

Sun, L., Wu, J., Du, F., Chen, X., & Chen, Z. J. (2013). Cyclic GMP-AMP synthase is a cytosolic DNA sensor that activates the type I interferon pathway. *Science*, *339*, 786–791. https://doi.org/10.1126/science.1232458.

Sutyagina, O. I., & Kisurina-Evgenieva, O. P. (2019). Morphofunctional differences of micronuclei in cultures of human p53-positive and p53-negative tumor cells. *Bulletin of Experimental Biology and Medicine*, *167*, 813–817. https://doi.org/10.1007/s10517-019-04629-3.

Suzuki, E., Shima, H., Toki, M., Hanew, K., Matsubara, K., Kurahashi, H., … Fukami, M. (2016). Complex X-chromosomal rearrangements in two women with ovarian dysfunction: Implications of chromothripsis/chromoanasynthesis-dependent and -independent origins of complex genomic alterations. *Cytogenetic and Genome Research*, *150*, 86–92. https://doi.org/10.1159/000455026.

Symer, D. E., Connelly, C., Szak, S. T., Caputo, E. M., Cost, G. J., Parmigiani, G., & Boeke, J. D. (2002). Human l1 retrotransposition is associated with genetic instability in vivo. *Cell*, *110*, 327–338. https://doi.org/10.1016/s0092-8674(02)00839-5.

Tan, E. H., Henry, I. M., Ravi, M., Bradnam, K. R., Mandakova, T., Marimuthu, M. P., … Chan, S.W. (2015). Catastrophic chromosomal restructuring during genome elimination in plants. *eLife*, *4*. https://doi.org/10.7554/eLife.06516.

Terradas, M., Martín, M., Tusell, L., & Genescà, A. (2010). Genetic activities in micronuclei: Is the DNA entrapped in micronuclei lost for the cell? *Mutation Research*, *705*, 60–67. https://doi.org/10.1016/j.mrrev.2010.03.004.

Terradas, M., Martín, M., & Genescà, A. (2016). Impaired nuclear functions in micronuclei results in genome instability and chromothripsis. *Archives of Toxicology*, *90*, 2657–2667. https://doi.org/10.1007/s00204-016-1818-4.

Terzoudi, G. I., Karakosta, M., Pantelias, A., Hatzi, V. I., Karachristou, I., & Pantelias, G. (2015). Stress induced by premature chromatin condensation triggers chromosome shattering and chromothripsis at DNA sites still replicating in micronuclei or multinucleate cells when primary nuclei enter mitosis. *Mutation Research, Genetic Toxicology and Environmental Mutagenesis*, *793*, 185–198. https://doi.org/10.1016/j.mrgentox.2015.07.014.

Thompson, S. L., & Compton, D. A. (2008). Examining the link between chromosomal instability and aneuploidy in human cells. *The Journal of Cell Biology*, *180*, 665–672. https://doi.org/10.1083/jcb.200712029.

Tubio, J. M. C., & Estivill, X. (2011). Cancer: When catastrophe strikes a cell. *Nature*, *470*, 476–477. https://doi.org/10.1038/470476a.

Umbreit, N. T., Zhang, C.-Z., Lynch, L. D., Blaine, L. J., Cheng, A. M., Tourdot, R., … Pellman, D. (2020). Mechanisms generating cancer genome complexity from a single cell division error. *Science*, *368*. https://doi.org/10.1126/science.aba0712.

Underwood, C. J., & Choi, K. (2019). Heterogeneous transposable elements as silencers, enhancers and targets of meiotic recombination. *Chromosoma*, *128*, 279–296. https://doi.org/10.1007/s00412-019-00718-4.

Ungelenk, M. (2021). Sequencing approaches. In T. Liehr (Ed.), *Cytogenomics* (pp. 87–122). Academic Press. Chapter 7 (in this book).

Ungricht, R., & Kutay, U. (2017). Mechanisms and functions of nuclear envelope remodelling. *Nature Reviews Molecular Cell Biology*, *18*, 229–245. https://doi.org/10.1038/nrm.2016.153.

Vargas, J. D., Hatch, E. M., Anderson, D. J., & Hetzer, M. W. (2012). Transient nuclear envelope rupturing during interphase in human cancer cells. *Nucleus*, *3*, 88–100. https://doi.org/10.4161/nucl.18954.

Venkatesan, S., Natarajan, A. T., & Hande, M. P. (2015). Chromosomal instability—mechanisms and consequences. *Mutation Research—Genetic Toxicology and Environmental Mutagenesis, 793*, 176–184. https://doi.org/10.1016/j.mrgentox.2015.08.008.

Voronina, N., Wong, J. K. L., Hübschmann, D., Hlevnjak, M., Uhrig, S., Heilig, C. E., … Ernst, A. (2020). The landscape of chromothripsis across adult cancer types. *Nature Communications, 11.* https://doi.org/10.1038/s41467-020-16134-7.

Wang, K., Wang, Y., & Collins, C. C. (2013). Chromoplexy: A new paradigm in genome remodeling and evolution. *Asian Journal of Andrology, 15*, 711–712. https://doi.org/10.1038/aja.2013.109.

Weckselblatt, B., Hermetz, K. E., & Rudd, M. K. (2015). Unbalanced translocations arise from diverse mutational mechanisms including chromothripsis. *Genome Research, 25*, 937–947. https://doi.org/10.1101/gr.191247.115.

Weise, A., & Liehr, T. (2021a). Cytogenetics. In T. Liehr (Ed.), *Cytogenomics* (pp. 25–34). Academic Press. Chapter 3 (in this book).

Weise, A., & Liehr, T. (2021b). Interchromosomal interactions with meaning for disease. In T. Liehr (Ed.), *Cytogenomics* (pp. 349–356). Academic Press. Chapter 17 (in this book).

Wilhelm, K., Pentzold, C., Schoener, S., Arakelyan, A., Hakobyan, A., Mrasek, K., & Weise, A. (2018). Fragile sites as drivers of gene and genome evolution. *Current Genetic Medicine Reports, 6*, 136–143.

Wilhelm, T., Said, M., & Naim, V. (2020). DNA replication stress and chromosomal instability: Dangerous liaisons. *Genes (Basel), 11.* https://doi.org/10.3390/genes11060642.

Willis, N. A., Rass, E., & Scully, R. (2015). Deciphering the code of the cancer genome: Mechanisms of chromosome rearrangement. *Trends Cancer, 1*, 217–230. https://doi.org/10.1016/j.trecan.2015.10.007.

Yang, J., Liu, J., Ouyang, L., Chen, Y., Liu, B., & Cai, H. (2016). CTLPScanner: A web server for chromothripsis-like pattern detection. *Nucleic Acids Research, 44*, W252–W258. https://doi.org/10.1093/nar/gkw434.

Yauy, K., Gatinois, V., Guignard, T., Sati, S., Puechberty, J., Gaillard, J. B., … Pellestor, F. (2018). Looking for broken TAD boundaries and changes on DNA interactions: Clinical guide to 3D chromatin change analysis in complex chromosomal rearrangements and chromothripsis. *Methods in Molecular Biology, 1769*, 353–361. https://doi.org/10.1007/978-1-4939-7780-2_22.

Ye, C. J., Sharpe, Z., Alemara, S., Mackenzie, S., Liu, G., Abdallah, B., … Heng, H.H. (2019). Micronuclei and genome chaos: Changing the system inheritance. *Genes (Basel), 10.* https://doi.org/10.3390/genes10050366.

Yumiceba, V., Souto Melo, U., & Spielmann, M. (2021). 3D cytogenomics: Structural variation in the three-dimensional genome. In T. Liehr (Ed.), *Cytogenomics* (pp. 247–266). Academic Press. Chapter 12 (in this book).

Zanardo, É. A., Piazzon, F. B., Dutra, R. L., Dias, A. T., Montenegro, M. M., Novo-Filho, G. M., … Kulikowski, L.D. (2014). Complex structural rearrangement features suggesting chromoanagenesis mechanism in a case of 1p36 deletion syndrome. *Molecular Genetics and Genomics, 289*, 1037–1043. https://doi.org/10.1007/s00438-014-0876-7.

Zepeda-Mendoza, C. J., & Morton, C. C. (2019). The iceberg under water: Unexplored complexity of chromoanagenesis in congenital disorders. *American Journal of Human Genetics, 104*, 565–577. https://doi.org/10.1016/j.ajhg.2019.02.024.

Zhang, F., Carvalho, C. M. B., & Lupski, J. R. (2009). Complex human chromosomal and genomic rearrangements. *Trends in Genetics*, *25*, 298–307. https://doi.org/10.1016/j.tig.2009.05.005.

Zhang, C.-Z., Spektor, A., Cornils, H., Francis, J. M., Jackson, E. K., Liu, S., … Pellman, D. (2015). Chromothripsis from DNA damage in micronuclei. *Nature*, *522*, 179–184. https://doi.org/10.1038/nature14493.

Zheng, Y. (2010). A membranous spindle matrix orchestrates cell division. *Nature Reviews Molecular Cell Biology*, *11*, 529–535. https://doi.org/10.1038/nrm2919.

3D cytogenomics: Structural variation in the three-dimensional genome

12

Veronica Yumiceba[a], Uirá Souto Melo[b], and Malte Spielmann[a,b]

[a]*Institute of Human Genetics, University of Lübeck, Lübeck, Germany*
[b]*Human Molecular Genomics Group, Max Planck Institute for Molecular Genetics, Berlin, Germany*

Chapter outline

Introduction

The human genome consists of three billion base pairs and measures approximately one-meter length when stretched out. Naturally, this genomic information is packed and accommodated in a micro-sized nucleus, yet it is still able to replicate and transcribe itself. Tight control of the level and timing of gene expression in different tissues maintains human health and homeostasis (Crutchley et al., 2010). Gene regulation is tightly controlled by cis-regulatory elements, such as enhancers, promoters, and insulators that reside in the 98% of the human DNA that is noncoding (Liehr, 2021a). Epigenetic marks, modified DNA wrapping proteins, and long noncoding RNA (Ichikawa & Saitoh, 2021) add another layer of complexity to gene regulation (ENCODE Project Consortium, 2012). However, the precise mechanism of how noncoding sequences can act as regulatory elements and how genetic variation in the noncoding genome can contribute to health and disease are still poorly understood and encompass an area for cytogenomic research (Liehr, 2021b, 2021c).

Human genome variation is classified into either structural variants (SVs) or single-nucleotide variants (SNVs). SVs include deletions, duplications, insertions, inversions, and translocations of more than 50 bp in size. Inversions and translocations are balanced rearrangements that change the genomic architecture without loss or gain of genomic information. Deletions and duplications, collectively known as copy number variants (CNVs), alter the diploid status of the genome (Weise & Liehr, 2021a). Although SVs occur less frequently than SNVs, they account for 60%–90% of the differences between individuals (Kosugi et al., 2019). It has been estimated that nearly 4%–7% of SVs impact gene expression, with larger effects compared to SNVs. Over 88% of SVs with transcriptional effects occur within enhancers and other regulatory elements, emphasizing their potential pathogenic effects (Chiang et al., 2017; Collins et al., 2021). The extent to which genomic and transcriptional activity perturbation can be tolerated or if there is a threshold between disease and heath remains elusive.

3D genome: Hierarchical organization of genetic information in the nucleus

The human DNA is a running code written with four characters (A-T-C-G). It is packed as a double-strand molecule wrapped around eight core histone proteins, two each of H2A, H2B, H3, and H4, forming nucleosomes. Nucleosomes, linked by 10–80 bp of DNA, form a linear bead-on-a-string structure referred to as 10 nm filaments. During cell replication, a 40-fold compaction of DNA length is reached by folding the 10 nm filaments into 30 nm chromatin fibers. During metaphase, a higher order of compaction of the 30 nm fibers takes place and forms the chromosomes. Chromosomes, which are visible under a microscope (Weise & Liehr, 2021b), enable the distribution of the cellular genomic information equally into their daughter cells (Fig. 12.1A; Coleman, 2018; Gross et al., 2015). Meanwhile, in interphase, the 10 nm filaments are organized into topologically associating domains (TADs) that define the three-dimensional (3D) genome architecture (Fig. 12.1B; see as well Daban, 2021).

As discussed further, TADs form the regulatory backbone of the genome. In a supra-nucleosomal level, TADs are grouped into two compartments, named A and B. The A compartment contains predominantly active chromatin (i.e., genes located in A compartment are more prone to be expressed) and is often located near the center of the nucleus, while B compartment mainly contains repressed genes and is found close to nuclear lamina. It is important to mention that the A/B compartmentalization is cell type-dependent, i.e., each cell type will have a specific A/B landscape for every chromosome. Although chromosomes are not densely packed during interphase, they have been shown to segregate into chromosome territories, which prevents interchromosomal intermingling (Liehr, 2021d; Szabo et al., 2019).

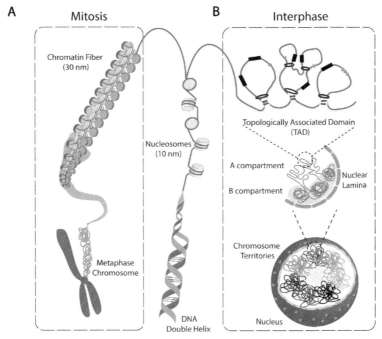

FIG. 12.1

Hierarchical 3D genome organization. The double helix of DNA compaction occurs at several levels. The first level is its wrapping around a core of 8 histones. The next levels of DNA compaction depend on the cellular cycle stage: (A) During metaphase, a mitotic phase, the chromosomes offer the most compacted form of DNA organization. This ensures a reliable genomic delivery to the resulting daughter cells. (B) In the interphase, the DNA is organized into less compact structures called TADs to allow controlled gene expression by providing insulated domains within which gene and cis-regulatory elements interact. At the next level, the genome is separated into transcriptionally active (A) and inactive (B) compartments. Inactive regions are preferentially bound to the nuclear lamina. At the scale of the nucleus, chromosomes (represented by different colors) occupy their own territory and do not physically interact with neighboring chromosomes.

Topologically associating domains

TADs are mega-base domains where regulatory elements (e.g., enhancers) and its cognate genes physically interact (Gross et al., 2015). They are insulated from one another by regions with low chromatin contacts, called TAD boundaries. Genes and enhancers located in the same TAD interact frequently with each other, but rarely with genes/enhancers from neighboring TADs (Dixon et al., 2012). The boundaries are defined by convergent CTCF binding sites, where CTCF protein binds, and cohesin protein complex accumulate (extrusion model).

These boundaries act as physical barriers blocking ectopic interactions between adjacent TADs (Spielmann et al., 2018). TADs and their boundaries are conserved across several tissues and preserved through mammalian evolution (Dixon et al., 2012). It has recently been shown that alteration of TAD structures caused by SVs or epimutations can cause abnormal gene expression (misexpression) and ultimately disease (Flavahan et al., 2016; Franke et al., 2016; Lupiáñez et al., 2015; Spielmann et al., 2018).

Techniques to detect chromatin interactions in the nuclear space

The rapid progress of chromosome conformation capture (3C) technologies coupled with next-generation sequencing has dramatically advanced our knowledge about chromatin organization (Ungelenk, 2021). In 2002, Dekker and his collaborators developed the 3C methodology using yeast genomic material (Dekker et al., 2002). In 3C, the genomic DNA is first fixed to freeze proteins and interacting DNA in the nucleus (Fig. 12.2).

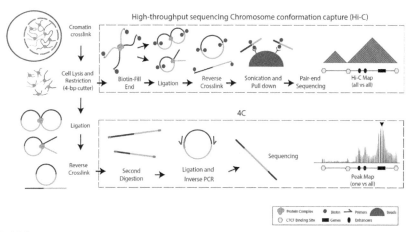

FIG. 12.2

Workflow of chromosome conformation capture assays. The basic workflow of 3C and related technologies are schematized on the left side. The starting material is segregated more less one million cells. First, the chromatin interactions mediated by proteins complexes inside the nucleus are stabilized by fixation. Second, lysed DNA is restricted by enzymatic processes. All segments containing proteins are protected from cutting enzymes. Thirdly, ligation of the interacting fragments from either ends or from one end stick fragments. Fourth, proteins are washed away from ligated fragments and now they are ready for downstream sequencing analysis. Inside boxes, Hi-C and 4C steps are depicted. Compared to the main sequence steps, in Hi-C restricted fragment ends are filled with biotin before ligation and reverse crosslink. Fragments are sonicated, as Hi-C is based on short-end sequencing. Biotin makes it easier to pull down biotinylated fragments

Second, cells are lysed to isolate DNA from cytoplasmic and nuclear content. Third genomic DNA devoid of proteins is cut out through enzymatic digestion (usually a 4-bp cutter, e.g., DpnII). Fourth, the DNA product is ligated, where the odds of ligating the DNA ends is higher when bound by proteins, due to increased proximity. Fourth, through a reverse-crosslinking process, proteins are detached from DNA for sequencing and analysis. For 4C assays, the fragments are redigested and self-circularized. The inverse PCR reaction allows the detection of all DNA interacting segments containing the target genomic site or viewpoint. The 4C results are usually depicted as peaks along the chromosomal position. Higher peaks represent preferred interactions or chromatin loops with the target genomic site (Simonis et al., 2006). Several advances to this concept have been made to achieve whole-genome interaction analysis in one experiment. For instance, in Hi-C, an important step during ligation is the biotinylation of digested chromatin ends by adding Biotin-dCTP or -dATP for later pull-down of chimeric fragments. After the reverse crosslink, fragments with biotinylated nucleotides are recovered with streptavidin-coated magnetic beads in preparation for pair-end sequencing (Lieberman-Aiden et al., 2009).

Features of a Hi-C map

While karyotyping offers a course view of the whole genome for cytogenetics (Liehr, 2021e; Weise & Liehr, 2021a), Hi-C heatmaps provide an equivalent, but finer and three-dimensional view in genomics (Fig. 12.3A). The Hi-C heatmaps represent the interacting 23 pairs of autosomes and sex chromosomes, similar to the banded karyotype. Higher interactions occur between segments that reside in the same chromosome (cis) than across chromosomes (trans). This is apparent in the higher interaction intensity along diagonal, where the chromosomes are matched in both axes (cis maps). The minimal interaction intensities away from the diagonal indicate chromosomal segregation under normal conditions. Thus, the various features of the 3D genome organization, such as compartments, TADs, and loops at various length scales are displayed visually in this two-dimensional Hi-C heatmap matrix (Fig. 12.3B and C). Chromatin loops appear as punctate marks at the vertices of triangular structures representing

FIG. 12.2, CONT'D

for sequencing (Dekker et al., 2002). For 4C assay, the basic 3C steps are performed until reverse crosslinked fragments are obtained. Later, a second digestion is performed, and the smaller fragments are self-circularized. 4C only tests the interaction between a particular bait or target segment with other genomic regions. Because the targeted fragment sequence is known, primers are designed to reverse amplify the circularized molecules. The results of contact maps are shown at the end of the boxes, the Hi-C mapped the whole-genome interaction as heatmaps (all vs. all), while 4C indicates all possible interactions with one target region as viewpoint (*arrow head* on top of the peak map) (one vs. all).

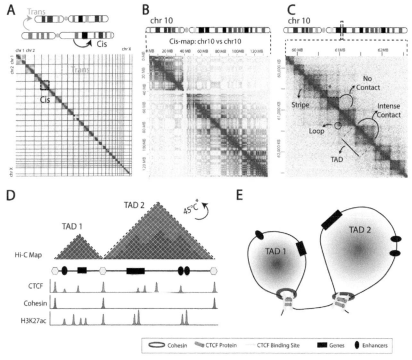

FIG. 12.3

Hi-C maps and topologically associating domains (TAD). (A) Cis and trans contacts are shown in the chromosome cartoon as well as in the Hi-C map, where cis-maps constitute the diagonal (Melo et al., 2020). (B) A Hi-C map of the chromosome 10 shows the alternating A and B compartmentalization, resembling a chess board, based on its transcriptional state. (C) A zoom-in into a 1 Mb section of the chromosome 10, where TADs emerge from the main diagonal. The interaction intensity is color coded in *red*. Several features such as loops or stripes, and TAD boundaries are highlighted. (D) Schematic representation of two TADs (rotated 45 degrees from (C)), highlighting the insulation of genes and regulatory elements across TADs. The *yellow* hexagon represents TAD boundaries. Below, the expected CTCF, cohesin and H3K27ac localization in relation to the TADs are depicted. (E) Hi-C maps (A–C) are representations of chromatin looping created by the extrusion model with architectural proteins bound to CTCF binding sites in convergent direction at the boundary.

TADs or sub-TADs, which in turn represent loci of pairwise interactions (Rao et al., 2014). The punctate signals are often associated with pairs of convergent CFTC binding sites, which mark the loop-anchors at TAD boundaries (Rao et al., 2014). The ring-shaped cohesin complex mediates the DNA loop formation and maintains the inter-TAD insulation (Fig. 12.3D and E; Kraft et al., 2019; Vian et al., 2018).

Position effects and TADs

Thirty years ago, it became evident that CNVs in the human genome cause genetic disorders. Since then, many examples of micro-deletion/duplications causing human disease have been described (Watson et al., 2014; Weise et al., 2012; Zarrei et al., 2015). One of the classic CNVs causing human disease is a duplication on the chromosome 17 associated with the Charcot-Marie-Tooth disease harboring *PMP22* (Lupski et al., 1991; Timmerman et al., 1992). Usually, the phenotypic effects of CNVs are explained by changes in gene dosage, i.e., the number of copies of a particular gene in the genome (Spielmann et al., 2018). However, in many cases, this explanation proved to be unsatisfactory. Early examples include reports on patients with aniridia, carrying translocations and inversions mapping 125 kb away from *PAX6*, a gene whose loss-of-function variants have been shown to cause aniridia. Therefore, the term "position effect" (Fantes et al., 1995) was introduced to describe mutations far away from the target gene. Position effects have also been described for balanced translocations associated with other known malformation syndromes as well as neurological and congenital diseases (Bugge et al., 2000).

Even before the discovery of TADs, it became clear that SVs can alter the long-range architecture of the noncoding genome. Several elegant experiments supported the idea of "position effects" in congenital malformation syndromes. For example, duplications of the *SHH* limb enhancer element, located 1-Mb upstream of *SHH*, were described as the cause of preaxial polydactyly (Lettice et al., 2003). Other examples include duplications, deletions, and translocations upstream and downstream (> 1 Mb) of *SOX9* that cause Pierre Robin sequence (PRS), a rare birth defect characterized by abnormal mandibular development (Benko et al., 2009).

Only after the discovery of TADs, it was evident that "position effects" were in fact SVs that crossed TAD boundaries, resulting in the fusion of two regulatory domains, that do not normally associate. Such position effects entail the reposition of TAD boundaries or the relocation of enhancer elements, causing misexpression and disease, and were termed "enhancer adoption" or "enhancer hijacking" (Spielmann et al., 2018). One of the earliest examples linking TADs to human diseases were large deletions including CTCF-associated boundary elements at the *EPHA4* locus (Lupiáñez et al., 2015). These deletions resulted in the ectopic interaction between an enhancer cluster and genes that are normally separated, causing enhancer adoption and congenital limb malformation.

Such findings have a direct impact on cytogenetics (Weise & Liehr, 2021b) and molecular genetics (Ungelenk, 2021). While high-throughput sequencing technologies have dramatically accelerated the discovery and characterization of complex SVs, their medical interpretation and the prediction of phenotypic consequences remain unsatisfactory. It is now certain that position effects can only be understood when taking into account the third dimension of genome, and the folding of the chromatin in the 3D space of the nucleus (Spielmann et al., 2018; Weise & Liehr, 2021c).

Intra-TAD structural variants

Structural variants located within a TAD are classified as intra-TAD SVs (Fig. 12.4A). Although intra-TAD SVs do not perturb the TAD architecture of the genome, they can alter gene expression. For example, deletion of regulatory elements may cause a regulatory loss of function (Fig. 12.4B; Spielmann et al., 2012; Tayebi et al., 2014). On the other hand, intra-TAD duplications may cause a tissue-specific increase in the expression of the associated genes (Fig. 12.4C; Dathe et al., 2009). Intra-TAD inversions, though conceptually possible, have not been described so far.

Even in the absence of gross TAD structure changes, as boundary elements remain intact, intra-TAD SVs may lead to diseases when the affected gene is dosage sensitive. In these cases, Hi-C maps are still useful, as they detect interaction intensity changes inside the TAD conserved topology.

Recent examples of intra-TAD SVs include skin, ear, and bone phenotypes. For instance, in several families affected with keratolytic winter erythema, a rare skin disorder, tandem duplications of an enhancer region upstream of *CTSB* were identified (Ngcungcu et al., 2017). Affected skin tissues showed a specific *CTSB* upregulation in the palmar epidermis. A similar mutational mechanism was described for recurrent duplications of a highly conserved enhancer downstream of *HMX1* in patients with isolated bilateral microtia (Fig. 12.5A). Mutations in *HMX1* have previously been associated with external ear malformations. Luciferase assays further confirmed the enhancer-activity of this conserved enhancer element (Si et al., 2020).

In addition to the noncoding enhancer elements, long noncoding RNAs (lncRNAs) have also attracted attention in the field of gene regulation. A particularly noteworthy example is that of an unannotated lncRNA implicated in a complex limb malformation syndrome associated with ventral fingernails (Fig. 12.5B; Allou et al.,

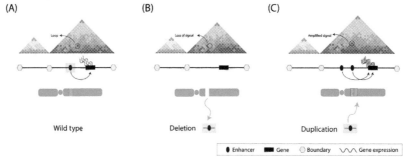

FIG. 12.4

Intra-TAD structural variants and its functional consequences. (A) A schematic of the Hi-C map of a wild-type reference portraying the interaction between a gene *(black box)* and its enhancer *(black oval, highlighted in light brown)* and the resultant gene expression (*wavy-lines*). The *light brown* segment in the chromosome below shows the enhancer element position. Enhancer dosage changes due to (B) deletion cause the lack of gene expression while (C) duplication overexpressed it in the same targeted tissue.

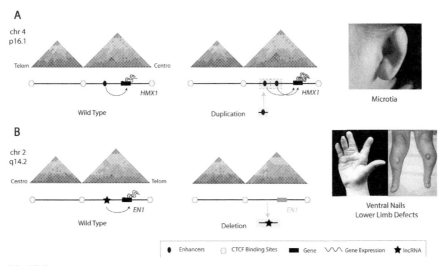

FIG. 12.5

Intra-TAD genomic alterations lead to diseases. (A) Several families with multiple duplications mapping to 4p16.1 have isolated bilateral-type microtia. The overlapping region of these duplications comprises an enhancer in the *HMX1*-TAD. Due to implication of *HMX1* in ear malformations, the overexpression of the gene has been proposed to be the disease mechanism (Si et al., 2020). (B) A lncRNA that controls the expression of *EN1* in cis by setting a permissive environment was identified. Families with a common homozygous deletion of lncRNA lost *EN1* expression, leading to abnormal limb development (Allou et al., 2021).

2021). Three individuals affected with this unique phenotype harbored homozygous deletions, 300 kb upstream of engrailed-1 (*EN1*). However, no limb enhancers were identified within the deletions. Deep sequencing of the region revealed that the deletion removed a new lncRNA called master activator of engrailed-1 (Maenli) inside *EN1*-TAD. Extensive mouse experiments demonstrated that *Maenli* transcription controls *En1* expression in limbs during embryonic development. The authors proposed that *Maenli*'s cis-acting mechanism could rely on the deposition of histone-activating marks (H3K4me3) at the *En1* promoter without any structural changes to the TAD (Allou et al., 2021).

Inter-TAD structural variants: Enhancer hijacking

Large structural variants that span beyond TAD boundaries are referred to as inter-TAD SVs (Fig. 12.6A). Inter-TAD SVs are more likely to cause enhancer adoption and subsequent human diseases because of the repositioning of TAD boundaries and/or the relocation of enhancer elements into other domains. TAD fusion occurs when the boundary sequences are deleted and neighboring TADs merge (Fig. 12.6B; Flöttmann et al., 2015; Ibn-Salem et al., 2014; Lupiáñez et al., 2015). Neo-TADs are

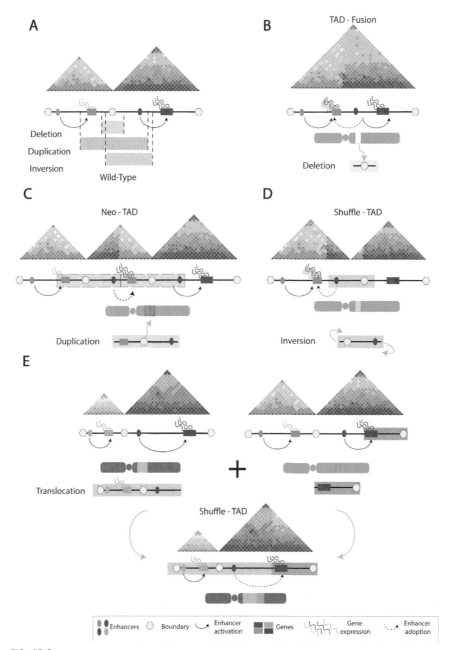

FIG. 12.6

Inter-TAD structural variants lead to enhancer adoption and misexpression. SVs that disrupt TAD architecture are readily identifiable on Hi-C maps, as new domains are created. (A) A graphical overview of the wild-type locus with two adjacent TADs with its gene and enhancers separated by CTCF binding sites. Each gene is regulated by the

the result of tandem duplications including a boundary element creating a new regulatory domain (Fig. 12.6C; Franke et al., 2016). Shuffle-TADs are the product of inversions or translocations when genomic information from two very distant locations come together (Fig. 12.6D and E). In shuffle-TADs, the regulatory elements can potentially activate other genes ectopically, causing enhancer adoption, or an enhancer can be disconnected from its target gene leading to regulatory loss of function.

Recent studies have described numerous inter-TAD SVs associated with diseases, causing a variety of different phenotypes. For example, whole-genome sequencing was used to resolve the breakpoints of de novo balanced rearrangements at a base pair level in a cohort of 248 patients with a wide range of congenital abnormalities: from organ-specific to multisystemic disorders. The breakpoints were mapped to the coding as well as intergenic regions. Recurrent intergenic breakpoints were mapped to the *MEF2C*-TAD, previously associated with a known microdeletion syndrome. Further exploration showed that patients with intergenic breakpoints in the *MEF2C*-TAD have a reduced *MEF2C* expression, indicating a regulatory loss of function (Redin et al., 2017).

A classic example of enhancer adoption is a set of noncoding deletions associated with a neurological disorder called the autosomal dominant adult-onset demyelinating leukodystrophy (ADLD), usually caused by the duplications of *LMNB1*. In a large family with ADLD, a heterozygous deletion of a TAD boundary upstream of *LMNB1* was identified (Fig. 12.7A; Giorgio et al., 2013). Further 4C analysis revealed that the deletion brings active forebrain enhancer elements in contact with *LMNB1*. This enhancer adoption most likely causes *LMNB1* overexpression and progressive demyelination, mimicking the classical *LMNB1* gene duplication effect (Giorgio et al., 2013, 2015). Since the initial study, three independent families with smaller deletions have been described, narrowing down the minimal critical region to the TAD boundary (Nmezi et al., 2019).

FIG. 12.6, CONT'D

tissue-specific enhancer (demarcated by matching colors of the enhancer and the wavy lines). The regions affected by structural variations (SVs) shown in A–D are depicted in different color squares below. Inter-TAD SVs may encompass various elements, and invariably the boundary region. (B) Deletions of boundary elements can cause TAD fusion, which allows the enhancer in one TAD to ectopically regulate the gene in the neighboring TAD (indicated by *turquoise and red wavy lines*). (C) Tandem duplications that lead to the creation of a new TAD (neo-TAD), where a gene *(turquoise box)* and a foreign enhancer *(red oval)* are confined between boundary elements. The result will be the ectopic regulation of the gene according to the enhancer's pattern of expression. (D) Inversions, that exclude protein coding genes, may drive abnormal expression when they relocate the boundary element and the enhancer. The gene expression in the TAD on the left is regulated by both the enhancers, either normally *(turquoise waves)* or in a different tissue *(red waves)*. (E) Translocations can replace elements in the genome, most likely generating shuffled-TAD, where a gene *(red box)* is located in the vicinity of an active enhancer that promotes expression at an abnormal anatomical location *(purple waves)*.

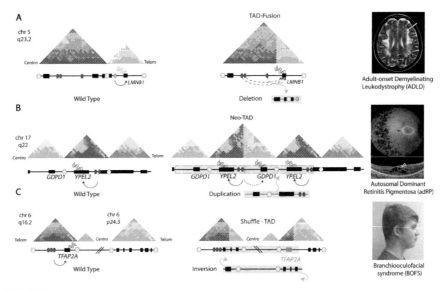

FIG. 12.7

Crossing boundaries and altered regulatory landscapes trigger human diseases.
(A) Overexpression of *LMNB1* has been found in patients with adult-onset demyelinating leukodystrophy (ADLD). Familial deletions, where the critical regions mapped to several elements among them the centromeric boundary element of LMNB1-TAD, can result in *LMNB1* overexpression. This occurs due to the fusion of the neighboring TADs and the activation of *LMNB1* via forebrain enhancer adoption (Giorgio et al., 2015). (B) Through a Hi-C map, a tandem duplication of several coding, noncoding and boundary elements creates a Neo-TAD. Inside this new domain, the retinal enhancers can activate its cognate *YPEL2* gene and *GDPD1* in the photoreceptors, which is the first tissue affected in the autosomal dominant retinitis pigmentosa (adRP) (de Bruijn et al., 2020). (C) A large inversion shuffles genetic content between two TADs in the either arms of chromosome 6. In the q-arm, the breakpoint separates *TFAP2A* from its enhancers and suppresses *TFAP2A* expression. The genes from the TAD disrupted in the p-arm are now next to the *TFAP2A* enhancers with no adoption (Laugsch et al., 2019).

A particular challenge for the medical interpretation of SVs are complex rearrangements, with multiple breakpoints. de Bruijn et al. (2020) demonstrated in an elegant study that complex rearrangements on chromosome 17 cause autosomal dominant retinitis pigmentosa (adRP) though neo-TAD formation (de Bruijn et al., 2020). The authors identified 22 families with duplications, triplications, and inversions close to *YPEL2*. Hi-C analysis of wild-type retinal organoids showed that *YPEL2* is located in a TAD rich in enhancers with binding sites for retinal transcription factors. The tandem duplication creates a neo-TAD with ectopic contacts between a distal gene *GDPD1* and several retinal enhancers causing misexpression of *GDPD1* in the retina (Fig. 12.7B; de Bruijn et al., 2020).

In comparison, copy number-neutral SVs, such as inversions, are difficult to be identified by short-read NGS data, but are key drivers of TAD shuffling. In patients with F-syndrome, a rare limb malformation, inversions have been shown to disconnect a set of limb enhancers from its target *EPHA4* and positioning them in front of *WNT6*, causing misexpression of *Wnt6* in a pattern similar to that of *Epha4* (Lupiáñez et al., 2015). In a more recent study, a patient diagnosed with branchio-oculo-facial syndrome (BOFS) carried a de novo, balanced inversion (89 Mb) that inverted a region between two chromosome arms (Fig. 12.7C). BOFS is known to be caused due to haploinsufficiency of *TFAP2A*, a neurocrestal regulator. One breakpoint of the inversion mapped between *TFAP2A* and its enhancers. Using human-induced pluripotent stem cells, the authors then demonstrated that this inversion separates *TFAP2A* from its neurocrestal enhancers, resulting in a regulatory loss of function (Laugsch et al., 2019). This study also beautifully demonstrates the use of human-induced pluripotent stem cells to model TAD shuffling in vitro.

Research of 3D genome architecture has also had a great impact in the oncology field. Cancer genomes have undergone numerous chromosomal rearrangements and somatic mutations, which make them attractive cells to investigate the effect of SVs. Hnisz et al. (2016) found that microdeletions in cells from T-cell acute lymphoblastic leukemia (T-ALL) patients remove TAD boundary elements near T-ALL proto-oncogenes such as *TAL1* and *LMO2*. Other cancer types also exhibited somatic SVs in the CTCF binding sites, followed by the activation of proto-oncogenes (Hnisz et al., 2016). Ectopic cross-boundary interactions can also be caused by epimutations. A loss of 30%–50% of CTCF binding at a hypermethylated TAD boundary has been shown to result in the overexpression of the glioma oncogene *PDGFRA* (Flavahan et al., 2016).

In a landmark study on over 7000 cancer genomes, Weischenfeldt et al. (2017) linked somatic copy-number alterations with enhancer hijacking at the *IRS4* and *IGF2* loci. In the wild-type genome, *IGF2* and its cognate enhancer are isolated from a set of superenhancers located in the neighboring TAD. However, in colorectal cancer genomes, a tandem duplication including a TAD boundary creates a neo-TAD, where the *IGF2* promoter and the superenhancer can strongly interact, resulting in a > 250-fold *IGF2* expression (Weischenfeldt et al., 2017).

SVs detection in the 3D genome

Since human genetic variation occurs at different levels, from nucleotides changes to chromosomal rearrangements, diagnostic tests should be able to capture all of them. Recent advances in the field apply both molecular and cytogenetic techniques to detect and interpret the genomic variants. Chromosomal microarray analysis (CMA) has been extensively used to detect CNVs (Weise & Liehr, 2021a). However, as high-throughput sequencing techniques have currently become cost-effective, they are preferred for genetic disorders diagnosis (Ungelenk, 2021). Whole-genome sequencing (WGS) has been suggested as the first-line test, as it has been able to confirm genetic variants detected with CMA and also other genetic variants as short tandem repeat previously missed. Indeed, WGS showed double the diagnostic rate efficiency compared to CMA (Lindstrand et al., 2019). Sequencing technologies can potentially resolve the balanced translocation breakpoints at a nucleotide resolution to determine the genes or the noncoding regions responsible for a patient's phenotype (Ordulu et al., 2016).

There are several genetic/genomic techniques to screen SVs and each has its advantages and drawbacks. While arrays can detect large CNVs, they fail to detect copy-neutral changes and do not offer precise breakpoint resolution. This latter issue can be compensated through WGS. Even though SVs can be properly mapped, their clinical interpretation can be challenging in particular within the noncoding genome (Liehr, 2021a). This could be due to the fact that until now most analyses have been restricted to the linear genome, ignoring the 3D genome architecture (Liehr, 2021d). In a recent study, Hi-C proved to be very informative in the diagnosis of patients with developmental diseases (Melo et al., 2020). Hi-C not only confirmed the SVs previously reported in those patients, but identified an additional layer of structural changes that were overlooked by other techniques. Hi-C also allowed resolving complex rearrangements with multiple breakpoints. The clinical impact of SVs was assessed based on the genes truncated at the breakpoints and genes whose expression could have been altered due to a remodeled regulatory landscape (Spielmann et al., 2018). The predictive 3D genome architecture in patients' samples puts forward the aberrant regulatory interactions that aid diagnosis in 55% of cases (Melo et al., 2020). In a similar study, Hi-C proved to be a beneficial tool for a precise detection of all types of structural variants and chromosome interaction maps in primary human tumor samples (Harewood et al., 2017). Hi-C protocol has been standardized for low input material (1000 cells) to directly analyze primary tissue derived from B-cell lymphoma patients. In this way, physicians can maximize the tissue availability for further testing and reduce the incidence of patients' resampling (Díaz et al., 2018). At a low coverage (1 M pair-end reads), Hi-C can detect reciprocal translocations, making this a valuable tool to look for balanced rearrangement if array CGH was not sufficiently informative. While SV callers overlooked small chromosomal changes, Hi-C was capable of unveiling them by divergent interactions when compared to control heatmaps (Melo et al., 2020). In summary, Hi-C holds a huge diagnostic potential for the field of cytogenetics and cytogenomics.

Clinical application and future perspectives

Our increasing knowledge about the noncoding genome has had a direct impact on human genetics and led to the recent development of an enhancer gene therapy in patients with β-thalassemia and sickle cell disease. In 2011, several deletions found in patients diagnosed with hereditary persistence of fetal hemoglobin revealed a 3.5 kb erythroid-specific enhancer. This noncoding region was implicated in the silencing of the γ-globin gene after birth. Hence, when this region was deleted, fetal hemoglobin was produced in adult cells (Sankaran et al., 2011). It has been reported that patients with β-thalassemia and sickle cell disease who retain fetal hemoglobin expression have a better disease prognosis or show no disease phenotype. CRISPR-Cas9 technology has been successfully used as a gene therapy in patient's derived hematopoietic stem and progenitor cells to block the *BCL11A*γ-globin gene represor) enhancer. The gene therapy reduced *BCL11A* expression in adult erythroid cells and fostered γ-globin synthesis reversing the disease phenotype (Frangoul et al., 2021). Similar approaches are being developed as molecular therapies targeting CTCF for acute lymphoblastic leukemia (Li et al., 2020). Another recent study used a CRISPR-based activation system to rescue a haploinsufficient phenotype by increasing the gene expression levels of the existing normal copy (Matharu et al., 2019). These studies are hopefully just a first step toward more enhancer gene therapies.

Genomics has been revolutionized by the analysis of 3D genome structure of different cell types to understand how alterations affect cell function and cause diseases. There are now web-based databases that make Hi-C data of various cell lines publicly available to compare disease-specific tissues (Wang et al., 2018). Even though there are still some cells not included in web databases, studies have reprogrammed patients' cells to hiPSCs and differentiate them into the different cell types such as photoreceptor cells or neurocrestal cells (de Bruijn et al., 2020; Laugsch et al., 2019). Hi-C has also been shown to be a cost-effective technique for resolving balanced rearrangements, when compared to WGS (Harewood et al., 2017). Despite all the benefits Hi-C has brought to diagnosis, its computational analysis remains challenging and has not been transferred to a clinical setting yet as has been done with microarrays or WGS tools (Melo et al., 2020). But many challenges remain and there are still many obstacles to be overcome until the effects of SVs can be predicted in a routine clinical setting. But we believe that in the coming years, new algorithms will be developed that could be used to predict the effect of all SVs in terms of a 3D genome landscape and gene expression.

The advances in sequencing technologies to characterize SVs at a base pair resolution, to measure their impact in 3D structure, and to gain an improved understanding of the transcriptional regulation will take 3D genome architecture from bench to bedside and nucleomics will become a meaningful tool for applied cytogenomics.

References

Allou, L., Balzano, S., Magg, A., Quinodoz, M., Royer-Bertrand, B., Schöpflin, R., … Chan, W. L. (2021). Non-coding deletions uncover Maenli lncRNA transcription as a major regulator of En1 during limb development. *Nature*. https://doi.org/10.1038/Q7 s41586-021-03208-9.

Benko, S., Fantes, J. A., Amiel, J., Kleinjan, D. J., Thomas, S., Ramsay, J., … Lyonnet, S. (2009). Highly conserved non-coding elements on either side of SOX9 associated with Pierre Robin sequence. *Nature Genetics*, *41*, 359–364. https://doi.org/10.1038/ng.329.

Bugge, M., Bruun-Petersen, G., Brøndum-Nielsen, K., Friedrich, U., Hansen, J., Jensen, G., … Tommerup, N. (2000). Disease associated balanced chromosome rearrangements: A resource for large scale genotype-phenotype delineation in man. *Journal of Medical Genetics*, *37*, 858–865. https://doi.org/10.1136/jmg.37.11.858.

Chiang, C., Scott, A. J., Davis, J. R., Tsang, E. K., Li, X., Kim, Y., ... Hall, I.M. (2017). The impact of structural variation on human gene expression. *Nature Genetics*, *49*, 692–699. https://doi.org/10.1038/ng.3834.

Coleman, W. B. (2018). The human genome: Understanding human disease in the post-genomic era. In W. B. Coleman, & G. J. Tsongalis (Eds.), *Molecular pathology* (2nd ed., pp. 121–134). Chapel Hill, NC: Academic Press. https://www.sciencedirect.com/science/article/pii/B9780128027615000067.

Collins, R. L., Glessner, J. T., Porcu, E., Niestroj, L.-M., Ulirsch, J., Kellaris, G., ... Talkowski, M.E. (2021). A cross-disorder dosage sensitivity map of the human genome. *medRxiv*. https://doi.org/10.1101/2021.01.26.21250098.

Crutchley, J. L., Wang, X. Q., Ferraiuolo, M. A., & Dostie, J. (2010). Chromatin conformation signatures: Ideal human disease biomarkers? *Biomarkers in Medicine*, *4*, 611–629. https://doi.org/10.2217/bmm.10.68.

Daban, J.-R. (2021). Multilayer organization of chromosomes. In T. Liehr (Ed.), *Cytogenomics* (pp. 267–296). Academic Press. Chapter 13 (in this book).

Dathe, K., Kjaer, K. W., Brehm, A., Meinecke, P., Nürnberg, P., Neto, J. C., ... Mundlos, S. (2009). Duplications involving a conserved regulatory element downstream of BMP2 are associated with brachydactyly type A2. *American Journal of Human Genetics*, *84*, 483–492. https://doi.org/10.1016/j.ajhg.2009.03.001.

de Bruijn, S. E., Fiorentino, A., Ottaviani, D., Fanucchi, S., Melo, U. S., Corral-Serrano, J. C., ... Hardcastle, A.J. (2020). Structural variants create new topological-associated domains and ectopic retinal enhancer-gene contact in dominant retinitis pigmentosa. *American Journal of Human Genetics*, *107*, 802–814. https://doi.org/10.1016/j.ajhg.2020.09.002.

Dekker, J., Rippe, K., Dekker, M., & Kleckner, N. (2002). Capturing chromosome conformation. *Science*, *295*, 1306–1311. https://doi.org/10.1126/science.1067799.

Díaz, N., Kruse, K., Erdmann, T., Staiger, A. M., Ott, G., Lenz, G., & Vaquerizas, J. M. (2018). Chromatin conformation analysis of primary patient tissue using a low input Hi-C method. *Nature Communications*, *9*, 4938. https://doi.org/10.1038/s41467-018-06961-0.

Dixon, J. R., Selvaraj, S., Yue, F., Kim, A., Li, Y., Shen, Y., ... Ren, B. (2012). Topological domains in mammalian genomes identified by analysis of chromatin interactions. *Nature*, *485*, 376–380. https://doi.org/10.1038/nature11082.

ENCODE Project Consortium. (2012). An integrated encyclopedia of DNA elements in the human genome. *Nature*, *489*, 57–74. https://doi.org/10.1038/nature11247.

Fantes, J., Redeker, B., Breen, M., Boyle, S., Brown, J., Fletcher, J., ... Hanson, I. (1995). Aniridia-associated cytogenetic rearrangements suggest that a position effect may cause the mutant phenotype. *Human Molecular Genetics*, *4*, 415–422. https://doi.org/10.1093/hmg/4.3.415.

Flavahan, W. A., Drier, Y., Liau, B. B., Gillespie, S. M., Venteicher, A. S., Stemmer-Rachamimov, A. O., ... Bernstein, B.E. (2016). Insulator dysfunction and oncogene activation in IDH mutant gliomas. *Nature*, *529*, 110–114. https://doi.org/10.1038/nature16490.

Flöttmann, R., Wagner, J., Kobus, K., Curry, C. J., Savarirayan, R., Nishimura, G., ... Spielmann, M. (2015). Microdeletions on 6p22.3 are associated with mesomelic dysplasia Savarirayan type. *Journal of Medical Genetics*, *52*, 476–483. https://doi.org/10.1136/jmedgenet-2015-103108.

Frangoul, H., Altshuler, D., Cappellini, M. D., Chen, Y. S., Domm, J., Eustace, B. K., ... Corbacioglu, S. (2021). CRISPR-Cas9 gene editing for sickle cell disease and β-thalassemia. *The New England Journal of Medicine*, *384*, 252–260. https://doi.org/10.1056/NEJMoa2031054.

Franke, M., Ibrahim, D. M., Andrey, G., Schwarzer, W., Heinrich, V., Schöpflin, R., … Mundlos, S. (2016). Formation of new chromatin domains determines pathogenicity of genomic duplications. *Nature, 538*, 265–269. https://doi.org/10.1038/nature19800.

Giorgio, E., Robyr, D., Spielmann, M., Ferrero, E., Di Gregorio, E., Imperiale, D., … Brusco, A. (2015). A large genomic deletion leads to enhancer adoption by the lamin B1 gene: A second path to autosomal dominant adult-onset demyelinating leukodystrophy (ADLD). *Human Molecular Genetics, 24*, 3143–3154. https://doi.org/10.1093/hmg/ddv065.

Giorgio, E., Rolyan, H., Kropp, L., Chakka, A. B., Yatsenko, S., Di Gregorio, E., … Padiath, Q.S. (2013). Analysis of LMNB1 duplications in autosomal dominant leukodystrophy provides insights into duplication mechanisms and allele-specific expression. *Human Mutation, 34*, 1160–1171. https://doi.org/10.1002/humu.22348.

Gross, D. S., Chowdhary, S., Anandhakumar, J., & Kainth, A. S. (2015). Chromatin. *Current Biology, 25*, R1158–R1163. https://doi.org/10.1016/j.cub.2015.10.059.

Harewood, L., Kishore, K., Eldridge, M. D., Wingett, S., Pearson, D., Schoenfelder, S., … Fraser, P. (2017). Hi-C as a tool for precise detection and characterisation of chromosomal rearrangements and copy number variation in human tumours. *Genome Biology, 18*, 125. https://doi.org/10.1186/s13059-017-1253-8.

Hnisz, D., Weintraub, A. S., Day, D. S., Valton, A. L., Bak, R. O., Li, C. H., … Young, R.A. (2016). Activation of proto-oncogenes by disruption of chromosome neighborhoods. *Science, 351*, 1454–1458. https://doi.org/10.1126/science.aad9024.

Ibn-Salem, J., Köhler, S., Love, M. I., Chung, H. R., Huang, N., Hurles, M. E., … Robinson, P.N. (2014). Deletions of chromosomal regulatory boundaries are associated with congenital disease. *Genome Biology, 15*, 423. https://doi.org/10.1186/s13059-014-0423-1.

Ichikawa, Y., & Saitoh, N. (2021). Shaping of genome by long noncoding RNAs. In T. Liehr (Ed.), *Cytogenomics* (pp. 357–372). Academic Press. Chapter 18 (in this book).

Kosugi, S., Momozawa, Y., Liu, X., Terao, C., Kubo, M., & Kamatani, Y. (2019). Comprehensive evaluation of structural variation detection algorithms for whole genome sequencing. *Genome Biology, 20*, 117. https://doi.org/10.1186/s13059-019-1720-5.

Kraft, K., Magg, A., Heinrich, V., Riemenschneider, C., Schöpflin, R., Markowski, J., … Mundlos, S. (2019). Serial genomic inversions induce tissue-specific architectural stripes, gene misexpression and congenital malformations. *Nature Cell Biology, 21*, 305–310. https://doi.org/10.1038/s41556-019-0273-x.

Laugsch, M., Bartusel, M., Rehimi, R., Alirzayeva, H., Karaolidou, A., Crispatzu, G., … Rada-Iglesias, A. (2019). Modeling the pathological long-range regulatory effects of human structural variation with patient-specific hiPSCs. *Cell Stem Cell, 24*, 736–752.e12. https://doi.org/10.1016/j.stem.2019.03.004.

Lettice, L. A., Heaney, S. J., Purdie, L. A., Li, L., de Beer, P., Oostra, B. A., … de Graaff, E. (2003). A long-range Shh enhancer regulates expression in the developing limb and fin and is associated with preaxial polydactyly. *Human Molecular Genetics, 12*, 1725–1735. https://doi.org/10.1093/hmg/ddg180.

Li, Y., Liao, Z., Luo, H., Benyoucef, A., Kang, Y., Lai, Q., … Huang, S. (2020). Alteration of CTCF-associated chromatin neighborhood inhibits TAL1-driven oncogenic transcription program and leukemogenesis. *Nucleic Acids Research, 48*, 3119–3133. https://doi.org/10.1093/nar/gkaa098.

Lieberman-Aiden, E., van Berkum, N. L., Williams, L., Imakaev, M., Ragoczy, T., Telling, A., … Dekker, J. (2009). Comprehensive mapping of long-range interactions reveals folding principles of the human genome. *Science, 326*, 289–293. https://doi.org/10.1126/science.1181369.

Liehr, T. (2021a). Repetitive elements, heteromorphisms, and copy number variants. In T. Liehr (Ed.), *Cytogenomics* (pp. 373–388). Academic Press. Chapter 19 (in this book).

Liehr, T. (2021b). A definition for cytogenomics - Which also may be called chromosomics. In T. Liehr (Ed.), *Cytogenomics* (pp. 1–7). Academic Press. Chapter 1 (in this book).

Liehr, T. (2021c). Overview of currently available approaches used in cytogenomics. In T. Liehr (Ed.), *Cytogenomics* (pp. 11–24). Academic Press. Chapter 2 (in this book).

Liehr, T. (2021d). Nuclear architecture. In T. Liehr (Ed.), *Cytogenomics* (pp. 297–305). Academic Press. Chapter 14 (in this book).

Liehr, T. (2021e). Molecular cytogenetics. In T. Liehr (Ed.), *Cytogenomics* (pp. 35–45). Academic Press. Chapter 4 (in this book).

Lindstrand, A., Eisfeldt, J., Pettersson, M., Carvalho, C. M. B., Kvarnung, M., Grigelioniene, G., … Nilsson, D. (2019). From cytogenetics to cytogenomics: Whole-genome sequencing as a first-line test comprehensively captures the diverse spectrum of disease-causing genetic variation underlying intellectual disability. *Genome Medicine, 11*, 68. https://doi.org/10.1186/s13073-019-0675-1.

Lupiáñez, D. G., Kraft, K., Heinrich, V., Krawitz, P., Brancati, F., Klopocki, E., … Mundlos, S. (2015). Disruptions of topological chromatin domains cause pathogenic rewiring of gene-enhancer interactions. *Cell, 161*, 1012–1025. https://doi.org/10.1016/j.cell.2015.04.004.

Lupski, J. R., de Oca-Luna, R. M., Slaugenhaupt, S., Pentao, L., Guzzetta, V., Trask, B. J., … Patel, P.I. (1991). DNA duplication associated with Charcot-Marie-Tooth disease type 1A. *Cell, 66*, 219–232. https://doi.org/10.1016/0092-8674(91)90613-4.

Matharu, N., Rattanasopha, S., Tamura, S., Maliskova, L., Wang, Y., Bernard, A., … Ahituv, N. (2019). CRISPR-mediated activation of a promoter or enhancer rescues obesity caused by haploinsufficiency. *Science, 363*, eaau0629. https://doi.org/10.1126/science.aau0629.

Melo, U. S., Schöpflin, R., Acuna-Hidalgo, R., Mensah, M. A., Fischer-Zirnsak, B., Holtgrewe, M., … Mundlos, S. (2020). Hi-C identifies complex genomic rearrangements and TAD-shuffling in developmental diseases. *American Journal of Human Genetics, 106*, 872–884. https://doi.org/10.1016/j.ajhg.2020.04.016.

Ngcungcu, T., Oti, M., Sitek, J. C., Haukanes, B. I., Linghu, B., Bruccoleri, R., … Ramsay, M. (2017). Duplicated enhancer region increases expression of CTSB and segregates with keratolytic winter erythema in south african and norwegian families. *American Journal of Human Genetics, 100*, 737–750. https://doi.org/10.1016/j.ajhg.2017.03.012.

Nmezi, B., Giorgio, E., Raininko, R., Lehman, A., Spielmann, M., Koenig, M. K., … Padiath, Q.S. (2019). Genomic deletions upstream of lamin B1 lead to atypical autosomal dominant leukodystrophy. *Neurology Genetics, 5*, e305. https://doi.org/10.1212/NXG.0000000000000305.

Ordulu, Z., Kammin, T., Brand, H., Pillalamarri, V., Redin, C. E., Collins, R. L., … Morton, C.C. (2016). Structural chromosomal rearrangements require nucleotide-level resolution: Lessons from next-generation sequencing in prenatal diagnosis. *American Journal of Human Genetics, 99*, 1015–1033. https://doi.org/10.1016/j.ajhg.2016.08.022.

Rao, S. S., Huntley, M. H., Durand, N. C., Stamenova, E. K., Bochkov, I. D., Robinson, J. T., … Aiden, E.L. (2014). A 3D map of the human genome at kilobase resolution reveals principles of chromatin looping. *Cell, 159*, 1665–1680. https://doi.org/10.1016/j.cell.2014.11.021.

Redin, C., Brand, H., Collins, R. L., Kammin, T., Mitchell, E., Hodge, J. C., … Talkowski, M.E. (2017). The genomic landscape of balanced cytogenetic abnormalities associated with human congenital anomalies. *Nature Genetics, 49*, 36–45. https://doi.org/10.1038/ng.3720.

Sankaran, V. G., Xu, J., Byron, R., Greisman, H. A., Fisher, C., Weatherall, D. J., … Bender, M.A. (2011). A functional element necessary for fetal hemoglobin silencing. *The New England Journal of Medicine*, *365*, 807–814. https://doi.org/10.1056/NEJMoa1103070.

Si, N., Meng, X., Lu, X., Liu, Z., Qi, Z., Wang, L., … Zhang, X. (2020). Duplications involving the long range HMX1 enhancer are associated with human isolated bilateral concha-type microtia. *Journal of Translational Medicine*, *18*, 244. https://doi.org/10.1186/s12967-020-02409-6.

Simonis, M., Klous, P., Splinter, E., Moshkin, Y., Willemsen, R., de Wit, E., … de Laat, W. (2006). Nuclear organization of active and inactive chromatin domains uncovered by chromosome conformation capture-on-chip (4C). *Nature Genetics*, *38*, 1348–1354. https://doi.org/10.1038/ng1896.

Spielmann, M., Brancati, F., Krawitz, P. M., Robinson, P. N., Ibrahim, D. M., Franke, M., … Mundlos, S. (2012). Homeotic arm-to-leg transformation associated with genomic rearrangements at the PITX1 locus. *American Journal of Human Genetics*, *91*, 629–635. https://doi.org/10.1016/j.ajhg.2012.08.014.

Spielmann, M., Lupiáñez, D. G., & Mundlos, S. (2018). Structural variation in the 3D genome. *Nature Reviews. Genetics*, *19*, 453–467. https://doi.org/10.1038/s41576-018-0007-0.

Szabo, Q., Bantignies, F., & Cavalli, G. (2019). Principles of genome folding into topologically associating domains. *Science Advances*, *5*, eaaw1668. https://doi.org/10.1126/sciadv.aaw1668.

Tayebi, N., Jamsheer, A., Flöttmann, R., Sowinska-Seidler, A., Doelken, S. C., Oehl-Jaschkowitz, B., … Spielmann, M. (2014). Deletions of exons with regulatory activity at the DYNC1I1 locus are associated with split-hand/split-foot malformation: Array CGH screening of 134 unrelated families. *Orphanet Journal of Rare Diseases*, *9*, 108. https://doi.org/10.1186/s13023-014-0108-6.

Timmerman, V., Nelis, E., Van Hul, W., Nieuwenhuijsen, B. W., Chen, K. L., Wang, S., … Van Broeckhoven, C. (1992). The peripheral myelin protein gene PMP-22 is contained within the Charcot-Marie-Tooth disease type 1A duplication. *Nature Genetics*, *1*, 171–175. https://doi.org/10.1038/ng0692-171.

Ungelenk, M. (2021). Sequencing approaches. In T. Liehr (Ed.), *Cytogenomics* (pp. 87–122). Academic Press. Chapter 7 (in this book).

Vian, L., Pękowska, A., Rao, S. S. P., Kieffer-Kwon, K. R., Jung, S., Baranello, L., … Casellas, R. (2018). The energetics and physiological impact of cohesin extrusion. *Cell*, *173*, 1165–1178.e20. https://doi.org/10.1016/j.cell.2018.03.072.

Wang, Y., Song, F., Zhang, B., Zhang, L., Xu, J., Kuang, D., … Yue, F. (2018). The 3D genome browser: A web-based browser for visualizing 3D genome organization and long-range chromatin interactions. *Genome Biology*, *19*, 151. https://doi.org/10.1186/s13059-018-1519-9.

Watson, C. T., Marques-Bonet, T., Sharp, A. J., & Mefford, H. C. (2014). The genetics of microdeletion and microduplication syndromes: An update. *Annual Review of Genomics and Human Genetics*, *15*, 215–244. https://doi.org/10.1146/annurev-genom-091212-153408.

Weischenfeldt, J., Dubash, T., Drainas, A. P., Mardin, B. R., Chen, Y., Stütz, A. M., … Korbel, J.O. (2017). Pan-cancer analysis of somatic copy-number alterations implicates IRS4 and IGF2 in enhancer hijacking. *Nature Genetics*, *49*, 65–74. https://doi.org/10.1038/ng.3722.

Weise, A., & Liehr, T. (2021a). Molecular karyotyping. In T. Liehr (Ed.), *Cytogenomics* (pp. 73–85). Academic Press. Chapter 6 (in this book).

Weise, A., & Liehr, T. (2021b). Cytogenetics. In T. Liehr (Ed.), *Cytogenomics* (pp. 25–34). Academic Press. Chapter 3 (in this book).

Weise, A., & Liehr, T. (2021c). Interchromosomal interactions with meaning for disease. In T. Liehr (Ed.), *Cytogenomics* (pp. 349–356). Academic Press. Chapter 17 (in this book).

Weise, A., Mrasek, K., Klein, E., Mulatinho, M., Llerena, J. C., Jr., Hardekopf, D., … Liehr, T. (2012). Microdeletion and microduplication syndromes. *The Journal of Histochemistry and Cytochemistry, 60*, 346–358. https://doi.org/10.1369/0022155412440001.

Zarrei, M., MacDonald, J. R., Merico, D., & Scherer, S. W. (2015). A copy number variation map of the human genome. *Nature Reviews. Genetics, 16*, 172–183. https://doi.org/10.1038/nrg3871.

Multilayer organization of chromosomes

13

Joan-Ramon Daban

Department of Biochemistry and Molecular Biology, Faculty of Biosciences, Autonomous University of Barcelona, Barcelona, Spain

Chapter outline

Cytogenomics. https://doi.org/10.1016/B978-0-12-823579-9.00010-2

Introduction

The genetic information of eukaryotic cells is stored in extremely long DNA molecules (Portugal & Cohen, 1977; Zimm, 1999). These thin and fragile one-dimensional structures are condensed at a high concentration in the mitotic chromosomes (Daban, 2003; Sumner, 2003). Since 1973 we know that all the genomic DNA is associated with histone proteins and forms long chromatin filaments containing many nucleosomes (Hewish & Burgoyne, 1973; Kornberg, 1974). The nucleosome cores are flat cylindrical particles (5.7 height and 11 nm diameter) formed by approximately two turns of DNA (146 bp) wrapped around a histone octamer (Luger et al., 1997), which are complemented with histone H1 and are connected between them by linker DNA. We also know the sequence of DNA in all human chromosomes (International Human Genome Sequencing Consortium, 2001; Liehr, 2021a). Nevertheless, the three-dimensional organization of the chromatin filament in interphase chromosomes and in condensed chromosomes in metaphase has been one of the most elusive problems in biology. In this chapter, diverse evidences showing that chromatin forms thin plates and that chromosomes are built by many layers of this two-dimensional structure will be reviewed. Furthermore, it will be shown that cytogenetic (Weise & Liehr, 2021) and genomic results (Ungelenk, 2021) provide important structural information that is compatible with a multilayered organization of chromatin in chromosomes. The dynamic properties and possible functional implications of this supramolecular structure will also be discussed as a major part of cytogenomics (Liehr, 2021b, 2021c).

Experimental approaches to study the internal structure of mitotic chromosomes

Scanning ion microscopy experiments (Strick et al., 2001) showed that the Mg^{2+} concentration in the interphase nucleus and in mitotic chromatin of cryopreserved cells is 2–4 and 5–17 mM, respectively; in isolated metaphase chromosomes, the observed concentration of Mg^{2+} is 12–22 mM. It is well known that this divalent cation induces chromatin condensation (Bartolomé et al., 1995; Bertin et al., 2007; Widom, 1986), and mitotic chromosomes are very compact in buffers containing Mg^{2+} (Caravaca et al., 2005; Eltsov et al., 2008; Poirier et al., 2002). As can be seen in the examples presented in the inset of Fig. 13.1B and in Fig. 13.1E, human mitotic chromosomes prepared in buffers containing Mg^{2+} have relatively thick chromatids (~0.6 μm diameter) and are very compact. They are highly electron-opaque, and their internal structure cannot be analyzed directly by transmission electron microscopy (TEM). This has been the major problem encountered in the study of the organization of chromatin within chromosomes. In order to try to overcome this problem, in early studies chromosomes were prepared in water without cations (Bahr, 1977; DuPraw, 1966, 1970) or in buffers containing the divalent cation chelator EDTA (Earnshaw & Laemmli, 1983). Fig. 13.1A and B show that under these conditions the chromatin

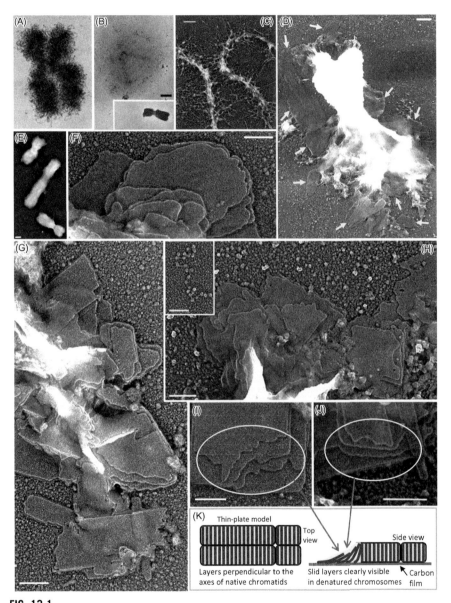

FIG. 13.1

TEM images of human metaphase chromosomes under different conditions.
Chromosomes with a fibrillar appearance: (A) prepared on a water-air interphase, (B)
treated with the cation chelator EDTA (0.2 mM) before the extension, and (C) incubated
with water without cations after extension on the carbon-coated grid. Native compact
chromosomes prepared in 5 mM Mg^{2+}: (B, inset) and (E). (D) Plates surrounding a
chromosome incubated at 37°C on the grid with a buffer containing 10 mM Mg^{2+} are

(Continued)

emanated from denatured chromosomes has a fibrillar structure. In another early study (Paulson & Laemmli, 1977), chromatin was depleted of histones and long DNA loops were observed surrounding a central remain of the initial chromosome. Since then, most of the structural models that have been proposed consider that mitotic chromosomes contain either chromatin fibers forming ordered loops (Gibcus et al., 2018; Paulson & Laemmli, 1977; Rattner & Lin, 1985; Saitoh & Laemmli, 1994) or irregularly folded chromatin fibers (Eltsov et al., 2008; Kireeva et al., 2004; Naumova et al., 2013; Ou et al., 2017; Poirier & Marko, 2002).

We developed experimental approaches to study the internal structure of mitotic chromosomes, in which the concentration of Mg^{2+} was maintained close to the physiological values and chromosomes were disturbed using soft treatments. In our early experiments (Caravaca et al., 2005), uncrosslinked metaphase chromosomes extended on carbon-coated grids of the electron microscope were incubated at 37°C for 30 min with a buffer containing Mg^{2+}. Unexpectedly, we observed chromosomes surrounded by laminar structures (Fig. 13.1D). These planar structures were also

---

FIG. 13.1, CONT'D

indicated with *yellow arrows*; (F,G) more examples of plates emanated from chromosomes under similar conditions. (H) Plates produced by the self-assembly, in buffers containing Mg^{2+}, of chromatin fragments obtained from metaphase chromosomes digested with micrococcal nuclease; a chromatin fragment in 1 mM EDTA (inset). Sliding of successive layers (indicated with *yellow rings*) in plates emanated from human (I) and chicken (J) chromosomes; (K) schematic representation of the thin-plate model proposed from the observed layer sliding *(red arrows)* and other data (see text). All images corresponding to rotary-shadowed preparations (C–J) are shown in reverse contrast; images in (A,B) correspond to negatively stained samples. Scale bars: 1 μm (B); 600 nm (C,D); 500 nm (E); 200 nm (F–J); 100 nm (inset in H).

Reproduced with permission from (A) Bahr, G. F. (1977). Chromosomes and chromatin structure. In J. J. Yunis (Ed.), Molecular structure of human chromosomes (pp. 143–203). Academic Press; (B) Earnshaw, W. C., & Laemmli, U. K. (1983). Architecture of metaphase chromosomes and chromosome scaffolds. Journal of Cell Biology, 96(1), 84–93. doi:10.1083/jcb.96.1.84; (C,D) Caravaca, J. M., Caño, S., Gállego, I., & Daban, J. R. (2005). Structural elements of bulk chromatin within metaphase chromosomes. Chromosome Research, 13(7), 725–743. doi:10.1007/s10577-005-1008-3; (E) Crosas, E., Chicano, A., Kamma-Longer, C., Martínez, J.C., Malfois, M., Svensson, A., & Daban, J. R. (2015). Structural analysis of condensed metaphase chromosomes by synchrotron small-angle x-ray scattering. ALBA Synchrotron User's Meeting, Cerdanyola del Vallès, Barcelona; (F) Daban, J.R., Castro-Hartmann, P., Gállego, I., Milla, M., Caño, S., & Caravaca, J.M. (2009). Stacked thin plates containing irregularly oriented nucleosomes in metaphase chromosomes. Joint Meeting of the Spanish and Portuguese Microscopy Societies, Segovia; (G,I) Castro-Hartmann, P., Milla, M., & Daban, J. R. (2010). Irregular orientation of nucleosomes in the well-defined chromatin plates of metaphase chromosomes. Biochemistry, 49(19), 4043–4050. doi:10.1021/bi100125f; (H) Milla, M., & Daban, J. R. (2012). Self-assembly of thin plates from micrococcal nuclease-digested chromatin of metaphase chromosomes. Biophysical Journal, 103(3), 567–575. doi:10.1016/j.bpj.2012.06.028; (J) Gállego, I., Castro-Hartmann, P., Caravaca, J. M., Caño, S., & Daban, J. R. (2009). Dense chromatin plates in metaphase chromosomes. European Biophysics Journal, 38(4), 503–522. doi:10.1007/s00249-008-0401-1.

observed (see Fig. 13.1F and G) when chromosomes (in buffers with Mg^{2+}) were disrupted using mechanical treatments and dilution with hyposmotic solutions (Castro-Hartmann et al., 2010; Gállego et al., 2009). Chromatin fibers were observed surrounding uncrosslinked plates denatured with EDTA-containing buffers (Gállego et al., 2009), indicating that the chromatin filament is folded within the plates. Plates were also observed (Fig. 13.1H) when chromatin fragments (obtained by micrococcal nuclease digestion of purified metaphase chromosomes) in 1 mM EDTA (inset of Fig. 13.1H) were dialyzed against solutions containing Mg^{2+} (Milla & Daban, 2012). As shown in Fig. 13.1F–H, the plates are multilayered. Many plates (see examples in Fig. 13.1I and J) emanated from chromosomes show a relative sliding between the successive layers, which facilitated the clear visualization of the edges of the layers in the rotary-shadowed preparations and allowed the discovery of multilayer planar chromatin. This easy sliding indicates that the interactions within layers are stronger than interactions between layers. All these results led to the proposal of the thin-plate model (Castro-Hartmann et al., 2010; Daban, 2011; Gállego et al., 2009), in which it is considered that native chromosomes are formed by many stacked layers of chromatin orthogonal to the chromosome axis (see simplified representation in Fig. 13.1K).

Chromatin plates in aqueous solution

The samples analyzed by TEM considered in the preceding section were extended on a carbon substrate, usually fixed by glutaraldehyde crosslinking, dehydrated, and metal-shadowed (or contrasted by negative staining) before imaging. In contrast, atomic force microscopy (AFM) allowed the analysis of uncrosslinked chromatin emanated from metaphase chromosomes in aqueous solutions at room temperature. As shown in Fig. 13.2A–C, plates adsorbed on mica in buffers containing different concentrations of Mg^{2+} can be visualized by tapping-mode (panels A and B) and contact-mode (panel C) AFM (Gállego et al., 2009). The topographical images obtained by scanning the sample with the AFM tip indicated that the surface of the plates is very smooth.

The results obtained with TEM and AFM can be criticized because the samples are adsorbed on flat substrates (carbon film and mica, respectively) that could alter the native structure of chromatin emanated from chromosomes. In contrast, cryo-electron tomography (cryo-ET) allows the study of the three-dimensional structure of uncrosslinked and unstained samples suspended in vitreous ice. The tomographic reconstructions (see examples in Fig. 13.2D and E) indicate that frozen-hydrated chromatin emanated from metaphase chromosomes has a planar structure (Chicano et al., 2019). The images obtained using all these techniques always show plates having different sizes and shapes. This is because plates have a large surface area but are very thin (see below) and can be easily broken and deformed during the preparation and deposition procedures. However, in some samples, the cryotomograms contained large multilayered plates with dimensions similar to the diameter of human

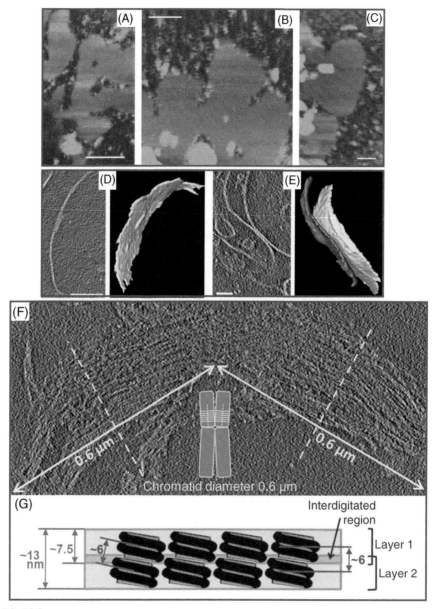

FIG. 13.2

AFM and cryo-ET studies of plates from human metaphase chromosomes. AFM images of monolayer plates in 10 mM (A), 15 mM (B), and 20 mM (C) Mg^{2+}; images acquired with unfixed samples in aqueous solution at room temperature. (D,E) Slices of two cryo-electron tomograms of frozen-hydrated plates in 5 mM Mg^{2+} *(left)*, and the corresponding three-dimensional reconstructions represented in *yellow* and *green (right)*. (F) Large multilayer plates perpendicular to a cryotomogram slice; their dimensions suggest that they metaphase chromosomes. The large multilayer structure in the cryotomogram slice shown in Fig. 13.2F was interpreted as a part of two sister chromatids formed by stacked chromatin layers (see simplified drawing in the inset of this figure).

The thickness of frozen-hydrated monolayer plates is ~7.5 nm (Chicano et al., 2019). Similar values were obtained using unidirectional shadowing in TEM experiments and from force curves in AFM experiments (Gállego et al., 2009). This width is compatible with the dimensions of a monolayer of nucleosomes slightly tilted with respect to the layer surface. Further measurements showed that the apparent thickness of each step in a multilayered plate is ~6 nm (Chicano et al., 2019; Gállego et al., 2009). This relatively small value suggests that in the multilaminar plates the successive layers are interdigitated. X-ray scattering of whole chromosomes prepared in buffers containing Mg^{2+} showed a main scattering peak at ~6 nm, which can be correlated with the distance between interdigitating layers and between nucleosomes interacting through their lateral faces. The schematic drawing in Fig. 13.2G represents the main structural elements and dimensions of two layers in close contact.

AFM-based friction measurements (nanotribology) have been another experimental approach applied to the study of chromatin plates in aqueous solution in buffers containing Mg^{2+} (Gállego et al., 2010). These friction measurements at the nanoscale showed that native plates have a relatively high friction coefficient ($\mu \approx 0.3$), which is markedly reduced when Mg^{2+} is removed from the plates with EDTA ($\mu \approx 0.1$). It was also found that protease digestion causes an increase in the friction coefficient of the plates ($\mu \approx 0.5$), but the highest friction was observed when DNA was cleaved with micrococcal nuclease ($\mu \approx 0.9$). Furthermore, plates digested with nuclease are irreversible denatured after the AFM scanning performed in the friction measurements, indicating that DNA is the main element responsible for the mechanical strength of the plates. Since native plates reversibly recover their original structure after the scanning performed in the friction measurements, but the lateral force exerted by the AFM tip during the friction measurements (up to 5 nN) is higher than the stretching force (~1 nN) required for the breakage of the covalent bonds in the DNA backbone (Bustamante et al., 2000), it was concluded that plates form a two-dimensional network with good elastic properties that protect DNA against mechanical damage.

FIG. 13.2, CONT'D

correspond to stacked layers of part of two sister chromatids (schematized in the inset). (G) Drawing showing the main structural elements of two layers of a multilaminar plate; the dimensions obtained from cryotomograms *(blue)* and synchrotron X-ray scattering *(red)* are indicated. Scale bars: 200 nm (A–D); 50 nm (E).

Reproduced with permission from (A-C) Gállego, I., Castro-Hartmann, P., Caravaca, J. M., Caño, S., & Daban, J. R. (2009). Dense chromatin plates in metaphase chromosomes. European Biophysics Journal, 38(4), 503–522. doi:10.1007/s00249-008-0401-1; (D-G) Chicano, A., Crosas, E., Otón, J., Melero, R., Engel, B. D., & Daban, J. R. (2019). Frozen-hydrated chromatin from metaphase chromosomes has an interdigitated multilayer structure. EMBO Journal, 38(7), e99769. doi:10.15252/embj.201899769.

Self-organization of multilayer chromatin

The discovery of multilayered chromatin was unexpected, but this structure should not be considered so surprising. The genomic DNA of dinoflagellates, which lack histones and nucleosomes, is packed within chromosomes as a multilayered liquid crystal with the layers orthogonal to the chromosome axis (Gornik et al., 2019; Livolant & Bouligand, 1978; Mitov, 2017). Furthermore, there is a widespread occurrence of multilaminar planar structures. Diverse two-dimensional natural materials such as mica and other layered silicates (Callister & Rethwisch, 2015), and graphene (Novoselov et al., 2004) and its analogs (Chen et al., 2015; Das et al., 2015) form multilaminar structures. All these materials slide and split easily, because they have interactions between layers that are weaker than those within layers, as it was observed for chromatin plates (see above; Fig. 13.1I and J). In the cell, the spontaneous assembly of phospholipid molecules gives rise to enormous two-dimensional structures: the membrane lipid bilayers. In vitro experiments showed that rod-shaped viruses self-assemble into two-dimensional membranes and large multilayered stacks (Adams et al., 1998; Gibaud et al., 2012). DNA of different sizes can also form multilayered structures (Koyfman et al., 2009; Martin & Dietz, 2012). Of note, it was found that the self-association of DNA origami elements in buffers containing Mg^{2+} produced large multilayered cylinders (Wagenbauer et al., 2017) having micrometer-scale dimensions.

Contemporary chemistry and nanotechnology research have been very interested in the study of many different types of self-organizing supramolecular assemblies (Lehn, 2002, 2007; Ozin et al., 2009). These structures are based on repetitive weak interactions between very diverse building blocks (Whitesides & Boncheva, 2002; Whitesides & Grzybowski, 2002). As indicated above (see Fig. 13.1H), it was found that multilayer chromatin plates can be self-assembled from small chromatin fragments (Milla & Daban, 2012). In this case, nucleosomes are the building blocks. It is known that nucleosomes can interact face-to-face giving rise to diverse structures (Adhireksan et al., 2020, 2021; Daban & Bermúdez, 1998; Dubochet & Noll, 1978; Ekudayo et al., 2017; Korolev et al., 2018; Leforestier et al., 1999; Luger et al., 2012; Robinson et al., 2006; Schalch et al., 2005; Song et al., 2014). In particular, it was found that purified nucleosome core particles interacting through their faces can form multilayer structures, in which each step consist of parallel columns of nucleosome cores (Leforestier et al., 2001; Mangenot et al., 2003). According to the thin-plate model (Fig. 13.1K), the lateral association of nucleosomes (Fig. 13.2G) is responsible for the stacking of the chromatin layers in metaphase chromosomes. In addition, a single molecule of DNA is covalently connecting all the nucleosomes in each chromatid.

The transformation of the long chromatin filaments into a multilayer planar structure changes completely the geometric and physical properties of chromatin and suggest that mitotic chromosomes could be a self-organizing supramolecular structure. This possibility could be criticized because it seems unlikely that large structures such as human chromosomes (with lengths of several micrometers) can be self-organized. However, nanotechnology research has demonstrated that diverse building blocks having nanometer-scale sizes can spontaneously form micrometric structures (Chung et al., 2011; Gibaud et al., 2012; Wagenbauer et al., 2017; Whitesides & Boncheva, 2002). Furthermore, there are structural similarities between metaphase chromosomes of diverse plant and animal species that strengthens the possibility that all chromosomes are self-organized following the same pattern: (i) chromosome volume depends on the size of the DNA molecule that it contains, but the DNA density is about the same (~ 166 Mb/μm^3) for diverse plant and animal chromosomes (Daban, 2000); (ii) in all cases, they are elongated smooth cylinders having a length-to-diameter ratio ≈ 13 (Daban, 2014). This morphology can be explained considering the energy components of the thin-plate model (Fig. 13.1K and Fig. 13.2G). Nucleosomes in the periphery of the chromosome are in contact with the medium, they cannot fully interact with nucleosomes within layers, and this generates a surface energy that destabilizes the structure. Chromosomes are smooth cylinders because this structure has a lower surface energy than other structures with irregular surfaces, and their elongated structure can be explained simply considering that the highly exposed nucleosomes in the telomere regions have a surface energy 13 times higher than that of the nucleosomes in the lateral surface (Daban, 2014).

Stretching experiments with micropipettes (Poirier et al., 2000) have demonstrated that mitotic chromosomes show reversible (i.e., elastic) extensions up to five times their native length and partially reversible extensions up to 80 times their initial length without suffering breakage. These outstanding mechanical properties can be interpreted considering that chromosomes are formed by stacked chromatin layers. Chromosomes can be easily stretched because the DNA covalent backbone is located within the layers, but there are only weak interactions between nucleosomes in the successive layers (Fig. 13.2G). During the extension produced by the stretching forces, a dissociation of face-to-face nucleosome contacts is produced; the reversible recovery of the initial chromosome length is due to the regeneration of these nucleosome-nucleosome contacts in adjacent layers (Daban, 2014). The energy of the internucleosome interactions is $\sim 9 k_B T$ (3.7×10^{-20} J) per nucleosome, according to values reported by several laboratories (Cui & Bustamante, 2000; De Jong et al., 2018; Kruithof et al., 2009; Moller et al., 2019; Norouzi & Zhurkin, 2018). The total energy of these interactions for a whole chromosome can justify the amount of work required for the elastic extension observed in stretching experiments (Daban, 2014). The higher amount of work required for larger extensions is probably absorbed by chromatin layers through a mechanism involving nucleosome unwrapping, which was studied by other authors using optical tweezers (Brower-Toland et al., 2002; Mihardja et al., 2006). If chromosomes suffer deformations, both the breakage of weak interactions between nucleosomes and the unwrapping of nucleosomes protect the covalent continuity of genomic DNA.

Considering early reports indicating that mitotic chromosomes have a helical shape (Boy de la Tour & Laemmli, 1988; Ohnuki, 1965; Rattner & Lin, 1985), it was suggested that the successive chromatin layers are connected between them forming a helicoid (Daban, 2014); in this structure each layer is equivalent to a helicoidal turn. This possibility is consistent with the observation of plates with a surface area larger than the cross-section of a chromatid (Castro-Hartmann et al., 2010). Other authors have found that the self-assembly of rod-shaped viruses and other materials can form large helical structures (Chung et al., 2011; Zhang et al., 2016). A continuous two-dimensional network folded as a helicoid can give a uniform protection to the entire DNA molecule within the chromosome. It can be considered that the chromosome is a hydrogel (in which the whole structure is crosslinked by the DNA backbone) having a lamellar liquid crystal order (Daban, 2014). Approximately one-third of the chromosome volume is occupied by water (Poirier et al., 2002), and this gives fluidity to this lyotropic liquid crystal, which can exist only in the presence of relatively high concentrations of Mg^{2+} (Strick et al., 2001) when the concentration of chromatin is high (Daban, 2003, 2011). In addition to the covalent bonds of the DNA backbone, the face-to-face nucleosome associations between layers are produced by electrostatic interactions (Luger et al., 1997, 2012; Mangenot et al., 2003; Moller et al., 2019; Schalch et al., 2005) that can be regenerated after chromosome stretching (see above). This self-healing capacity has been observed in nanotechnology studies of other hydrogels stabilized by electrostatic interactions in addition to covalent crosslinks (Sun et al., 2012). In the cell, all these structural and mechanical properties explain chromosome condensation and the maintenance of the genomic DNA integrity during mitosis.

Stacked thin layers explain the morphology of bands and chromosome rearrangements

Diverse cytogenetic and genomic findings provide information on the organization of chromatin within chromosomes (Daban, 2015). This structural information and its relationship to the multilayered structure of chromosomes will be reviewed in this section. Each data set will be presented separately.

Cytogenetic map and genome sequence

Fluorescence in situ hybridization (FISH) (Liehr, 2021d) allowed the positioning of genes with respect to the chromosome bands (Trask, 2002). In FISH experiments with metaphase chromosomes, the resolution of probe ordering is 1–3 Mb (Trask et al., 1993). It was found that there is a close correlation between the cytogenetic maps of human chromosomes and the corresponding genomic sequences (Fig. 13.3A, left; Cheung et al., 2001), which implies that DNA fills the chromatids progressively from one end to the other (Fig. 13.3A, right). This fundamental cytogenomic finding generates a three-dimensional structural constraint that has to be satisfied by any feasible model for metaphase chromosomes. The chromatin filament must be folded without overlapping along the chromosome axis. Otherwise, the genes would have a random zigzag order with respect to the chromosome axis. The continuous helicoidal model

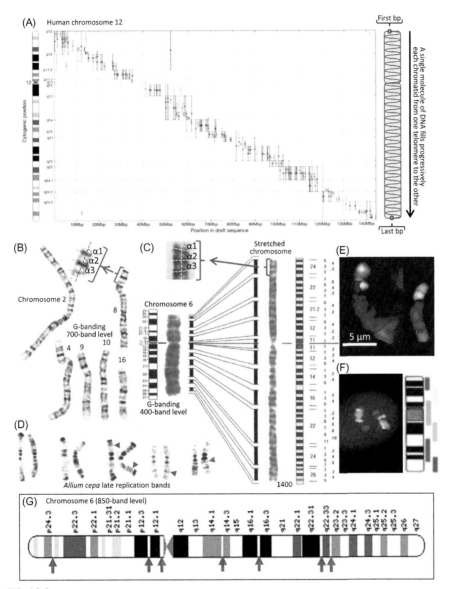

FIG. 13.3

Cytogenomic results containing important structural information. (A) Correspondence between cytogenetic locations and positions on the sequence of chromosome 12; these results indicate that DNA occupies the chromatids progressively from one end to the other (schematized at *right*). The orientation angles α1, α2, α3... were measured for typical G-bands such as those shown as an example in (B) and for the split G-bands produced by chromosome stretching (C); in all cases the mean value of the orientation angle α

(Continued)

considered in the preceding section is fully consistent with this structural constraint and justifies a progressive placement of the successive genes in the DNA sequence along the chromosome axis.

Chromosome bands

Each staining method produces a continuous pattern of light and dark bands that can be correlated with a multilayered structure of chromosomes. The quantitative measurement of the orientation angle α with respect to the chromosome axis (as indicated in Fig. 13.3B) for G-bands (at different resolution levels) and R-bands showed that the mean value of this angle is ~90 degrees (Daban, 2015). This orthogonal orientation of the bands is implicitly considered in the idiograms of the chromosome banding patterns (McGowan-Jordan et al., 2020), in which bands are represented exactly perpendicular to the longitudinal axis of the chromosomes. The sequence of the human genome (International Human Genome Sequencing Consortium, 2001) indicated a correlation between regions of low GC content and dark G-bands (light R-bands) and between regions of high GC content and light G-bands (dark R-bands). Since the orientation of the bands is the same than that observed for chromatin layers, it can be proposed that chromosome bands are produced by selective staining of chromatin plate clusters having the DNA base composition adequate for the interaction with

---

FIG. 13.3, CONT'D

is ~90 degrees (Daban, 2015). (D) Example of late-replication banding; *red arrowheads* point to thin bands. Examples of the maintenance of chromosome bands during interphase: (E) the multicolor banding pattern of metaphase chromosome 5 *(right)* does not disappear in interphase *(left)*; (F) the multicolor banding of two copies of chromosome 9 in interphase is equivalent to that observed in metaphase (idiogram at *right*). (G) Thin bands (containing 0.9–1.3 Mb) in human chromosome 6 are indicated with *red arrows*; the DNA content of the bands was obtained from www.genome.ucsc.edu; the idiogram was obtained from www.ensembl.org/Homo_sapiens/Location/Genome.

Reproduced with permission from (A) Cheung, V.G., Nowak, N., Jang, W., Kirsch, I.R., Zhao, S., Chen, X.N., ..., The BAC Resource Consortium. (2001). Integration of cytogenetic landmarks into the draft sequence of the human genome. Nature, 409, 953–958. doi:10.1038/35057192; (B) Levy, S., Sutton, G., Ng, P.C., Feuk, L., Halpern, A.L., Walenz, B.P., ..., Venter. J.C. (2007). The diploid genome sequence of an individual human. PLoS Biology, 5(10), e254. doi: 10.1371/journal.pbio.0050254; (C) Hliscs, R., Mühlig, P., & Claussen, U. (1997). The nature of G-bands analyzed by chromosome stretching. Cytogenetic and Genome Research, 79 (1–2), 162–166. doi:10.1159/000134710; (D) Cortés, F., & Escalza, P. (1986). Analysis of different banding patterns and late replicating regions in chromosomes of Allium cepa, A. sativum and A. nigrum, Genetica, 71(1), 39–46. doi:10.1007/BF00123231; (E) Lemke, J., Claussen, J., Michel, S., Chudoba, I., Mühlig, P., Westermann, M., ..., Claussen, U. (2002). The DNA-based structure of human chromosome 5 in interphase. American Journal of Human Genetics, 71(5), 1051–1059. doi:10.1086/344286; (F) Yurov, Y. B., Iourov, I. Y., Vorsanova, S. G., Liehr, T., Kolotii, A. D., Kutsev, S. I., ..., Soloviev, I. V. (2007). Aneuploidy and confined chromosomal mosaicism in the developing human brain. PLoS One, 2(6), e558. doi:10.1371/journal.pone.0000558.

different dyes. The number of chromatin layers in a specific band is dependent on the amount of DNA contained in the band. Therefore, the chromosome structure proposed in the thin-plate model is also consistent with these cytogenomic observations.

Thin bands

Even the thinnest bands in the karyotypes (Fig. 13.3B) and in the corresponding idiograms are transverse and occupy the entire cross-section of the two sister chromatids. Since in the human genome the thinnest bands correspond to sequences containing ~1 Mb [Fig. 13.3G (International Human Genome Sequencing Consortium, 2001)], these results indicate that a chromatin filament containing short stretches of DNA can fill completely the cross-section of the chromosome. Thus, in metaphase chromosomes, the chromatin filament must fold forming disc-like structures with the diameter of the chromosome and the height of the thinnest bands. In agreement with this, early FISH results showed that sequences separated by 0.5–1 Mb frequently produce hybridization spots on either side of the chromatid, suggesting that short stretches of DNA can cross the width of the chromatid (Lawrence et al., 1990; see also Weise et al., 2002). All these observations are compatible with the thin-plate model, in which each layer contains ~0.5 Mb (Daban, 2014) and has a small thickness (~6 nm; Fig. 13.2G); the thinnest bands could be formed by one or two chromatin layers. As discussed elsewhere (see Fig. 3 of Daban, 2015), the fibrillar models proposed by other authors (Eltsov et al., 2008; Kireeva et al., 2004; Naumova et al., 2013; Ou et al., 2017; Paulson & Laemmli, 1977; Poirier & Marko, 2002; Rattner & Lin, 1985; Saitoh & Laemmli, 1994) require large amounts of DNA to cover the chromosome cross-section and cannot justify the existence of very thin orthogonal bands.

Band splitting

The mechanical stretching of G-banded chromosomes causes band splitting into several subbands (Hliscs et al., 1997). In the example presented in Fig. 13.3C, it can be seen that after stretching chromosome 6 (at ~400-band level) is transformed into a highly extended chromosome (~1400-band level), in which each initial band is split into 3–4 subbands. The resulting split bands are well defined and relatively small; on average each band contains ~2.4 Mb (calculated considering 3.3 Gb in the whole genome distributed into 1400 bands). The orientation angle α of the split bands with respect to the chromosome axis is ~90 degrees (Daban, 2015). The easy sliding of layers in chromatin plates (see above; Fig. 13.1I and J) indicates that there is no topological entanglement between the DNA of adjacent layers and justifies the observed band splitting and the maintenance of the orthogonal orientation of the resulting thinner subbands. In contrast, in the fibrillar models considered above, DNA goes up and down with respect to the chromosome axis, and, consequently, these models cannot explain the appearance of well-defined thin subbands in the chromosome stretching experiments.

Replication bands

In the example presented in Fig. 13.3D, dark bands correspond to late-replication regions (Cortés & Escalza, 1986). The measurements of the orientation angle of replication bands of different species indicate that they are orthogonal to the chromosome axis (Daban, 2015). The well-defined thin bands observed in these experiments (some of them are indicated with red arrowheads in Fig. 13.3D) contain short segments of DNA corresponding to different late-replicating sequences. In agreement with the structural interpretation of the thin bands observed in typical karyotypes, this observation indicates that short stretches of DNA can form disks that occupy the whole cross-section of the chromatid. These observations are compatible with the small thickness of the layers in the thin-plate model.

Sister chromatid exchanges

The metaphase spread presented in Fig. 13.4A (reproduced from Friebe & Cortés, 1996) shows several chromosomes with reciprocal exchanges between sister chromatids. The boundaries between the original chromatids and the interchanged segments are sharp, indicating that the connection of the chromatids occurs in a very narrow zone. The mean value of the orientation angle α of the connection boundaries is ~90 degrees (Daban, 2015). These results suggest that the resulting mixed cylindrical chromatids are connected in thin planar surfaces oriented perpendicular to the chromosome axis. Well-defined flat connection surfaces cannot be justified by fibrillar models because, as indicated above, the chromatin filament goes up and down with respect to the chromatid axis, but if chromosomes are formed by thin stacked layers of chromatin, the ends of all exchanged segments are expected to be planar surfaces orthogonal to the chromosome axis.

Chromosome translocations

Examples of multiple translocations occurring in different cancer cells are presented in Fig. 13.4B–E (Gisselsson et al., 2000; Landry et al., 2013; Roschke et al., 2003; Schröck et al., 1996), in which chromosomes are identified with different painting techniques. It can be seen that many junctions are sharp and well defined, indicating that the connection between the original chromatid and the translocated part is a planar surface. As observed in sister chromatid exchanges, the orientation angle α of the connection surfaces in different carcinomas and hematological malignancies is ~90 degrees (see Table 1 in Daban, 2015). Even when the translocated parts are very thin (some examples are indicated with yellow asterisks in Fig. 13.4C–E), the boundaries between the translocated and original chromatids are well defined, indicating that relatively short sequences of genomic DNA are confined within thin discoidal structures. The simplified drawing in Fig. 13.4F show schematically the direct structural relationship between the thin-plate model and the morphology of

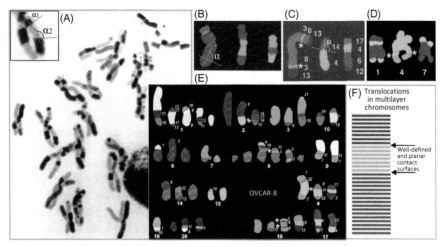

FIG. 13.4

Structural properties of chromosomes inferred from translocations and other rearrangements. The orientation angles α, $\alpha 1$, $\alpha 2$... of the connection surfaces were measured in sister chromatid exchanges (A) and in translocations (B–E). Examples of translocations visualized by multiplex-FISH (B,D) and multicolor spectral karyotyping (C,E); *yellow asterisks* indicate translocated thin chromosome segments. In all rearrangements, the mean value of the orientation angle α is ~90 degrees (Daban, 2015). (F) Scheme showing that a multilayered organization of chromosomes can explain the planar geometry of the connection surfaces and their orthogonal orientation with respect to the chromosome axis.

Reproduced with permission from (A) Friebe, B., & Cortés, F. (1996). Sister chromatid exchange and replication banding. In K. Fukui & S. Nakayama (Eds.), Plant chromosomes: Laboratory methods (pp. 171–186). CRC Press; (B,D) Gisselsson, D., Pettersson, L., Höglund, M., Heidenblad, M., Gorunova, L., Wiegant, J., Mertens, F., Dal Cin, P., Mitelman, F., & Mandahl, N. (2000). Chromosomal breakage-fusion-bridge events cause genetic intratumor heterogeneity. Proceedings of the National Academy of Sciences of the United States of America, 97 (10), 5357–5362. doi:10.1073/pnas.090013497; Landry, J. J. M., Pyl, P. T., Rausch, T., Zichner, T., Tekkedil, M. M., Stütz, A. M., Jauch, A., Aiyar, R. S., Pau, G., Delhomme, N., Gagneur, J., Korbel, J. O., Huber, W., & Steinmetz, L. M. (2013). The genomic and transcriptomic landscape of a HeLa cell line, G3: Genes, Genomes, Genetics, 3(8), 1213–1224. doi:10.1534/g3.113.005777; (C,E) Schröck, E., Du Manoir, S., Veldman, T., Schoell, B., Wienberg, J., Ferguson-Smith, M. A., Ning, Y., Ledbetter, D. H., Bar-Am, I., Soenksen, D., Garini, Y., & Ried, T. (1996). Multicolor spectral karyotyping of human chromosomes. Science, 273(5274), 494–497. doi:10.1126/science.273.5274.494; Cell Line NCI60 Drug Discovery Panel (OVCAR-8 cell line): http://home.ncifcrf.gov/CCR/60SKY/new/demo1.asp.

chromosome translocations. In particular, according to this model, the thickness of the chromatin sheets stacked in chromosomes is very small because it corresponds to a monolayer of nucleosomes (see above) and therefore, as observed experimentally, the junction between the original chromosome and the translocated part is produced in a very narrow zone.

The topological organization of DNA is preserved in fixed chromosomes

Isolated metaphase chromosomes in aqueous solution, even in the presence of structuring buffers containing physiological concentrations of Mg^{2+}, are extremely soft structures that very frequently appear to be distorted in TEM images (Caravaca et al., 2005). As described in the preceding sections, the study of chromosomes under these conditions has allowed the discovery of their internal multilaminar structure. In contrast, in the typical cell spreads used in cytogenetic studies, chromosomes are fixed with methanol-acetic acid, and apparently their global structure is completely preserved (McGowan-Jordan et al., 2020). Since early studies revealed that this organic fixative removes part of the histones (Comings, 1978), it could be considered that the structural information inferred from cytogenetic results is not adequate for the understanding of the chromatin organization in native chromosomes. However, the observation of well-defined and orthogonal thin bands and smooth surface connections in sister chromatid exchanges and translocations indicates that the backbone of the multilayered structure is maintained in the chromosome fixed with organics solvents. Considering all these observations, it was proposed that the topological organization of DNA in metaphase chromosomes is preserved when methanol-acetic acid fixatives are used (Daban, 2015). The resulting model is self-consistent because it allows explaining both the results obtained in aqueous media and in the presence of strong fixatives.

Functional implications of multilayered chromosomes

All the previous sections have been devoted exclusively to the study of the structure of chromatin in mitotic chromosomes. In the last part of this chapter, the structure of chromatin along the cell cycle will be considered. The multilayered structure of chromosomes may have multiple functional roles (Daban, 2020) that will be reviewed in this section.

Planar chromatin in interphase

As indicated above, in early experiments chromatin fibers were observed when mitotic chromosomes were treated with water without cations (Fig. 13.1A–C). In the initial studies about the structure of chromatin during interphase, cell nuclei were swollen using similar conditions and chromatin fibers were observed too (see Fig. 13.5A and B; Rattner & Hamkalo, 1979). In buffers containing EDTA, chromatin forms the typical beads-on-a-string filaments with nucleosomes clearly visible (Fig. 13.5C; Thoma et al., 1979). However, it is known that even in interphase there are significant amounts of cations in the cell nucleus (see above; Strick et al., 2001). In buffers containing Mg^{2+}, it was found that the chromatin emanated from mechanically disrupted nuclei was laminar rather than fibrillar (Fig. 13.5D; Chicano & Daban, 2019); this chromatin morphology was observed in all the stages of the interphase. Furthermore, the chromatin fragments produced by micrococcal nuclease

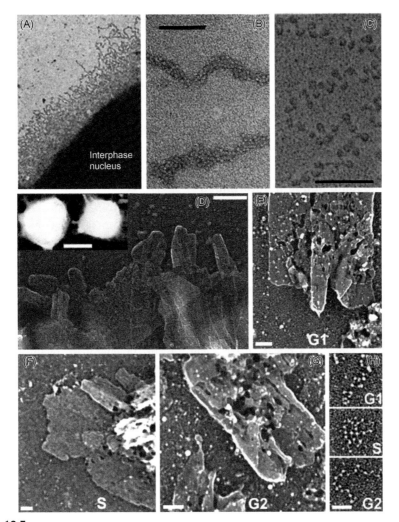

FIG. 13.5

Structure of chromatin in the interphase nucleus. (A) TEM image of chromatin fibers emanating from a mouse cell lysed by hypotonic shock with water; (B) fibers at a higher magnification. (C) Rat liver chromatin in 0.2 mM EDTA. (D) Plates emanated from disrupted human nuclei (vortexed in the presence of glass beads) in a buffer containing 2.5 mM Mg^{2+}; native nuclei are compact and electron-opaque (inset). (E–G) Self-assembled plates observed when chromatin fragments (obtained from micrococcal nuclease digestion of nuclei in G1, S, and G2 phases) were dialyzed against a buffer containing 5 mM Mg^{2+}. (H) Chromatin fragments of G1, S, and G2 nuclei in 1 mM EDTA before self-assembly. Preparations in (A–C) were negatively stained; rotary-shadowed preparations (D–H) are shown in reverse contrast. Scale bars: 100 nm (B,C,H); 200 nm (D–G); 5 μm (inset in D).

Reproduced with permission from (A,B) Rattner, J. B., & Hamkalo, B. A. (1979). Nucleosome packing in interphase chromatin. Journal of Cell Biology, 81(2), 453–457. doi:10.1083/jcb.81.2.453; (C) Thoma, F., Koller, T., & Klug, A. (1979). Involvement of histone H1 in the organization of the nucleosome and of the salt-dependent superstructures of chromatin. Journal of Cell Biology, 83(2), 403–427. doi:10.1083/jcb.83.2.403; (D–H) Chicano, A., & Daban, J. R. (2019). Chromatin plates in the interphase nucleus. FEBS Letters, 593(8), 810–819. doi:10.1002/1873-3468.13370.

digestion of G1, S, and G2 nuclei form extended beads-on-a-string filaments in buffers containing EDTA (Fig. 13.5H) but, in the presence of Mg^{2+}, they self-assemble into plate-like structures (Fig. 13.5E–G). Planar chromatin from interphase nuclei do not form thick plates, indicating that it has a lower tendency to form multilayered structures than metaphase chromatin.

Chromosome bands in interphase

Banding is a characteristic property of mitotic chromosomes. In fact, the typical banding procedures were developed for mitotic chromosomes (McGowan-Jordan et al., 2020). However, cytogenetic experiments using microdissection-based multicolor banding showed that the band pattern of metaphase chromosomes is essentially maintained during all the stages of interphase (Fig. 13.3E; Lemke et al., 2002; Weise et al., 2002). Of note, this method allowed to obtain a high-resolution banding of interphase chromosomes in human brain cells (Fig. 13.3F; Iourov et al., 2006, 2021; Yurov et al., 2007). These results indicate that, although chromosomes are slightly more extended in interphase than in metaphase, the chromosome territories (Cremer & Cremer, 2010; Liehr, 2021a) in interphase retain essential structural elements of the structure of metaphase chromosomes. There are many experimental evidences indicating that bands of mitotic chromosomes are directly related with the multilayered planar structure of chromatin observed in mitosis (see the preceding section); therefore, the maintenance of the chromosome bands in interphase observed in multicolor banding experiments may be directly related with the maintenance of planar chromatin during interphase considered in the preceding paragraph.

Topologically associating domains

Chromosome conformation capture experiments (including genome-wide Hi-C analyses) allow the identification of contacts between different regions of the chromatin filament within the cell nucleus by chemical crosslinking (Lieberman-Aiden et al., 2009; Sajan & Hawkins, 2012). These studies revealed the existence of topologically associating domains (TADs) having a high frequency of internal contacts at distances ≤ 1 Mb (Dixon et al., 2012; Nora et al., 2012; Yumiceba et al., 2021). These results were interpreted considering that chromatin fibers are flexible polymers (Parmar et al., 2019), and the resulting structural models for TADs consisted of loosely packed fibers forming many loops (Bonev & Cavalli, 2016; Dekker et al., 2013). However, according to the observations considered above the contacts corresponding to TADs in Hi-C experiments can be interpreted as contacts produced by the folding of the chromatin filament in multilayered chromosomes. Since the size of a layer in a human chromosome (~0.5 Mb) is comparable to the size of a TAD, it was suggested that each layer may correspond to a TAD (Daban, 2020). From a physical point of view, the capability of self-assembly, high compaction, and good mechanical properties of planar chromatin (see above) could justify the structural stability of TADs, which are considered to be the universal building blocks of chromosomes and are conserved in different species and cell types (Dixon et al., 2016).

TADs were not detected during mitosis in Hi-C studies of synchronous cell cultures (Naumova et al., 2013). Using Hi-C methods at single-cell resolution (Nagano et al., 2017), it was found that the dominant contacts ($\leq$1 Mb, corresponding to TADs) detected in S and G2 cells are replaced by the dominant contacts at much larger distances ($\sim$10 Mb) in mitotic and early-G1 cells. These results can be interpreted considering that TADs disappear during mitosis (Dekker, 2014), but more recent Hi-C studies (Gibcus et al., 2018) have shown that in mitotic chromosomes there are also contacts in the chromatin filament at short distances (crosslinks at $\sim$400 and 80 kb). An easy interpretation of these results was proposed considering that the multilayer organization of chromosomes have different compaction degrees during the cell cycle (Chicano & Daban, 2019). It was suggested that, in addition to the crosslinks at short distances ($\sim$0.5 Mb) corresponding to intralayer contacts, the close stacking of chromatin layers in mitotic cells favors long-distance interlayer contacts and consequently the dominant crosslinks are observed at long distances. During interphase the layers have a low tendency to be stacked (see above) and the dominant crosslinks are produced at short distances within the layers.

Gene expression

The two compaction states of chromatin considered in the preceding paragraph led to the proposal of a hypothesis about the function of multilayer chromosomes in gene expression (Daban, 2020). The compact structure of metaphase chromosomes can justify the general gene inhibition observed during mitosis; in this state the nucleosomes of the successive layers are interacting through their faces (Fig. 13.2G) and chromatin is completely inaccessible to the proteins of the transcription machinery. During interphase, in many regions (A compartments) there are unstacked layers with chromatin accessible for gene expression, and in heterochromatic regions (B compartments) layers are stacked and inactive (see schematic drawing in Fig. 13.6, left panel; see also Liehr, 2021e). The CCCTC-binding factor (CTCF), cohesin, and other proteins may be involved in the configuration of these two states (Bonev & Cavalli, 2016; Nora et al., 2017; Rao et al., 2017). It has been suggested that cohesin and the insulator protein CTCF could generate boundaries between layers leading to well-defined domains for gene expression (Daban, 2020). This is in agreement with the equivalent size of TADs and layers considered above.

Instead of the complex looping mechanisms required to generate functional contacts between enhancer and promoters that are currently considered in the fibrillar models for TADs (Bohn & Heermann, 2010; Dekker, 2008; Dekker & Mirny, 2016), these contacts may simply be due to the folding of the chromatin filament within the layers. This possibility is consistent with data indicating that enhancer-promoter contacts exist before gene activation (Ghavi-Helm et al., 2014; Jin et al.,

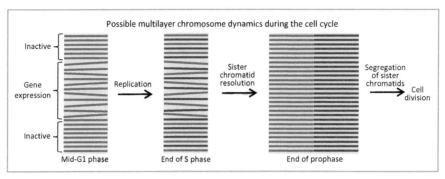

FIG. 13.6

Hypothetical function of multilayered chromosomes in gene regulation and DNA replication. The figure is a simplified representation of part of an original multilayered chromosome and the resulting replicated chromatids (see text and Daban, 2020).

2013). Furthermore, presumably even the noncoding DNA regions are essential to promote the correct enhancer-promoter contacts in the chromatin layers, and this can explain results showing that disruptions of noncoding regions cause misexpression and disease (Hnisz et al., 2016; Lupiáñez et al., 2015; Spielmann et al., 2018). Transcription factor binding requires nucleosome recognition (Ballaré et al., 2013). Since the thickness of a single chromatin layer corresponds to a monolayer of nucleosomes (see above; Chicano et al., 2019), chromatin accessibility is very high on both sides of each unstacked layer. Therefore, in active regions the elements of the basic transcription machinery, as well as specific transcription factors and mediators, can interact with promoters and regulatory sequences in the layers. The relationship of TADs with cell differentiation (Dixon et al., 2015; Zheng & Xie, 2019) can be interpreted considering that different chromosomal regions become unstacked and active during the successive stages of the development.

DNA replication

As indicated above, the easy sliding of layers (Fig. 13.1I and J) indicates the absence of entanglements in the chromatin filament between adjacent layers. This well-defined topology may justify recent superresolution FISH and Hi-C results showing a lack of knots and entanglements within and between chromosomal domains during interphase (Goundaroulis et al., 2020; Tavares-Cadete et al., 2020). Furthermore, this chromatin organization may facilitate DNA replication and repair in multilayered chromosomes (Daban, 2020). Currently, it is considered that DNA replication occurs in many replication bubbles in the linear chromatin filament of each chromosome giving rise to two daughter filaments at the end of S phase (Alabert & Groth, 2012). Planar chromatin must be locally disrupted by the replication machinery to build active replication bubbles. In mammals, domains containing 5–10 bubbles replicate simultaneously. The size of these clusters [0.4–0.8 Mb (Pope et al., 2014)] suggests that each chromatin layer could be a replication domain. The late replication regions correspond to inactive chromosome compartments. It is expected that replication will produce unentangled replicated layers on top of each other; if all the layers are connected between them forming a helicoid (see above), eventually the complete replication will give rise to a double helicoid (Fig. 13.6, middle panel). Finally, an unknown mechanism will separate the two sister chromatids (Fig. 13.6, right panel). Sister chromatid resolution takes place at the end of prophase (Liang et al., 2015; Nagasaka et al., 2016). Interphase cohesin is replaced by condensins during mitosis (Abramo et al., 2019); presumably these proteins and topoisomerase II (Hirano, 2015) are required for the separation of the two helicoids. The resulting multilayered helicoidal sister chromatids have a compact structure and good mechanical properties to be safely transferred to the daughter cells at the end of mitosis.

DNA repair

The compactness of planar chromatin could be related with DNA repair (Daban, 2020). There is no covalent continuity of DNA in plates self-assembled from small chromatin fragments (see above; Chicano & Daban, 2019; Milla & Daban, 2012), indicating that chromatin can maintain its planar structure when DNA is broken. This self-consistent two-dimensional network may explain results showing that in living cells, after double-strand breaks, the resulting DNA ends remain in close proximity (Misteli & Soutoglou, 2009; Soutoglou et al., 2007). Such a structural stability, which is difficult to be justified by the loosely packed chromatin fibers considered in the current literature, may facilitate DNA repair. Furthermore, the hypothesized intermediate double helicoids produced during DNA replication (Fig. 13.6, middle panel) contain two copies of any genomic DNA sequence in close proximity between them. This may facilitate the repair, during late S-G2 phases (Misteli & Soutoglou, 2009), of double-stranded DNA breaks by homologous recombination in any region of the genome.

Concluding remarks

Considering the intrinsic linear structure of DNA and the chromatin filament, the existence of multilayered chromatin plates is counterintuitive. However, if a given structure, such as chromatin plates, can be spontaneously self-assembled from its native components, it is reasonable to consider that in the cell it may have a biological role based on its physicochemical properties. The results reviewed in this chapter demonstrates that it is possible to build a bridge between four independent fields: structural and biophysical research about planar chromatin, cytogenomic observations, chromosome conformation capture results, and functional studies of chromatin. The multilayer chromosome model is compatible with the dynamic structure and multiple biological functions of chromatin and provides for the first time a consistent explanation of the chromosome structural elements on which diverse clinical cytogenetic analyses are based.

References

Abramo, K., Valton, A. L., Venev, S. V., Ozadam, H., Fox, A. N., & Dekker, J. (2019). A chromosome folding intermediate at the condensin-to-cohesin transition during telophase. *Nature Cell Biology*, *21*(11), 1393–1402. https://doi.org/10.1038/s41556-019-0406-2.

Adams, M., Dogic, Z., Keller, S. L., & Fraden, S. (1998). Entropically driven microphase transitions in mixtures of colloidal rods and spheres. *Nature*, *393*, 349–352. https://doi.org/10.1038/30700.

Adhireksan, Z., Sharma, D., Lee, P. L., Bao, Q., Padavattan, S., Shum, W. K., ... Davey, C. A. (2021). Engineering nucleosomes for generating diverse chromatin assemblies. *Nucleic Acids Research*, gkab070. https://doi.org/10.1093/nar/gkab070.

Adhireksan, Z., Sharma, D., Lee, P. L., & Davey, C. A. (2020). Near-atomic resolution structures of interdigitated nucleosome fibres. *Nature Communications*, *11*(1), 4747. https://doi.org/10.1038/s41467-020-18533-2.

Alabert, C., & Groth, A. (2012). Chromatin replication and epigenome maintenance. *Nature Reviews Molecular Cell Biology*, *13*(3), 153–167. https://doi.org/10.1038/nrm3288.

Bahr, G. F. (1977). Chromosomes and chromatin structure. In J. J. Yunis (Ed.), *Molecular structure of human chromosomes* (pp. 143–203). New York: Academic Press.

Ballaré, C., Castellano, G., Gaveglia, L., Althammer, S., González-Vallinas, J., Eyras, E., ... Beato, M. (2013). Nucleosome-driven transcription factor binding and gene regulation. *Molecular Cell*, *49*(1), 67–79. https://doi.org/10.1016/j.molcel.2012.10.019.

Bartolomé, S., Bermúdez, A., & Daban, J. R. (1995). Electrophoresis of chromatin on non-denaturing agarose gels containing Mg^{2+}: Self-assembly of small chromatin fragments and folding of the 30-nm fiber. *Journal of Biological Chemistry*, *270*(38), 22514–22521. https://doi.org/10.1074/jbc.270.38.22514.

Bertin, A., Mangenot, S., Renouard, M., Durand, D., & Livolant, F. (2007). Structure and phase diagram of nucleosome core particles aggregated by multivalent cations. *Biophysical Journal*, *93*(10), 3652–3663. https://doi.org/10.1529/biophysj.107.108365.

Bohn, M., & Heermann, D. W. (2010). Diffusion-driven looping provides a consistent framework for chromatin organization. *PLoS One*, *5*(8), e12218. https://doi.org/10.1371/journal.pone.0012218.

Bonev, B., & Cavalli, G. (2016). Organization and function of the 3D genome. *Nature Reviews Genetics*, *17*(11), 661–678. https://doi.org/10.1038/nrg.2016.112.

Boy de la Tour, E., & Laemmli, U. K. (1988). The metaphase scaffold is helically folded: Sister chromatids have predominantly opposite helical handedness. *Cell*, *55*, 937–944. https://doi.org/10.1016/0092-8674(88)90239-5.

Brower-Toland, B. D., Smith, C. L., Yeh, R. C., Lis, J. T., Peterson, C. L., & Wang, M. D. (2002). Mechanical disruption of individual nucleosomes reveals a reversible multistage release of DNA. *Proceedings of the National Academy of Sciences of the United States of America*, *99*(4), 1960–1965. https://doi.org/10.1073/pnas.022638399.

Bustamante, C., Smith, S. B., Liphardt, J., & Smith, D. (2000). Single-molecule studies of DNA mechanics. *Current Opinion in Structural Biology*, *10*(3), 279–285. https://doi.org/10.1016/S0959-440X(00)00085-3.

Callister, W. D., & Rethwisch, D. G. (2015). *Materials science and engineering*. New York: Wiley.

Caravaca, J. M., Caño, S., Gállego, I., & Daban, J. R. (2005). Structural elements of bulk chromatin within metaphase chromosomes. *Chromosome Research*, *13*(7), 725–743. https://doi.org/10.1007/s10577-005-1008-3.

Castro-Hartmann, P., Milla, M., & Daban, J. R. (2010). Irregular orientation of nucleosomes in the well-defined chromatin plates of metaphase chromosomes. *Biochemistry*, *49*(19), 4043–4050. https://doi.org/10.1021/bi100125f.

Chen, Y., Tan, C., Zhang, H., & Wang, L. (2015). Two-dimensional graphene analogues for biomedical applications. *Chemical Society Reviews*, *44*(9), 2681–2701. https://doi.org/10.1039/c4cs00300d.

Cheung, V. G., Nowak, N., Jang, W., Kirsch, I. R., Zhao, S., Chen, X. N., … The BAC Resource Consortium. (2001). Integration of cytogenetic landmarks into the draft sequence of the human genome. *Nature*, *409*, 953–958. https://doi.org/10.1038/35057192.

Chicano, A., Crosas, E., Otón, J., Melero, R., Engel, B. D., & Daban, J. R. (2019). Frozen-hydrated chromatin from metaphase chromosomes has an interdigitated multilayer structure. *EMBO Journal*, *38*(7), e99769. https://doi.org/10.15252/embj.201899769.

Chicano, A., & Daban, J. R. (2019). Chromatin plates in the interphase nucleus. *FEBS Letters*, *593*(8), 810–819. https://doi.org/10.1002/1873-3468.13370.

Chung, W. J., Oh, J. W., Kwak, K., Lee, B. Y., Meyer, J., Wang, E., … Lee, S.W. (2011). Biomimetic self-templating supramolecular structures. *Nature*, *478*(7369), 364–368. https://doi.org/10.1038/nature10513.

Comings, D. E. (1978). Mechanisms of chromosome banding and implications for chromosome structure. *Annual Review of Genetics*, *12*, 25–46. https://doi.org/10.1146/annurev.ge.12.120178.000325.

Cortés, F., & Escalza, P. (1986). Analysis of different banding patterns and late replicating regions in chromosomes of *Allium cepa*, *A. sativum* and *A. nigrum*. *Genetica*, *71*(1), 39–46. https://doi.org/10.1007/BF00123231.

Cremer, T., & Cremer, M. (2010). Chromosome territories. *Cold Spring Harbor Perspectives in Biology*, *2*(1), a003889. https://doi.org/10.1101/cshperspect.a003889.

Cui, Y., & Bustamante, C. (2000). Pulling a single chromatin fiber reveals the forces that maintain its higher-order structure. *Proceedings of the National Academy of Sciences of the United States of America*, *97*(1), 127–132. https://doi.org/10.1073/pnas.97.1.127.

Daban, J. R. (2000). Physical constraints in the condensation of eukaryotic chromosomes. Local concentration of DNA versus linear packing ratio in higher order chromatin structures. *Biochemistry*, *39*(14), 3861–3866. https://doi.org/10.1021/bi992628w.

Daban, J. R. (2003). High concentration of DNA in condensed chromatin. *Biochemistry and Cell Biology*, *81*(3), 91–99. https://doi.org/10.1139/o03-037.

Daban, J. R. (2011). Electron microscopy and atomic force microscopy studies of chromatin and metaphase chromosome structure. *Micron*, *42*(8), 733–750. https://doi.org/10.1016/j.micron.2011.05.002.

Daban, J. R. (2014). The energy components of stacked chromatin layers explain the morphology, dimensions and mechanical properties of metaphase chromosomes. *Journal of the Royal Society, Interface*, *11*, 20131043.

Daban, J. R. (2015). Stacked thin layers of metaphase chromatin explain the geometry of chromosome rearrangements and banding. *Scientific Reports*, *5*, 14891. https://doi.org/10.1038/srep14891.

Daban, J. R. (2020). Supramolecular multilayer organization of chromosomes: Possible functional roles of planar chromatin in gene expression and DNA replication and repair. *FEBS Letters*, *594*(3), 395–411. https://doi.org/10.1002/1873-3468.13724.

Daban, J. R., & Bermúdez, A. (1998). Interdigitated solenoid model for compact chromatin fibers. *Biochemistry*, *37*(13), 4299–4304. https://doi.org/10.1021/bi973117h.

Das, S., Robinson, J. A., Dubey, M., Terrones, H., & Terrones, M. (2015). Beyond graphene: Progress in novel two-dimensional materials and van der Waals solids. *Annual Review of Materials Research, 45,* 1–27. https://doi.org/10.1146/annurev-matsci-070214-021034.

De Jong, B. E., Brouwer, T. B., Kaczmarczyk, A., Visscher, B., & van Noort, J. (2018). Rigid basepair Monte Carlo simulations of one-start and two-start chromatin fiber unfolding by force. *Biophysical Journal, 115*(10), 1848–1859. https://doi.org/10.1016/j.bpj.2018.10.007.

Dekker, J. (2008). Gene regulation in the third dimension. *Science, 319*(5871), 1793–1794. https://doi.org/10.1126/science.1152850.

Dekker, J. (2014). Two ways to fold the genome during the cell cycle: Insights obtained with chromosome conformation capture. *Epigenetics & Chromatin, 7,* 25. https://doi.org/10.1186/1756-8935-7-25.

Dekker, J., Marti-Renom, M. A., & Mirny, L. A. (2013). Exploring the three-dimensional organization of genomes: Interpreting chromatin interaction data. *Nature Reviews Genetics, 14*(6), 390–403. https://doi.org/10.1038/nrg3454.

Dekker, J., & Mirny, L. (2016). The 3D genome as moderator of chromosomal communication. *Cell, 164*(6), 1110–1121. https://doi.org/10.1016/j.cell.2016.02.007.

Dixon, J. R., Gorkin, D. U., & Ren, B. (2016). Chromatin domains: The unit of chromosome organization. *Molecular Cell, 62*(5), 668–680. https://doi.org/10.1016/j.molcel.2016.05.018.

Dixon, J. R., Jung, I., Selvaraj, S., Shen, Y., Antosiewicz-Bourget, J. E., Lee, A. Y., … Ren, B. (2015). Chromatin architecture reorganization during stem cell differentiation. *Nature, 518*(7539), 331–336. https://doi.org/10.1038/nature14222.

Dixon, J. R., Selvaraj, S., Yue, F., Kim, A., Li, Y., Shen, Y., … Ren, B. (2012). Topological domains in mammalian genomes identified by analysis of chromatin interactions. *Nature, 485*(7398), 376–380. https://doi.org/10.1038/nature11082.

Dubochet, J., & Noll, M. (1978). Nucleosome arcs and helices. *Science, 202*(4365), 280–286. https://doi.org/10.1126/science.694532.

DuPraw, E. J. (1966). Evidence for a "folded-fibre" organization in human chromosomes. *Nature, 209*(5023), 577–581. https://doi.org/10.1038/209577a0.

DuPraw, E. J. (1970). *DNA and chromosomes.* New York: Holt, Rinehart and Winston.

Earnshaw, W. C., & Laemmli, U. K. (1983). Architecture of metaphase chromosomes and chromosome scaffolds. *Journal of Cell Biology, 96*(1), 84–93. https://doi.org/10.1083/jcb.96.1.84.

Ekudayo, B., Richmond, T. J., & Schalch, T. (2017). Capturing structural heterogeneity in chromatin fibers. *Journal of Molecular Biology, 429*(20), 3031–3042. https://doi.org/10.1016/j.jmb.2017.09.002.

Eltsov, M., MacLellan, K. M., Maeshima, K., Frangakis, A. S., & Dubochet, J. (2008). Analysis of cryo-electron microscopy images does not support the existence of 30-nm chromatin fibers in mitotic chromosomes in situ. *Proceedings of the National Academy of Sciences of the United States of America, 105*(50), 19732–19737. https://doi.org/10.1073/pnas.0810057105.

Friebe, B., & Cortés, F. (1996). Sister chromatid exchange and replication banding. In K. Fukui, & S. Nakayama (Eds.), *Plant chromosomes: Laboratory methods* (pp. 171–186). Cleveland: CRC Press.

Gállego, I., Castro-Hartmann, P., Caravaca, J. M., Caño, S., & Daban, J. R. (2009). Dense chromatin plates in metaphase chromosomes. *European Biophysics Journal, 38*(4), 503–522. https://doi.org/10.1007/s00249-008-0401-1.

Gállego, I., Oncins, G., Sisquella, X., Fernàndez-Busquets, X., & Daban, J. R. (2010). Nanotribology results show that DNA forms a mechanically resistant 2D network in metaphase chromatin plates. *Biophysical Journal*, *99*(12), 3951–3958. https://doi.org/10.1016/j.bpj.2010.11.015.

Ghavi-Helm, Y., Klein, F. A., Pakozdi, T., Ciglar, L., Noordermeer, D., Huber, W., & Furlong, E. E. M. (2014). Enhancer loops appear stable during development and are associated with paused polymerase. *Nature*, *512*(1), 96–100. https://doi.org/10.1038/nature13417.

Gibaud, T., Barry, E., Zakhary, M. J., Henglin, M., Ward, A., Yang, Y., … Dogic, Z. (2012). Reconfigurable self-assembly through chiral control of interfacial tension. *Nature*, *481*(7381), 348–351. https://doi.org/10.1038/nature10769.

Gibcus, J. H., Samejima, K., Goloborodko, A., Samejima, I., Naumova, N., Nuebler, J., … Dekker, J. (2018). A pathway for mitotic chromosome formation. *Science*, *359*(6376), eaao6135. https://doi.org/10.1126/science.aao6135.

Gisselsson, D., Pettersson, L., Höglund, M., Heidenblad, M., Gorunova, L., Wiegant, J., … Mandahl, N. (2000). Chromosomal breakage-fusion-bridge events cause genetic intratumor heterogeneity. *Proceedings of the National Academy of Sciences of the United States of America*, *97*(10), 5357–5362. https://doi.org/10.1073/pnas.090013497.

Gornik, S. G., Hu, I., Lassadi, I., & Waller, R. F. (2019). The biochemistry and evolution of dinoflagellate nucleus. *Microorganisms*, *7*(8), 245.

Goundaroulis, D., Lieberman Aiden, E., & Stasiak, A. (2020). Chromatin is frequently unknotted at the megabase scale. *Biophysical Journal*, *118*, 2268–2279. https://doi.org/10.1016/j.bpj.2019.11.002.

Hewish, D. R., & Burgoyne, L. A. (1973). Chromatin sub-structure. The digestion of chromatin DNA at regularly spaced sites by a nuclear deoxyribonuclease. *Biochemical and Biophysical Research Communications*, *52*(2), 504–510. https://doi.org/10.1016/0006-291X(73)90740-7.

Hirano, T. (2015). Chromosome dynamics during mitosis. *Cold Spring Harbor Perspectives in Biology*, *7*, a015792. https://doi.org/10.1101/cshperspect.a015792.

Hliscs, R., Mühlig, P., & Claussen, U. (1997). The nature of G-bands analyzed by chromosome stretching. *Cytogenetic and Genome Research*, *79*(1–2), 162–166. https://doi.org/10.1159/000134710.

Hnisz, D., Weintraub, A. S., Day, D. S., Valton, A. L., Bak, R. O., Li, C. H., … Young, R.A. (2016). Activation of proto-oncogenes by disruption of chromosome neighborhoods. *Science*, *351*(6280), 1454–1458. https://doi.org/10.1126/science.aad9024.

International Human Genome Sequencing Consortium. (2001). Initial sequencing and analysis of the human genome. *Nature*, *409*, 860–921. https://doi.org/10.1038/35057062.

Iourov, I. Y., Liehr, T., Vorsanova, S. G., Kolotii, A. D., & Yurov, Y. B. (2006). Visualization of interphase chromosomes in postmitotic cells of the human brain by multicolour banding (MCB). *Chromosome Research*, *14*(3), 223–229. https://doi.org/10.1007/s10577-006-1037-6.

Iourov, I. Y., Vorsanova, S. G., & Yurov, Y. B. (2021). Cytogenomic landscape of the human brain. In T. Liehr (Ed.), *Cytogenomics* (pp. 327–348). Academic Press. Chapter 16 (in this book).

Jin, F., Li, Y., Dixon, J. R., Selvaraj, S., Ye, Z., Lee, A. Y., … Ren, B. (2013). A high-resolution map of the three-dimensional chromatin interactome in human cells. *Nature*, *503*(7475), 290–294. https://doi.org/10.1038/nature12644.

Kireeva, N., Lakonishok, M., Kireev, I., Hirano, T., & Belmont, A. S. (2004). Visualization of early chromosome condensation: A hierarchical folding, axial glue model of chromosome structure. *Journal of Cell Biology*, *166*(6), 775–785. https://doi.org/10.1083/jcb.200406049.

Kornberg, R. D. (1974). Chromatin structure: A repeating unit of histones and DNA. *Science, 184*(4139), 868–871. https://doi.org/10.1126/science.184.4139.868.

Korolev, N., Lyubartsev, A. P., & Nordenskiöld, L. (2018). A systematic analysis of nucleosome core particle and nucleosome-nucleosome stacking structure. *Scientific Reports, 8,* 1543. https://doi.org/10.1038/s41598-018-19875-0.

Koyfman, A. Y., Magonov, S. N., & Reich, N. O. (2009). Self-assembly of DNA arrays into multilayer stacks. *Langmuir, 25*(2), 1091–1096. https://doi.org/10.1021/la801306j.

Kruithof, M., Chien, F. T., Routh, A., Logie, C., Rhodes, D., & Van Noort, J. (2009). Single-molecule force spectroscopy reveals a highly compliant helical folding for the 30-nm chromatin fiber. *Nature Structural and Molecular Biology, 16*(5), 534–540. https://doi.org/10.1038/nsmb.1590.

Landry, J. J. M., Pyl, P. T., Rausch, T., Zichner, T., Tekkedil, M. M., Stütz, A. M., ... Steinmetz, L. M. (2013). The genomic and transcriptomic landscape of a HeLa cell line. *G3: Genes, Genomes, Genetics, 3*(8), 1213–1224. https://doi.org/10.1534/g3.113.005777.

Lawrence, J. B., Singer, R. H., & McNeil, J. A. (1990). Interphase and metaphase resolution of different distances within the human dystrophin gene. *Science, 249*(4971), 928–932. https://doi.org/10.1126/science.2203143.

Leforestier, A., Dubochet, J., & Livolant, F. (2001). Bilayers of nucleosome core particles. *Biophysical Journal, 81*(4), 2414–2421. https://doi.org/10.1016/S0006-3495(01)75888-2.

Leforestier, A., Fudaley, S., & Livolant, F. (1999). Spermidine-induced aggregation of nucleosome core particles: Evidence for multiple liquid crystalline phases. *Journal of Molecular Biology, 290*(2), 481–494. https://doi.org/10.1006/jmbi.1999.2895.

Lehn, J. M. (2002). Toward self-organization and complex matter. *Science, 295*(5564), 2400–2403. https://doi.org/10.1126/science.1071063.

Lehn, J. M. (2007). From supramolecular chemistry towards constitutional dynamic chemistry and adaptive chemistry. *Chemical Society Reviews, 36,* 151–160. https://doi.org/10.1039/b616752g.

Lemke, J., Claussen, J., Michel, S., Chudoba, I., Mühlig, P., Westermann, M., ... Claussen, U. (2002). The DNA-based structure of human chromosome 5 in interphase. *American Journal of Human Genetics, 71*(5), 1051–1059. https://doi.org/10.1086/344286.

Liang, Z., Zickler, D., Prentiss, M., Chang, F. S., Witz, G., Maeshima, K., & Kleckner, N. (2015). Chromosomes progress to metaphase in multiple discrete steps via global compaction/expansion cycles. *Cell, 161*(5), 1124–1137. https://doi.org/10.1016/j.cell.2015.04.030.

Lieberman-Aiden, E., Van Berkum, N. L., Williams, L., Imakaev, M., Ragoczy, T., Telling, A., ... Dekker, J. (2009). Comprehensive mapping of long-range interactions reveals folding principles of the human genome. *Science, 326*(5950), 289–293. https://doi.org/10.1126/science.1181369.

Liehr, T. (2021a). Nuclear architecture. In T. Liehr (Ed.), *Cytogenomics* (pp. 297–305). Academic Press. Chapter 14 (in this book).

Liehr, T. (2021b). A definition for cytogenomics - Which also may be called chromosomics. In T. Liehr (Ed.), *Cytogenomics* (pp. 1–7). Academic Press. Chapter 1 (in this book).

Liehr, T. (2021c). Overview of currently available approaches used in cytogenomics. In T. Liehr (Ed.), *Cytogenomics* (pp. 11–24). Academic Press. Chapter 2 (in this book).

Liehr, T. (2021d). Molecular cytogenetics. In T. Liehr (Ed.), *Cytogenomics* (pp. 35–45). Academic Press. Chapter 4 (in this book).

Liehr, T. (2021e). Repetitive elements, heteromorphisms, and copy number variants. In T. Liehr (Ed.), *Cytogenomics* (pp. 373–388). Academic Press. Chapter 19 (in this book).

Livolant, F., & Bouligand, Y. (1978). New observations on the twisted arrangement of dinoflagellate chromosomes. *Chromosoma*, *68*(1), 21–44. https://doi.org/10.1007/BF00330370.

Luger, K., Dechassa, M. L., & Tremethick, D. J. (2012). New insights into nucleosome and chromatin structure: An ordered state or a disordered affair? *Nature Reviews Molecular Cell Biology*, *13*(7), 436–447. https://doi.org/10.1038/nrm3382.

Luger, K., Mäder, A. W., Richmond, R. K., Sargent, D. F., & Richmond, T. J. (1997). Crystal structure of the nucleosome core particle at 2.8 Å resolution. *Nature*, *389*(6648), 251–260. https://doi.org/10.1038/38444.

Lupiáñez, D. G., Kraft, K., Heinrich, V., Krawitz, P., Brancati, F., Klopocki, E., & Mundlos, S. (2015). Disruptions of topological domains cause pathogenic rewiring of gene-enhancer interactions. *Cell*, *161*, 1012–1025. https://doi.org/10.1016/j.cell.2015.04.004.

Mangenot, S., Leforestier, A., Durand, D., & Livolant, F. (2003). Phase diagram of nucleosome core particles. *Journal of Molecular Biology*, *333*(5), 907–916. https://doi.org/10.1016/j.jmb.2003.09.015.

Martin, T. G., & Dietz, H. (2012). Magnesium-free self-assembly of multi-layer DNA objects. *Nature Communications*, *3*, 1103. https://doi.org/10.1038/ncomms2095.

McGowan-Jordan, J., Hastings, R. J., & Moore, S. (2020). *ISCN 2020 - An international system for human cytogenomic nomenclature (2020)*. Basel: Karger.

Mihardja, S., Spakowitz, A. J., Zhang, Y., & Bustamante, C. (2006). Effect of force on mononucleosomal dynamics. *Proceedings of the National Academy of Sciences of the United States of America*, *103*(43), 15871–15876. https://doi.org/10.1073/pnas.0607526103.

Milla, M., & Daban, J. R. (2012). Self-assembly of thin plates from micrococcal nuclease-digested chromatin of metaphase chromosomes. *Biophysical Journal*, *103*(3), 567–575. https://doi.org/10.1016/j.bpj.2012.06.028.

Misteli, T., & Soutoglou, E. (2009). The emerging role of nuclear architecture in DNA repair and genome maintenance. *Nature Reviews Molecular Cell Biology*, *10*(4), 243–254. https://doi.org/10.1038/nrm2651.

Mitov, M. (2017). Cholesteric liqid crystals in soft matter. *Soft Matter*, *13*, 4176–4209.

Moller, J., Lequieu, J., & De Pablo, J. J. (2019). The free energy landscape of internucleosome interactions and its relation to chromatin fiber structure. *ACS Central Science*, *5*(2), 341–348. https://doi.org/10.1021/acscentsci.8b00836.

Nagano, T., Lubling, Y., Várnai, C., Dudley, C., Leung, W., Baran, Y., … Tanay, A. (2017). Cell-cycle dynamics of chromosomal organization at single-cell resolution. *Nature*, *547*(7661), 61–67. https://doi.org/10.1038/nature23001.

Nagasaka, K., Hossain, M. J., Roberti, M. J., Ellenberg, J., & Hirota, T. (2016). Sister chromatid resolution is an intrinsic part of chromosome organization in prophase. *Nature Cell Biology*, *18*(6), 692–699. https://doi.org/10.1038/ncb3353.

Naumova, N., Imakaev, M., Fudenberg, G., Zhan, Y., Lajoie, B. R., Mirny, L. A., & Dekker, J. (2013). Organization of the mitotic chromosome. *Science*, *342*(6161), 948–953. https://doi.org/10.1126/science.1236083.

Nora, E. P., Goloborodko, A., Valton, A. L., Gibcus, J. H., Uebersohn, A., Abdennur, N., … Bruneau, B.G. (2017). Targeted degradation of CTCF decouples local insulation of chromosome domains from genomic compartmentalization. *Cell*, *169*(5), 930–944.e22. https://doi.org/10.1016/j.cell.2017.05.004.

Nora, E. P., Lajoie, B. R., Schulz, E. G., Giorgetti, L., Okamoto, I., Servant, N., … Heard, E. (2012). Spatial partitioning of the regulatory landscape of the X-inactivation centre. *Nature*, *485*(7398), 381–385. https://doi.org/10.1038/nature11049.

Norouzi, D., & Zhurkin, V. B. (2018). Dynamics of chromatin fibers: Comparison of Monte Carlo simulations with force spectroscopy. *Biophysical Journal*, *115*(9), 1644–1655. https://doi.org/10.1016/j.bpj.2018.06.032.

Novoselov, K. S., Geim, A. K., Morozov, S. V., Jiang, D., Zhang, Y., Duvonos, S. V., ... Firsov, A.A. (2004). Electric field effect in atomically thin carbon films. *Science*, *306*, 666–669. https://doi.org/10.1126/science.1102896.

Ohnuki, Y. (1965). Demonstration of the spiral structure of human chromosomes. *Nature*, *208*(5013), 916–917. https://doi.org/10.1038/208916a0.

Ou, H. D., Phan, S., Deerinck, T. J., Thor, A., Ellisman, M. H., & O'Shea, C. C. (2017). Chrom EMT: Visualizing 3D chromatin structure and compaction in interphase and mitotic cells. *Science*, *357*(6349), eaag0025. https://doi.org/10.1126/science.aag0025.

Ozin, G. A., Arsenault, A. C., & Cademartiri, L. (2009). *Nanochemistry. A chemical approach to nanomaterials*. Cambridge: RSC Publishing.

Parmar, J. J., Woringer, M., & Zimmer, C. (2019). How the genome folds: The biophysics of four-dimensional chromatin organization. *Annual Review of Biophysics*, *48*, 231–253. https://doi.org/10.1146/annurev-biophys-052118-115638.

Paulson, J. R., & Laemmli, U. K. (1977). The structure of histone-depleted metaphase chromosomes. *Cell*, *12*(3), 817–828. https://doi.org/10.1016/0092-8674(77)90280-X.

Poirier, M., Eroglu, S., Chatenay, D., & Marko, J. F. (2000). Reversible and irreversible unfolding of mitotic newt chromosomes by applied force. *Molecular Biology of the Cell*, *11*(1), 269–276. https://doi.org/10.1091/mbc.11.1.269.

Poirier, M. G., & Marko, J. F. (2002). Mitotic chromosomes are chromatin networks without a mechanically contiguous protein scaffold. *Proceedings of the National Academy of Sciences of the United States of America*, *99*(24), 15393–15397. https://doi.org/10.1073/pnas.232442599.

Poirier, M. G., Monhait, T., & Marko, J. F. (2002). Reversible hypercondensation and decondensation of mitotic chromosomes studied using combined chemical-micromechanical techniques. *Journal of Cellular Biochemistry*, *85*(2), 422–434. https://doi.org/10.1002/jcb.10132.

Pope, B. D., Ryba, T., Dileep, V., Yue, F., Wu, W., Denas, O., & Gilbert, D. M. (2014). Topologically associating domains are stable units of replication-timing regulation. *Nature*, *515*(7527), 402–405. https://doi.org/10.1038/nature13986.

Portugal, F. H., & Cohen, J. S. (1977). *A century of DNA. A history of the discovery of the structure and function of the genetic substance*. Cambridge: The MIT Press.

Rao, S. S. P., Huang, S. C., St Hilaire, B. G., Engreitz, J. M., Perez, E. M., Kieffer-Kwon, K. R., & Aiden, E. L. (2017). Cohesin loss eliminates all loop domains. *Cell*, *171*(2), 305–320. e24. https://doi.org/10.1016/j.cell.2017.09.026.

Rattner, J. B., & Hamkalo, B. A. (1979). Nucleosome packing in interphase chromatin. *Journal of Cell Biology*, *81*(2), 453–457. https://doi.org/10.1083/jcb.81.2.453.

Rattner, J. B., & Lin, C. C. (1985). Radial loops and helical coils coexist in metaphase chromosomes. *Cell*, *42*(1), 291–296. https://doi.org/10.1016/S0092-8674(85)80124-0.

Robinson, P. J. J., Fairall, L., Huynh, V. A. T., & Rhodes, D. (2006). EM measurements define the dimensions of the "30-nm" chromatin fiber: Evidence for a compact, interdigitated structure. *Proceedings of the National Academy of Sciences of the United States of America*, *103*(17), 6506–6511. https://doi.org/10.1073/pnas.0601212103.

Roschke, A. V., Tonon, G., Gehlhaus, K. S., McTyre, N., Bussey, K. J., Lababidi, S., ... Kirsch, I. R. (2003). Karyotypic complexity of the NCI-60 drug-screening panel. *Cancer Research*, *63*, 8634–8647.

Saitoh, Y., & Laemmli, U. K. (1994). Metaphase chromosome structure: Bands arise from a differential folding path of the highly AT-rich scaffold. *Cell, 76*(4), 609–622. https://doi.org/10.1016/0092-8674(94)90502-9.

Sajan, S. A., & Hawkins, R. D. (2012). Methods for identifying higher-order chromatin structure. *Annual Review of Genomics and Human Genetics, 13,* 59–82. https://doi.org/10.1146/annurev-genom-090711-163818.

Schalch, T., Duda, S., Sargent, D. F., & Richmond, T. J. (2005). X-ray structure of a tetranucleosome and its implications for the chromatin fibre. *Nature, 436*(7047), 138–141. https://doi.org/10.1038/nature03686.

Schröck, E., Du Manoir, S., Veldman, T., Schoell, B., Wienberg, J., Ferguson-Smith, M. A., … Ried, T. (1996). Multicolor spectral karyotyping of human chromosomes. *Science, 273*(5274), 494–497. https://doi.org/10.1126/science.273.5274.494.

Song, F., Chen, P., Sun, D., Wang, M., Dong, L., Liang, D., … Li, G. (2014). Cryo-EM study of the chromatin fiber reveals a double helix twisted by tetranucleosomal units. *Science, 344*(6182), 376–380. https://doi.org/10.1126/science.1251413.

Soutoglou, E., Dorn, J. F., Sengupta, K., Jasin, M., Nussenzweig, A., Ried, T., … Misteli, T. (2007). Positional stability of single double-strand breaks in mammalian cells. *Nature Cell Biology, 9*(6), 675–682. https://doi.org/10.1038/ncb1591.

Spielmann, M., Lupiáñez, D. G., & Mundlos, S. (2018). Structural variation in the 3D genome. *Nature Reviews Genetics, 19*(7), 453–467. https://doi.org/10.1038/s41576-018-0007-0.

Strick, R., Strissel, P. L., Gavrilov, K., & Levi-Setti, R. (2001). Cation-chromatin binding as shown by ion microscopy is essential for the structural integrity of chromosomes. *Journal of Cell Biology, 155*(6), 899–910. https://doi.org/10.1083/jcb.200105026.

Sumner, A. T. (2003). *Chromosomes: Organization and function.* Oxford: Blackwell Publishing.

Sun, J. Y., Zhao, X., Illeperuma, W. R. K., Chaudhuri, O., Oh, K. H., Mooney, D. J., … Suo, Z. (2012). Highly stretchable and tough hydrogels. *Nature, 489*(7414), 133–136. https://doi.org/10.1038/nature11409.

Tavares-Cadete, F., Norouzi, D., Dekker, B., Liu, Y., & Dekker, J. (2020). Multi-contact 3C reveals that the human genome during interphase is largely not entangled. *Nature Structural & Molecular Biology, 27,* 1105–1114. https://doi.org/10.1038/s41594-020-0506-5.

Thoma, F., Koller, T., & Klug, A. (1979). Involvement of histone H1 in the organization of the nucleosome and of the salt-dependent superstructures of chromatin. *Journal of Cell Biology, 83*(2), 403–427. https://doi.org/10.1083/jcb.83.2.403.

Trask, B. J. (2002). Human cytogenetics: 46 chromosomes, 46 years and counting. *Nature Reviews Genetics, 3*(10), 769–778. https://doi.org/10.1038/nrg905.

Trask, B. J., Allen, S., Massa, H., Fertitta, A., Sachs, R., Van den Engh, G., & Wu, M. (1993). Studies of metaphase and interphase chromosomes using fluorescence in situ hybridization. In *Vol. 58. Cold spring harbor symposia on quantitative biology* (pp. 767–775). Cold Spring Harbor: Cold Spring Harbor Laboratory Press.

Ungelenk, M. (2021). Sequencing approaches. In T. Liehr (Ed.), *Cytogenomics* (pp. 87–122). Academic Press. Chapter 7 (in this book).

Wagenbauer, K. F., Sigl, C., & Dietz, H. (2017). Gigadalton-scale shape-programmable DNA assemblies. *Nature, 552*(7683), 78–83. https://doi.org/10.1038/nature24651.

Weise, A., & Liehr, T. (2021). Cytogenetics. In T. Liehr (Ed.), *Cytogenomics* (pp. 25–34). Academic Press. Chapter 3 (in this book).

Weise, A., Starke, H., Heller, A., Claussen, U., & Liehr, T. (2002). Evidence for interphase DNA decondensation transverse to the chromosome axis: A multicolor banding analysis. *International Journal of Molecular Medicine, 9*(4), 359–361. https://doi.org/10.3892/ijmm.9.4.359.

Whitesides, G. M., & Boncheva, M. (2002). Beyond molecules: Self-assembly of mesoscopic and macroscopic components. *Proceedings of the National Academy of Sciences of the United States of America*, 99(8), 4769–4774. https://doi.org/10.1073/pnas.082065899.

Whitesides, G. M., & Grzybowski, B. (2002). Self-assembly at all scales. *Science*, 295(5564), 2418–2421. https://doi.org/10.1126/science.1070821.

Widom, J. (1986). Physicochemical studies of the folding of the 100 Å nucleosome filament into the 300 Å filament. Cation dependence. *Journal of Molecular Biology*, 190(3), 411–424. https://doi.org/10.1016/0022-2836(86)90012-4.

Yumiceba, V., Souto Melo, U., & Spielmann, M. (2021). 3D cytogenomics: Structural variation in the three-dimensional genome. In T. Liehr (Ed.), *Cytogenomics* (pp. 247–266). Academic Press. Chapter 12 (in this book).

Yurov, Y. B., Iourov, I. Y., Vorsanova, S. G., Liehr, T., Kolotii, A. D., Kutsev, S. I., … Soloviev, I. V. (2007). Aneuploidy and confined chromosomal mosaicism in the developing human brain. *PLoS One*, 2(6), e558. https://doi.org/10.1371/journal.pone.0000558.

Zhang, L., Wang, T., Shen, Z., & Liu, M. (2016). Chiral nanoarchitectonics: Towards the design, self-assembly, and function of nanoscale chiral twists and helices. *Advanced Materials*, 28(6), 1044–1059. https://doi.org/10.1002/adma.201502590.

Zheng, H., & Xie, W. (2019). The role of 3D genome organization in development and cell differentiation. *Nature Reviews Molecular Cell Biology*, 20(9), 535–550. https://doi.org/10.1038/s41580-019-0132-4.

Zimm, B. H. (1999). One chromosome: One DNA molecule. *Trends in Biochemical Sciences*, 24(3), 121–123. https://doi.org/10.1016/S0968-0004(99)01361-4.

Nuclear architecture

14

Thomas Liehr

*Jena University Hospital, Friedrich Schiller University, Institute of Human Genetics,
Jena, Germany*

Chapter outline

Background

Considerations about the nuclear architecture of euchromatic cells are quite old; the first models of how nuclei and comprised chromosomes could be arranged had already arisen by the end of the 19th century. Surprisingly, some of those first ideas were, as it turned out during last 25 years, less far from the truth than those developed later, after the middle of the 20th century (Cremer et al., 2016).

The researchers who deserve most of the credit for establishing in-depth insights into the euchromatic nucleus are the retired Thomas Cremer, his brother Christoph Cremer, and Thomas Cremer's erstwhile group at Munich Ludwig-Maximilians-University (Cremer et al., 2016). They were among the leading European pioneers who paved the way towards the launch of the "4D nucleome project/network" by National Institutes of Health (NIH) in 2015. This project

"aims to develop and apply approaches to map the structure and dynamics of the human and mouse genomes in space and time with the goal of gaining deeper mechanistic understanding of how the nucleus is organized and functions. The project will develop and benchmark experimental and computational approaches for measuring genome conformation and nuclear organization and investigate

Cytogenomics. https://doi.org/10.1016/B978-0-12-823579-9.00011-4

how these contribute to gene regulation and other genome functions. Validated experimental approaches will be combined with biophysical modeling to generate quantitative models of spatial genome organization in different biological states, both in cell populations and in single cells"

(Dekker et al., 2017)

The four major lines of modern cytogenomic research (Liehr, 2021a, 2021b) are: (1) cytogenetics (Weise & Liehr, 2021a) and molecular cytogenetics (Bisht & Avarello, 2021; Delpu et al., 2021; Iourov et al., 2021; Liehr, 2021c); (2) molecular genetics and genomics (Ichikawa & Saitoh, 2021; Ungelenk, 2021; Weise & Liehr, 2021b); (3) epigenetics and epigenomics (Eggermann, 2021; Harutyunyan & Hovhannisyan, 2021); and (4) nucleomics (see also Baranov & Kuznetzova, 2021; Pellestor et al., 2021); of these, the latter is the still least understood scientific field.

Short history of nucleomics
1888 to 1950s

Long before designations like genes, genomes, cytogenomics, or nucleomics were established, the microscopically visible equivalents of gene-carrying units were postulated. Indeed, it was Gregor Mendel who suggested that there are "coupling groups" where the underlying information for phenotypes must be located; in other words, he postulated the existence of what Wilhelm H. Waldeyer, back in 1888, called for the first time "chromosomes." Chromosomes had been seen in the microscope already, about 15–20 years earlier, by Walther Flemming, the founder of cytogenetics (Liehr, 2020). Interestingly, the first model for interphase architecture of chromosomes dates back to 1885; it was Carl Rabl who then suggested that chromosomes in interphase are threadlike structures, however, separated into individual chromosomes (Rabl, 1885). He did not use the term "chromosome territories" to describe what he was talking about; this designation was shaped in 1909 by Theodor Boveri, and the related idea of "chromosome domains" was suggested by Eduard Strasburger at about the same time (Cremer et al., 2016; Liehr, 2020).

From the end of the Victorian era until the 1950s, neither were corresponding approaches to study the euchromatic nucleus at hand, nor were breakthroughs achieved in understanding the structure and real nature of genetic material; nucleomics research in essence stopped for almost 50 years.

1950s to 1980s

The decades following were even worse, as studies on nuclear architecture were then practically discredited. This was most likely initially owing to the rise of electron microscopy; these studies led to a concentration of science on the DNA fiber rather than the chromosomes. In addition, it turned out to be impossible to distinguish anything like chromosome territories by electron microscopy, which led to a denial of their existence (Wischnitzer, 1973); as summarized in another chapter of this book (Daban, 2021), it is in part a matter of preparation methods to bring chromosomes to reveal their secrets in electron microscopy.

One exception to mention for these times were the studies by Murray Barr published in the early 1950s, which paved the way for the general finding that genetic inactive DNA (like the inactive X-chromosome in females, or Barr-body) is predominantly located at the nuclear periphery (Barr et al., 1950) (see also Harutyunyan & Hovhannisyan, 2021).

Later on, the great success of applying molecular genetic approaches for cytogenomic research and human genetic diagnostics led the mainstream of science in the same direction as electron microscopy, previously; i.e., away from nucleomics (Waters, 2013). Amusingly, at the same time, molecular genetics and electron microscopy gave deep insights into the structure, packing, and folding of DNA, which was later important for the understanding of chromosome structure in meta- and interphase.

1980s to present

The work on nuclear architecture was inspired again in the 1970s, when radioactive in situ hybridization became possible, and much more from 1986 onward, with the introduction of fluorescence in situ hybridization (Liehr, 2020, 2021c). In addition, fluorescence microscopy had meanwhile made much progress, and already in 1984, well-elaborated immunohistochemical studies could confirm that chromosomes are not spaghetti-like, decondensed in interphase, but arranged in chromosome territories or domains (Cremer & Cremer, 2006, 2010). One year later, Günther Blobel additionally hypothesized that these chromosome domains change their shapes and structures in time-dependent manners during cell differentiation; an idea now widely supported by available cytogenomic data (Blobel, 1985).

In the 21st century, nucleomics has received more and more attention. Technical possibilities to study the genome have never been as numerous. Insights into interphase architecture came from high-resolution fluorescence microscopes, like laser scanning, or microscopes specifically constructed for nucleomic research (Cremer & Cremer, 1978; Cremer & Masters, 2013). Furthermore, technical approaches for retaining the original shape and size of interphase nuclei were established (Steinhaeuser et al., 2002) (Fig. 14.1), new approaches enabled life-cell imaging (Ishii et al., 2021; Potlapalli et al., 2020), and, during the last decade, surprisingly, the chromosome came back into focus in molecular genetic techniques; the next generation and long read sequencing approaches. Here the description of topologically associating domains (TADs) and the proof that genes do not only interact in cis on a DNA level, but that there are things happening like chromosome kissing, or the fact that interaction of DNA is necessary for a proper expression of proteins, were unexpected catalyzers of nucleomic research (Krumm & Duan, 2019; Lemke et al., 2002; Maass et al., 2019; Roy et al., 2018; Weise & Liehr, 2021c; Yu & Ren, 2017; Yumiceba et al., 2021).

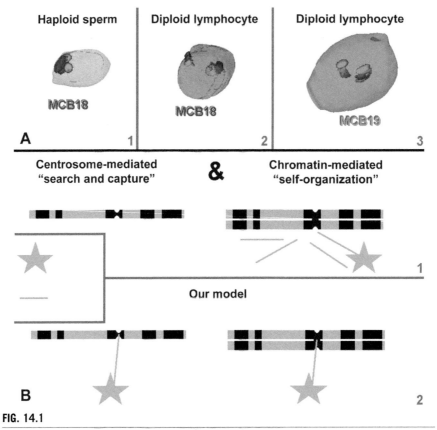

FIG. 14.1

Chromosomes in interphase-nuclei—all the time connected by a centrosome.
(A) Multicolor banding (MCB) in 3D-preserved interphase nuclei clearly showing
that chromosomes keep their basic structure as visible in metaphase chromosomes.
Chromosomes also keep their preferred positions, as, e.g., chromosome 18 in the
periphery (A-1 and A2) and chromosome 19 in the central part of the nucleus (A-
3), irrespective if they are detected in a haploid (sperm) nucleus (A-1), or a diploid
(lymphocyte) nucleus (A-2). (B) Textbooks state that after cell division centrosome and
spindle apparatus are degraded, and only build up again after DNA has doubled. There
are different theories how this is done (B-1). However, own studies showed that each cell
keeps the parental haploid sets separated in each body cell. This can only mean that, as
shown in B-2, the centrosome is always connected to the centromeres of each of the 23
chromosomes of one haploid set, during all cell cycles. To postulate, this is also much
more economic for the cell than the suggested degradation and rebuilding of whole spindle
apparatus for each mitosis. The latter would also encompass an intrinsic major risk not
to include all chromosomes in nuclear and cellular division. Cytogenomic research will
also help here to break old dogmas based on light microscopy, such as it having already
disproved the presence of "spaghetti-like chromosomes" in interphase.

Present insights into nuclear architecture

Based on the abovementioned historical ideas and molecular genetic, molecular cytogenetic, and cytogenomic data foundated on modern tools and approaches (Liehr, 2021a, 2021b), nowadays we can design the following picture of the eukaryotic nuclear architecture.

The nucleus with its DNA content is considered to be the center of information and regulation of all what goes on in a eukaryotic cell, including things like division, RNA- and protein-production, reaction on environmental changes, and cell death via apoptosis. The nucleus is normally depicted in textbooks as spherical and round; however, it can also have various other shapes. In the DNA of the nucleus covered with proteins, there is a never-ending transcription, de- and refolding, and repair going on (Cremer et al., 2016; Liehr, 2020). The picture of what a nucleus and cytoplasm look like developed during the last decades from a more static view to a flexible, ever-changing, fluent one. Nonetheless, DNA compartments of chromosomes stay together in the nucleus and are something like subunits of the nucleus, which are never really separated (Cremer et al., 2016; Lemke et al., 2002; Liehr, 2020). Even the maternal and paternal genomes seem to be connected for life via the spindle-pole and its spindles (Weise et al., 2016), which, in contrast to the textbooks, most likely are never completely degraded during the cell cycle (Fig. 14.1).

Two principles of nuclear organization at a chromosomal level have been shown by numerous authors in many different cell types and species: (1) larger chromosomes are located in the periphery, and smaller ones in the center; and (2) gene-poor chromosomes can be found in the periphery, and gene-rich chromosomes in the center (Cremer et al., 2016; Manvelyan, Hunstig, Bhatt, et al., 2008; Manvelyan, Hunstig, Mrasek, et al., 2008). The second principle can also concern chromosomal subunits (see below). Yet only one exception from this rule could be found: rod cells in retina of nocturnal mammals, where an "inverted" arrangement is present to support night vision (Solovei et al., 2009).

In a human diploid cell, the genome consists of 2×3 billion base pairs of DNA, separated into 46 portions (the chromosomes), forming overall 4m of genetic material in length. It must be mentioned that outside of the nucleus there are in each cell additionally hundreds of circular DNAs in the mitochondria, ~17,000bp in length. Due to packing with histone and nonhistone proteins, the DNA is condensed and packed in the nucleus of a cell, being on average 10–20μm in diameter. By winding the double-stranded DNA around nucleosomes, ~10nm thick chromatin threads are formed and the length of the DNA is shortened by a factor of 10. The 10nm chromatin threads are considered as basics of chromatin domains. This means that the cell accomplishes the feat of pressing 46 overall 40cm long and 10nm in diameter chromatin threads, in a self-organized way, and for all proteins and enzymes available at a given time point, into a nucleus of about 5–10μm (Fig. 14.2) (see also Daban, 2020, 2021). Thomas Cremer showed that the basic units of the chromosome territories are chromatin domains of ~1 Mbp in size, the latter being formed by ~100 kbp subdomains. Interestingly, active genes are at the periphery of these

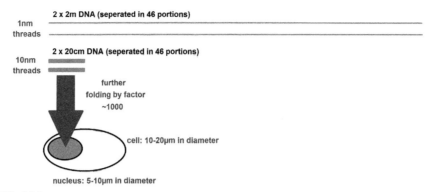

FIG. 14.2

Length of human DNA and small size of the interphase nucleus.
Schematic depiction highlighting the fact that 2×2 m long DNA fits into a nucleus 5–10 μm in diameter.

subdomains and inactive ones in their center. The subdomains are in contact with the interchromatin-compartment, which delivers RNA transcripts through the nuclear pores into the cytoplasm. In addition, parts of interchromatin-compartment are lacunae with splice speckles and other nuclear bodies with multiple functions (Cremer et al., 2016; Liehr, 2020).

Conclusion

Cytogenomic research about the interphase nucleus' nature has already been very successful during the last two decades. An important gap in the ongoing puzzle was recently closed by the paper of Joan-Ramon Daban, who greatly reviewed his own data and the literature on chromosome structure in 2020, and is also coauthor of this book (Daban, 2020, 2021). As written by him:

> *"Experimental evidence indicates that the chromatin filament is self-organized into a multilayer planar structure that is densely stacked in metaphase and unstacked in interphase. This chromatin organization is unexpected, but it is shown that diverse supramolecular assemblies are multilayered. The mechanical strength of planar chromatin protects the genome integrity, even when double-strand breaks are produced. Here, it is hypothesized that the chromatin filament in the loops and topologically associating domains is folded within the thin layers of the multilaminar chromosomes. It is also proposed that multilayer chromatin has two states: inactive when layers are stacked and active when layers are unstacked. Importantly, the well-defined topology of planar chromatin may facilitate DNA replication without entanglements and DNA repair by homologous recombination"*

(Daban, 2020)

As the puzzle of chromosome and interphase structure of DNA now seems to be resolved, it is time to study changes of nuclear architecture in human diseases and especially in cancer in much more detail, as already started in small pioneering studies (Klonisch et al., 2010; Manvelyan et al., 2009; Timme et al., 2011).

References

Baranov, V. S., & Kuznetzova, T. V. (2021). Nuclear stability in early embryo. Chromosomal aberrations. In T. Liehr (Ed.), *Cytogenomics* (pp. 307–325). Academic Press. Chapter 15 (in this book).

Barr, M. L., Bertram, L. F., & Lindsay, H. A. (1950). The morphology of the nerve cell nucleus, according to sex. *The Anatomical Record, 107*(3), 283–297. https://doi.org/10.1002/ar.1091070307.

Bisht, P., & Avarello, M. D. M. (2021). Molecular combing solutions to characterize replication kinetics and genome rearrangements. In T. Liehr (Ed.), *Cytogenomics* (pp. 47–71). Academic Press. Chapter 5 (in this book).

Blobel, G. (1985). Gene gating: A hypothesis. *Proceedings of the National Academy of Sciences of the United States of America, 82*(24), 8527–8529. https://doi.org/10.1073/pnas.82.24.8527.

Cremer, C., & Cremer, T. (1978). Considerations on a laser-scanning-microscope with high resolution and depth of field. *Microscopica Acta, 81*(1), 31–44.

Cremer, T., & Cremer, C. (2006). Rise, fall and resurrection of chromosome territories: A historical perspective. Part I. The rise of chromosome territories. *European Journal of Histochemistry, 50*(3), 161–176. http://www.ejh.it/pdf/2006_3/Cremer.pdf.

Cremer, T., & Cremer, M. (2010). Chromosome territories. *Cold Spring Harbor Perspectives in Biology, 2*, a003889. https://doi.org/10.1101/cshperspect.a003889.

Cremer, T., Cremer, M., & Cremer, C. (2016). Der Zellkern – eine Stadt in der Zelle: Teil 1: Chromosomenterritorien und Chromatindomänen. *Biologie in unserer Zeit, 46*(5), 290–299. https://doi.org/10.1002/biuz.201610601.

Cremer, C., & Masters, B. R. (2013). Resolution enhancement techniques in microscopy. *European Physical Journal, 38*(3), 281–344. https://doi.org/10.1140/epjh/e2012-20060-1.

Daban, J. R. (2020). Supramolecular multilayer organization of chromosomes: Possible functional roles of planar chromatin in gene expression and DNA replication and repair. *FEBS Letters, 594*(3), 395–411. https://doi.org/10.1002/1873-3468.13724.

Daban, J.-R. (2021). Multilayer organization of chromosomes. In T. Liehr (Ed.), *Cytogenomics* (pp. 267–296). Academic Press. Chapter 13 (in this book).

Dekker, J., Belmont, A. S., Guttman, M., Leshyk, V. O., Lis, J. T., Lomvardas, S., ... Zhong, S. (2017). The 4D nucleome project. *Nature*, 219–226. https://doi.org/10.1038/nature23884.

Delpu, Y., Barseghyan, H., Bocklandt, S., Hastie, A., & Chaubey, A. (2021). Next-generation cytogenomics: High-resolution structural variation detection by optical genome mapping. In T. Liehr (Ed.), *Cytogenomics* (pp. 123–146). Academic Press. Chapter 8 (in this book).

Eggermann, T. (2021). Epigenetics. In T. Liehr (Ed.), *Cytogenomics* (pp. 389–401). Academic Press. Chapter 20 (in this book).

Harutyunyan, T., & Hovhannisyan, G. (2021). Approaches for studying epigenetic aspects of the human genome. In T. Liehr (Ed.), *Cytogenomics* (pp. 155–209). Academic Press. Chapter 10 (in this book).

Ichikawa, Y., & Saitoh, N. (2021). Shaping of genome by long noncoding RNAs. In T. Liehr (Ed.), *Cytogenomics* (pp. 357–372). Academic Press. Chapter 18 (in this book).

Iourov, I. Y., Vorsanova, S. G., & Yurov, Y. B. (2021). Cytogenomic landscape of the human brain. In T. Liehr (Ed.), *Cytogenomics* (pp. 327–348). Academic Press. Chapter 16 (in this book).

Ishii, T., Nagaki, K., & Houben, A. (2021). Application of CRISPR/Cas9 to visualize defined genomic sequences in fixed chromosomes and nuclei. In T. Liehr (Ed.), *Cytogenomics* (pp. 147–153). Academic Press. Chapter 9 (in this book).

Klonisch, T., Wark, L., Hombach-Klonisch, S., & Mai, S. (2010). Nuclear imaging in three dimensions: A unique tool in cancer research. *Annals of Anatomy, 192*(5), 292–301. https://doi.org/10.1016/j.aanat.2010.07.007.

Krumm, A., & Duan, Z. (2019). Understanding the 3D genome: Emerging impacts on human disease. *Seminars in Cell and Developmental Biology, 90,* 62–77. https://doi.org/10.1016/j.semcdb.2018.07.004.

Lemke, J., Claussen, J., Michel, S., Chudoba, I., Mühlig, P., Westermann, M., … Claussen, U. (2002). The DNA-based structure of human chromosome 5 in interphase. *American Journal of Human Genetics, 71*(5), 1051–1059. https://doi.org/10.1086/344286.

Liehr, T. (2020). *Human genetics—Edition 2020: A basic training package.* Epubli.

Liehr, T. (2021a). A definition for cytogenomics - Which also may be called chromosomics. In T. Liehr (Ed.), *Cytogenomics* (pp. 1–7). Academic Press. Chapter 1 (in this book).

Liehr, T. (2021b). Overview of currently available approaches used in cytogenomics. In T. Liehr (Ed.), *Cytogenomics* (pp. 11–24). Academic Press. Chapter 2 (in this book).

Liehr, T. (2021c). Molecular cytogenetics. In T. Liehr (Ed.), *Cytogenomics* (pp. 35–45). Academic Press. Chapter 4 (in this book).

Maass, P. G., Barutcu, A. R., & Rinn, J. L. (2019). Interchromosomal interactions: A genomic love story of kissing chromosomes. *Journal of Cell Biology, 218*(1), 27–38. https://doi.org/10.1083/jcb.201806052.

Manvelyan, M., Hunstig, F., Bhatt, S., Mrasek, K., Pellestor, F., Weise, A., … Liehr, T. (2008). Chromosome distribution in human sperm—A 3D multicolor banding-study. *Molecular Cytogenetics, 25.* https://doi.org/10.1186/1755-8166-1-25.

Manvelyan, M., Hunstig, F., Mrasek, K., Bhatt, S., Pellestor, F., Weise, A., & Liehr, T. (2008). Position of chromosomes 18, 19, 21 and 22 in 3D-preserved interphase nuclei of human and gorilla and white hand gibbon. *Molecular Cytogenetics, 9.* https://doi.org/10.1186/1755-8166-1-9.

Manvelyan, M., Kempf, P., Weise, A., Mrasek, K., Heller, A., Lier, A., … Mkrtcyhan, H. (2009). Preferred co-localization of chromosome 8 and 21 in myeloid bone marrow cells detected by three dimensional molecular cytogenetics. *International Journal of Molecular Medicine, 24*(3), 335–341. https://doi.org/10.3892/ijmm_00000237.

Pellestor, F., Gaillard, J.-B., Schneider, A., Puechberty, J., & Gatinois, V. (2021). Chromoanagenesis phenomena and their formation mechanisms. In T. Liehr (Ed.), *Cytogenomics* (pp. 213–245). Academic Press. Chapter 11 (in this book).

Potlapalli, B. P., Schubert, V., Metje-Sprink, J., Liehr, T., & Houben, A. (2020). Application of Tris-HCl allows the specific labeling of regularly prepared chromosomes by CRISPR-FISH. *Cytogenetic and Genome Research, 160*(3), 156–165. https://doi.org/10.1159/000506720.

Rabl, C. (1885). Über Zelltheilung. In C. Gegenbaur (Ed.), *10. Morphologisches Jahrbuch* (pp. 214–330).

Roy, S. S., Mukherjee, A. K., & Chowdhury, S. (2018). Insights about genome function from spatial organization of the genome. *Human Genomics, 12*(1), 8. https://doi.org/10.1186/s40246-018-0140-z.

Solovei, I., Kreysing, M., Lanctôt, C., Kösem, S., Peichl, L., Cremer, T., … Joffe, B. (2009). Nuclear architecture of rod photoreceptor cells adapts to vision in mammalian evolution. *Cell*, *137*(2), 356–368. https://doi.org/10.1016/j.cell.2009.01.052.

Steinhaeuser, U., Starke, H., Nietzel, A., Lindenau, J., Ullmann, P., Claussen, U., & Liehr, T. (2002). Suspension (S)-FISH, a new technique for interphase nuclei. *Journal of Histochemistry and Cytochemistry*, *50*(12), 1697–1698. https://doi.org/10.1177/002215540205001216.

Timme, S., Schmitt, E., Stein, S., Schwarz-Finsterle, J., Wagner, J., Walch, A., … Wiech, T. (2011). Nuclear position and shape deformation of chromosome 8 territories in pancreatic ductal adenocarcinoma. *Analytical Cellular Pathology*, *34*(1–2), 21–33. https://doi.org/10.3233/ACP-2011-0004.

Ungelenk, M. (2021). Sequencing approaches. In T. Liehr (Ed.), *Cytogenomics* (pp. 87–122). Academic Press. Chapter 7 (in this book).

Waters, K. (2013). Molecular genetics. In *The Stanford encyclopedia of philosophy* (Fall 2013 ed.). https://plato.stanford.edu/archives/fall2013/entries/molecular-genetics/.

Weise, A., Bhatt, S., Piaszinski, K., Kosyakova, N., Fan, X., Altendorf-Hofmann, A., … Chaudhuri, J.P. (2016). Chromosomes in a genome-wise order: Evidence for metaphase architecture. *Molecular Cytogenetics*, *9*(1), 36. https://doi.org/10.1186/s13039-016-0243-y.

Weise, A., & Liehr, T. (2021a). Cytogenetics. In T. Liehr (Ed.), *Cytogenomics* (pp. 25–34). Academic Press. Chapter 3 (in this book).

Weise, A., & Liehr, T. (2021b). Molecular karyotyping. In T. Liehr (Ed.), *Cytogenomics* (pp. 73–85). Academic Press. Chapter 6 (in this book).

Weise, A., & Liehr, T. (2021c). Interchromosomal interactions with meaning for disease. In T. Liehr (Ed.), *Cytogenomics* (pp. 349–356). Academic Press. Chapter 17 (in this book).

Wischnitzer, S. (1973). The submicroscopic morphology of the interphase nucleus. *International Review of Cytology*, *34*(C), 1–48. https://doi.org/10.1016/S0074-7696(08)61933-6.

Yu, M., & Ren, B. (2017). The three-dimensional organization of mammalian genomes. *Annual Review of Cell and Developmental Biology*, *33*, 265–289. https://doi.org/10.1146/annurev-cellbio-100616-060531.

Yumiceba, V., Souto Melo, U., & Spielmann, M. (2021). 3D cytogenomics: Structural variation in the three-dimensional genome. In T. Liehr (Ed.), *Cytogenomics* (pp. 247–266). Academic Press. Chapter 12 (in this book).

Nuclear stability in early embryo. Chromosomal aberrations

15

Vladislav S. Baranov and Tatiana V. Kuznetzova

D.O. Ott Research Institute of Obstetrics, Gynecology and Reproductology,
Saint Petersburg, Russia

Chapter outline

Introduction

Almost all cellular DNA is localized in the nucleus and less than 1% in the mitochondria. The genes, hidden in the DNA spirals, are unevenly packaged along different chromosomes of the haploid genome, where they are transmitted from cell to cell, from one generation to the other. The continuity of heritability, its accuracy, depends on many factors and is regulated by complex mechanisms at the level of genes, chromosomes, cells, and the whole organism. Deviations of regulatory mechanisms are very diverse and can occur at any stage of development (see also Liehr, 2021a). In particular, the occurrence of these aberrations in gametes and early embryos can be the cause of developmental anomalies and intrauterine death (see also Pellestor et al., 2021).

Cytogenomics. https://doi.org/10.1016/B978-0-12-823579-9.00012-6

Here, we summarize and discuss up-to-date data on zygote nucleus formation and molecular genetic and epigenetic mechanisms of embryo development at cleavage stages, during compaction, and primary differentiation into trophectoderm cells (TEs) and inner cell mass (ICM). Frequency and mechanisms of aneuploidy, chromosomal mosaicism, genomic imprinting (GI), and complex chromosomal rearrangements – chromothripsis (CT) are also covered (see also (Pellestor et al., 2021)), as being an important part of the cytogenomic research field (see also Liehr, 2021a, 2021b).

The nucleus of the zygote

The formation of the nucleus of a new organism begins at the zygote stage, that is, after fertilization, when the sperm cell entered the egg. Cardinal changes in the structure of the nucleus and chromosomes occur during the formation of the zygote, during the cleavage of blastomeres, the compaction of the morula, and its differentiation into TE and ICM cells, during the formation of a blastocyst, and implantation of the embryo (Baranov et al., 2012; Baranov & Kuznetzova, 2007, 2012).

Recent years have been marked by the rapid development of research in the field of early human embryogenesis. Despite the official ban on work on human embryos, in 2015 in the UK, permission was issued to clone human embryos to the blastocyst stage in order to analyze the molecular genetic mechanisms that regulate the preimplantation stages of embryogenesis. It is assumed that the results of such studies will improve the efficiency of in vitro fertilization (IVF) and gene therapy of hereditary diseases (Luke, 2017). At present, such in vitro scientific research on human embryos is being actively carried out in many countries. Their high scientific and practical significance is beyond doubt. They not only allow obtaining unique scientific data on the biology of early human development, but also significantly expand the possibilities of using assisted reproductive technologies.

Zygotes and blastomeres

The possibilities of cultivating human embryos in vitro from the zygote to the blastocyst stage, and according to the latest data even to the stage of the primitive streak (13 days) (Morris, 2017), made it possible to shed light on many processes occurring in early human embryogenesis and compare it with other mammals, primarily laboratory mice. The chronology of preimplantation human development and the genes that control it have been discussed in detail by us earlier (Baranov et al., 2012; Baranov & Kuznetzova, 2012; Carlson, 2008). It is known that the male and female pronuclei formed a few hours after fertilization in humans, in contrast to mice, have a similar size, but differ significantly in their chromatin configuration. Using modern methods studying chromatin filaments, the presence of the human pronucleus was established (Racko et al., 2019), which also forms local clusters, the so-called topologically associated domains (TADs) (see also Yumiceba et al., 2021). There are many TADs in the male pronucleus; they cluster near the nuclear membrane, while in the female pronucleus, there are much less of them, and they rarely form clusters. The biological meaning of these structures and their functions are poorly understood, yet. To some extent, they reflect the different degrees of spiralization of the chromosomes in male and female pronuclei; however, the differences in their functional activity remain unclear, yet. The biological meaning of TADs and their functions are still not completely resolved. Most likely, they are a consequence of the peculiarities of the spiralization of the chromosomes in male and female pronuclei and, to a certain extent, indicate the differences in their functional activity. It is believed that TAD compartments reflect the features of the structural and functional organization of the embryonal genome. They determine the spatial arrangement of functional domains, facilitate the convergence of enhancer and promoter DNA sequences of structural genes, and form the program of the initial stages of genome operation (Hug & Vaquerizas, 2018). It is pertinent to note that the study of TADs in health and disease is now considered a new direction of cytogenetics (Weise & Liehr, 2021a), included in the cytogenomics field (Daban, 2021; Lebedev, 2018; Liehr, 2021c). The ability to study TAD patterns and the dynamics of their gene expression already at the earliest stages of embryogenesis opens up great options for understanding the etiology and pathogenesis of chromosomal diseases, searching for ways to prevent and treat them.

There is evidence that the genes of the male pronucleus begin to be expressed earlier than the female ones. These observations are in good agreement with the features of the morphology of chromosomes in the metaphase of the first cleavage division. Chromosomes from the male pronucleus are low-coiled, filamentous, and differ from short, highly coiled maternal chromosomes. Consequently, the heterocyclicity of chromosomes of the first cleavage division, previously established in experimental models, is also valid for humans and, as recent studies show (see below), it most likely indicates an earlier functional activity of the genes of the paternal genome.

Other global changes in pronuclei relate to the processes of total demethylation-remethylation of the embryo genome, corresponding to the processes of genome reprogramming as a part of individual development program. In human embryos obtained under the in vitro fertilization/assisted reproductive technology (IVF/ART) program, we have studied in detail the dynamics of the processes of methylation-demethylation of the genome from fertilization to the blastocyst stage. In a series of studies using specific fluorescent antibodies to 5-methylcytosine (m^5C) and 5-hydroxymethylcytosine (Hm^5C), an uneven distribution of the m^5C label in the pronuclei and a different pattern of antibody fluorescence on homologous chromosomes in the metaphase of the first cleavage division were shown, which indicates an uneven demethylation process (Efimova et al., 2015; Pendina et al., 2011; Petrussa et al., 2016).

It has recently been established that, in addition to the mechanism of active demethylation of chromatin in the male pronucleus using special enzymes (demethylases), another variant of demethylation is possible - excision of methylated DNA fragments with their subsequent replacement by homologous regions devoid of a methyl tag. The switching of the individual genetic program from the maternal RNA to the genome of the embryo itself occurs gradually and begins with the genes of the male pronucleus (for RNA in cytogenomics see also Ichikawa and Saitoh (2021)). Indeed, the premature activation of hundreds of genes of the paternal genome at the zygote stage has been shown by the method of whole-genome sequencing (Yamamoto & Aoki, 2017).

The fragmentation of the human zygote is completely asynchronous. The first division of cleavage occurs 30 h after the penetration of the sperm into the egg, and each subsequent division occurs after about 9 h. At the stage of metaphase of the second cleavage division, all chromosomes are characterized by asymmetric methylation of sister chromatids (one of the chromatids is brightly colored (methylated), and the other practically does not contain the m5C label). In the future, the demethylation process progressively increases. From the eight-cell stage, chromosomes begin to acquire a characteristic segment-specific pattern, which is finally established at the blastocyst stage (for epigenetics see also (Eggermann, 2021; Harutyunyan & Hovhannisyan, 2021)).

Primary differentiation and blastocyst formation

The cleavage of the human fertilized egg after 2.5–3 days is followed by compaction of blastomeres and their further differentiation into ICM and TE cells. It has recently been established by molecular genetic methods that primary differentiation begins already at the stage of four blastomeres. The key role in this is played by the transcription factor SOX2, which binds to the DNA of dividing cells (Du et al., 2017). Depending on the presence or absence of the amino acid arginine in histone - 3 (H3R26), the strength of such binding and its duration vary. With prolonged binding, germ cells begin to express the transcription factor SOX21, which inhibits differentiation. Such cells later become ICM cells, which subsequently give rise to all the tissues of the fetus itself. Cells with a short contact time with SOX2 begin to differentiate and turn into TE cells. The enzyme CARM1, which is responsible for the methylation of histone H3, triggers the process of primary differentiation. The factors that induce the synthesis of this key enzyme are still unknown. It is assumed that this may be the product of the *DUX4* gene, which is initially expressed in the zygote and, as shown by experiments with targeted shutdown of this gene using the CRISPR/Cas9 technology, it induces the activation of the entire embryonic genome (De Iaco et al., 2017; White et al., 2016). Surprisingly, the primary activator of the human genome, the *DUX4* gene, has previously been identified as the main gene whose mutations lead to severe neuromuscular disease of the facioscapulohumeral muscular dystrophy (FSHD) (Geng et al., 2012).

Thanks to the ability to work directly on human embryos under the IVF and ART programs, key information have been obtained about the main morphogenetic processes that are implemented in the preimplantation period of human development and the factors that regulate them. During these studies, several other interesting facts were revealed. In particular, it was found that when the human oocyte is cleaved, heterochromatin is activated, which is accompanied by the expression of transposons located in it (see also Liehr, 2021d). It is assumed that the expression of transposons, especially the retrotransposon LINE1, is required to initiate the work of the entire genome after fertilization, and chromatin remodeling in the pronuclei is necessary to activate transcription processes (Munch et al., 2016). In the process of chromatin remodeling and blastomere division, breakdowns often occur, which are repaired using a special enzyme system (Godini & Fallahi, 2019). It is important to emphasize that breakdowns associated with defects in the DNA repair system and impaired epigenetic reprogramming of the genome lead to developmental arrest (Fogarty et al., 2017) and are one of the possible mechanisms for the emergence of complex structural rearrangements (see "Chromothripsis" section).

Chromosomal abnormalities
Methods of karyotype analysis

The main genetic diagnostic method remains karyotyping (Weise & Liehr, 2021a), i.e., analysis of the number and structure of differentially stained chromosomes in a cell at the metaphase stage. Cytogenetic analysis diagnoses all types of numerical anomalies of the karyotype and their form (complete and/or mosaic), and with an appropriate level of resolution of differential staining, all types of chromosomal rearrangements can be identified. The other methods (molecular cytogenetics (Liehr, 2021e), array comparative genomic hybridization (aCGH - see Weise & Liehr, 2021b), and other molecular ones - (see e.g., Ungelenk, 2021) can either reveal or clarify the imbalance in individual chromosomes or their regions, but do not give an idea about the karyotype itself. Even the so-called "molecular karyotyping" using the aCGH method does not always allow diagnosing changes in the number of chromosomes (polyploidy, low-clonal mosaicism for individual chromosomes), identifying balanced rearrangements, and analyzing the structural basis of the imbalance in the resulting "virtual karyotype." Thus, different methods of analyzing cytogenomic abnormalities complement each other; therefore, they are not fundamentally alternative (Liehr, 2021a, 2021b).

The chapter deals with current problems of human developmental biology, the solution of which has become possible due to the introduction of new technologies and methods of molecular genetics for the study of human embryos obtained through IVF and ART programs. Accordingly, the most significant scientific achievements in the study of the structure of the nucleus and chromosomal instability of early human embryos include: assessment of the contribution of chromosomal aberrations, GI (see as well Eggermann, 2021; Harutyunyan & Hovhannisyan, 2021) and chromosome mosaicism to the pathology of early stages of embryogenesis, molecular genetic mechanisms of expression of the zygote genome of primary embryonic induction, structural and functional organization of chromosomes, and epigenetic evolution of the human genome before implantation.

Chromosomal abnormalities in early embryogenesis

Cytogenetic studies of human embryos and, mainly, the material of spontaneous abortions showed a large contribution of chromosomal abnormalities to the pathology of human embryonic development (Pendina et al., 2014; Wu et al., 2016).

The main mechanisms of aneuploidy in meiosis and early human embryogenesis are considered to be lagging of chromatids in anaphase and, less often, true nondisjunction of chromosomes (Griffin & Ogur, 2018). Abnormal chromosome segregation is explained by the lack of control of the cell cycle in the first divisions of cleavage, which are characterized by abrupt changes in cell size and time parameters, as well as peculiarities of kinetochore composition, and their orientation to the division poles (van de Werken et al., 2015; Vázquez-Diez & FitzHarris, 2018). The likelihood of disorganization of the fission spindle, overduplication of centrosomes, and the formation of three-pole spindles is also not excluded, which leads to an unequal distribution of homologous chromosomes, anomalies of nuclei (multiple nuclei, micronuclei) (Wu et al., 2016).

The formation of micronuclei containing chromatids lagging behind in anaphase is considered as a mechanism explaining not only the occurrence of aneuploidy during the cleavage of the zygote, but also the "self-correction" of trisomy (i.e., restoration of the euploid chromosome set) (for more details, see Lebedev, 2008).

Using the aCGH and next generation sequencing (NGS) methods, it was found that more than 25% of cases of aneuploidy occur after fertilization, as a result of abnormal segregation of chromosomes in cleavage divisions (McCoy, 2017). These mitotic errors can occur during cleavage of zygotes, both with initially normal and abnormal sets of chromosomes, and do not depend on the age of the mother. They can affect one or several chromosomes of both maternal and paternal origin, in one or several blastomeres, or occur simultaneously or sequentially in different cleavage divisions.

When karyotyping early spontaneous abortions (5–10 weeks of gestation), chromosomal aberrations were recorded in almost 80% of cases (Hug & Vaquerizas, 2018; Lebedev, 2008). These observations were confirmed by karyotyping of preimplantation embryos obtained by IVF programs (Su et al., 2016). Moreover, it has been established that it is chromosomal abnormalities that lead to a delay in blastocyst formation, impaired implantation, and developmental arrest (Fogarty et al., 2017). At the same time, the spectrum of chromosomal abnormalities in preimplantation embryos was more diverse than in embryos of later stages (Lee & Kiessling, 2017; Wu et al., 2016). Typical abnormalities were not only trisomy (three copies of one chromosome), possibly involving almost all chromosomes, including rather rare ones at later stages, but multiple trisomies and even monosomies (absence of one copy of a chromosome), which, with few exceptions (chromosome 21, 22, gonosomes X and Y), practically do not occur at the postimplantation stages (Baranov & Kuznetzova, 2007). In many embryos with monosomy, the absence of mitotic divisions and a delay in the formation of a vesicle (blastocyst) were noted. The presence of monosomy was confirmed on interphase nuclei by hybridization with labeled DNA probes (fluorescence in situ hybridization=FISH method) (Baranov & Kuznetzova, 2007). It was also found that chromosomal abnormalities are found in almost 80% of cases in human embryos at the blastocyst stage (Sermon, 2017).

It has been suggested that aneuploidy in preimplantation human embryos, in contrast to other mammals (mice), is normal (Lee & Kiessling, 2017). Aneuploid cells are more common in the trophectoderm than in ICM, from which all embryonic tissues actually develop. The aneuploid cells themselves undergo strict selection and die off, as evidenced by the high frequency of chromosomal abnormalities in DNA samples, from the blastocoel (blastocyst cavity) (Skryabin et al., 2015; Tšuiko et al., 2018). There are also data proving the presence of special processes of repair of a set of chromosomes in the karyotype of the embryo. Aneuploid human embryos deliberately transplanted into the uterus in several cases with further development revealed a completely normal balanced karyotype (Su et al., 2016). Embryos with a developmental delay in vitro for almost a day can form blastocysts capable of implantation (Lee & Kiessling, 2017). Nevertheless, preliminary screening of aneuploid embryos obtained during IVF using PGS (preimplantation genetic screening) methods significantly increases (by more than 10%) the effectiveness of ART, as indicated by higher numbers of karyotypically balanced embryos (Lee & Kiessling, 2017).

It should be noted that in addition to the visible disturbances of the karyotype determined under the microscope, various variants of microchromosomal rearrangements are encountered relatively often (in about 1%) in embryos of IVF programs, mainly insertion/deletions (indels) with a length ranging from several hundred to several thousand and even millions of base pairs (Baranov & Kuznetzova, 2007). Habitual miscarriage (more than two spontaneous abortions with a balanced karyotype in the number of chromosomes in the parents) allows one to suspect the presence of cryptic chromosomal rearrangements in the parents. Microarray analysis, the method of aCGH, is successfully used for their detection in unbalanced offspring (Baranov & Kuznetzova, 2007).

Thus, as studies of embryos obtained in IVF/ART programs have shown, chromosomal abnormalities that occur during fertilization or during the first divisions of egg cell cleavage play an important role in the pathology of preimplantation human development. Selection of embryos using PGS significantly increases its efficiency. Improvement of cytogenetic methods leads to a gradual erasure of the boundaries of the resolving power of molecular and cytogenetic methods for studying the genome. Data on the contribution of chromosomal abnormalities to the pathology of embryos in the early stages of development are constantly being refined. Further progress in this area can be achieved thanks to the introduction of a new technology into the PGS algorithm based on the use of the karyomapping method, which makes it possible to identify and select preimplantation embryos with high efficiency, not only with numerical, structural, and cryptic chromosomal abnormalities but also with mutations leading to many common hereditary diseases, the so-called PGD-2 technology (Handyside et al., 2010; Natesan et al., 2014).

Genomic imprinting

GI - nonrandom, sex-linked homogeneous expression of some genes or even entire genomes - also makes a significant contribution to the pathogenesis of various hereditary diseases (Dyban & Baranov, 1987; Ferguson-Smith & Bourc'his, 2018). A classic example of GI disorders, which is not uncommon in conditions of IVF, is diandria (**diandria**, the embryo genome is represented only by paternal genes, and - **digynia**, the presence of two haploid maternal genomes). Dispersed fertilization or delayed separation of the second polar body may be the cause of such disorders. In the case of division of triploid zygotes, the chromosome set of one pronucleus may be lost, while the restored diploid set and the embryo's genome may turn out to be normal if obtained from both parents or pathological if it is represented by two paternal (diandria) or two maternal (digynia) haploid sets. The peculiarities of embryogenesis under conditions of diandria and digynia were previously studied in detail in experiments on laboratory mice (Solter, 2006; Surani, 1998). It has been shown experimentally that normal development occurs only when the embryo has both parental genomes. In the case of digynia, the formation of the placenta is completely disturbed, while in diandria, on the contrary, the chorionic tissue grows in excess. The patterns of GI, which are laid down in gametogenesis, have significant sex differences with respect to gene expression, which is the main reason for the disruption of the development of such embryos. The morphogenetic effect of GI is manifested not only at the level of the whole genome, but also at the level of individual chromosomes (full uniparental disomy - two homologous chromosomes are obtained from the same parent) and are explained by the presence of special - imprinted - genes in them. In humans, about 90 imprinted genes have been found, the products of most of which are involved in control of placental growth and proliferation of embryonic cells and regulation of cell differentiation in the central nervous system (Lepshin et al., 2014). Errors in the methylation of such genes (the so-called epimutation) arising in the process of gametogenesis, thus leading to various dysfunctions of the genome. The consequence of epimutations during the maturation of oocytes in the mother may be a serious disease - blighted ova (Sazhenova & Lebedev, 2008). In studies conducted on the material of spontaneous abortions, multiple epimutations of imprinted genes were found in 11.7% of cases (Sazhenova et al., 2017). Moreover, the total frequency of imprinted paternal genes was two times higher than that of maternal genes (34.4% and 16%). However, the share of inactivated maternal alleles accounts for 16% of epimutations, while the share of hypermethylated paternal genes accounts for 18.4%. In most infertile married couples, hypomethylation of imprinted genes on the chromosomes of maternal origin was present in the germ cells, which could be due to the quality of the nutrient media used for the cultivation of gametes and embryos. It is assumed that the disturbance of GI pattern in immature oocytes may be one of the causes of developmental pathology and embryonic mortality under IVF/ART conditions (Takahashi et al., 2019) - see also Liehr (2020).

It is known that selective methylation in the gametogenesis is the basis of GI that results in the expression of certain genes based on the parental origin of the chromosomes on which they are localized (Babariya et al., 2017; Treff & Franasiak, 2017).

Thus, the disorder of GI (epimutation) is associated with errors in methylation of genes of the paternal and maternal chromosomes in gametogenesis, which manifests itself already in the early stages of development and can be the cause of early embryonic death.

Chromosomal mosaicism

The term *"chromosome mosaicism" is a combination in one embryo of cell lines with different chromosomal sets, identified during cytogenetic analysis*. It should be emphasized that genetic mosaics are only those individuals that arose from one zygote, and, accordingly, the presence of two or more populations of cells with different genotypes in one organism is due to somatic mutations that occurred during development. Real mosaicism must be distinguished from **chimerism**. Let us recall that a chimera is an individual that develops from different zygotes after their fusion.

The mosaic variant of the chromosome set with normal and trisomic cell lines is the result of a series of successive mitotic errors in the early stages of embryonic development in individual cells. They can affect one or more chromosomes of both maternal and paternal origin, in one or more blastomeres, occurring simultaneously or sequentially in different cleavage divisions. According to the combination of karyologically different blastomeres, four main groups of mosaic embryos are distinguished: (i) aneuploid mosaics (there is no single euploid cell), (ii) aneuploid/euploid (a combination of aneuploid and euploid cells), (iii) mixoploid (a combination of haploid, di-, tri-, and tetraploid, each blastomere has a random set of chromosomes) (Taylor et al., 2014), and (iv) mitotic mosaicism occurs due to abnormal segregation of chromosomes during division of a normal diploid cell to form cells with trisomy (an additional maternal or paternal chromosome) and monosomy and the resting pool of diploid cells. The formed monosomic cells are usually eliminated, and trisomic cells, along with euploid cells, form cell clones.

Meiotic mosaicism $2n/2n+1$ occurs during division of the trisomic zygote, formed as a result of meiotic nondisjunction in gametogenesis in one of the parents. After the loss of an extra chromosome (due to mitotic nondisjunction or lagging in anaphase), euploid cells are formed, which form a normal cell lineage along with a trisomic cell.

It should be noted that in addition to disorders of the number of chromosomes, segmental aneuploidies are observed relatively often (on average, 7.5%) in preimplantation embryos (i.e., duplication/deletion of chromosome regions of various lengths and localization). Segmental aneuploidies can be mosaic (so-called segmental mosaicism) and be combined with mosaicism along whole chromosomes (Sachdev et al., 2017; Taylor et al., 2014). It has been established that segmental aneuploidies can have both meiotic and zygotic origins and more often affect chromosomes of groups A, B, and C (chromosomes 1–12). Hot spots of their formation are associated with known fragile sites of chromosomes or are specific for gametogenesis and/or embryogenesis (Taylor et al., 2014).

The total frequency of mosaics with complete and partial forms of aneuploidy on different chromosomes varies widely at the morula stage (15%–90%), and at the blastocyst stage, it averages about 30% (Taylor et al., 2014). To a large extent, such variations are explained by the different sensitivities of the chromosome analysis methods and characteristics of the cells under study. Hence, at the cleavage stage, no more than two blastomeres are available for analysis, and using the FISH method, one can suspect the presence of mosaicism on several (from 5 to 12) chromosomes, but not determine its level (i.e., the ratio of cells with different chromosome sets). At the blastocyst stage, DNA samples from several cells of TE are usually examined, which makes it possible to determine mosaicism for all chromosomes of the set at the level of 40% (by aCGH method) and 20% (NGS) (Sachdev et al., 2017). It has been established that in the blastocyst, abnormal cells are distributed randomly, both in the TE and ICM, but not necessarily in the TE and ICM simultaneously (COGEN Position Statement on Chromosomal Mosaicism Detected in Preimplantation Blastocyst Biopsies, n.d.; PGDIS Newsletter, 2016). Therefore, information on the presence and level of mosaicism, determined by analyzing several neighboring TE cells, cannot be used to judge the karyotype of ICM cells (Sachdev et al., 2017).

The proportion of an abnormal cell clone and its localization in tissues and organs depend on the stage of embryogenesis at which the mitotic mutation occurred as well as the viability and proliferative potencies of cells with an abnormal karyotype. In the study of cleaving human embryos, it was proven that the frequency of many trisomies increases with age of the mother and is mainly due to errors in the segregation of chromosomes in the first division of meiosis in oogenesis. They can affect one or more chromosomes of both maternal and paternal origin, in one or more blastomeres, occurring simultaneously or sequentially in different divisions of splitting (Taylor et al., 2014).

It was noted that the molecular mechanisms of genomic instability in TEs are close or identical to those in carcinogenesis (Carlson, 2008; Pendina et al., 2014). The main mechanisms of aneuploidy in meiosis and early human embryogenesis are lagging chromatids in anaphase and, less often, true nondisjunction of chromosomes. Abnormal segregation of chromosomes is explained by the lack of control of the cell cycle in the first divisions of cleavage, which are characterized by abrupt changes in cell size and time parameters of the cell cycle, peculiarities of the kinetochore composition, and their orientation toward the division poles. It was found that in the blastocyst, abnormal TE and ICM cells are distributed randomly, but not necessarily evenly. A decrease in the frequency of mosaic embryos, as well as a decrease in the level of mosaicism at the blastocyst stage, indicate selection against chromosomal imbalance at the level of the organism (developmental arrest and embryo death) or individual cells (decrease in the rate of proliferation, apoptosis). Elimination of aneuploid cells by apoptosis is more pronounced in ICMs than in TEs and, apparently, is the main mechanism of "self-correction" of aneuploidy in human embryos before implantation and the cause of placenta-limited chromosome mosaicism recorded at postimplantation stages. The death of aneuploid cells in mosaics can lead to a delay in blastocyst formation, but their presence in TE does not prevent implantation and even promotes trophoblast invasion into the uterine wall (Taylor et al., 2014).

A combination of two approaches - cytogenetic and FISH with a locus- and/or centromere-specific DNA probe for a particular chromosome - is considered optimal for the diagnosis of mosaicism (Iourov et al., 2019). The presence of 7–10 to 20% of nuclei with an aneuploid chromosome number in preparations with FISH staining indicates the presence of mosaicism. In this case, it is recommended to conduct a cytogenetic study on metaphase chromosomes. Preliminary background values should help in analyzing the results of hybridization on interphase nuclei and in deciding on the proportion of trisomic cells. Recall that the percentage of abnormal cells according to the results of the analysis of metaphases and interphase nuclei will almost always be different, which is associated with different proliferative potencies of diploid and aneuploid cells in vivo and, moreover, in vitro. Thus, chromosomal mosaicism on preparations made from any cell can be either artifact, or true (generalized), or limited to the cells under study. Obviously, it is not possible to determine the cell type using metaphase chromosomes and interphase nuclei on cytogenetic preparations from heterogeneous cell samples. Placental mosaicism receives increased attention in prenatal diagnostics. Taking into account the extraembryonic origin of TE and the likelihood of placental mosaicism, the predictive value of the results of prenatal karyotyping on spontaneously dividing cytotrophoblast (CTB) cells is considered reduced in comparison with the analysis of cells stimulated to divide in vitro, which are already differentiated from epiblast implantation (mesoderm of chorion, amniotic fluid cells, and fetal cord blood lymphocytes). At the same time, rejection of metaphase analysis for CTB cells in favor of screening for frequent aneuploidies using interphase FISH or quantitative fluorescence PCR (QF-PCR) does not allow solving the problems of intercellular and intertissue mosaicism. It is interesting to note that when testing chromosomal abnormalities using high-resolution molecular methods (NGS) within the framework of preimplantation testing at the blastocyst stage, as a rule, samples from the TE are used, and within the framework of NIPT (noninvasive prenatal testing) - the so-called "free fetal" (cell-free fetal DNA) DNA, which is actually placental DNA, mainly of trophoblastic origin (Liehr et al., 2017). At the same time, the high incidence of aneuploidy in trophoblast cells is well known both in preimplantation embryos and in invasive trophoblasts (this is the main source of extracellular DNA in the mother's bloodstream for NIPT). To verify the detected chromosomal imbalance in preimplementation genetic testing (PGT) or NIPT, chorionic or placenta biopsy is often recommended, followed by analysis of metaphase chromosomes or DNA from chorion/placenta. Obviously, information will be obtained only in relation to the chorion/placenta, but not the fetus, and in cases of discordance of the chromosome set in the chorionic cells and in the fetal cells, it will lead to false-positive and false-negative results. The presence of these and other paradoxes significantly affects the accuracy of the PD of chromosomal diseases, complicates the interpretation of the results, and often leads to annoying errors (Liehr et al., 2017).

Chromothripsis

Specific features of gametogenesis and early development may contribute to emergence of complex chromosome rearrangements (CCRs) including massive chromosome fragmentation followed by random joining of fragments of one or more chromosomes. The resulting acentric fragments are lost and CT could not account for all phenomena of chaotic and quick genome rearrangements.

Depending on the mechanisms of chromosome fragmentation, types, and joining of their fragments, there are three variants of CCR: CT, chromoanasynthesis (CAS), chromoplexy (CP). CT and CAS are single catastrophic events, and CP is the result of errors in several consecutive cycles (Madan, 2013; Pellestor, 2019; Pellestor et al., 2021). CT and CAS are found in somatic cells, while CP is found only in tumor cells. CT is the fragmentation of chromosomes and chaotic reunification of chromosomal fragments with the possible loss of some regions (Madan, 2013). CAS is the result of multiple errors in chromosome replication and repair (Pellestor, 2019). CP is mainly represented by deleted and translocated fragments of chromosomes in tumor cells. A distinctive feature of CP is a smaller number of breakpoints and a larger number of those involved in chromosome rearrangements. The cytogenetic identification of various types of structural rearrangements of chromosomes is significantly complicated and requires special methods.

Along with the standard FISH analysis, spectral karyotyping, the method of comparative genomic hybridization on microarrays and various variants of DNA sequencing are widely used to detect CT, CAS, or CP. The most commonly used is multicolor FISH (multicolor spectral karyotyping (SKY or M-FISH), hybridization on metaphase chromosomes with labeled whole chromosomal DNA probes, and MCB-FISH (multicolor banding FISH) method, which is a DNA-specific analogue of differential chromosome staining. Recently, for the analysis of various constitutional CT (CCT) variants, genome-wide sequencing of paired DNA ends (mate-pair sequencing, MPseq) has been widely used to study the DNA sequence at the junctions of various fragments, where genes whose mutations are cell-lethal are often located. With an almost complete absence of genetic imbalance, constitutional CT can be accompanied by damage to genes localized at the sites of chromosome breaks, or dysregulation of their expression (Houge et al., 2003; Liu et al., 2011; Stephens et al., 2011).

The causes, frequency, and significance of these CCRs in the normal and pathological development of multicellular organisms and humans are not fully understood and are areas of active study by cytogeneticists and molecular biologists. It is believed that the cause of their occurrence can be chromatin condensation and mutations in the repair genes (Liu et al., 2011; van Roosmalen et al., 2015). These disturbances can be caused by various damaging external factors (viral infections, reactive oxygen species, ionizing radiation, etc.), as well as by processes associated with the metamorphosis of the chromosomes themselves, which occur during replication, homologous and nonhomologous recombination, premature separation of sister chromatids, shortening telomeric ends, loss of centromeres, divergence errors and chromosome rearrangements (Koltsova et al., 2019). The arising chromosomal abnormalities are considered as CCT. All of these processes are concentrated and especially critical in gametogenesis and in the early stages of embryogenesis. Studies on model objects have established that CCT occurs especially often in the processes of spermatogenesis due to the high frequency of replication errors and meiotic recombination and the poorly developed system for repairing random double-strand breaks. During recombination in meiosis, a disorder of the repair processes of double-strand breaks can also be observed (Chavez et al., 2012). Having undergone incomplete (abortive) apoptosis, such cells can continue to differentiate and participate in fertilization (Bertelsen et al., 2016; Chavez et al., 2012). Mutant spermatids with a haploid number of chromosomes form spermatozoa that are capable of fertilization and transmit CCT to the embryo, which can continue to develop.

The rarity of detection of embryos with CCT after implantation suggests that embryos with CCT are eliminated at earlier stages of development. In this case, the process of formation of the nucleus of the embryo during the unification (syngamy) of the chromosomes of the female and male pronuclei is considered to be an important cause of the onset of CCT. Differences in replication time (the chromosomes of the male pronucleus are replicated earlier than the female), the state and capabilities of the repair systems (all replication errors are corrected by the oocyte enzyme systems) are additional stress that provokes the onset of CCT (Middelkamp et al., 2017). Unlike male germ cells, oocytes retain the ability to repair breaks by homologous recombination and nonhomologous joining of free ends (Pellestor et al., 2014). The cytological manifestation of these disorders is micronuclei, which are often found in the analysis of preimplantation embryos in ART programs. The "pulverization" of chromosomes in micronuclei is considered the most recognized hypothesis for the emergence of CT (Pellestor et al., 2014). Micronuclei are considered as a passive indicator of chromosome instability (Tang et al., 2012).

It is believed that, in addition to gametogenesis, DNA damage of the KX type can also occur de novo after fertilization, as a result of micronucleus formation, blastomere fragmentation, and mitotic disruption (Marchetti et al., 2007). At the same time, the asynchronous development of the female and male pronuclei and, as a consequence, the under-replication of the paternal genetic material, can lead to "pulverization" of chromosomes in the zygote (Marchetti et al., 2007). CT can arise via the telomere attrition, deregulation, and loss of *TP53* gene.

As already noted, CCT is practically not detected during cytogenetic analysis of chorionic cells when pregnancy stops in the first trimester (Liehr et al., 2017; Madan, 2013; Pellestor, 2019; Victor et al., 2019). However, despite the widespread use of PGT technologies, its true frequency in gametes and early embryos is still unknown.

Conclusion

Complex cytogenetic and molecular genetics, i.e., chromosomic studies of the structural and functional organization of the genome and nucleus of gametes and early human embryos, are of decisive importance not only for solving the fundamental problem of human biology - the features of the implementation of hereditary information in development, but also significantly expand the possibilities of using assisted reproductive technologies to solve problems of human reproduction. Fundamentally, new cytogenomic data were obtained on the structure of the genome of the male and female pronuclei, on the features of the formation of the nucleus of the embryo, the dynamics of epigenetic changes in the preimplantation period, and the features of the expression of early genes at the initial stages of development immediately preceding the primary embryonic differentiation were clarified. For the first time, the contribution of chromosomal abnormalities arising in gametogenesis and during cleavage to the pathology of human intrauterine development was determined. Evaluation of the contribution of chromosomal aberrations and GI to the pathology of early stages of embryogenesis, elucidation of the molecular genetic mechanisms of primary embryonic induction and the first phase of gastrulation, features of the structural and functional organization of chromosomes, and epigenetic evolution of the genome before implantation have been discussed so far in this work. Significant progress has been achieved in understanding the mechanisms of occurrence and in the diagnosis of chromosomal mosaicism as well as complex chromosomal rearrangements, including constitutional CT.

Acknowledgments

This work was supported by the Ministry of Science and Higher Education of the Russian Federation, Basic Research Program number AAAA-A19-119021290033-1.

References

Babariya, D., Fragouli, E., Alfarawati, S., Spath, K., & Wells, D. (2017). The incidence and origin of segmental aneuploidy in human oocytes and preimplantation embryos. *Human Reproduction (Oxford, England)*, *32*(12), 2549–2560. https://doi.org/10.1093/humrep/dex324.

Baranov, V. S., & Kuznetzova, T. V. (2007). *Cytogenetics of human embryonic development: Scientific & practical aspects*. St. Petersburg: N-L.

Baranov, V. S., & Kuznetzova, T. V. (2012). In N. P. Bochkov, E. K. Ginter, & V. P. Puzyrev (Eds.), *Genetics of human development* (pp. 81–125). GEOTAR-Media.

Baranov, V. S., Kuznetzova, T. V., Pendina, A. A., Efimova, O. A., Fedorova, I. D., & Trofimova, I. L. (2012). In S. M. Zakijan, S. M. Vlasov, & E. V. Dement'eva (Eds.), *Epigenetic mechanisms of normal and pathological human development* (pp. 225–266). SD RAN.

Bertelsen, B., Nazaryan-Petersen, L., Sun, W., Mehrjouy, M. M., Xie, G., Chen, W., ... Tümer, Z. (2016). A germline chromothripsis event stably segregating in 11 individuals through three generations. *Genetics in Medicine*, *18*(5), 494–500. https://doi.org/10.1038/gim.2015.112.

Carlson, B. (2008). *Human embryology and developmental biology* (4th ed.). Mosby Elsevier.

Chavez, S. L., Loewke, K. E., Han, J., Moussavi, F., Colls, P., Munne, S., ... Reijo Pera, R.A. (2012). Dynamic blastomere behaviour reflects human embryo ploidy by the four-cell stage. *Nature Communications*, *3*, 1251. https://doi.org/10.1038/ncomms2249.

COGEN Position Statement on Chromosomal Mosaicism Detected in Preimplantation Blastocyst Biopsies. (2020). *COGEN position statement on chromosomal mosaicism detected in preimplantation blastocyst biopsies.* Retrieved 31 August 2020 from: https://ivf-worldwide.com/cogen/general/cogen-statement.html.

Daban, J.-R. (2021). Multilayer organization of chromosomes. In T. Liehr (Ed.), *Cytogenomics* (pp. 267–296). Academic Press. Chapter 13 (in this book).

De Iaco, A., Planet, E., Coluccio, A., Verp, S., Duc, J., & Trono, D. (2017). DUX-family transcription factors regulate zygotic genome activation in placental mammals. *Nature Genetics, 49*(6), 941–945. https://doi.org/10.1038/ng.3858.

de Pagter, M. S., van Roosmalen, M. J., Baas, A. F., Renkens, I., Duran, K. J., van Binsbergen, E., … Kloosterman, W.P. (2015). Chromothripsis in healthy individuals affects multiple protein-coding genes and can result in severe congenital abnormalities in offspring. *American Journal of Human Genetics, 96*(4), 651–656. https://doi.org/10.1016/j.ajhg.2015.02.005.

Du, Z., Zheng, H., Huang, B., Ma, R., Wu, J., Zhang, X., … Xie, W. (2017). Allelic reprogramming of 3D chromatin architecture during early mammalian development. *Nature, 547*(7662), 232–235. https://doi.org/10.1038/nature23263.

Dyban, A. P., & Baranov, V. S. (1987). *Cytogenetics of mammalian embryonic development.* Oxford University Press.

Efimova, O. A., Pendina, A. A., Tikhonov, A. V., Fedorova, I. D., Krapivin, M. I., Chiryaeva, O. G., … Baranov, V.S. (2015). Chromosome hydroxymethylation patterns in human zygotes and cleavage-stage embryos. *Reproduction, 149*(3), 223–233. https://doi.org/10.1530/REP-14-0343.

Eggermann, T. (2021). Epigenetics. In T. Liehr (Ed.), *Cytogenomics* (pp. 389–401). Academic Press. Chapter 20 (in this book).

Ferguson-Smith, A. C., & Bourc'his, D. (2018). The discovery and importance of genomic imprinting. *eLife, 7.* https://doi.org/10.7554/eLife.42368, e42368.

Fogarty, N. M. E., McCarthy, A., Snijders, K. E., Powell, B. E., Kubikova, N., Blakeley, P., … Niakan, K.K. (2017). Genome editing reveals a role for OCT4 in human embryogenesis. *Nature, 550*(7674), 67–73. https://doi.org/10.1038/nature24033.

Geng, L. N., Yao, Z., Snider, L., Fong, A. P., Cech, J. N., Young, J. M., … Tapscott, S.J. (2012). DUX4 activates germline genes, retroelements, and immune mediators: Implications for facioscapulohumeral dystrophy. *Developmental Cell, 22*(1), 38–51. https://doi.org/10.1016/j.devcel.2011.11.013.

Godini, R., & Fallahi, H. (2019). Dynamics changes in the transcription factors during early human embryonic development. *Journal of Cellular Physiology, 234*(5), 6489–6502. https://doi.org/10.1002/jcp.27386.

Griffin, D. K., & Ogur, C. (2018). Chromosomal analysis in IVF: Just how useful is it? *Reproduction, 156*(1), F29–F50. https://doi.org/10.1530/REP-17-0683.

Handyside, A. H., Harton, G. L., Mariani, B., Thornhill, A. R., Affara, N., Shaw, M.-A., & Griffin, D. K. (2010). Karyomapping: A universal method for genome wide analysis of genetic disease based on mapping crossovers between parental haplotypes. *Journal of Medical Genetics, 47*(10), 651–658. https://doi.org/10.1136/jmg.2009.069971.

Harutyunyan, T., & Hovhannisyan, G. (2021). Approaches for studying epigenetic aspects of the human genome. In T. Liehr (Ed.), *Cytogenomics* (pp. 155–209). Academic Press. Chapter 10 (in this book).

Houge, G., Liehr, T., Schoumans, J., Ness, G. O., Solland, K., Starke, H., … Vermeulen, S. (2003). Ten years follow up of a boy with a complex chromosomal rearrangement: Going from a >5 to 15-breakpoint CCR. *American Journal of Medical Genetics, 118A*(3), 235–240. https://doi.org/10.1002/ajmg.a.10106.

Hug, C. B., & Vaquerizas, J. M. (2018). The birth of the 3D genome during early embryonic development. *Trends in Genetics, 34*(12), 903–914. https://doi.org/10.1016/j.tig.2018.09.002.

Iourov, I. Y., Vorsanova, S. G., Yurov, Y. B., & Kutsev, S. I. (2019). Ontogenetic and pathogenetic views on somatic chromosomal mosaicism. *Genes (Basel), 10*(5), 379. https://doi.org/10.3390/genes10050379.

Ichikawa, Y., & Saitoh, N. (2021). Shaping of genome by long noncoding RNAs. In T. Liehr (Ed.), *Cytogenomics* (pp. 357–372). Academic Press. Chapter 18 (in this book).

Koltsova, A. S., Pendina, A. A., Efimova, O. A., Chiryaeva, O. G., Kuznetzova, T. V., & Baranov, V. S. (2019). On the complexity of mechanisms and consequences of chromothripsis: An update. *Frontiers in Genetics, 10*, 393. https://doi.org/10.3389/fgene.2019.00393.

Lebedev, B. N. (2008). Cytogenetics of human embryogenesis: Historical overview and current conceptions. In A. B. Maslennikov (Ed.), *Vol. 12. Molecular-biology technology in medical practice* (pp. 127–140). Alfa Vista N.

Lebedev, I. N. (2018). Human cytogenetics in genome and postgenome ERA: From genome architecture to novel chromosomal diseases. *Tsitologiya, 60*(7), 499–502. https://doi.org/10.31116/tsitol.2018.07.02.

Lee, A., & Kiessling, A. A. (2017). Early human embryos are naturally aneuploid-can that be corrected? *Journal of Assisted Reproduction and Genetics, 34*(1), 15–21. https://doi.org/10.1007/s10815-016-0845-7.

Lepshin, M. V., Sazhenova, E. A., & Lebedev, I. N. (2014). Multiple epimutations in imprinted genes in the human genome and congenital disorders. *Russian Journal of Genetics, 50*(3), 221–236. https://doi.org/10.7868/S0016675814030059.

Liehr, T. (2020). *Cases with uniparental disomy.* http://cs-tl.de/DB/CA/UPD/0-Start.html. (Accessed 20 December 2020).

Liehr, T. (2021a). A definition for cytogenomics - Which also may be called chromosomics. In T. Liehr (Ed.), *Cytogenomics* (pp. 1–7). Academic Press. Chapter 1 (in this book).

Liehr, T. (2021b). Overview of currently available approaches used in cytogenomics. In T. Liehr (Ed.), *Cytogenomics* (pp. 11–24). Academic Press. Chapter 2 (in this book).

Liehr, T. (2021c). Nuclear architecture. In T. Liehr (Ed.), *Cytogenomics* (pp. 297–305). Academic Press. Chapter 14 (in this book).

Liehr, T. (2021d). Repetitive elements, heteromorphisms, and copy number variants. In T. Liehr (Ed.), *Cytogenomics* (pp. 373–388). Academic Press. Chapter 19 (in this book).

Liehr, T. (2021e). Molecular cytogenetics. In T. Liehr (Ed.), *Cytogenomics* (pp. 35–45). Academic Press. Chapter 4 (in this book).

Liehr, T., Lauten, A., Schneider, U., Schleussner, E., & Weise, A. (2017). Noninvasive prenatal testing—When is it advantageous to apply. *Biomedicine Hub, 2*(1), 1–11. https://doi.org/10.1159/000458432.

Liu, P., Erez, A., Nagamani, S. C. S., Dhar, S. U., Kołodziejska, K. E., Dharmadhikari, A. V., … Bi, W. (2011). Chromosome catastrophes involve replication mechanisms generating complex genomic rearrangements. *Cell, 146*(6), 889–903. https://doi.org/10.1016/j.cell.2011.07.042. In this issue.

Luke, B. (2017). Pregnancy and birth outcomes in couples with infertility with and without assisted reproductive technology: With an emphasis on US population-based studies. *American Journal of Obstetrics and Gynecology, 217*(3), 270–281. https://doi.org/10.1016/j.ajog.2017.03.012.

Madan, K. (2013). What is a complex chromosome rearrangement? *American Journal of Medical Genetics. Part A, 161A*(5), 1181–1184. https://doi.org/10.1002/ajmg.a.35834.

Marchetti, F., Essers, J., Kanaar, R., & Wyrobek, A. J. (2007). Disruption of maternal DNA repair increases sperm-derived chromosomal aberrations. *Proceedings of the National Academy of Sciences of the United States of America*, *104*(45), 17725–17729.

McCoy, R. C. (2017). Mosaicism in preimplantation human embryos: When chromosomal abnormalities are the norm. *Trends in Genetics*, *33*(7), 448–463. https://doi.org/10.1016/j.tig.2017.04.001.

Middelkamp, S., van Heesch, S., Braat, A. K., de Ligt, J., van Iterson, M., Simonis, M., … Cuppen, E. (2017). Molecular dissection of germline chromothripsis in a developmental context using patient-derived iPS cells. *Genome Medicine*, *9*(1), 9. https://doi.org/10.1186/s13073-017-0399-z.

Morris, S. A. (2017). Human embryos cultured in vitro to 14 days. *Open Biology*, *7*(1), 170003. https://doi.org/10.1098/rsob.170003.

Munch, E. M., Sparks, A. E., Gonzalez Bosquet, J., Christenson, L. K., Devor, E. J., & Van Voorhis, B. J. (2016). Differentially expressed genes in preimplantation human embryos: Potential candidate genes for blastocyst formation and implantation. *Journal of Assisted Reproduction and Genetics*, *33*(8), 1017–1025. https://doi.org/10.1007/s10815-016-0745-x.

Natesan, S. A., Handyside, A. H., Thornhill, A. R., Ottolini, C. S., Sage, K., Summers, M. C., … Griffin, D.K. (2014). Live birth after PGD with confirmation by a comprehensive approach (karyomapping) for simultaneous detection of monogenic and chromosomal disorders. *Reproductive Biomedicine Online*, *29*(5), 600–605. https://doi.org/10.1016/j.rbmo.2014.07.007.

Pellestor, F. (2019). Chromoanagenesis: Cataclysms behind complex chromosomal rearrangements. *Molecular Cytogenetics*, *12*, 6. https://doi.org/10.1186/s13039-019-0415-7.

Pellestor, F., Gatinois, V., Puechberty, J., Geneviève, D., & Lefort, G. (2014). Chromothripsis: Potential origin in gametogenesis and preimplantation cell divisions. A review. *Fertility and Sterility*, *102*(6), 1785–1796. https://doi.org/10.1016/j.fertnstert.2014.09.006.

Pellestor, F., Gaillard, J.-B., Schneider, A., Puechberty, J., & Gatinois, V. (2021). Chromoanagenesis phenomena and their formation mechanisms. In T. Liehr (Ed.), *Cytogenomics* (pp. 213–245). Academic Press. Chapter 11 (in this book).

Pendina, A. A., Efimova, O. A., Chiryaeva, O. G., Tikhonov, A. V., Petrova, L. I., Dudkina, V. S., … Baranov, V.S. (2014). A comparative cytogenetic study of miscarriages after IVF and natural conception in women aged under and over 35 years. *Journal of Assisted Reproduction and Genetics*, *31*(2), 149–155. https://doi.org/10.1007/s10815-013-0148-1.

Pendina, A., Efimova, O. A., Fedorova, I. D., Leont'eva, O. A., Shilnikova, E. M., Lezhnina, J. G., … Baranov, V.S. (2011). DNA methylation patterns of metaphase chromosomes in human preimplantation embryos. *Cytogenetic and Genome Research*, *132*(1–2), 1–7. https://doi.org/10.1159/000318673.

Petrussa, L., Van de Velde, H., & De Rycke, M. (2016). Similar kinetics for 5-methylcytosine and 5-hydroxymethylcytosine during human preimplantation development in vitro. *Molecular Reproduction and Development*, *83*(7), 594–605. https://doi.org/10.1002/mrd.22656.

PGDIS Newsletter. (2016). *Position statement on chromosome mosaicism and preimplantation aneuploidy testing at the blastocyst stage*. http://www.pgdis.org/docs/newsletter_071816.html.

Racko, D., Benedetti, F., Dorier, J., & Stasiak, A. (2019). Are TADs supercoiled? *Nucleic Acids Research*, *47*(2), 521–532. https://doi.org/10.1093/nar/gky1091.

Sachdev, N. M., Maxwell, S. M., Besser, A. G., & Grifo, J. A. (2017). Diagnosis and clinical management of embryonic mosaicism. *Fertility and Sterility*, *107*(1), 6–11. https://doi.org/10.1016/j.fertnstert.2016.10.006.

Sazhenova, E. A., & Lebedev, I. N. (2008). Cytogenetic and epigenetic aspects of blighted ova. In A. B. Maslennikov (Ed.), *Vol. 12. Molecular-biology technology in medical practice* (pp. 151–161). Alfa Vista N.

Sazhenova, E. A., Nikitina, T. V., Skryabin, N. A., Minaycheva, L. I., Ivanova, T. V., Nemtseva, T. N., … Lebedev, I.N. (2017). Epigenetic status of imprinted genes in placenta during recurrent pregnancy loss. *Russian Journal of Genetics*, *53*(3), 376–387.

Sermon, K. (2017). Novel technologies emerging for preimplantation genetic diagnosis and preimplantation genetic testing for aneuploidy. *Expert Review of Molecular Diagnostics*, *17*(1), 71–82.

Skryabin, N. A., Lebedev, I. N., Artukhova, V. G., Zhigalina, D. I., Stepanov, I. A., Krivoschekova, G. V., & Svetlakov, A. V. (2015). Molecular karyotyping of cell-free DNA from blastocoele fluid as a basis for noninvasive preimplantation genetic screening of aneuploidy. *Genetika*, *51*(11), 1301–1307. https://doi.org/10.7868/S0016675815110156.

Solter, D. (2006). Imprinting today: End of the beginning or beginning of the end? *Cytogenetic and Genome Research*, *113*(1–4), 12–16.

Stephens, P. J., Greenman, C. D., Fu, B., Yang, F., Bignell, G. R., Mudie, L. J., … Campbell, P.J. (2011). Massive genomic rearrangement acquired in a single catastrophic event during cancer development. *Cell*, *144*(1), 27–40. https://doi.org/10.1016/j.cell.2010.11.055.

Su, Y., Li, J.-J., Wang, C., Haddad, G., & Wang, W.-H. (2016). Aneuploidy analysis in day 7 human blastocysts produced by in vitro fertilization. *Reproductive Biology and Endocrinology*, *14*, 20. https://doi.org/10.1186/s12958-016-0157-x.

Surani, M. A. (1998). Imprinting and the initiation of gene silencing in the germ line. *Cell*, *93*(3), 309–312.

Takahashi, N., Coluccio, A., Thorball, C. W., Planet, E., Shi, H., Offner, S., … Trono, D. (2019). ZNF445 is a primary regulator of genomic imprinting. *Genes & Development*, *33*(1–2), 49–54. https://doi.org/10.1101/gad.320069.118.

Tang, H. L., Tang, H. M., Mak, K. H., Hu, S., Wang, S. S., Wong, K. M., … Fung, M.C. (2012). Cell survival, DNA damage, and oncogenic transformation after a transient and reversible apoptotic response. *Molecular Biology of the Cell*, *23*(12), 2240–2252. https://doi.org/10.1091/mbc.E11-11-0926.

Taylor, T. H., Gitlin, S. A., Patrick, J. L., Crain, J. L., Wilson, J. M., & Griffin, D. K. (2014). The origin, mechanisms, incidence and clinical consequences of chromosomal mosaicism in humans. *Human Reproduction Update*, *20*(4), 571–581. https://doi.org/10.1093/humupd/dmu016.

Treff, N. R., & Franasiak, J. M. (2017). Detection of segmental aneuploidy and mosaicism in the human preimplantation embryo: Technical considerations and limitations. *Fertility and Sterility*, *107*(1), 27–31. https://doi.org/10.1016/j.fertnstert.2016.09.039.

Tšuiko, O., Zhigalina, D. I., Jatsenko, T., Skryabin, N. A., Kanbekova, O. R., Artyukhova, V. G., … Lebedev, I.N. (2018). Karyotype of the blastocoel fluid demonstrates low concordance with both trophectoderm and inner cell mass. *Fertility and Sterility*, *109*(6), 1127–1134.e1. https://doi.org/10.1016/j.fertnstert.2018.02.008.

Ungelenk, M. (2021). Sequencing approaches. In T. Liehr (Ed.), *Cytogenomics* (pp. 87–122). Academic Press. Chapter 7 (in this book).

van de Werken, C., Avo Santos, M., Laven, J. S. E., Eleveld, C., Fauser, B. C. J. M., Lens, S. M. A., & Baart, E. B. (2015). Chromosome segregation regulation in human zygotes: Altered mitotic histone phosphorylation dynamics underlying centromeric targeting of the chromosomal passenger complex. *Human Reproduction (Oxford, England)*, *30*(10), 2275–2291. https://doi.org/10.1093/humrep/dev186.

Vázquez-Diez, C., & FitzHarris, G. (2018). Causes and consequences of chromosome segregation error in preimplantation embryos. *Reproduction*, *155*(1), R63–R76. https://doi.org/10.1530/REP-17-0569.

Victor, A. R., Griffin, D. K., Brake, A. J., Tyndall, J. C., Murphy, A. E., Lepkowsky, L. T., … Viotti, M. (2019). Assessment of aneuploidy concordance between clinical trophectoderm biopsy and blastocyst. *Human Reproduction (Oxford, England)*, *34*(1), 181–192. https://doi.org/10.1093/humrep/dey327.

Weise, A., & Liehr, T. (2021a). Cytogenetics. In T. Liehr (Ed.), *Cytogenomics* (pp. 25–34). Academic Press. Chapter 3 (in this book).

Weise, A., & Liehr, T. (2021b). Molecular karyotyping. In T. Liehr (Ed.), *Cytogenomics* (pp. 73–85). Academic Press. Chapter 6 (in this book).

White, M. D., Angiolini, J. F., Alvarez, Y. D., Kaur, G., Zhao, Z. W., Mocskos, E., … Plachta, N. (2016). Long-lived binding of Sox2 to DNA predicts cell fate in the four-cell mouse embryo. *Cell*, *165*(1), 75–87. https://doi.org/10.1016/j.cell.2016.02.032.

Wu, T., Yin, B., Zhu, Y., Li, G., Ye, L., Chen, C., … Liang, D. (2016). Molecular cytogenetic analysis of early spontaneous abortions conceived from varying assisted reproductive technology procedures. *Molecular Cytogenetics*, *9*, 79. https://doi.org/10.1186/s13039-016-0284-2.

Yamamoto, R., & Aoki, F. (2017). A unique mechanism regulating gene expression in 1-cell embryos. *The Journal of Reproduction and Development*, *63*(1), 9–11. https://doi.org/10.1262/jrd.2016-133.

Yumiceba, V., Souto Melo, U., & Spielmann, M. (2021). 3D cytogenomics: Structural variation in the three-dimensional genome. In T. Liehr (Ed.), *Cytogenomics* (pp. 247–266). Academic Press. Chapter 12 (in this book).

Cytogenomic landscape of the human brain

16

Ivan Y. Iourov[a,b,c], Svetlana G. Vorsanova[a,b], and Yuri B. Yurov[a,b]

[a]Yurov's Laboratory of Molecular Genetics and Cytogenomics of the Brain, Mental Health Research Center, Moscow, Russia
[b]Laboratory of Molecular Cytogenetics of Neuropsychiatric Diseases, Veltischev Research and Clinical Institute for Pediatrics of the Pirogov Russian National Research Medical University, Moscow, Russia
[c]Russian Medical Academy of Continuous Postgraduate Education, Moscow, Russia

Chapter outline

Introduction to molecular neurocytogenetics/ neurocytogenomics

Molecular neurocytogenetics, which can also be termed neurocytogenomics, encompasses all studies on chromosomal abnormalities and chromosome organization in neural cells of the mammalian brain (Iourov et al., 2006b; Yurov et al., 2018a), and are a subfield of cytogenomics (Liehr 2021a, b). These studies are designed to describe the genomic or, more precisely, cytogenomic landscape of the brain (genomic variations at the chromosomal/subchromosomal level and chromosome arrangement in interphase). Generally, three topics are addressed during neurocytogenetic or cytogenomic analyses of the human brain: somatic mosaicism, the chromosomal organization in interphase, and molecular/cellular pathways.

Firstly, the availability of interphase molecular cytogenetic techniques (e.g., fluorescence in situ hybridization (FISH)-based methods) and single-cell whole-genome assays (Gupta et al., 2020; Heng et al., 2018; Iourov et al., 2012, 2013b; Liehr 2021c; Vorsanova et al., 2010c) has led to studies defining somatic mosaicism as a significant contributor to neural diversity and as a mechanism for brain diseases. Accordingly, somatic mosaicism has been recognized as an underlying mechanism for a variety of neurological and psychiatric diseases (Campbell et al., 2015; D'Gama & Walsh 2018; Iourov et al., 2019a, 2010a, b; Jourdon et al., 2020; Thorpe et al., 2020; Yurov et al., 2019b). Secondly, elucidating genome organization at chromosomal level in the human brain was repeatedly noted to be required for understanding molecular and cellular pathways to neural diversity and brain disorders (Daban, 2020, 2021; Finn & Misteli, 2019; Iourov, 2006a, 2012, 2019a, b, c). Thirdly, postgenomic (systems biology) techniques provide pathway-based analysis of functional consequences of cytogenomic variations and chromosomal organization in interphase nuclei of neural cells (Iourov, 2019; Iourov et al., 2019b, c).

Here, the aforementioned topics are described in the light of studies on cytogenomic landscape of the human brain. To succeed, cytogenomic variations at different ontogenetic stages are reviewed. Furthermore, the contribution of brain-specific somatic chromosomal mosaicism to neuropsychiatric diseases is addressed. Chromosomal organization in the human brain is considered in the postgenomic perspective. The latter logically implies the evaluation of the role played by pathway-based analysis in describing the neurocytogenomic landscape.

Cytogenomic variation in the human brain: Neuronal diversity versus disease

Since regular and mosaic chromosomal imbalances (e.g., aneuploidy) possess devastating effects on humans, brain-specific genomic variations at the chromosomal level (i.e., neurocytogenomic variations) have been systematically highlighted as a mechanism for neurological and psychiatric disorders (Iourov et al., 2006a, 2008, 2019a; Jourdon et al., 2020; Thorpe et al., 2020). However, cytogenomic variations are an integral part of prenatal development of the human central nervous system, as well (Rohrback et al., 2018a, b; Yurov et al., 2007b, 2010). Moreover, these variations are observed in the unaffected human brain and are involved in brain aging (Andriani et al., 2017; Faggioli et al., 2012; Fischer et al., 2012; Iourov et al., 2019a; Yurov et al., 2009, 2010). Thus, to describe cytogenomic landscape of the human brain from the perspective of canonical genomics, it is to consider ontogenetic variations of the neural genomes at the chromosomal level in health and diseases. In this instance, cyto(onto)genomic variations in the unaffected brain are those contributing to neural diversity, whereas increased neurocytogenomic variations and chromosomal instability are likely to be contributors to brain diseases and aging.

Cyto(onto)genomic variations in the unaffected brain

The developing human brain (gestational age 8–11 weeks) is characterized by high levels of chromosomal instability and mosaicism (aneuploidy) affecting 30%–35% or aneuploidy rates ranging from 1.25% to 1.45% per individual homologous chromosome pair (Yurov et al., 2005, 2007b). Similarly, high rates of tissue-specific genomic variation at the subchromosomal (submicroscopic) level, i.e., copy number variants/variations (CNVs with < 1 Mb in size), are detectable in the developing human brain (McConnell et al., 2013; Rohrback et al., 2018a, b) - for technical aspects of CNV detection see as well (Weise & Liehr, 2021). Tissue-specific genomic variation at the sequence level has not been observed in the developing human brain (Knouse et al., 2014; McConnell et al., 2017; Rohrback et al., 2018a, b).

So far, the developing human brain is the sole embryonic tissue, which may be affected by confined chromosomal mosaicism in contrast to extraembryonic chromosomal mosaicism confined to the placenta (Iourov et al., 2019a; Yurov et al., 2007a). High rates of (sub)chromosomal instability and mosaicism in the developing brain suggest two alternative scenarios for the chromosomally abnormal cell populations in postnatal period: (i) decrease via programmed cell death (Fricker et al., 2018; Pfisterer & Khodosevich, 2017) or (ii) increase/lack of decrease leading to risks for brain functioning and development after birth (Bushman & Chun, 2013; Iourov et al., 2006a, 2008; Vorsanova et al., 2017). In total, one may conclude the developing human brain to have a specific cytogenomic landscape due to neurocytogenomic variations at chromosomal and subchromosomal levels.

Devastating effects of mosaic aneuploidy and chromosome instability within cellular populations are likely to underlie the rate decrease during late prenatal and early postnatal development of the human central nervous system. Studies on neurocytogenomic variations in the postnatal brain have shown the rates of chromosomal (subchromosomal) mosaicism and/or instability to be below 10%–12% (Iourov et al., 2006a, 2009a, b; Knouse et al., 2014; McConnell et al., 2013; Mosch et al., 2007; Yurov et al., 2005, 2010). The decrease is likely to be mediated by programmed cell death, inasmuch as aneuploidy and/or chromosome instability are involved in cascade cell clearance processes (Arendt et al., 2015; Fielder et al., 2017; Peterson et al., 2012; Yang et al., 2003). Still, the remainder of chromosomally abnormal (e.g., aneuploid) cells is likely to be an integral component of the human brain, being functionally active and integrated into brain circuitry (Kingsbury et al., 2005). These observations suggest that low-level mosaic aneuploidy may contribute to normal neuronal diversity.

Cytogenomic landscape of the aging brain remains enigmatic. However, there are evidences that aneuploidy rate changes with age in late brain ontogeny (Iourov et al., 2019a; Yurov et al., 2009, 2010; Zhang & Vijg, 2018). Notably, a number of studies demonstrate an increase in rates of aneuploidy/chromosome instability during brain aging. These data have allowed hypothesizing that accumulation of chromosomally abnormal (aneuploid) cells is a mechanism of brain aging (Andriani et al., 2017; Faggioli et al., 2012; Fischer et al., 2012; Vorsanova et al., 2020; Yurov et al., 2009,

2010, 2018b, 2019a, b; Zhang & Vijg, 2018). However, there are reports indicating that brain aging is not associated with variations in aneuploidy/chromosome instability rates (Chronister et al., 2019; Iourov et al., 2019a; Shepherd et al., 2018). Probably, studies of cyto(onto)genomic variation in the human brain by molecular cytogenetic (FISH) techniques, whole-genome (single-cell) assays, and bioinformatic (systems biology) methods could clarify the nature of this disparity in the future.

Neurocytogenomic variations and brain diseases

Somatic genomic variations importantly contribute to the etiology of brain diseases, among which are Alzheimer's disease (AD), autism spectrum disorders, epilepsy, intellectual disability, schizophrenia, and other psychiatric and neurological disorders. Indeed, brain-specific chromosomal abnormalities and instability require special attention, inasmuch as directly affecting the diseased brain (Bushman & Chun, 2013; Graham et al., 2019; Hochstenbach et al., 2011; Iourov et al., 2006a, 2010a, b, 2013b, 2019a; Liehr and Al-Rikabi 2019; Paquola et al., 2017; Potter et al., 2019; Smith et al., 2010; Vorsanova et al., 2007, 2010a; Ye et al., 2019; Yurov et al., 2001).

Neurodegenerative diseases have been repeatedly associated with chromosomal instability, aneuploidy, and related phenomena (Iourov et al., 2009a, b; Leija-Salazar et al., 2018; Potter et al., 2019; Shepherd et al., 2018; Yurov et al., 2019a). Neurological parallels between AD and Down syndrome (trisomy of chromosome 21) have led to a hypothesis suggesting the AD brain to be affected by chromosome 21 aneuploidy (Potter et al., 2016; Rao et al., 2020). Indeed, chromosome 21-specific instability has been found associated with AD pathogenesis (Fig. 16.1). It is to note that cells exhibiting trisomy 21 have been detected in the AD brain in contrast to control samples (Iourov et al., 2009a, b) (Fig. 16.2). Another cytogenetic marker of human aging, X-chromosome-loss (mosaic aneuploidy (monosomy) of X-chromosome), is commonly detected in female AD brains (i.e., X-chromosome loss is more frequent in the AD brain than in controls) (Bajic et al., 2020; Yurov et al., 2014). The AD brain has also demonstrated suchromosomal instability manifesting as nonspecific CNVs of the *APP* gene (Kaeser & Chun, 2020).

Chromosome instability in the brain seems to be an important element of the AD pathogenic cascade (Iourov et al., 2010a, b; Yurov et al., 2019a). This may be further supported by the following observations:

- Aneuploidy causes neuronal cell death in the AD brain similarly to developing brain (Arendt, 2012; Arendt et al., 2015).
- LDL/cholesterol induces altered chromosome segregation in AD, Niemann-Pick C1, and atherosclerosis (Granic and Potter 2013).
- The AD brain demonstrates altered DNA damage response resulting in chromosome instability mediating neurodegeneration (Coppedè & Migliore, 2015; Fielder et al., 2017; Lin et al., 2020).

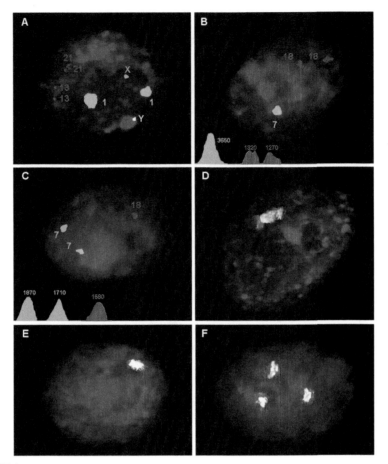

FIG. 16.1

Molecular cytogenetic analysis of aneuploidy in the normal and diseased human brain tissue. (A) 5-color FISH with five DNA probes on nuclei of the normal brain: chromosome 21(red, 21q22–q22.22), chromosome 13 (pink, 13q14), chromosome 1 (blue, 1q12), chromosome X (green, Xp11–q11), and chromosome Y (yellow, Yp11–q11). (B) FISH with chromosome enumeration probes for chromosome 7 (one green signal, relative intensity is 3650 pixels) and 18 (two red signals, relative intensities are 1320 and 1270 pixels), characterized by somatic pairing of two homologous chromosomes 7, but not a monosomy 7 in the AT brain. (C) FISH with chromosome enumeration probes for chromosome 7 (two green signals, relative intensities are 1870 and 1710 pixels) and 18 (one red signal, relative intensity is 1680 pixels), characterized by true monosomy 18 in the AT (Ataxia telangiectasia) brain. (D) True monosomy 14 revealed by interphase chromosome-specific multicolor banding (ICS-MCB) with chromosome 14-specific probe in the AT brain. (E) True monosomy 21 revealed by ICS-MCB with chromosome 21-specific probe in the AD (Alzheimer disease) brain. (F) True trisomy 21 revealed by ICS-MCB with chromosome 21-specific probe in the AD brain.

Reprinted from Iourov, I. Y., Vorsanova, S. G., Liehr, T., & Yurov, Y. B., Aneuploidy in the normal, Alzheimer's disease and ataxia-telangiectasia brain: Differential expression and pathological meaning. Neurobiology of Disease, 34, 212–220, Copyright (2009), with permission from Elsevier.

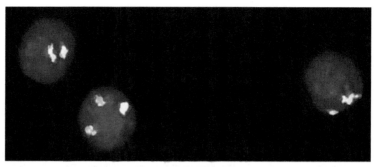

FIG. 16.2

Interphase chromosome-specific multicolor banding (ICS-MCB) for chromosome 21 in the human diseased brain. Two nuclei with disomy 21 and a nucleus with true trisomy 21 revealed by ICS-MCB with chromosome 21-specific probe in the AD brain.

Reprinted from I. Y., Vorsanova, S. G., Liehr, T., & Yurov, Y. B., Aneuploidy in the normal,
Alzheimer's disease and ataxia-telangiectasia brain: Differential expression and pathological meaning.
Neurobiology of Disease, 34, 212–220, Copyright (2009), with permission from Elsevier.

- Replication stress is likely to produce chromosome instability in the AD brain (Wilhelm et al., 2020; Yurov et al., 2011).
- Cellular senescence associated with genome instability hallmarks neurodegeneration (Fielder et al., 2017; Martínez-Cué and Rueda 2020; Monika et al., 2020).

There are other neurodegenerative diseases associated with chromosome instability and/or aneuploidy in the brain, as well. Chromosome instability/aneuploidy has been firstly found to cause neurodegeneration in the ataxia-telangiectasia (AT) brain (Fig. 16.3) (Iourov et al., 2009a, b). AT is an autosomal recessive syndrome caused by mutations in the *ATM* gene and characterized, among other, by chromosome instability and cerebellar degeneration (Rothblum-Oviatt et al., 2016). Chromosome-14 instability manifesting as interphase chromosomal breaks and additional rearranged chromosomes affects > 40% of cells in the degenerating AT cerebellum (Iourov et al., 2009a, b). Neural aneuploidy can be seen at high rates in the brains of individuals with Lewy body (Yang et al., 2015). Mutations in the *MAPT* gene are associated with mitotic defects, which may lead to neuronal aneuploidy and apoptosis in frontotemporal lobar degeneration (Caneus et al., 2018). CNVs/gains of α-synuclein (*SNCA*) gene have been associated with Parkinson's disease and multiple system atrophy (Mokretar et al., 2018).

Chromosome instability is generally recognized as a hallmark of cancer (Simonetti et al., 2019; Valind et al., 2013; Vishwakarma & McManus, 2020). Alternatively, as mentioned before, neurodegeneration is mediated by chromosome instability, as well (Bajic et al., 2015; Nudelman et al., 2019). However, cancerous chromosome instability results from clonal evolution, whereas neurodegenerative

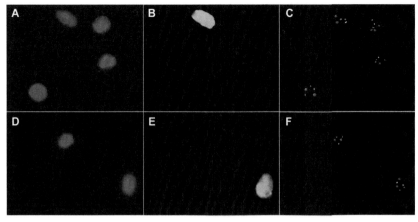

FIG. 16.3

NeuN immunophenotyping of neural cell in the normal (A–C) and AT (Ataxia telangiectasia) brain samples (D–F) and simultaneous triple-color FISH with chromosome enumeration probes forchromosomes 1, 18 and X. . (A) and (D) DAPI staining nuclei. (B) and (F) NeuN immunophenotyping demonstrating one NeuN+ neuronal nucleus and three NeuN – nonneuronal nuclei in (A) and one NeuN+ neuronal nucleus and one NeuN – nonneuronal nucleus in (F); (C) Two red signals corresponding to chromosome 1 in each nuclei; one green signal corresponding to X-chromosome in each nuclei; chromosome 18-specific probe demonstrating two magenta signals in three nuclei with a fourth upper left nucleus of nonneuronal cell demonstrating three signals or trisomy 18. (E) Two red signals corresponding to chromosome 1 in each nucleus; one green signal specific for X-chromosome in nonneuronal nucleus and two signals in neuronal nucleus (disomy of X-chromosome); and chromosome 18-specific probe demonstrates two magenta signals in both nuclei.

Reprinted from Iourov, I. Y., Vorsanova, S. G., Liehr, T., & Yurov, Y. B., Aneuploidy in the normal,
Alzheimer's disease and ataxia-telangiectasia brain: Differential expression and pathological meaning.
Neurobiology of Disease, 34, 212–220, Copyright (2009), with permission from Elsevier.

chromosome instability is the result of genetic-environment interactions with altered cellular genomes triggering progressive neuronal cell loss through programmed cell death (Iourov et al., 2013a; Yurov et al., 2019b).

Individual cases of schizophrenia have been associated with mosaic aneuploidy (trisomy X and 18) in the brain (Yurov et al., 2001). The schizophrenia brain has been further found to demonstrate chromosome 1-specific instability (Fig. 16.4) and gonosomal instability manifested as aneuploidy (Yurov et al., 2008, 2016, 2018a). Tissue-specific CNVs have been also found to be a feature for the schizophrenia brain (Sakai et al., 2015).

Unfortunately, studies of cytogenomic landscape of the brain in neurodevelopmental diseases (e.g., autism, epilepsy, intellectual disability) are scarce.

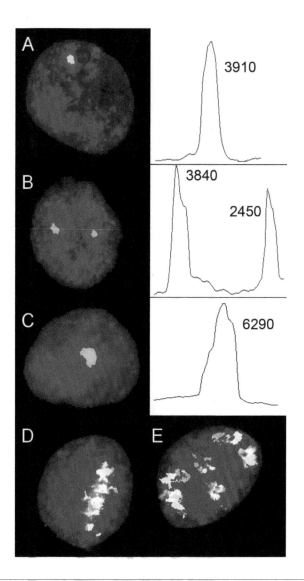

FIG. 16.4

Molecular-cytogenetic analysis of aneuploidy in the postmortem schizophrenia brain. Interphase FISH with chromosome enumeration DNA probes. (A) a nucleus with monosomy involving chromosome 1 (one white signal, relative intensity: 3910) and disomy X (two red signals); (B) a nucleus with disomy 1 (two white signals, relative intensities: 3840 and 2450) and disomy X (two red signals); (C) a nucleus with disomy 1 (one large white signal composed from two paired signals, relative intensity: 6290) and disomy X (two red signals). Interphase chromosome-specific MCB: nuclei with monosomy (D) and trisomy (E) involving chromosome 1.

Reprinted from Yurov, Y. B., Iourov, I. Y., Vorsanova, S. G., Demidova, I. A., Kravetz, V. S., Beresheva, A. K., Kolotii, A. D., Monakchov, V. V., Uranova, N. A., Vostrikov, V. M., Soloviev, I. V., & Liehr, T., The schizophrenia brain exhibits low-level aneuploidy involving chromosome 1. Schizophrenia Research, 98, 139–147, Copyright (2008), with permission from Elsevier.

Nonetheless, some potential characteristics of the brain in neurodevelopmental diseases are assessable. Thus, somatic aneuploidy is common in autistic individuals affecting ~16% of males, 10% of which are low-level 47,XXY/46,XY mosaics (Yurov et al., 2007a). Furthermore, segregation of X-chromosome aneuploidy and X-chromosome instability is common in families with autistic children (Vorsanova et al., 2007, 2017). Interestingly, there is a line of evidences that pathways regulating cell cycle, programmed cell death, and genome stability maintenance (i.e., pathways preventing chromosome instability) are altered in the brain of individuals suffering from neurodevelopmental and neurodegenerative disorders (Bajic et al., 2015; Crawley et al., 2016; Gordon & Geschwind, 2020; Herrup & Yang, 2007; Kumar et al., 2019; Okazaki et al., 2016; Vorsanova et al., 2017; Ye et al., 2019). In a number of cases, chromosome instability (chromothripsis) has been uncovered in the autistic brain (Iourov et al., 2017). Cellular models of autistic spectrum disorders caused by CNV affecting chromosome 15 seem to demonstrate related pathological processes (Casamassa et al., 2020). Finally, chromosome instability in the brain is hypothesized to shape behavior in neurobehavioral diseases and gulf war illness (Liu et al., 2018; Vorsanova et al., 2018). To this end, despite these empirical and theoretical data, direct studies of cytogenomic landscape of the brain in neurodevelopmental and neurobehavioral diseases are still required.

Chromosomal organization in the human brain: Is there a postgenomic perspective?

Chromosome organization in the interphase nucleus determines chromatin remodeling, genome activity (transcription), genome safeguarding (DNA damage response, proper chromosome segregation, mitotic checkpoint, etc.), DNA repair and replication, and programmed cell death at the supramolecular level (Fritz et al., 2019; Liehr, 2021d; Ravi et al., 2020; Rouquette et al., 2010; Tortora et al., 2020; Yu & Ren, 2017). Neurocytogenetic studies have rarely addressed chromosome organization in interphase nuclei of the human brain (Iourov et al., 2006a; Yurov et al., 2013, 2018a). However, molecular neurocytogenetic analyses of somatic mosaicism and chromosome instability have indicated a number of behavioral peculiarities of interphase chromosomes in the human brain.

Available data on chromosome organization in interphase nuclei of the human brain show the following: (i) homologous interphase chromosomes are paired in the human brain; (ii) chromosomes are arranged specifically in a "band-by-band" manner (Iourov, 2017; Yurov et al., 2017a). These data have been acquired by interphase chromosome-specific multicolor banning, which is the unique opportunity to see an interphase chromosome in the human brain (Fig. 16.5) (Iourov et al., 2006b, 2007). In our opinion, these data would be more informative if considered in the systems biology (postgenomic) context. This idea may be supported by a number of theoretic analyses of molecular cytogenetic data using systems biology methodology (Heng, 2020; Iourov, 2019; Yurov et al., 2017b). Interphase chromosome organization in the

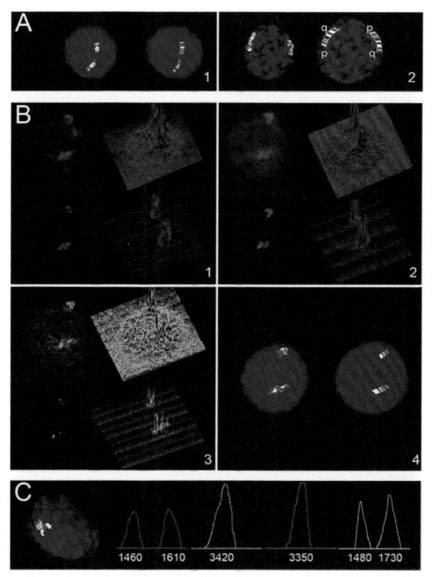

FIG. 16.5

ICS-MCB on human interphase nuclei from peripheral blood lymphocytes. (A) (1) ICS-MCB probe-set for chromosome 14 characterizing the localization and orientation of the two homologues (right: pseudocolor-results; left: schematically depicted results as G-banding ideograms); (2) corresponding results obtained with an ICS-MCB with probe-set for chromosome 7. (B) ICS-MCB probe-set for chromosome 21 depicted as 3D-intensity profiles (parts 1–3) and as depicted in part 4. 3D-intensity profiles can be used to define chromosome integrity, the original results can be evaluated by equalizing the background

(Continued)

human brain is a missing link in current neuroscience. It is highly likely that postgenomic approaches to single cell genomes would bring new insights into molecular neurocytogenetics and cytogenomics.

Pathway-based analysis of neurocytogenomic landscape

Neurocytogenomic variations are most likely to be a result of alterations to pathways preventing chromosome instability. Recently, several pathways were described to produce chromosome instability and reviewed in details (Iourov et al., 2015; Okazaki et al., 2016; Rai et al., 2019; Ravi et al., 2020; Sugaya, 2019; Tan et al., 2019; Xiong et al., 2020). These are likely to be candidate pathways for shaping neurocytogenomic landscape of the human brain. Accordingly, pathway-based analysis might help to complement our understanding of molecular and cellular processes determining the neural genome behavior.

Recently, we have developed a concept for evaluating functional outcomes of the whole set of CNVs in an individual genome (Fig. 16.6). In the technical context, this concept suggests that one has to evaluate all the genome variations (including sequence variations) to define the individual genomic landscape. The definition would be the classification of genomic variations according to cellular and molecular pathways, which are likely to be changed through the involvement of genes affected by the variations. This concept is applicable to neurocytogenomic landscape through the connection between regular genome variations somatic, chromosomal mosaicism, chromosome instability (Fig. 16.7), and identification of functional meaning of chromosomal arrangement in interphase nuclei of the human brain.

FIG. 16.5, CONT'D

within nuclear area to zero: SpectrumOrange signals specific for 21q11.1–q21 (1); TexasRed signals specific for 21q21-qter (2), and SpectrumGreen signals located in 21q21–q22 (3) were treated in this manner; original results are shown in the upper and equalized ones in the lower lines, each. In (4) results corresponding to these are shown in figure obtained with an ICS-MCB with probe-set for chromosome 21. (C) Results of ICS-MCB obtained with a probe-set for chromosome 16 in pseudocolor-depiction (left) together with the corresponding quantitative FISH (QFISH9) results (right). For the latter the relative intensity for each color channel used (SpectrumOrange, SpectrumGreen, TexasRed, and Cyanine5) values are given in fluorescence intensity curves. According to pseudocolor-depiction the presence of one or two chromosomes 16 on this nucleus could have been suggested. QFISH clearly shows that two overlapping chromosomes are present, since intensities of two discrete signals of SpectrumOrange and Cyanine5 are approximately two times lower than those of single signals in SpectrumGreen and TexasRed.

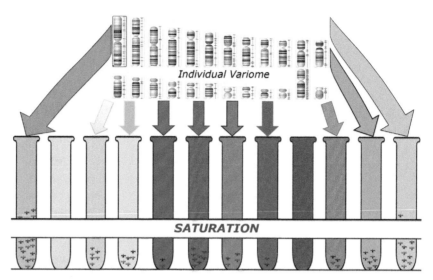

FIG. 16.6

Schematic representation of the variome (CNVariome) concept. A set of genomic variations (e.g., CNVs) may contribute to functional variability of molecular/cellular pathways depicted as colored laboratory tubes; when the number of genomic variations achieves a critical level (i.e., saturation in genomic variations), an alteration to the pathway occurs (e.g., left-most and right-most tubes). A screenshot of CNV analysis by Chromosome Analysis Suite (ChAS) 3.10.15© 2015 Affymetrix Inc. was used to depict individual variome (CNVariome).

From Iourov, I. Y., Vorsanova, S. G., & Yurov, Y. B. (2019c). The variome concept: Focus on CNVariome. Molecular Cytogenetics, 12(1), 52. https://doi:10.1186/s13039-019-0467-8, an open-access article distributed under the terms of the Creative Commons Attribution License.

Concluding remarks: What is and what should be
What is

Currently, following knowledge about cytogenomic landscape of the human brain is available: cytoontogenomic variations are an integral part of brain development; somatic chromosomal mosaicism contributes to neuronal diversity (however, the scale of the contribution is a matter of debates); somatic chromosomal mosaicism and chromosome instability are mechanisms for brain diseases, mediating neurodegeneration and neurodysfunction.

What should be

Currently, following knowledge about cytogenomic landscape of the human brain is to be gained: interindividual and intercellular genome variability in the postnatal and aging brain; the role of chromosomal mosaicism and chromosome instability in

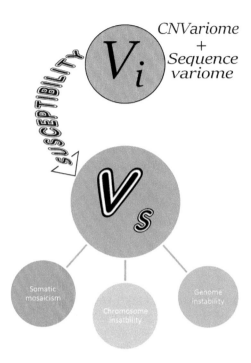

FIG. 16.7

Interplay between individual variome (Vi) and somatic variome (Vs). CNVariome and sequence variome forming Vi may alter pathways critical for maintaining genome stability, cell cycle, and/or programmed cell death; the result of such alterations is likely to be a susceptibility for specific Vs, which encompass somatic mosaism as well as chromosome and genome instability.

From Iourov, I. Y., Vorsanova, S. G., & Yurov, Y. B. (2019c). The variome concept: Focus on CNVariome. Molecular Cytogenetics, 12(1), 52. https://doi:10.1186/s13039-019-0467-8, an open-access article distributed under the terms of the Creative Commons Attribution License.

the etiology of brain diseases, which have not been as yet a focus of neurocytogenetic/neurocytogenomic studies; chromosome arrangement in interphase nuclei of the unaffected and diseased brain; pathway-based analysis of functional meaning of chromosome instability and specific chromosome arrangement in the human brain.

Finally, there are a couple of practical issues in current molecular neurocytogenetics/cytogenomics. Firstly, data on neurocytogenomic variations are to be put in the diagnostic context. Fortunately, there is a basis for related efforts in current cytogenomics and medical genetics (Iourov et al., 2014; Vorsanova et al., 2010b). Secondly, neurocytogenomic landscape is important for the development of therapeutic interventions in brain diseases. Certainly, these interventions would be more effective and less dangerous if developed using pathway-based (systems biology) analysis (Iourov et al., 2019a, b, c; Tan et al., 2019; Thadathil et al., 2019).

Moreover, one may suggest that similar data would become even more valuable applied to the 3D genomic landscape (Razin et al., 2019; Yu & Ren, 2017) of the human brain encompassing genome behavior and variability from DNA sequence to whole chromosomes.

Acknowledgments

The research in authors' labs is partially supported by RFBR and CITMA according to the research project №18-515-34005; prof. IY Iourov's lab is supported by the Government Assignment of the Russian Ministry of Science and Higher Education, Assignment no. AAAA-A19-119040490101-6; prof. SG Vorsanova's lab is supported by the Government Assignment of the Russian Ministry of Health, Assignment no. 121031000238-1.

References

Andriani, G. A., Vijg, J., & Montagna, C. (2017). Mechanisms and consequences of aneuploidy and chromosome instability in the aging brain. *Mechanisms of Ageing and Development, 161*, 19–36. https://doi.org/10.1016/j.mad.2016.03.007.

Arendt, T. (2012). Cell cycle activation and aneuploid neurons in Alzheimer's disease. *Molecular Neurobiology, 46*(1), 125–135. https://doi.org/10.1007/s12035-012-8262-0.

Arendt, T., Brückner, M. K., & Lösche, A. (2015). Regional mosaic genomic heterogeneity in the elderly and in Alzheimer's disease as a correlate of neuronal vulnerability. *Acta Neuropathologica, 130*(4), 501–510. https://doi.org/10.1007/s00401-015-1465-5.

Bajic, V., Spremo-Potparevic, B., Zivkovic, L., Isenovic, E. R., & Arendt, T. (2015). Cohesion and the aneuploid phenotype in Alzheimer's disease: A tale of genome instability. *Neuroscience and Biobehavioral Reviews, 55*, 365–374. https://doi.org/10.1016/j.neubiorev.2015.05.010.

Bajic, V. P., Essack, M., Zivkovic, L., Stewart, A., Zafirovic, S., Bajic, V. B., … Spremo-Potparevic, B. (2020). The X files: "The mystery of X chromosome instability in Alzheimer's disease". *Frontiers in Genetics, 10*(1), 1368. https://doi.org/10.3389/fgene.2019.01368.

Bushman, D. M., & Chun, J. (2013). The genomically mosaic brain: Aneuploidy and more in neural diversity and disease. *Seminars in Cell and Developmental Biology, 24*(4), 357–369. https://doi.org/10.1016/j.semcdb.2013.02.003.

Campbell, I. M., Shaw, C. A., Stankiewicz, P., & Lupski, J. R. (2015). Somatic mosaicism: Implications for disease and transmission genetics. *Trends in Genetics, 31*(7), 382–392. https://doi.org/10.1016/j.tig.2015.03.013.

Caneus, J., Granic, A., Rademakers, R., Dickson, D. W., Coughlan, C. M., Chial, H. J., & Potter, H. (2018). Mitotic defects lead to neuronal aneuploidy and apoptosis in frontotemporal lobar degeneration caused by MAPT mutations. *Molecular Biology of the Cell, 29*(5), 575–586. https://doi.org/10.1091/mbc.E17-01-0031.

Casamassa, A., Ferrari, D., Gelati, M., Carella, M., Vescovi, A. L., & Rosati, J. (2020). A link between genetic disorders and cellular impairment, using human induced pluripotent stem cells to reveal the functional consequences of copy number variations in the central nervous system – A close look at chromosome 15. *International Journal of Molecular Sciences*, *21*(5), 1860. https://doi.org/10.3390/ijms21051860.

Chronister, W. D., Burbulis, I. E., Wierman, M. B., Wolpert, M. J., Haakenson, M. F., Smith, A. C. B., … McConnell, M.J. (2019). Neurons with complex karyotypes are rare in aged human neocortex. *Cell Reports*, *26*(4). https://doi.org/10.1016/j.celrep.2018.12.107. 825–835.e7.

Coppedè, F., & Migliore, L. (2015). DNA damage in neurodegenerative diseases. *Mutation Research, Fundamental and Molecular Mechanisms of Mutagenesis*, *776*, 84–97. https://doi.org/10.1016/j.mrfmmm.2014.11.010.

Crawley, J. N., Heyer, W. D., & LaSalle, J. M. (2016). Autism and cancer share risk genes, pathways, and drug targets. *Trends in Genetics*, *32*(3), 139–146. https://doi.org/10.1016/j.tig.2016.01.001.

Daban, J. R. (2020). Supramolecular multilayer organization of chromosomes: Possible functional roles of planar chromatin in gene expression and DNA replication and repair. *FEBS Letters*, *594*(3), 395–411. https://doi.org/10.1002/1873-3468.13724.

Daban, J.-R. (2021). Multilayer organization of chromosomes. In T. Liehr (Ed.), *Cytogenomics* (pp. 267–296). Academic Press. Chapter 13 (in this book).

D'Gama, A. M., & Walsh, C. A. (2018). Somatic mosaicism and neurodevelopmental disease. *Nature Neuroscience*, *21*(11), 1504–1514. https://doi.org/10.1038/s41593-018-0257-3.

Faggioli, F., Wang, T., Vijg, J., & Montagna, C. (2012). Chromosome-specific accumulation of aneuploidy in the aging mouse brain. *Human Molecular Genetics*, *21*(24), 5246–5253. https://doi.org/10.1093/hmg/dds375.

Fielder, E., Von Zglinicki, T., & Jurk, D. (2017). The DNA damage response in neurons: Die by apoptosis or survive in a senescence-like state? *Journal of Alzheimer's Disease*, *60*(1), S107–S131. https://doi.org/10.3233/JAD-161221.

Finn, E. H., & Misteli, T. (2019). Molecular basis and biological function of variability in spatial genome organization. *Science*, *365*(6457). https://doi.org/10.1126/science.aaw9498, eaaw9498.

Fischer, H. G., Morawski, M., Brückner, M. K., Mittag, A., Tarnok, A., & Arendt, T. (2012). Changes in neuronal DNA content variation in the human brain during aging. *Aging Cell*, *11*(4), 628–633. https://doi.org/10.1111/j.1474-9726.2012.00826.x.

Fricker, M., Tolkovsky, A. M., Borutaite, V., Coleman, M., & Brown, G. C. (2018). Neuronal cell death. *Physiological Reviews*, *98*(2), 813–880. https://doi.org/10.1152/physrev.00011.2017.

Fritz, A. J., Sehgal, N., Pliss, A., Xu, J., & Berezney, R. (2019). Chromosome territories and the global regulation of the genome. *Genes, Chromosomes and Cancer*, *58*(7), 407–426. https://doi.org/10.1002/gcc.22732.

Gordon, A., & Geschwind, D. H. (2020). Human in vitro models for understanding mechanisms of autism spectrum disorder. *Molecular Autism*, *11*(1), 26. https://doi.org/10.1186/s13229-020-00332-7.

Graham, E. J., Vermeulen, M., Vardarajan, B., Bennett, D., De Jager, P., Pearse, R. V., … Mostafavi, S. (2019). Somatic mosaicism of sex chromosomes in the blood and brain. *Brain Research*, *1721*(1), 146345. https://doi.org/10.1016/j.brainres.2019.146345.

Granic, A., & Potter, H. (2013). Mitotic spindle defects and chromosome mis-segregation induced by LDL/cholesterol-implications for Niemann-Pick C1, Alzheimer's disease, and atherosclerosis. *PLoS One*, *8*(4). https://doi.org/10.1371/journal.pone.0060718, e60718.

Gupta, P., Balasubramaniam, N., Chang, H. Y., Tseng, F. G., & Santra, T. S. (2020). A single-neuron: Current trends and future prospects. *Cell*, *9*(6), 1528. https://doi.org/10.3390/cells9061528.

Heng, H. (2020). New data collection priority: Focusing on genome-based bioinformation. *ResearchResultsinBiomedicine,6*(1),5–8.https://doi.org/10.18413/2658-6533-2020-6-1-0-1.

Heng, H. H., Horne, S. D., Chaudhry, S., Regan, S. M., Liu, G., Abdallah, B. Y., & Ye, C. J. (2018). A postgenomic perspective on molecular cytogenetics. *Current Genomics*, *19*(3), 227–239. https://doi.org/10.2174/1389202918666170717145716.

Herrup, K., & Yang, Y. (2007). Cell cycle regulation in the postmitotic neuron: Oxymoron or new biology? *Nature Reviews Neuroscience*, *8*(5), 368–378. https://doi.org/10.1038/nrn2124.

Hochstenbach, R., Buizer-Voskamp, J. E., Vorstman, J. A. S., & Ophoff, R. A. (2011). Genome arrays for the detection of copy number variations in idiopathic mental retardation, idiopathic generalized epilepsy and neuropsychiatric disorders: Lessons for diagnostic workflow and research. *Cytogenetic and Genome Research*, *135*(3–4), 174–202. https://doi.org/10.1159/000332928.

Iourov, I. Y. (2012). To see an interphase chromosome or: How a disease can be associated with specific nuclear genome organization. *BioDiscovery*, *4*(1). https://doi.org/10.7750/BioDiscovery.2012.4.5, e8932.

Iourov, I. Y. (2017). Quantitative fluorescence in situ hybridization (QFISH). *Methods in Molecular Biology*, *1541*, 143–149. https://doi.org/10.1007/978-1-4939-6703-2_13.

Iourov, I. Y. (2019). Cytopostgenomics: What is it and how does it work? *Current Genomics*, *20*(2), 77–78. https://doi.org/10.2174/1389202920021904221220524.

Iourov, I. Y., Vorsanova, S. G., & Yurov, Y. B. (2006a). Chromosomal variation in mammalian neuronal cells: Known facts and attractive hypotheses. *International Review of Cytology*, *249*, 143–191. https://doi.org/10.1016/S0074-7696(06)49003-3.

Iourov, I. Y., Liehr, T., Vorsanova, S. G., Kolotii, A. D., & Yurov, Y. B. (2006b). Visualization of interphase chromosomes in postmitotic cells of the human brain by multicolour banding (MCB). *Chromosome Research*, *14*(3), 223–229. https://doi.org/10.1007/s10577-006-1037-6.

Iourov, I. Y., Liehr, T., Vorsanova, S. G., & Yurov, Y. B. (2007). Interphase chromosome-specific multicolor banding (ICS-MCB): A new tool for analysis of interphase chromosomes in their integrity. *Biomolecular Engineering*, *24*(4), 415–417. https://doi.org/10.1016/j.bioeng.2007.05.003.

Iourov, I. Y., Vorsanova, S., & Yurov, Y. (2008). Chromosomal mosaicism goes global. *Molecular Cytogenetics*, *1*(1), 26. https://doi.org/10.1186/1755-8166-1-26.

Iourov, I. Y., Vorsanova, S. G., Liehr, T., Kolotii, A. D., & Yurov, Y. B. (2009a). Increased chromosome instability dramatically disrupts neural genome integrity and mediates cerebellar degeneration in the ataxia-telangiectasia brain. *Human Molecular Genetics*, *18*(14), 2656–2669. https://doi.org/10.1093/hmg/ddp207.

Iourov, I. Y., Vorsanova, S. G., Liehr, T., & Yurov, Y. B. (2009b). Aneuploidy in the normal, Alzheimer's disease and ataxia-telangiectasia brain: Differential expression and pathological meaning. *Neurobiology of Disease*, *34*(2), 212–220. https://doi.org/10.1016/j.nbd.2009.01.003.

Iourov, I. Y., Vorsanova, S. G., & Yurov, Y. B. (2010a). Genomic landscape of the Alzheimer's disease brain: Chromosome instability—aneuploidy, but not tetraploidy—mediates neurodegeneration. *Neurodegenerative Diseases*, *8*(1–2), 35–37. https://doi.org/10.1159/000315398.

Iourov, I. Y., Vorsanova, S. G., & Yurov, Y. B. (2010b). Somatic genome variations in health and disease. *Current Genomics*, *11*(6), 387–396. https://doi.org/10.2174/138920210793176065.

Iourov, I. Y., Vorsanova, S. G., & Yurov, Y. B. (2012). Single cell genomics of the brain: Focus on neuronal diversity and neuropsychiatric diseases. *Current Genomics*, *13*(6), 477–488. https://doi.org/10.2174/138920212802510439.

Iourov, I. Y., Vorsanova, S. G., Voinova, V. Y., Kurinnaia, O. S., Zelenova, M. A., Demidova, I. A., & Yurov, Y. B. (2013a). Xq28 (MECP2) microdeletions are common in mutation-negative females with Rett syndrome and cause mild subtypes of the disease. *Molecular Cytogenetics*, *6*(1), 53. https://doi.org/10.1186/1755-8166-6-53.

Iourov, I. Y., Vorsanova, S. G., & Yurov, Y. B. (2013b). Somatic cell genomics of brain disorders: A new opportunity to clarify genetic-environmental interactions. *Cytogenetic and Genome Research*, *139*(3), 181–188. https://doi.org/10.1159/000347053.

Iourov, I. Y., Vorsanova, S. G., Liehr, T., & Yurov, Y. B. (2014). Mosaike im Gehirn des Menschen: Diagnostische Relevanz in der Zukunft? *Medizinische Genetik*, *26*(3), 342–345. https://doi.org/10.1007/s11825-014-0010-6.

Iourov, I. Y., Vorsanova, S. G., Zelenova, M. A., Korostelev, S. A., & Yurov, Y. B. (2015). Genomic copy number variation affecting genes involved in the cell cycle pathway: Implications for somatic mosaicism. *International Journal of Genomics*, *2015*(1), 757680. https://doi.org/10.1155/2015/757680.

Iourov, I., Vorsanova, S., Liehr, T., Zelenova, M., Kurinnaia, O., Vasin, K., & Yurov, Y. (2017). Chromothripsis as a mechanism driving genomic instability mediating brain diseases. *Molecular Cytogenetics*, *10*(Suppl. 1), 20.

Iourov, I. Y., Vorsanova, S. G., Yurov, Y. B., & Kutsev, S. I. (2019a). Ontogenetic and pathogenetic views on somatic chromosomal mosaicism. *Genes*, *10*(5), 379. https://doi.org/10.3390/genes10050379.

Iourov, I. Y., Vorsanova, S. G., & Yurov, Y. B. (2019b). Pathway-based classification of genetic diseases. *Molecular Cytogenetics*, *12*(1), 4. https://doi.org/10.1186/s13039-019-0418-4.

Iourov, I. Y., Vorsanova, S. G., & Yurov, Y. B. (2019c). The variome concept: Focus on CNVariome. *Molecular Cytogenetics*, *12*(1), 52. https://doi.org/10.1186/s13039-019-0467-8.

Jourdon, A., Fasching, L., Scuderi, S., Abyzov, A., & Vaccarino, F. M. (2020). The role of somatic mosaicism in brain disease. *Current Opinion in Genetics and Development*, *65*, 84–90. https://doi.org/10.1016/j.gde.2020.05.002.

Kaeser, G. E., & Chun, J. (2020). Mosaic somatic gene recombination as a potentially unifying hypothesis for Alzheimer's disease. *Frontiers in Genetics*, *11*(1), 390. https://doi.org/10.3389/fgene.2020.00390.

Kingsbury, M. A., Friedman, B., McConnell, M. J., Rehen, S. K., Yang, A. H., Kaushal, D., & Chun, J. (2005). Aneuploid neurons are functionally active and integrated into brain circuitry. *Proceedings of the National Academy of Sciences of the United States of America*, *102*(17), 6143–6147. https://doi.org/10.1073/pnas.0408171102.

Knouse, K. A., Wu, J., Whittaker, C. A., & Amon, A. (2014). Single cell sequencing reveals low levels of aneuploidy across mammalian tissues. *Proceedings of the National Academy of Sciences of the United States of America*, *111*(37), 13409–13414. https://doi.org/10.1073/pnas.1415287111.

Kumar, S., Reynolds, K., Ji, Y., Gu, R., Rai, S., & Zhou, C. J. (2019). Impaired neurodevelopmental pathways in autism spectrum disorder: A review of signaling mechanisms and crosstalk. *Journal of Neurodevelopmental Disorders*, *11*(1), 10. https://doi.org/10.1186/s11689-019-9268-y.

Leija-Salazar, M., Piette, C., & Proukakis, C. (2018). Review: Somatic mutations in neurodegeneration. *Neuropathology and Applied Neurobiology*, *44*(3), 267–285. https://doi.org/10.1111/nan.12465.

Liehr, T. (2021a). A definition for cytogenomics - Which also may be called chromosomics. In T. Liehr (Ed.), *Cytogenomics* (pp. 1–7). Academic Press. Chapter 1 (in this book).

Liehr, T. (2021b). Overview of currently available approaches used in cytogenomics. In T. Liehr (Ed.), *Cytogenomics* (pp. 11–24). Academic Press. Chapter 2 (in this book).

Liehr, T. (2021c). Molecular cytogenetics. In T. Liehr (Ed.), *Cytogenomics* (pp. 35–45). Academic Press. Chapter 4 (in this book).

Liehr, T. (2021d). Nuclear architecture. In T. Liehr (Ed.), *Cytogenomics* (pp. 297–305). Academic Press. Chapter 14 (in this book).

Liehr, T., & Al-Rikabi, A. (2019). Mosaicism: Reason for normal phenotypes in carriers of small supernumerary marker chromosomes with known adverse outcome. A systematic review. *Frontiers in Genetics, 10*(1), 1131. https://doi.org/10.3389/fgene.2019.01131.

Lin, X., Kapoor, A., Gu, Y., Chow, M. J., Peng, J., Zhao, K., & Tang, D. (2020). Contributions of DNA damage to Alzheimer's disease. *International Journal of Molecular Sciences, 21*(5), 1666. https://doi.org/10.3390/ijms21051666.

Liu, G., Ye, C. J., Chowdhury, S. K., Abdallah, B. Y., Horne, S. D., Nichols, D., & Heng, H. H. (2018). Detecting chromosome condensation defects in gulf war illness patients. *Current Genomics, 19*(3), 200–206. https://doi.org/10.2174/1389202918666170705150819.

Martínez-Cué, C., & Rueda, N. (2020). Cellular senescence in neurodegenerative diseases. *Frontiers in Cellular Neuroscience, 14*(1). https://doi.org/10.3389/fncel.2020.00016.

McConnell, M. J., Lindberg, M. R., Brennand, K. J., Piper, J. C., Voet, T., Cowing-Zitron, C., … Gage, F.H. (2013). Mosaic copy number variation in human neurons. *Science, 342*(6158), 632–637. https://doi.org/10.1126/science.1243472.

McConnell, M. J., Moran, J. V., Abyzov, A., Akbarian, S., Bae, T., Cortes-Ciriano, I., … Rosenbluh, C. (2017). Intersection of diverse neuronal genomes and neuropsychiatric disease: The brain somatic mosaicism network. *Science, 356*(6336). https://doi.org/10.1126/science.aal1641, eaal1641.

Mokretar, K., Pease, D., Taanman, J. W., Soenmez, A., Ejaz, A., Lashley, T., … Proukakis, C. (2018). Somatic copy number gains of α-synuclein (SNCA) in Parkinson's disease and multiple system atrophy brains. *Brain, 141*(8), 2419–2431. https://doi.org/10.1093/brain/awy157.

Monika, B., Macedo, J., Reis, M., Warren, J., Compton, D., & Logarinho, E. (2020). Small-molecule inhibition of aging-associated chromosomal instability delays cellular senescence. *EMBO Reports, 21*. https://doi.org/10.15252/embr.201949248, e49248.

Mosch, B., Morawski, M., Mittag, A., Lenz, D., Tarnok, A., & Arendt, T. (2007). Aneuploidy and DNA replication in the normal human brain and Alzheimer's disease. *Journal of Neuroscience, 27*(26), 6859–6867. https://doi.org/10.1523/JNEUROSCI.0379-07.2007.

Nudelman, K. N. H., McDonald, B. C., Lahiri, D. K., & Saykin, A. J. (2019). Biological hallmarks of cancer in Alzheimer's disease. *Molecular Neurobiology, 56*(10), 7173–7187. https://doi.org/10.1007/s12035-019-1591-5.

Okazaki, S., Boku, S., Otsuka, I., Mouri, K., Aoyama, S., Shiroiwa, K., … Hishimoto, A. (2016). The cell cycle-related genes as biomarkers for schizophrenia. *Progress in Neuro-Psychopharmacology and Biological Psychiatry, 70*, 85–91. https://doi.org/10.1016/j.pnpbp.2016.05.005.

Paquola, A. C. M., Erwin, J. A., & Gage, F. H. (2017). Insights into the role of somatic mosaicism in the brain. *Current Opinion in Systems Biology, 1*, 90–94. https://doi.org/10.1016/j.coisb.2016.12.004.

Peterson, S. E., Yang, A. H., Bushman, D. M., Westra, J. W., Yung, Y. C., Barral, S., … Chun, J. (2012). Aneuploid cells are differentially susceptible to caspase- mediated death during embryonic cerebral cortical development. *Journal of Neuroscience*, *32*(46), 16213–16222. https://doi.org/10.1523/JNEUROSCI.3706-12.2012.

Pfisterer, U., & Khodosevich, K. (2017). Neuronal survival in the brain: Neuron type-specific mechanisms. *Cell Death & Disease*, *8*(3). https://doi.org/10.1038/cddis.2017.64, e2643.

Potter, H., Granic, A., & Caneus, J. (2016). Role of trisomy 21 mosaicism in sporadic and familial Alzheimer's disease. *Current Alzheimer Research*, *13*(1), 7–17. https://doi.org/10.2174/1567205013011512071006116.

Potter, H., Chial, H. J., Caneus, J., Elos, M., Elder, N., Borysov, S., & Granic, A. (2019). Chromosome instability and mosaic aneuploidy in neurodegenerative and neurodevelopmental disorders. *Frontiers in Genetics*, *10*(1), 1092. https://doi.org/10.3389/fgene.2019.01092.

Rai, S. N., Dilnashin, H., Birla, H., Singh, S. S., Zahra, W., Rathore, A. S., … Singh, S.P. (2019). The role of PI3K/Akt and ERK in neurodegenerative disorders. *Neurotoxicity Research*, *35*(3), 775–795. https://doi.org/10.1007/s12640-019-0003-y.

Rao, C. V., Asch, A. S., Carr, D. J. J., & Yamada, H. Y. (2020). "Amyloid-beta accumulation cycle" as a prevention and/or therapy target for Alzheimer's disease. *Aging Cell*, *19*(3). https://doi.org/10.1111/acel.13109, e13109.

Ravi, M., Ramanathan, S., & Krishna, K. (2020). Factors, mechanisms and implications of chromatin condensation and chromosomal structural maintenance through the cell cycle. *Journal of Cellular Physiology*, *235*(2), 758–775. https://doi.org/10.1002/jcp.29038.

Razin, S. V., Ulianov, S. V., & Gavrilov, A. A. (2019). 3D genomics. *Molekuliarnaia Biologiia*, *53*(6), 911–923. https://doi.org/10.1134/S0026898419060156.

Rohrback, S., April, C., Kaper, F., Rivera, R. R., Liu, C. S., Siddoway, B., & Chun, J. (2018a). Submegabase copy number variations arise during cerebral cortical neurogenesis as revealed by single-cell whole-genome sequencing. *Proceedings of the National Academy of Sciences of the United States of America*, *115*(42), 10804–10809. https://doi.org/10.1073/pnas.1812702115.

Rohrback, S., Siddoway, B., Liu, C. S., & Chun, J. (2018b). Genomic mosaicism in the developing and adult brain. *Developmental Neurobiology*, *78*(11), 1026–1048. https://doi.org/10.1002/dneu.22626.

Rothblum-Oviatt, C., Wright, J., Lefton-Greif, M., McGrath-Morrow, S., Crawford, T., & Howard, L. (2016). Ataxia telangiectasia: A review. *Orphanet Journal of Rare Diseases*, *11*(1), 159. https://doi.org/10.1186/s13023-016-0543-7.

Rouquette, J., Cremer, C., Cremer, T., & Fakan, S. (2010). Functional nuclear architecture studied by microscopy: Present and future. *International Review of Cell and Molecular Biology*, *282*(C), 1–90. https://doi.org/10.1016/S1937-6448(10)82001-5.

Sakai, M., Watanabe, Y., Someya, T., Araki, K., Shibuya, M., Niizato, K., … Nawa, H. (2015). Assessment of copy number variations in the brain genome of schizophrenia patients. *Molecular Cytogenetics*, *8*(1), 46. https://doi.org/10.1186/s13039-015-0144-5.

Shepherd, C. E., Yang, Y., & Halliday, G. M. (2018). Region- and cell-specific aneuploidy in brain aging and neurodegeneration. *Neuroscience*, *374*(1), 326–334. https://doi.org/10.1016/j.neuroscience.2018.01.050.

Simonetti, G., Bruno, S., Padella, A., Tenti, E., & Martinelli, G. (2019). Aneuploidy: Cancer strength or vulnerability? *International Journal of Cancer*, *144*(1), 8–25. https://doi.org/10.1002/ijc.31718.

Smith, C. L., Bolton, A., & Nguyen, G. (2010). Genomic and epigenomic instability, fragile sites, schizophrenia and autism. *Current Genomics*, *11*(6), 447–469. https://doi.org/10.2174/138920210793176001.

Sugaya, K. (2019). Chromosome instability caused by mutations in the genes involved in transcription and splicing. *RNA Biology*, *16*(11), 1521–1525. https://doi.org/10.1080/1547 6286.2019.1652523.

Tan, S. H., Karri, V., Tay, N. W. R., Chang, K. H., Ah, H. Y., Ng, P. Q., … Candasamy, M. (2019). Emerging pathways to neurodegeneration: Dissecting the critical molecular mechanisms in Alzheimer's disease, Parkinson's disease. *Biomedicine and Pharmacotherapy*, *111*, 765–777. https://doi.org/10.1016/j.biopha.2018.12.101.

Thadathil, N., Hori, R., Xiao, J., & Khan, M. M. (2019). DNA double-strand breaks: A potential therapeutic target for neurodegenerative diseases. *Chromosome Research*, *27*(4), 345–364. https://doi.org/10.1007/s10577-019-09617-x.

Thorpe, J., Osei-Owusu, I., Avigdor, B. E., Tupler, R., & Pevsner, J. (2020). Mosaicism in human health and disease. *Annual Review of Genetics*. https://doi.org/10.1146/annurev-genet-041720-093403.

Tortora, M. M., Salari, H., & Jost, D. (2020). Chromosome dynamics during interphase: A biophysical perspective. *Current Opinion in Genetics and Development*, *61*, 37–43. https://doi.org/10.1016/j.gde.2020.03.001.

Valind, A., Jin, Y., Baldetorp, B., & Gisselsson, D. (2013). Whole chromosome gain does not in itself confer cancer-like chromosomal instability. *Proceedings of the National Academy of Sciences of the United States of America*, *110*(52), 21119–21123. https://doi.org/10.1073/pnas.1311163110.

Vishwakarma, R., & McManus, K. J. (2020). Chromosome instability; Implications in cancer development, progression, and clinical outcomes. *Cancers*, *12*(4), 824. https://doi.org/10.3390/cancers12040824.

Vorsanova, S. G., Yurov, I. Y., Demidova, I. A., Voinova-Ulas, V. Y., Kravets, V. S., Solov'ev, I. V., … Yurov, Y.B. (2007). Variability in the heterochromatin regions of the chromosomes and chromosomal anomalies in children with autism: Identification of genetic markers of autistic spectrum disorders. *Neuroscience and Behavioral Physiology*, *37*(6), 553–558. https://doi.org/10.1007/s11055-007-0052-1.

Vorsanova, S. G., Yurov, Y. B., Soloviev, I. V., & Iourov, I. Y. (2010a). Molecular cytogenetic diagnosis and somatic genome variations. *Current Genomics*, *11*(6), 440–446. https://doi.org/10.2174/138920210793176010.

Vorsanova, S. G., Voinova, V. Y., Yurov, I. Y., Kurinnaya, O. S., Demidova, I. A., & Yurov, Y. B. (2010b). Cytogenetic, molecular-cytogenetic, and clinical-genealogical studies of the mothers of children with autism: A search for familial genetic markers for autistic disorders. *Neuroscience and Behavioral Physiology*, *40*(7), 745–756. https://doi.org/10.1007/s11055-010-9321-5.

Vorsanova, S. G., Yurov, Y. B., & Iourov, I. Y. (2010c). Human interphase chromosomes: A review of available molecular cytogenetic technologies. *Molecular Cytogenetics*, *3*(1), 1. https://doi.org/10.1186/1755-8166-3-1.

Vorsanova, S. G., Yurov, Y. B., & Iourov, I. Y. (2017). Neurogenomic pathway of autism spectrum disorders: Linking germline and somatic mutations to genetic-environmental interactions. *Current Bioinformatics*, *12*(1), 19–26. https://doi.org/10.2174/1574893611666160606164849.

Vorsanova, S. G., Zelenova, M. A., Yurov, Y. B., & Iourov, I. Y. (2018). Behavioral variability and somatic mosaicism: A cytogenomic hypothesis. *Current Genomics*, *19*(3), 158–162. https://doi.org/10.2174/1389202918666170719165339.

Vorsanova, S. G., Yurov, Y. B., & Iourov, I. Y. (2020). Dynamic nature of somatic chromosomal mosaicism, genetic-environmental interactions and therapeutic opportunities in disease and aging. *Molecular Cytogenetics*, *13*(1), 16. https://doi.org/10.1186/s13039-020-00488-0.

Weise, A., & Liehr, T. (2021). Molecular karyotyping. In T. Liehr (Ed.), *Cytogenomics* (pp. 73–85). Academic Press. Chapter 6 (in this book).

Wilhelm, T., Said, M., & Naim, V. (2020). DNA replication stress and chromosomal instability: Dangerous liaisons. *Genes, 11*(6), 1–35. https://doi.org/10.3390/genes11060642.

Xiong, Y., Zhang, Y., Xiong, S., & Williams-Villalobo, A. E. (2020). A glance of p53 functions in brain development, neural stem cells, and brain cancer. *Biology, 9*(9), 1–13. https://doi.org/10.3390/biology9090285.

Yang, A. H., Kaushal, D., Rehen, S. K., Kriedt, K., Kingsbury, M. A., McConnell, M. J., & Chun, J. (2003). Chromosome segregation defects contribute to aneuploidy in normal neural progenitor cells. *Journal of Neuroscience, 23*(32), 10454–10462. https://doi.org/10.1523/jneurosci.23-32-10454.2003.

Yang, Y., Shepherd, C., & Halliday, G. (2015). Aneuploidy in Lewy body diseases. *Neurobiology of Aging, 36*(3), 1253–1260. https://doi.org/10.1016/j.neurobiolaging.2014.12.016.

Ye, Z., McQuillan, L., Poduri, A., Green, T. E., Matsumoto, N., Mefford, H. C., … Hildebrand, M.S. (2019). Somatic mutation: The hidden genetics of brain malformations and focal epilepsies. *Epilepsy Research, 155*(1), 106161. https://doi.org/10.1016/j.eplepsyres.2019.106161.

Yu, M., & Ren, B. (2017). The three-dimensional organization of mammalian genomes. *Annual Review of Cell and Developmental Biology, 33*, 265–289. https://doi.org/10.1146/annurev-cellbio-100616-060531.

Yurov, Y. B., Vostrikov, V. M., Vorsanova, S. G., Monakhov, V. V., & Iourov, I. Y. (2001). Multicolor fluorescent in situ hybridization on post-mortem brain in schizophrenia as an approach for identification of low-level chromosomal aneuploidy in neuropsychiatric diseases. *Brain and Development, 23*(Suppl 1), S186–S190. https://doi.org/10.1016/S0387-7604(01)00363-1.

Yurov, Y. B., Iourov, I. Y., Monakhov, V. V., Soloviev, I. V., Vostrikov, V. M., & Vorsanova, S. G. (2005). The variation of aneuploidy frequency in the developing and adult human brain revealed by an interphase FISH study. *Journal of Histochemistry and Cytochemistry, 53*(3), 385–390. https://doi.org/10.1369/jhc.4A6430.2005.

Yurov, Y. B., Vorsanova, S. G., Iourov, I. Y., Demidova, I. A., Beresheva, A. K., Kravetz, V. S., … Gorbachevskaya, N.L. (2007a). Unexplained autism is frequently associated with low-level mosaic aneuploidy. *Journal of Medical Genetics, 44*(8), 521–525. https://doi.org/10.1136/jmg.2007.049312.

Yurov, Y. B., Iourov, I. Y., Vorsanova, S. G., Liehr, T., Kolotii, A. D., Kutsev, S. I., … Soloviev, I.V. (2007b). Aneuploidy and confined chromosomal mosaicism in the developing human brain. *PLoS One, 2*(6). https://doi.org/10.1371/journal.pone.0000558, e558.

Yurov, Y. B., Iourov, I. Y., Vorsanova, S. G., Demidova, I. A., Kravetz, V. S., Beresheva, A. K., … Liehr, T. (2008). The schizophrenia brain exhibits low-level aneuploidy involving chromosome 1. *Schizophrenia Research, 98*(1–3), 139–147. https://doi.org/10.1016/j.schres.2007.07.035.

Yurov, Y., Vorsanova, S. G., & Iourov, I. Y. (2009). GIN'n'CIN hypothesis of brain aging: Deciphering the role of somatic genetic instabilities and neural aneuploidy during ontogeny. *Molecular Cytogenetics, 2*(1), 23. https://doi.org/10.1186/1755-8166-2-23.

Yurov, Y., Vorsanova, S. G., & Iourov, I. Y. (2010). Ontogenetic variation of the human genome. *Current Genomics, 11*(6), 420–425. https://doi.org/10.2174/138920210793175958.

Yurov, Y. B., Vorsanova, S. G., & Iourov, I. Y. (2011). The DNA replication stress hypothesis of Alzheimer's disease. *The Scientific World Journal, 11*, 2602–2612. https://doi.org/10.1100/2011/625690.

Yurov, Y., Vorsanova, S., & Iourov, I. Y. (2013). *Human interphase chromosomes: Biomedical aspects*. Springer. https://doi.org/10.1007/978-1-4614-6558-4.

Yurov, Y. B., Vorsanova, S. G., Liehr, T., Kolotii, A. D., & Iourov, I. Y. (2014). X chromosome aneuploidy in the Alzheimer's disease brain. *Molecular Cytogenetics, 7*(1), 20. https://doi.org/10.1186/1755-8166-7-20.

Yurov, Y. B., Vorsanova, S. G., Demidova, I. A., Kravets, V. S., Vostrikov, V. M., Soloviev, I. V., … Iourov, I.Y. (2016). Genomic instability in the brain: Chromosomal mosaicism in schizophrenia. *Zhurnal Nevrologii i Psihiatrii imeni S.S. Korsakova, 116*(11), 86–91. https://doi.org/10.17116/jnevro201611611186-91.

Yurov, Y. B., Vorsanova, S. G., Soloviev, I. V., Ratnikov, A. M., & Iourov, I. Y. (2017a). FISH-based assays for detecting genomic (chromosomal) mosaicism in human brain cells. In *Vol. 131. Neuromethods* (pp. 27–41). Humana Press Inc. https://doi.org/10.1007/978-1-4939-7280-7_2.

Yurov, Y. B., Vorsanova, S. G., & Iourov, I. Y. (2017b). Network-based classification of molecular cytogenetic data. *Current Bioinformatics, 12*(1), 27–33. https://doi.org/10.2174/1574893611666160606165119.

Yurov, Y. B., Vorsanova, S. G., Demidova, I. A., Kolotii, A. D., Soloviev, I. V., & Iourov, I. Y. (2018a). Mosaic brain aneuploidy in mental illnesses: An association of low-level post-zygotic aneuploidy with schizophrenia and comorbid psychiatric disorders. *Current Genomics, 19*(3), 163–172. https://doi.org/10.2174/1389202918666170717154340.

Yurov, Y., Vorsanova, S., & Iourov, I. (2018b). Human molecular neurocytogenetics. *Current Genetic Medicine Reports, 6*(4), 155–164. https://doi.org/10.1007/s40142-018-0152-y.

Yurov, Y. B., Vorsanova, S. G., & Iourov, I. Y. (2019a). Chromosome instability in the neurodegenerating brain. *Frontiers in Genetics, 10*(1), 892. https://doi.org/10.3389/fgene.2019.00892.

Yurov, Y., Vorsanova, S., & Iourov, I. (2019b). FISHing for unstable cellular genomes in the human brain. *OBM Genetics, 3*(2), 11. https://doi.org/10.21926/obm.genet.1902076.

Zhang, L., & Vijg, J. (2018). Somatic mutagenesis in mammals and its implications for human disease and aging. *Annual Review of Genetics, 52*, 397–419. https://doi.org/10.1146/annurev-genet-120417-031501.

Interchromosomal interactions with meaning for disease

17

Anja Weise and Thomas Liehr

Jena University Hospital, Friedrich Schiller University, Institute of Human Genetics,
Jena, Germany

Chapter outline

Background

During last decades, it became obvious that spatial organization of chromosomes within the transcriptionally active interphase nucleus is a key to understand gene regulation and cell-specific gene expression (Cremer & Cremer, 2019; Daban, 2020, 2021; Liehr, 2021a; Maass et al., 2018; Pederson, 2011). Thus, nucleomics (see also Baranov & Kuznetzova, 2021; Liehr, 2021a; Pellestor et al., 2021) is one of the four major lines of modern cytogenomic research (Liehr, 2021b, 2021c), together with cytogenetics (Weise & Liehr, 2021a) and molecular cytogenetics (Bisht & Avarello, 2021; Delpu et al., 2021; Iourov et al., 2021; Ishii et al., 2021; Liehr, 2021d, 2021e), molecular genetics and genomics (Ichikawa & Saitoh, 2021; Ungelenk, 2021; Weise & Liehr, 2021b), and epigenetics and epigenomics (Eggermann, 2021; Harutyunyan & Hovhannisyan, 2021). The description of topologically associating domains (TADs) and associated extensive intrachromosomal and interchromosomal interactions involved in gene regulation was one of the major insights in genome biology during the last 10 years (Dixon et al., 2015; Yumiceba et al., 2021). It is clear by now that structural chromosomal rearrangements may lead to TAD alterations being associated with human disease (Franke et al., 2016; Lupiáñez et al., 2015). As also outlined in Liehr's Cytogenomics book (2021a), that is, this book, transcriptionally active gene loci show the tendency to be in a more central position (Liehr, 2021a), while inactive genes are located more peripherally, i.e., at the nuclear border as lamina-associated domains (LADs) (Briand & Collas, 2020).

Interchromosomal interactions in the normal human interphase nucleus have been suggested for quite some time (Bickmore, 2013); recent studies on chromosome architecture supported this idea in general terms. Work performed in the laboratory of Thomas Cremer and others (Bickmore, 2013; Cremer & Cremer, 2019; Daban, 2020, 2021; Dekker et al., 2017; Liehr, 2021a; Su et al., 2020; Yu & Ren, 2017) showed that chromosomes are organized in clearly separated chromosome territories, but at the same time, have the ability to intermingle at the borders of these territories (Branco & Pombo, 2006; Su et al., 2020). Biological functions of these contacts in general are still not well understood. One idea, coming from the observation of "chromosome kissing" (also interchromosomal interactions or nonhomologous chromosomal contacts=NHCCs) (Bastia & Singh, 2011; Cavalli, 2007; Kioussis, 2005; Spilianakis et al., 2005) seen in cells of early stages of female embryonal development includes that this contact of X-chromosomes is necessary to determine the number of these gonosomes per cell; according to the chromosome counting-result, nothing special happens (cell with one X-chromosome), or all but one X-chromosome is inactivated (cell with two or more X-chromosome) (Rinčić et al., 2016). The idea of general chromosome counting by contact between two (or more) homologues chromosomes was deduced from that as early embryonic cells have the ability to perform targeted monosomic or trisomic rescue (Liehr, 2014; Pellestor, 2014; Pellestor et al., 2021; Rinčić et al., 2016). *SLC16A2* gene family was proposed to play hereby a crucial role (Rinčić et al., 2016). Also, recently, long noncoding RNA (lncRNA) loci have been suggested to play a role in formation of "chromosome kissing" (Hacisuleyman et al., 2014; Maass et al., 2012; Rinn & Guttman, 2014; Ichikawa & Saitoh, 2021).

Methods used in interchromosomal nucleomics

Available approaches to study interchromosomal interactions are summarized by Su et al. (2020) as two ways to address the problem. First there are "high-throughput chromosome conformation capture methods, such as Hi-C, genome architecture mapping, and other sequencing-based methods, which have revealed chromatin structures such as domains and compartments with a genome-wide view. Although these high-throughput sequencing-based approaches are powerful and have greatly enriched our knowledge of 3D genome organization, they provide contact information for pairs of chromatin loci but not direct spatial position information of individual loci. Furthermore, most of the genome-wide insights into chromatin organization are built on population-averaged contact maps across millions of cells" (Su et al., 2020). Besides, there exists the second way to approach the question for interchromosomal interactions, i.e., to do investigations of three-dimensional genome organization at single-cell level. This is possible by single-cell Hi-C techniques or by imaging-based approaches. The latter enable "a direct measure of the spatial positions of chromatin loci in individual cells with high detection efficiency. In particular, fluorescence in situ hybridization (FISH) technique provides highly specific detection of chromatin loci in fixed cells, and the clustered regularly interspaced short palindromic repeats (CRISPR) system substantially enhances our ability to image specific chromatin loci in live cells. However, current imaging methods have limited throughput in sequence space, traditionally allowing the study of only a few genomic loci at a time" (Su et al., 2020).

Accordingly, even though being laborious, FISH, optimally in three-dimensionally preserved interphase nuclei (Steinhaeuser et al., 2002), is presently the best and most mature single-cell oriented approach to study chromosomal interactions. While there are since recently also multiscale multiplexed FISH imaging technologies available, enabling "simultaneous imaging of hundreds to more than 1,000 distinct genomic loci at various resolutions and genomic coverages in single cells" (Su et al., 2020), it can also be worthwhile to concentrate on single loci, especially to answer questions for the meaning of a specific interchromosomal interaction for a certain disease (Lakadamyali & Cosma, 2020; Maass et al., 2018). While in 2016, Krijger and de Laat still stated: "The actual existence of cooperative 3D interactions between enhancers and genes is still debated" (Krijger & De Laat, 2016), recent own studies proved that such interactions obviously exist and have, if disrupted, deleterious effects for the carrier (Maass et al., 2018; Manvelyan et al., 2009; Othman et al., 2012).

Insights from two own studies
Manvelyan et al. (2009) and Othman et al. (2012)

In the first here summarized study, FISH in 3-D preserved bone marrow cells of healthy and persons diseased with leukemia was studied (Manvelyan et al., 2009). It could be shown that in both kinds of bone marrow cells studied, chromosomes 8 and 21 were colocalized. Deduced from that study, it was suggested that this colocalization could drive acute myelogenous leukemia type M2 (AML-M2) by promoting the typical translocation t(8;21)(q22;q22). Our initial research in 2009 was done using multicolor banding probes; a second study applying locus-specific probes for the corresponding involved genes *RUNX1* and *RUNXT1* in 8q22 and 21q22, respectively, confirmed this suggestion (Othman et al., 2012, Fig. 17.1). However, the reason for colocalization yet remains enigmatic and it is unclear why this interchromosomal interaction is taking place. Here also not necessarily RUNX1 and RUNXT1, but other nearby genes may indeed interact and the translocation t(8;21) is just an unfortunate by-product.

Maass et al. (2018)

"Here, we investigated the interchromosomal interactions between chromosomes 2q, 12, and 17 in human mesenchymal stem cells (MSCs) and MSC-derived cell types by DNA-FISH. We compared our findings in normal karyotypes with a three-generation

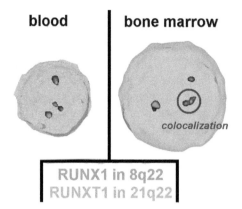

FIG. 17.1

Colocalization of 8q22 and 21q22 in bone marrow but not in lymphocytes. Interphase-FISH results applying locus-specific probes for *RUNX1* in 8q22 and *RUNXT1* in 21q22 in normal nuclei derived from blood and bone marrow. Only in bone marrow cells, colocalization of 8q22 and 21q22 were observed.

family harboring a 2q37-deletion syndrome, featuring a heterozygous partial deletion of histone deacetylase 4 (*HDAC4*) on chr2q37. In normal karyotypes, we detected stable, recurring arrangements and interactions between the three chromosomal territories with a tissue-specific interaction bias at certain loci. These interchromosomal interactions were confirmed by Hi-C. Interestingly, the disease-related HDAC4 deletion resulted in displaced interchromosomal arrangements and altered interactions between the deletion-affected chromosome 2 and chromosome 12 and/or 17 in 2q37-deletion syndrome patients. Our findings provide evidence for a direct link between a structural chromosomal aberration and altered interphase architecture that results in a nuclear configuration, supporting a possible molecular pathogenesis" (Maass et al., 2018). Fig. 17.2 shows a typical FISH result from that study.

Outlook

As nicely stated by Krijger and De Laat (2016): "Studying disease-associated mutations and chromosomal rearrangements in the context of the 3D genome will enable the identification of dysregulated target genes and aid the progression from descriptive genetic association results to discovering molecular mechanisms underlying disease" (Krijger & De Laat, 2016). For sure there is still much more out there in our genome and in that of other species, where interchromosomal interactions play a role, in normal cell regulation and in disease, as "intact nuclear locus positioning in chromatin is important to ensure gene regulation either by intrachromosomal, interchromosomal regulation, or by associated proteins" (Maass et al., 2018).

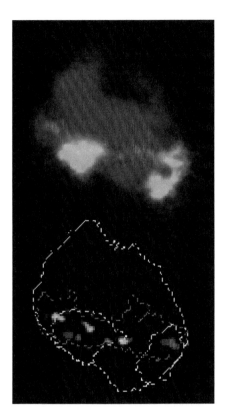

FIG. 17.2

Multicolor FISH to classify NHCCs. In whole chromosome painting probes, chromosome 12 and 17 (in upper part in green and red) were cohybridized with locus-specific probes (in lower part). Probes are specific for *PTHLH* (in 12p11.22, blue), *CISTR-ACT* (in 12q13.13, red), and *ETV4, HOXB, NOG,* and *SOX9* (in 17q24.3, green). Results like shown here were used to study the colocalization of chromosome territories 2q, 12, and 17 in human mesenchymal stem cells.

References

Baranov, V. S., & Kuznetzova, T. V. (2021). Nuclear stability in early embryo. Chromosomal aberrations. In T. Liehr (Ed.), *Cytogenomics* (pp. 307–325). Academic Press. Chapter 15 (in this book).

Bastia, D., & Singh, S. K. (2011). "Chromosome kissing" and modulation of replication termination. *BioArchitecture, 1*(1), 24–28. https://doi.org/10.4161/bioa.1.1.14664.

Bickmore, W. A. (2013). The spatial organization of the human genome. *Annual Review of Genomics and Human Genetics, 14,* 67–84. https://doi.org/10.1146/annurev-genom-091212-153515.

Bisht, P., & Avarello, M. D. M. (2021). Molecular combing solutions to characterize replication kinetics and genome rearrangements. In T. Liehr (Ed.), *Cytogenomics* (pp. 47–71). Academic Press. Chapter 5 (in this book).

Branco, M. R., & Pombo, A. (2006). Intermingling of chromosome territories in interphase suggests role in translocations and transcription-dependent associations. *PLoS Biology*, *4*(5), 780–788. https://doi.org/10.1371/journal.pbio.0040138.

Briand, N., & Collas, P. (2020). Lamina-associated domains: Peripheral matters and internal affairs. *Genome Biology*, *21*(1), 85. https://doi.org/10.1186/s13059-020-02003-5.

Cavalli, G. (2007). Chromosome kissing. *Current Opinion in Genetics and Development*, *17*(5), 443–450. https://doi.org/10.1016/j.gde.2007.08.013.

Cremer, M., & Cremer, T. (2019). Nuclear compartmentalization, dynamics, and function of regulatory DNA sequences. *Genes, Chromosomes and Cancer*, *58*(7), 427–436. https://doi.org/10.1002/gcc.22714.

Daban, J. R. (2020). Supramolecular multilayer organization of chromosomes: Possible functional roles of planar chromatin in gene expression and DNA replication and repair. *FEBS Letters*, *594*(3), 395–411. https://doi.org/10.1002/1873-3468.13724.

Daban, J.-R. (2021). Multilayer organization of chromosomes. In T. Liehr (Ed.), *Cytogenomics* (pp. 267–296). Academic Press. Chapter 13 (in this book).

Dekker, J., Belmont, A. S., Guttman, M., Leshyk, V. O., Lis, J. T., Lomvardas, S., … Zhong, S. (2017). The 4D nucleome project. *Nature*, *549*(7671), 219–226. https://doi.org/10.1038/nature23884.

Delpu, Y., Barseghyan, H., Bocklandt, S., Hastie, A., & Chaubey, A. (2021). Next-generation cytogenomics: High-resolution structural variation detection by optical genome mapping. In T. Liehr (Ed.), *Cytogenomics* (pp. 123–146). Academic Press. Chapter 8 (in this book).

Dixon, J. R., Jung, I., Selvaraj, S., Shen, Y., Antosiewicz-Bourget, J. E., Lee, A. Y., … Ren, B. (2015). Chromatin architecture reorganization during stem cell differentiation. *Nature*, *518*(7539), 331–336. https://doi.org/10.1038/nature14222.

Eggermann, T. (2021). Epigenetics. In T. Liehr (Ed.), *Cytogenomics* (pp. 389–401). Academic Press. Chapter 20 (in this book).

Franke, M., Ibrahim, D. M., Andrey, G., Schwarzer, W., Heinrich, V., Schöpflin, R., … Mundlos, S. (2016). Formation of new chromatin domains determines pathogenicity of genomic duplications. *Nature*, *538*(7624), 265–269. https://doi.org/10.1038/nature19800.

Hacisuleyman, E., Goff, L. A., Trapnell, C., Williams, A., Henao-Mejia, J., Sun, L., … Rinn, J.L. (2014). Topological organization of multichromosomal regions by the long intergenic noncoding RNA firre. *Nature Structural and Molecular Biology*, *21*(2), 198–206. https://doi.org/10.1038/nsmb.2764.

Harutyunyan, T., & Hovhannisyan, G. (2021). Approaches for studying epigenetic aspects of the human genome. In T. Liehr (Ed.), *Cytogenomics* (pp. 155–209). Academic Press. Chapter 10 (in this book).

Ichikawa, Y., & Saitoh, N. (2021). Shaping of genome by long noncoding RNAs. In T. Liehr (Ed.), *Cytogenomics* (pp. 357–372). Academic Press. Chapter 18 (in this book).

Iourov, I. Y., Vorsanova, S. G., & Yurov, Y. B. (2021). Cytogenomic landscape of the human brain. In T. Liehr (Ed.), *Cytogenomics* (pp. 327–348). Academic Press. Chapter 16 (in this book).

Ishii, T., Nagaki, K., & Houben, A. (2021). Application of CRISPR/Cas9 to visualize defined genomic sequences in fixed chromosomes and nuclei. In T. Liehr (Ed.), *Cytogenomics* (pp. 147–153). Academic Press. Chapter 9 (in this book).

Kioussis, D. (2005). Gene regulation: Kissing chromosomes. *Nature*, *435*(7042), 579–580. https://doi.org/10.1038/435579a.

Krijger, P. H. L., & De Laat, W. (2016). Regulation of disease-associated gene expression in the 3D genome. *Nature Reviews Molecular Cell Biology*, *17*(12), 771–782. https://doi.org/10.1038/nrm.2016.138.

Lakadamyali, M., & Cosma, M. P. (2020). Visualizing the genome in high resolution challenges our textbook understanding. *Nature Methods*, *17*(4), 371–379. https://doi.org/10.1038/s41592-020-0758-3.

Liehr, T. (2014). *Uniparental disomy (UPD) in clinical genetics. A guide for clinicians and patients* (1st ed.). Springer.

Liehr, T. (2021a). Nuclear architecture. In T. Liehr (Ed.), *Cytogenomics* (pp. 297–305). Academic Press. Chapter 14 (in this book).

Liehr, T. (2021b). A definition for cytogenomics - Which also may be called chromosomics. In T. Liehr (Ed.), *Cytogenomics* (pp. 1–7). Academic Press. Chapter 1 (in this book).

Liehr, T. (2021c). Overview of currently available approaches used in cytogenomics. In T. Liehr (Ed.), *Cytogenomics* (pp. 11–24). Academic Press. Chapter 2 (in this book).

Liehr, T. (2021d). Molecular cytogenetics. In T. Liehr (Ed.), *Cytogenomics* (pp. 35–45). Academic Press. Chapter 4 (in this book).

Liehr, T. (2021e). Repetitive elements, heteromorphisms, and copy number variants. In T. Liehr (Ed.), *Cytogenomics* (pp. 373–388). Academic Press. Chapter 19 (in this book).

Lupiáñez, D. G., Kraft, K., Heinrich, V., Krawitz, P., Brancati, F., Klopocki, E., … Mundlos, S. (2015). Disruptions of topological chromatin domains cause pathogenic rewiring of gene-enhancer interactions. *Cell*, *161*(5), 1012–1025. https://doi.org/10.1016/j.cell.2015.04.004.

Maass, P. G., Rump, A., Schulz, H., Stricker, S., Schulze, L., Platzer, K., … Bähring, S. (2012). A misplaced lncRNA causes brachydactyly in humans. *Journal of Clinical Investigation*, *122*(11), 3990–4002. https://doi.org/10.1172/JCI65508.

Maass, P. G., Weise, A., Rittscher, K., Lichtenwald, J., Barutcu, A. R., Liehr, T., … Bähring, S. (2018). Reorganization of inter-chromosomal interactions in the 2q37-deletion syndrome. *EMBO Journal*, *37*(15). https://doi.org/10.15252/embj.201696257, e96257.

Manvelyan, M., Kempf, P., Weise, A., Mrasek, K., Heller, A., Lier, A., … Mkrtcyhan, H. (2009). Preferred co-localization of chromosome 8 and 21 in myeloid bone marrow cells detected by three dimensional molecular cytogenetics. *International Journal of Molecular Medicine*, *24*(3), 335–341. https://doi.org/10.3892/ijmm_00000237.

Othman, M. A. K., Lier, A., Junker, S., Kempf, P., Dorka, F., Gebhart, E., … Manvelyan, M. (2012). Does positioning of chromosomes 8 and 21 in interphase drive t(8;21) in acute myelogenous leukemia? *BioDiscovery*, *4*(1), 4.

Pederson, T. (2011). The nucleus introduced. *Cold Spring Harbor Perspectives in Biology*, *3*(5), 1–16. https://doi.org/10.1101/cshperspect.a000521.

Pellestor, F. (2014). Chromothripsis: How does such a catastrophic event impact human reproduction? *Human Reproduction*, *29*(3), 388–393. https://doi.org/10.1093/humrep/deu003.

Pellestor, F., Gaillard, J.-B., Schneider, A., Puechberty, J., & Gatinois, V. (2021). Chromoanagenesis phenomena and their formation mechanisms. In T. Liehr (Ed.), *Cytogenomics* (pp. 213–245). Academic Press. Chapter 11 (in this book).

Rinčić, M., Iourov, I. Y., & Liehr, T. (2016). Thoughts about SLC16A2, TSIX and XIST gene like sites in the human genome and a potential role in cellular chromosome counting. *Molecular Cytogenetics*, *9*(1), 56. https://doi.org/10.1186/s13039-016-0271-7.

Rinn, J., & Guttman, M. (2014). RNA and dynamic nuclear organization. *Science*, *345*(1), 1240–1241. https://doi.org/10.1126/science.1252966.

Spilianakis, C. G., Lalioti, M. D., Town, T., Lee, G. R., & Flavell, R. A. (2005). Interchromosomal associations between alternatively expressed loci. *Nature, 435*(7042), 637–645. https://doi.org/10.1038/nature03574.

Steinhaeuser, U., Starke, H., Nietzel, A., Lindenau, J., Ullmann, P., Claussen, U., & Liehr, T. (2002). Suspension (S)-FISH, a new technique for interphase nuclei. *Journal of Histochemistry and Cytochemistry, 50*(12), 1697–1698. https://doi.org/10.1177/002215540205001216.

Su, J. H., Zheng, P., Kinrot, S. S., Bintu, B., & Zhuang, X. (2020). Genome-scale imaging of the 3D organization and transcriptional activity of chromatin. *Cell, 182*(6). https://doi.org/10.1016/j.cell.2020.07.032. 1641–1659.e26.

Ungelenk, M. (2021). Sequencing approaches. In T. Liehr (Ed.), *Cytogenomics* (pp. 87–122). Academic Press. Chapter 7 (in this book).

Weise, A., & Liehr, T. (2021a). Cytogenetics. In T. Liehr (Ed.), *Cytogenomics* (pp. 25–34). Academic Press. Chapter 3 (in this book).

Weise, A., & Liehr, T. (2021b). Molecular karyotyping. In T. Liehr (Ed.), *Cytogenomics* (pp. 73–85). Academic Press. Chapter 6 (in this book).

Yu, M., & Ren, B. (2017). The three-dimensional organization of mammalian genomes. *Annual Review of Cell and Developmental Biology, 33*, 265–289. https://doi.org/10.1146/annurev-cellbio-100616-060531.

Yumiceba, V., Souto Melo, U., & Spielmann, M. (2021). 3D cytogenomics: Structural variation in the three-dimensional genome. In T. Liehr (Ed.), *Cytogenomics* (pp. 247–266). Academic Press. Chapter 12 (in this book).

Shaping of genome by long noncoding RNAs

18

Yuichi Ichikawa and Noriko Saitoh

Division of Cancer Biology, The Cancer Institute of JFCR, Tokyo, Japan

Chapter outline

Introduction.. 357
Technologies to determine genome-wide RNA-chromatin interactions......................... 360
    Chromatin isolation by RNA purification (ChIRP), followed by
    high-throughput sequencing (ChIRP-seq).. 360
    Capture hybridization analysis of RNA targets 360
    Proximity ligation-based methods like RADICL-seq 361
Nuclear lncRNAs that recruit chromatin modifiers.. 362
Nuclear lncRNAs that promote long-range chromatin interactions............................... 363
Interchromosomal interactions via subnuclear structures... 364
Breast cancer-specific lncRNAs regulate a long-range chromatin interaction.............. 366
Conclusion 367
Acknowledgments.. 368
References.. 368

Introduction

In eukaryotes, genomic DNA is packaged in chromatin, a DNA-protein complex (Fig. 18.1) being the major topic of cytogenomics (Liehr, 2021a, 2021b). The basic unit of the chromatin architecture is the nucleosome, which consists of approximately 147 bp of DNA wrapped around a core histone octamer including two copies of each of the four histone proteins H2A, H2B, H3, and H4 (Luger et al., 1997). Nucleosome arrays, which resemble beads on a string, are associated with various nuclear factors and form chromatin fibers. In living cells, the chromatin fiber is not uniform. The lightly packed form of chromatin is called euchromatin, whereas the tightly packed form is called heterochromatin (Heitz, 1928; Janssen et al., 2018; Liehr, 2021c).

In the cell nucleus, chromatin fibers are folded into multiple layers with a hierarchical order: chromatin loops, topologically associating domains (TADs), A and B compartments, and chromosome territories (Fig. 18.1) (Daban, 2021; Dekker et al., 2013;

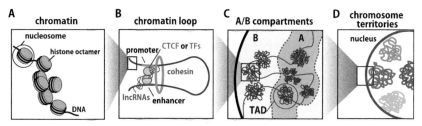

FIG. 18.1

Hierarchies in chromatin structures in the nucleus. The eukaryotic genomic DNA is folded into multiple layers and stored in the cell nucleus. (A) Genomic DNA forms nucleosomes with a histone octamer, and the chromatin structure is composed of a series of nucleosomes. (B) Chromatin forms a loop structure. One example is an enhancer-promoter interaction. Various factors, such as CCCTC-binding factor (CTCF), cohesin complex, transcription factors (TFs), and long noncoding RNAs (lncRNAs) are involved in chromatin loop formation. (C) Megabase-sized chromatin domains, topologically associating domains (TADs), and A/B compartments are formed. (D) Chromosome territories are regions occupied by each chromosome in the nucleus.

Liehr, 2021d; Weise & Liehr, 2021a; Yumiceba et al., 2021). These structural and spatial organizations are closely linked to gene expression regulation.

Transcriptionally active genes are often enriched in euchromatin, while silenced genes are usually present in heterochromatin. Chromatin loop formation may compartmentalize genes to separate them from other genes with a different expression program. Local chromatin loops are also formed for enhancer-promoter interactions and transcription activation (Rao et al., 2014; Weise & Liehr, 2021a). TAD is a megabase sized chromatin domain that was detected by a technology for genome-wide chromatin interactions, Hi-C. TAD is relatively conserved among species and different cell types, and the genes in the same TAD tend to be co-regulated. A principal component analysis of the Hi-C data revealed two kinds of TADs: A and B compartments, corresponding to active chromatin (euchromatin) and inactive chromatin (heterochromatin), respectively (Dixon et al., 2012; Lieberman-Aiden et al., 2009). Many kinds of nuclear bodies consist of RNAs and proteins, including the nucleolus and paraspeckles, which are associated with chromatin. Each chromosome occupies its distinct space in the nucleus, called the chromosome territory. Chromosomes with high gene density and strong transcription activities tend to be localized in the center of the nucleus, while those with low gene density and minimal transcription activity tend to be at the nuclear periphery, which is called radial positioning (Liehr, 2021d; Tanabe et al., 2002).

What shapes the genome for gene regulation? Over 50 years ago, it was reported that chromosomes contain RNA, which is associated with chromosomal proteins such as histones (Bonner & Widholm, 1967). However, the functions of nuclear noncoding RNAs have remained elusive for a long time.

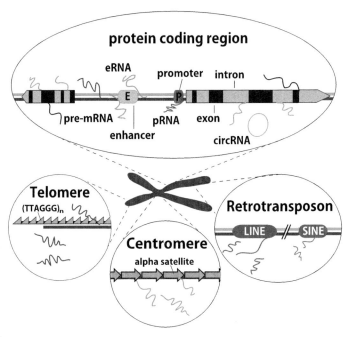

FIG. 18.2

Genomic regions that produce long noncoding RNAs. Almost all (more than 70%) of the genome is transcribed, and various types of lncRNAs are produced. For convenience, we have used a diagram of a mitotic chromosome (see also Weise & Liehr, 2021b) to show the positions where representative types of lncRNAs are transcribed.

Recent studies have revealed that almost all parts of the genome are transcribed, and estimated that over 100,000 noncoding RNAs lacking protein coding potential are produced in cells (Fig. 18.2) (Zhao et al., 2016). For example, enhancer-associated RNA (eRNA) and promoter-associated RNA (pRNA) are transcribed from enhancer and promoter regions, respectively (Kung et al., 2013). Introns in pre-mRNAs are usually removed and degraded quickly by splicing, but sometimes circular RNAs (circRNAs) containing intron sequences are produced by a mechanism called back-splicing. About 40% of the human genome consists of DNA sequences derived from retrotransposons, including long interspersed elements (LINEs) and short interspersed elements (SINEs). These sequences are also transcribed as noncoding RNAs and contribute to transcriptional regulation (Kung et al., 2013; Liehr, 2021c; Statello et al., 2020). Repetitive DNA sequences, such as centromeres and telomeres, are also transcribed and the produced noncoding RNAs are involved in the maintenance of chromatin structure (Kung et al., 2013; Liehr, 2021c; McNulty et al., 2017; Statello et al., 2020).

Noncoding RNAs (ncRNAs) are technically grouped into two types, small ncRNAs and long ncRNAs (lncRNAs), which are shorter and longer than 200 nt, respectively. LncRNAs are transcribed in developmental stage-, tissue- and disease-specific manners, and many of them are localized in the cell nucleus, suggesting that they are important factors for gene regulation in organisms (Statello et al., 2020; Tachiwana et al., 2020; Yamamoto & Saitoh, 2019).

In this chapter, we first introduce recently developed technologies to investigate genome-wide RNA-chromatin interactions. These were critical for facilitating the discovery of the molecular mechanisms of lncRNAs, as described in the following section. We then focus on some nuclear lncRNAs and their various functions in gene regulation, mainly through modulating epigenetic marks and chromatin architecture. Furthermore, we focus on roles of a specific lncRNA, ELEANOR, in breast cancer and discuss the potential of lncRNAs as diagnostic and therapeutic targets for disease.

Technologies to determine genome-wide RNA-chromatin interactions

As described above, lncRNAs tend to be retained in the cell nucleus, where they interact with chromatin to regulate its functions, such as transcription. To investigate where the lncRNAs bind on the genome, and to determine their molecular mechanisms, several technologies have been developed (Bonetti et al., 2020; Chu et al., 2011; Kato & Carninci, 2020; Simon et al., 2011). The methods have identified genome-wide lncRNA binding sites and they can be divided into two main categories, according to their methodological differences: hybridization-based methods or proximity ligation-based methods (Fig. 18.3).

Chromatin isolation by RNA purification (ChIRP), followed by high-throughput sequencing (ChIRP-seq)

The hybridization-based methods use oligonucleotide probes containing complementary sequences to pull-down the RNA of interest and its genomic binding sites. Among the methods in this category, the most widely used may be Chromatin Isolation by RNA Purification (ChIRP), followed by high-throughput sequencing (ChIRP-seq) (Chu et al., 2011; for sequencing, see also Ungelenk, 2021). In this method, cells are cross-linked with glutaraldehyde or formaldehyde to stabilize RNA-chromatin complexes, and then sonicated to fragment the chromatin. The resulting solubilized target RNA-chromatin complexes are captured with specific probes. For ChIRP-seq, multiple DNA probes, tiling the RNA of interest, are required and split into two-probe sets. Performing ChIRP-seq with different probe sets is recommended to ensure specificity. The multiple probes allow ChIRP-seq to be performed without prior information about the secondary structure of the target RNA, which can inhibit the hybridization step. ChIRP is also useful for detecting RNA-protein interactions in combination with immunoblotting or mass spectrometry analyses.

Capture hybridization analysis of RNA targets

Capture hybridization analysis of RNA targets (CHART) (Simon et al., 2011), which is conceptually similar to ChIRP. The major difference is that a ribonuclease

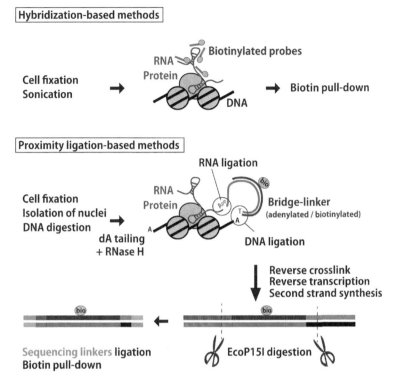

Hybridization-based methods

Cell fixation
Sonication

RNA
Protein
Biotinylated probes
DNA

Biotin pull-down

Proximity ligation-based methods

RNA ligation

Cell fixation
Isolation of nuclei
DNA digestion
dA tailing
+ RNase H

RNA
Protein

Bridge-linker
(adenylated / biotinylated)

DNA ligation

Reverse crosslink
Reverse transcription
Second strand synthesis

Sequencing linkers ligation
Biotin pull-down

EcoP15I digestion

FIG. 18.3

Methods for investigating RNA-chromatin interactions. As a hybridization-based method, a schematic representation of the ChIRP protocol is shown (upper) (Chu et al., 2011). As a proximity ligation-based method, a schematic representation of RADICL-seq is shown (bottom) (Bonetti et al., 2020). Abbreviations: dA: deoxyadenosine, A: 3′-adenine overhang, T: 3′-thymine overhang, App: 5′-adenyl pyrophosphoryl moiety, bio: biotin.

H (RNaseH) sensitivity assay is included to identify the regions accessible to the target RNA, to design better CHART probes.

Proximity ligation-based methods like RADICL-seq

The proximity ligation-based methods are effective to investigate the chromatin-associated RNAs in all chromatin interactions. Before 2020, several reported methods used a bivalent linker which can bridge lncRNAs and genomic DNA (Bonetti et al., 2020; Kato & Carninci, 2020; Li et al., 2017; Sridhar et al., 2017). The latest method is RNA and DNA interacting complexes ligated and sequenced (RADICL-seq) (Bonetti et al., 2020). In this method, cells are cross-linked with formaldehyde and lysed, and then nuclei are isolated. To maintain the integrity of the nuclei and the spatial information of the RNA-chromatin complex, the chromatin is digested with deoxyribonuclease I (DNase I), rather than sonicated. After chromatin end-repair, deoxyadenosine-tailing, and RNase H treatment, the preadenylated side of the bridge-linker is ligated to the 3′ end of the RNAs and the other side is ligated to the genomic DNA ends. The RNA-DNA chimera is converted into a fully double-stranded DNA molecule by reverse transcription, followed by second strand synthesis. The resulting products are digested by the restriction enzyme *Eco*P15I to a designated length from the two recognition sites within the bridge-linker, and ligated to sequencing linkers, followed by biotin-pull down. A unique feature of RADICL-seq is the RNase H treatment which can reduce the amount of abundant nascent transcripts by digesting RNA-DNA hybrids, thus improving the capture rates of other types of RNA–DNA interactions.

Nuclear lncRNAs that recruit chromatin modifiers

LncRNAs play important roles in both gene repression and activation through recruiting chromatin modifiers. The most intensively studied case of a repressive lncRNA is the X-inactive specific transcript (*Xist*/*XIST* in mice and humans, respectively) which inactivates the X-chromosome (Borsani et al., 1991; Brown, 1991; Brown et al., 1992). Mammalian females have two X-chromosomes, carrying over 1000 genes essential for development and cell viability. Early in development, one of the X-chromosomes becomes highly condensed and transcriptionally inactivated (Fig. 18.4). In this way, the transcript levels of X-linked genes are adjusted to those in males with only one X chromosome (Boumil & Lee, 2001; Liehr, 2021d; Lyon, 1961). In this process, *Xist* is transcribed from the region called Xic (X chromosome inactivation center), spreads along the X-chromosome in cis, and eventually covers the entire X-chromosome to inactivate it (Fig. 18.4).

Xist interacts with heterogeneous nuclear ribonucleoprotein K (HNRNPK) and recruits the polycomb repressive complex 1 (PRC) for the ubiquitination of histone H2A at lysine 119, leading to the accumulation of PRC2 for the trimethylation of histone H3 at lysine 27 (H3K27me3) (Fig. 18.4) (Almeida et al., 2017; Boumil & Lee, 2001; Bousard et al., 2019; Pintacuda et al., 2017; Zhao et al., 2008). One of the hybridization-based methods, the RNA antisense purification (RAP) method, revealed that Xist also interacts with SMRT/HDAC1- associated repressor protein (SHARP) and recruits histone deacetylase 3 (HDAC3) to remove active histone marks such as the acetylation of histone H3 at lysine 27 (H3K27ac) (Fig. 18.4) (McHugh et al., 2015). These epigenetic regulations (see also Eggermann, 2021; Harutyunyan & Hovhannisyan, 2021) contribute to the formation of repressive heterochromatin. The inactive X-chromosome is often observed near the nuclear membrane or the nucleolus, suggesting that these spatial features are also related to the inactivation (Fritz et al., 2019; Zhang et al., 2007).

Some long noncoding RNAs contribute to the activation of transcription. For example, *HOXA* transcript at the distal tip (HOTTIP) is transcribed from the 5′ end of the homeobox A cluster (HOXA gene cluster) and plays a role in the activation of the target *HOXA* genes (Wang et al., 2011). HOTTIP recruits the mixed-lineage

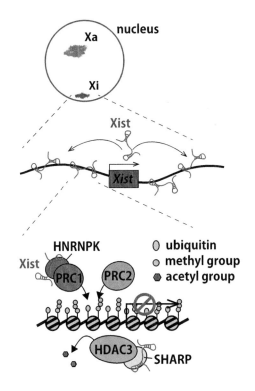

FIG. 18.4

X-chromosome inactivation mediated by Xist lncRNA. The inactive X-chromosome (Xi) has a peripheral location within the nucleus relative to the active X-chromosome (Xa) counterpart. *Xist* is transcribed from Xi, spreads along the Xi in cis, and recruits chromatin modifiers to form a silenced chromatin state.

leukemia protein complex (MLL complex) by a physical interaction with the adaptor protein WDR5 and the MLL complex catalyzes the trimethylation of histone H3 at lysine 4 (H3K4me3) to activate the target genes (Yang et al., 2014).

Nuclear lncRNAs that promote long-range chromatin interactions

LncRNAs regulate gene expression by mediating higher order chromatin structures such as chromatin loops. For example, HOTTIP promotes the formation of a chromosomal loop, which enables the delivery of HOTTIP from its site of transcription over a long distance to the target HOXA genes (Kung et al., 2013; Statello et al., 2020). Like HOTTIP, enhancer-associated RNAs (eRNAs), which are lncRNAs transcribed from enhancer regions, are thought to promote such long-range chromosomal interactions between enhancers and promoters (Kung et al., 2013; Statello et al., 2020).

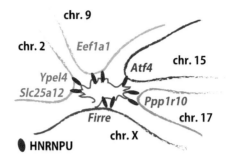

FIG. 18.5

Trans-chromosomal assembly mediated by Firre lncRNA. The Firre RNA is transcribed from the inactive X-chromosome (chr.X) and mediates interactions of multiple gene loci located on different chromosomes via physical association with HNRNPU.

HOTTIP and eRNAs promote long-range chromatin interactions within the same chromosome in cis, while some lncRNAs serve as a platform for trans-chromosomal associations. In mice, functional intergenic repeating RNA element (Firre), encoded on the X-chromosome, escapes X-chromosome inactivation, and the Firre lncRNA promotes interactions between its own transcription site and other genomic loci, encoding *Ypel4*, *Slc25a12*, *Eef1a1*, *Ppp1r10*, and *Aft4* through physical binding with heterogeneous nuclear ribonucleoprotein U (HNRNPU) (Fig. 18.5) (Hacisuleyman et al., 2014; Lewandowski et al., 2019).

In human cells, the Firre locus also interacts with *ATF4* and *YPEL4* on other chromosomes indicating that the trans-acting feature of the Firre lncRNA is conserved between mice and humans (Maass et al., 2018, 2019). In addition to its role in chromosomal interactions, Firre also participates in anchoring the inactive X-chromosome to the nucleolus or nuclear periphery to maintain its repressed status (Yang et al., 2015).

Interchromosomal interactions via subnuclear structures

LncRNAs have been proposed to establish seeding platforms to trigger the recruitment and retention of components of subnuclear structures (Dundr, 2011; Mao et al., 2011; Shevtsov & Dundr, 2011). Subnuclear structures contain various factors involved in transcription, replication, and repair, and regulate chromatin functions by providing nuclear microenvironments. Among the diverse subnuclear structures, the most prominent is the nucleolus, which can be observed with an optical microscope and functions as a site for ribosome assembly (Lafontaine et al., 2021). In the nucleolus, repeats of the ribosomal RNA (rRNA) genes, called the nucleolar organizer regions (NORs), are transcribed by RNA polymerase I (Fig. 18.6A), and rRNAs are processed and assembled with ribosome proteins into mature ribosome particles, which are then exported to the cytoplasm (Ide et al., 2020; Lafontaine et al., 2021).

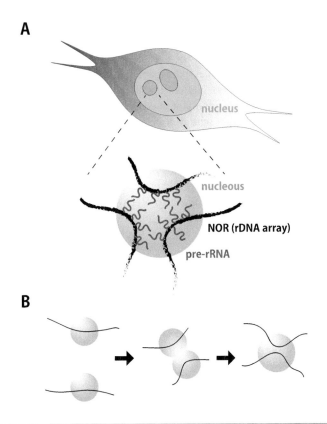

FIG. 18.6

The nucleolus serves as a platform for trans-chromosomal associations. (A) Several NORs (rDNA array) are assembled within the nucleolus. (B) Nucleoli exhibit liquid droplet-like properties in vitro and in *Xenopus*. Small nucleoli can fuse with each other and form larger nucleoli.

In humans, NORs are located on the short arms of five different loci on chromosomes 13, 14, 15, 21, and 22, and come into physical proximity in the nucleolus (Fig. 18.6A), suggesting interchromosomal interactions through the nucleolus. In addition, larger nucleoli are reportedly formed by the fusion of smaller nucleoli (Fig. 18.6B), suggesting that the fusion process may drive the clustering of multiple NORs on different chromosomes (Maass et al., 2019; Sawyer et al., 2019).

The nucleolus is a multiphase liquid condensate, which is produced through a physical phenomenon called liquid-liquid phase separation (LLPS). Typically, LLPS is driven by weak, multivalent interactions between proteins with low complexity domains or intrinsically disordered regions that have biased amino acid compositions or repetitive sequences (Banani et al., 2017; Sawyer et al., 2019). RNA itself and RNA-protein complexes are also known to form liquid droplets by LLPS and contribute to the regulation of protein viscosity and solubility (Banani et al., 2017; Sawyer et al., 2019).

Indeed, subnuclear structures with liquid droplet-like features contain lncRNA as a main component (Nozawa et al., 2020). For example, nuclear speckles and paraspeckles contain the lncRNA metastasis-associated in lung adenocarcinoma transcript 1 (MALAT1) and nuclear paraspeckle assembly transcript 1 (NEAT1), respectively (Sawyer et al., 2019). NEAT1 is essential for the formation of paraspeckles, whereas nuclear speckles are maintained in the absence of MALAT1. MALAT1 and NEAT1 reportedly interact with transcriptionally active chromatin regions, suggesting their involvement in the regulation of gene expression (Statello et al., 2020; West et al., 2014). Both of these RNAs are aberrantly expressed in cancer and have been proposed to function as oncogenes or tumor suppressor genes depending on the type of cancer (Li et al., 2018; Mello et al., 2017; Pisani & Baron, 2020). Cajal bodies and histone locus bodies are also RNA-containing subnuclear structures, and both regulate genome organization, gene transcription, and RNA processing (Sawyer et al., 2019).

Breast cancer-specific lncRNAs regulate a long-range chromatin interaction

Estrogen receptor (ER)-positive breast cancer is the most common tumor subtype; it depends on the hormone estrogen for growth, and accounts for 60%–70% of all breast cancer patients. ER is a nuclear receptor that functions as a transcription factor by binding to estrogen. Although endocrine therapy, which inhibits estrogen activity, is effective for ER-positive breast cancer, there is a serious problem that the cancer may acquire resistance to therapy during long-term treatment and recur at a high frequency. Long-term estrogen deprivation (LTED) cells are a model for recurrent breast cancer and are established by culturing ER-positive breast cancer cell lines in estrogen-depleted media for several months. An analysis of total RNA expression in LTED cells revealed that a cluster of lncRNAs, termed *ESR1* locus enhancing and activating noncoding RNAs (ELEANOR), are transcribed from the neighbor of the *ESR1* locus, which encodes the ER (Tomita et al., 2015). Once ELEANOR was knocked down, the transcription of *ESR1* and its neighboring genes was suppressed. Thus ELEANOR contributes to the transcriptional activation of the *ESR1* locus (Tomita et al., 2015). Fluorescence in situ hybridization (FISH) experiments (see also Liehr (2021e)) revealed that ELEANOR forms a clump, called an ELEANOR cloud surrounding the *ESR1* locus in the cell nucleus (Tomita et al., 2015). Inhibition of the ELEANOR cloud formation by antisense oligonucleotides or small molecule compounds suppressed the proliferation of LTED cells (Abdalla et al., 2019; Tomita et al., 2015; Yamamoto et al., 2018).

A Hi-C analysis of the LTED cells revealed that the region where ELEANOR RNAs are transcribed corresponds to approximately 0.7 Mb of TAD including the *ESR1* locus, named the ELEANOR TAD. Furthermore, the ELEANOR TAD is physically close to the TAD containing the apoptotic transcription factor forkhead box O3 (*FOXO3*) gene, which is approximately 42.9 Mb away from the ELEANOR TAD in

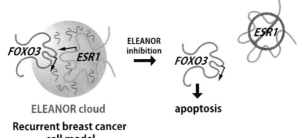

FIG. 18.7

Long-range chromatin interactions mediated by ELEANOR lncRNAs. ELEANOR lncRNAs are transcribed from a location near the *ESR1* locus, form an RNA cloud (ELEANOR cloud) in LTED cells and modulate long-range chromatin interactions between *ESR1* and *FOXO3*. *ESR1* is responsible for cell growth, and *FOXO3* participates in cell apoptosis. Upon ELEANOR inhibition, *ESR1* is suppressed, and the *ESR1-FOXO3* interaction becomes weak. Regardless of this disconnection, the *FOXO3* expression is maintained intact, thus leading to apoptosis of LTED cells.

the genome, and both the ELEANOR TAD and *FOXO3* TAD are in the transcriptionally active A-compartment (Abdalla et al., 2019). When ELEANOR was inhibited by small molecule compounds, the interaction between the two loci was weakened (Fig. 18.7). Both the *ESR1* and *FOXO3* genes were actively transcribed in LTED cells, but only *ESR1* transcription was decreased upon ELEANOR inhibition. These results indicate that the RNA cloud functions as a mediator of long-range chromatin interactions, and that the inhibition of ELEANOR function induces cell death, predominantly by balancing out the expression of *FOXO3* and *ESR1* (Fig. 18.7) (Abdalla et al., 2019).

These results suggest the potential of lncRNAs as novel therapeutic targets for cancers. Like other subnuclear structures containing lncRNAs, the ELEANOR cloud may also have droplet-like properties and use RNA-mediated droplet interactions to bring different parts of a chromosome into spatial proximity.

Conclusion

An atlas of human long noncoding RNAs was published in 2017, and it reported that 19,175 lncRNAs are potentially functional, with 1970 of them possibly involved in diseases (Hon et al., 2017). In this chapter, we focused on the representative lncRNAs that are involved in gene expression regulation through chromatin and nuclear structures, although many more have been identified both inside and outside of the nucleus. Interestingly, most of the genomic mutations found in cancer are located in noncoding regions (Rheinbay et al., 2020). Thus the elucidation of the functions of lncRNAs and their relationships to diseases is an important goal, and it may lead to

RNA-targeted therapeutics and diagnostic markers in cytogenomics. To date, RNA-sequencing at the single-cell level has been used for patient-derived tissues, and transcriptomic alteration signatures have been identified (Newman et al., 2019; Nomura, 2021). Since many lncRNAs function as epigenetic regulators through associations with chromatin, the use of RNA-chromatin interactome analyses, such as ChIRP-seq and RADICL-seq for patient-derived samples is expected to pave the way for disease treatments in the future.

Acknowledgments

We thank all of the members of Saitoh's laboratory at JFCR. This work was supported in part by JSPS KAKENHI grant numbers JP18H05531, JP18K19310, JP20H0352 (to N.S.), and JP20K15722 (to Y.I.). This work was supported by grants from the Takeda Science Foundation (to N.S.), The Vehicle Racing Commemorative Foundation (to N.S.), and a Research Grant of the Princess Takamatsu Cancer Research Fund (to N.S.).

References

Abdalla, M. O. A., Yamamoto, T., Maehara, K., Nogami, J., Ohkawa, Y., Miura, H., … Saitoh, N. (2019). The *Eleanor* ncRNAs activate the topological domain of the *ESR1* locus to balance against apoptosis. *Nature Communications, 10*, 3778. https://doi.org/10.1038/s41467-019-11378-4.

Almeida, M., Pintacuda, G., Masui, O., Koseki, Y., Gdula, M., Cerase, A., … Brockdorff, N. (2017). PCGF3/5-PRC1 initiates polycomb recruitment in X chromosome inactivation. *Science, 356*(6342), 1081–1084. https://doi.org/10.1126/science.aal2512.

Banani, S. F., Lee, H. O., Hyman, A. A., & Rosen, M. K. (2017). Biomolecular condensates: Organizers of cellular biochemistry. *Nature Reviews. Molecular Cell Biology, 18*(5), 285–298. https://doi.org/10.1038/nrm.2017.7.

Bonetti, A., Agostini, F., Suzuki, A. M., Hashimoto, K., Pascarella, G., Gimenez, J., … Carninci, P. (2020). RADICL-seq identifies general and cell type-specific principles of genome-wide RNA-chromatin interactions. *Nature Communications, 11*(1), 1018. https://doi.org/10.1038/s41467-020-14337-6.

Bonner, J., & Widholm, J. (1967). Molecular complementarity between nuclear DNA and organ-specific chromosomal RNA. *Proceedings of the National Academy of Sciences of the United States of America, 57*(5), 1379–1385. https://doi.org/10.1073/pnas.57.5.1379.

Borsani, G., Tonlorenzi, R., Simmler, M. C., Dandolo, L., Arnaud, D., Capra, V., … Ballabio, A. (1991). Characterization of a murine gene expressed from the inactive X chromosome. *Nature, 351*(6324), 325–329. https://doi.org/10.1038/351325a0.

Boumil, R. M., & Lee, J. T. (2001). Forty years of decoding the silence in X-chromosome inactivation. *Human Molecular Genetics, 10*(20), 2225–2232. https://doi.org/10.1093/hmg/10.20.2225.

Bousard, A., Raposo, A. C., Żylicz, J. J., Picard, C., Pires, V. B., Qi, Y., … da Rocha, S. T. (2019). The role of *Xist*-mediated polycomb recruitment in the initiation of X-chromosome inactivation. *EMBO Reports, 20*(10). https://doi.org/10.15252/embr.201948019, e48019.

Brown, S. D. (1991). XIST and the mapping of the X chromosome inactivation centre. *BioEssays, 13*(11), 607–612. https://doi.org/10.1002/bies.950131112.

Brown, C. J., Hendrich, B. D., Rupert, J. L., Lafrenière, R. G., Xing, Y., Lawrence, J., & Willard, H. F. (1992). The human *XIST* gene: Analysis of a 17 kb inactive X-specific RNA that contains conserved repeats and is highly localized within the nucleus. *Cell, 71*(3), 527–542. https://doi.org/10.1016/0092-8674(92)90520-m.

Chu, C., Qu, K., Zhong, F. L., Artandi, S. E., & Chang, H. Y. (2011). Genomic maps of long noncoding RNA occupancy reveal principles of RNA-chromatin interactions. *Molecular Cell, 44*(4), 667–678. https://doi.org/10.1016/j.molcel.2011.08.027.

Daban, J.-R. (2021). Multilayer organization of chromosomes. In T. Liehr (Ed.), *Cytogenomics* (pp. 267–296). Academic Press. Chapter 13 (in this book).

Dekker, J., Marti-Renom, M. A., & Mirny, L. A. (2013). Exploring the three-dimensional organization of genomes: Interpreting chromatin interaction data. *Nature Reviews. Genetics, 14*(6), 390–403. https://doi.org/10.1038/nrg3454.

Dixon, J. R., Selvaraj, S., Yue, F., Kim, A., Li, Y., Shen, Y., … Ren, B. (2012). Topological domains in mammalian genomes identified by analysis of chromatin interactions. *Nature, 485*(7398), 376–380. https://doi.org/10.1038/nature11082.

Dundr, M. (2011). Seed and grow: A two-step model for nuclear body biogenesis. *The Journal of Cell Biology, 193*(4), 605–606. https://doi.org/10.1083/jcb.201104087.

Eggermann, T. (2021). Epigenetics. In T. Liehr (Ed.), *Cytogenomics* (pp. 389–401). Academic Press. Chapter 20 (in this book).

Fritz, A. J., Sehgal, N., Pliss, A., Xu, J., & Berezney, R. (2019). Chromosome territories and the global regulation of the genome. *Genes, Chromosomes & Cancer, 58*(7), 407–426. https://doi.org/10.1002/gcc.22732.

Hacisuleyman, E., Goff, L. A., Trapnell, C., Williams, A., Henao-Mejia, J., Sun, L., … Rinn, J.L. (2014). Topological organization of multichromosomal regions by the long intergenic noncoding RNA firre. *Nature Structural & Molecular Biology, 21*(2), 198–206. https://doi.org/10.1038/nsmb.2764.

Harutyunyan, T., & Hovhannisyan, G. (2021). Approaches for studying epigenetic aspects of the human genome. In T. Liehr (Ed.), *Cytogenomics* (pp. 155–209). Academic Press. Chapter 10 (in this book).

Heitz, E. (1928). Das Heterochromatin der Moose. *Jahrbucher fur Wissenschaftliche Botanik, 69*, 762–818.

Hon, C. C., Ramilowski, J. A., Harshbarger, J., Bertin, N., Rackham, O. J., Gough, J., … Forrest, A.R. (2017). An atlas of human long non-coding RNAs with accurate 5′ ends. *Nature, 543*(7644), 199–204. https://doi.org/10.1038/nature21374.

Ide, S., Imai, R., Ochi, H., & Maeshima, K. (2020). Transcriptional suppression of ribosomal DNA with phase separation. *Science Advances, 6*(42). https://doi.org/10.1126/sciadv.abb5953, eabb5953.

Janssen, A., Colmenares, S. U., & Karpen, G. H. (2018). Heterochromatin: Guardian of the genome. *Annual Review of Cell and Developmental Biology, 34*, 265–288. https://doi.org/10.1146/annurev-cellbio-100617-062653.

Kato, M., & Carninci, P. (2020). Genome-wide technologies to study RNA-chromatin interactions. *Noncoding RNA, 6*(2), 20. https://doi.org/10.3390/ncrna6020020.

Kung, J. T., Colognori, D., & Lee, J. T. (2013). Long noncoding RNAs: Past, present, and future. *Genetics, 193*(3), 651–669. https://doi.org/10.1534/genetics.112.146704.

Lafontaine, D. L. J., Riback, J. A., Bascetin, R., & Brangwynne, C. P. (2021). The nucleolus as a multiphase liquid condensate. *Nature Reviews. Molecular Cell Biology*. https://doi.org/10.1038/s41580-020-0272-6.

Lewandowski, J. P., Lee, J. C., Hwang, T., Sunwoo, H., Goldstein, J. M., Groff, A. F., … Rinn, J. L. (2019). The *firre* locus produces a *trans*-acting RNA molecule that functions in hematopoiesis. *Nature Communications*, *10*(1), 5137. https://doi.org/10.1038/s41467-019-12970-4.

Li, X., Zhou, B., Chen, L., Gou, L. T., Li, H., & Fu, X. D. (2017). GRID-seq reveals the global RNA-chromatin interactome. *Nature Biotechnology*, *35*(10), 940–950. https://doi.org/10.1038/nbt.3968.

Li, Z. X., Zhu, Q. N., Zhang, H. B., Hu, Y., Wang, G., & Zhu, Y. S. (2018). MALAT1: A potential biomarker in cancer. *Cancer Management and Research*, *10*, 6757–6768. https://doi.org/10.2147/CMAR.S169406.

Lieberman-Aiden, E., van Berkum, N. L., Williams, L., Imakaev, M., Ragoczy, T., Telling, A., … Dekker, J. (2009). Comprehensive mapping of long-range interactions reveals folding principles of the human genome. *Science*, *326*(5950), 289–293. https://doi.org/10.1126/science.1181369.

Liehr, T. (2021a). A definition for cytogenomics - Which also may be called chromosomics. In T. Liehr (Ed.), *Cytogenomics* (pp. 1–7). Academic Press. Chapter 1 (in this book).

Liehr, T. (2021b). Overview of currently available approaches used in cytogenomics. In T. Liehr (Ed.), *Cytogenomics* (pp. 11–24). Academic Press. Chapter 2 (in this book).

Liehr, T. (2021c). Repetitive elements, heteromorphisms, and copy number variants. In T. Liehr (Ed.), *Cytogenomics* (pp. 373–388). Academic Press. Chapter 19 (in this book).

Liehr, T. (2021d). Nuclear architecture. In T. Liehr (Ed.), *Cytogenomics* (pp. 297–305). Academic Press. Chapter 14 (in this book).

Liehr, T. (2021e). Molecular cytogenetics. In T. Liehr (Ed.), *Cytogenomics* (pp. 35–45). Academic Press. Chapter 4 (in this book).

Luger, K., Mäder, A. W., Richmond, R. K., Sargent, D. F., & Richmond, T. J. (1997). Crystal structure of the nucleosome core particle at 2.8 Å resolution. *Nature*, *389*(6648), 251–260. https://doi.org/10.1038/38444.

Lyon, M. F. (1961). Gene action in the X-chromosome of the mouse (*Mus musculus* L.). *Nature*, *190*, 372–373. https://doi.org/10.1038/190372a0.

Maass, P. G., Barutcu, A. R., & Rinn, J. L. (2019). Interchromosomal interactions: A genomic love story of kissing chromosomes. *The Journal of Cell Biology*, *218*(1), 27–38. https://doi.org/10.1083/jcb.201806052.

Maass, P. G., Barutcu, A. R., Weiner, C. L., & Rinn, J. L. (2018). Inter-chromosomal contact properties in live-cell imaging and in hi-C. *Molecular Cell*, *70*(1), 188–189. https://doi.org/10.1016/j.molcel.2018.03.021.

Mao, Y. S., Sunwoo, H., Zhang, B., & Spector, D. L. (2011). Direct visualization of the co-transcriptional assembly of a nuclear body by noncoding RNAs. *Nature Cell Biology*, *13*(1), 95–101. https://doi.org/10.1038/ncb2140.

McHugh, C. A., Chen, C. K., Chow, A., Surka, C. F., Tran, C., McDonel, P., … Guttman, M. (2015). The *Xist* lncRNA interacts directly with SHARP to silence transcription through HDAC3. *Nature*, *521*(7551), 232–236. https://doi.org/10.1038/nature14443.

McNulty, S. M., Sullivan, L. L., & Sullivan, B. A. (2017). Human centromeres produce chromosome-specific and array-specific alpha satellite transcripts that are complexed with CENP-A and CENP-C. *Developmental Cell*, *42*(3), 226–240.e6. https://doi.org/10.1016/j.devcel.2017.07.001.

Mello, S. S., Sinow, C., Raj, N., Mazur, P. K., Bieging-Rolett, K., Broz, D. K., … Attardi, L. D. (2017). *Neat1* is a p53-inducible lincRNA essential for transformation suppression. *Genes & Development*, *31*(11), 1095–1108. https://doi.org/10.1101/gad.284661.116.

Newman, A. M., Steen, C. B., Liu, C. L., Gentles, A. J., Chaudhuri, A. A., Scherer, F., … Alizadeh, A.A. (2019). Determining cell type abundance and expression from bulk tissues with digital cytometry. *Nature Biotechnology*, *37*(7), 773–782. https://doi.org/10.1038/s41587-019-0114-2.

Nomura, S. (2021). Single-cell genomics to understand disease pathogenesis. *Journal of Human Genetics*, *66*(1), 75–84. https://doi.org/10.1038/s10038-020-00844-3.

Nozawa, R. S., Yamamoto, T., Takahashi, M., Tachiwana, H., Maruyama, R., Hirota, T., & Saitoh, N. (2020). Nuclear microenvironment in cancer: Control through liquid-liquid phase separation. *Cancer Science*, *111*(9), 3155–3163. https://doi.org/10.1111/cas.14551.

Pintacuda, G., Wei, G., Roustan, C., Kirmizitas, B. A., Solcan, N., Cerase, A., … Brockdorff, N. (2017). hnRNPK recruits PCGF3/5-PRC1 to the Xist RNA B-repeat to establish polycomb-mediated chromosomal silencing. *Molecular Cell*, *68*(5), 955–969.e10. https://doi.org/10.1016/j.molcel.2017.11.013.

Pisani, G., & Baron, B. (2020). NEAT1 and paraspeckles in cancer development and chemoresistance. *Noncoding RNA*, *6*(4), 43. https://doi.org/10.3390/ncrna6040043.

Rao, S. S., Huntley, M. H., Durand, N. C., Stamenova, E. K., Bochkov, I. D., Robinson, J. T., … Aiden, E.L. (2014). A 3D map of the human genome at kilobase resolution reveals principles of chromatin looping. *Cell*, *159*(7), 1665–1680. https://doi.org/10.1016/j.cell.2014.11.021.

Rheinbay, E., Nielsen, M. M., Abascal, F., Wala, J. A., Shapira, O., Tiao, G., … PCAWG Consortium. (2020). Analyses of non-coding somatic drivers in 2,658 cancer whole genomes. *Nature*, *578*(7793), 102–111. https://doi.org/10.1038/s41586-020-1965-x. In this issue.

Sawyer, I. A., Sturgill, D., & Dundr, M. (2019). Membraneless nuclear organelles and the search for phases within phases. *Wiley Interdisciplinary Reviews RNA*, *10*(2). https://doi.org/10.1002/wrna.1514, e1514.

Shevtsov, S. P., & Dundr, M. (2011). Nucleation of nuclear bodies by RNA. *Nature Cell Biology*, *13*(2), 167–173. https://doi.org/10.1038/ncb2157.

Simon, M. D., Wang, C. I., Kharchenko, P. V., West, J. A., Chapman, B. A., Alekseyenko, A. A., … Kingston, R.E. (2011). The genomic binding sites of a noncoding RNA. *Proceedings of the National Academy of Sciences of the United States of America*, *108*(51), 20497–20502. https://doi.org/10.1073/pnas.1113536108.

Sridhar, B., Rivas-Astroza, M., Nguyen, T. C., Chen, W., Yan, Z., Cao, X., … Zhong, S. (2017). Systematic mapping of RNA-chromatin interactions in vivo. *Current Biology*, *27*(4), 610–612. https://doi.org/10.1016/j.cub.2017.01.068.

Statello, L., Guo, C. J., Chen, L. L., & Huarte, M. (2020). Gene regulation by long non-coding RNAs and its biological functions. *Nature Reviews. Molecular Cell Biology*, *22*, 1–23. https://doi.org/10.1038/s41580-020-00315-9.

Tachiwana, H., Yamamoto, T., & Saitoh, N. (2020). Gene regulation by non-coding RNAs in the 3D genome architecture. *Current Opinion in Genetics & Development*, *61*, 69–74. https://doi.org/10.1016/j.gde.2020.03.002.

Tanabe, H., Müller, S., Neusser, M., von Hase, J., Calcagno, E., Cremer, M., … Cremer, T. (2002). Evolutionary conservation of chromosome territory arrangements in cell nuclei from higher primates. *Proceedings of the National Academy of Sciences of the United States of America*, *99*(7), 4424–4429. https://doi.org/10.1073/pnas.072618599.

Tomita, S., Abdalla, M. O. A., Fujiwara, S., Matsumori, H., Maehara, K., Ohkawa, Y., … Nakao, M. (2015). A cluster of noncoding RNAs activates the *ESR1* locus during breast cancer adaptation. *Nature Communications*, *6*, 6966. https://doi.org/10.1038/ncomms7966.

Ungelenk, M. (2021). Sequencing approaches. In T. Liehr (Ed.), *Cytogenomics* (pp. 87–122). Academic Press. Chapter 7 (in this book).

Wang, K. C., Yang, Y. W., Liu, B., Sanyal, A., Corces-Zimmerman, R., Chen, Y., … Chang, H.Y. (2011). A long noncoding RNA maintains active chromatin to coordinate homeotic gene expression. *Nature, 472*(7341), 120–124. https://doi.org/10.1038/nature09819.

Weise, A., & Liehr, T. (2021a). Interchromosomal interactions with meaning for disease. In T. Liehr (Ed.), *Cytogenomics* (pp. 349–356). Academic Press. Chapter 17 (in this book).

Weise, A., & Liehr, T. (2021b). Cytogenetics. In T. Liehr (Ed.), *Cytogenomics* (pp. 25–34). Academic Press. Chapter 3 (in this book).

West, J. A., Davis, C. P., Sunwoo, H., Simon, M. D., Sadreyev, R. I., Wang, P. I., … Kingston, R.E. (2014). The long noncoding RNAs NEAT1 and MALAT1 bind active chromatin sites. *Molecular Cell, 55*(5), 791–802. https://doi.org/10.1016/j.molcel.2014.07.012.

Yamamoto, T., & Saitoh, N. (2019). Non-coding RNAs and chromatin domains. *Current Opinion in Cell Biology, 58*, 26–33. https://doi.org/10.1016/j.ceb.2018.12.005.

Yamamoto, T., Sakamoto, C., Tachiwana, H., Kumabe, M., Matsui, T., Yamashita, T., … Nakao, M. (2018). Endocrine therapy-resistant breast cancer model cells are inhibited by soybean glyceollin I through *Eleanor* non-coding RNA. *Scientific Reports, 8*(1), 15202. https://doi.org/10.1038/s41598-018-33227-y.

Yang, F., Deng, X., Ma, W., Berletch, J. B., Rabaia, N., Wei, G., … Disteche, C.M. (2015). The lncRNA firre anchors the inactive X chromosome to the nucleolus by binding CTCF and maintains H3K27me3 methylation. *Genome Biology, 16*(1), 52. https://doi.org/10.1186/s13059-015-0618-0.

Yang, Y. W., Flynn, R. A., Chen, Y., Qu, K., Wan, B., Wang, K. C., … Chang, H.Y. (2014). Essential role of lncRNA binding for WDR5 maintenance of active chromatin and embryonic stem cell pluripotency. *eLife, 3*. https://doi.org/10.7554/eLife.02046, e02046.

Yumiceba, V., Souto Melo, U., & Spielmann, M. (2021). 3D cytogenomics: Structural variation in the three-dimensional genome. In T. Liehr (Ed.), *Cytogenomics* (pp. 247–266). Academic Press. Chapter 12 (in this book).

Zhang, L. F., Huynh, K. D., & Lee, J. T. (2007). Perinucleolar targeting of the inactive X during S phase: Evidence for a role in the maintenance of silencing. *Cell, 129*(4), 693–706. https://doi.org/10.1016/j.cell.2007.03.036.

Zhao, Y., Li, H., Fang, S., Kang, Y., Wu, W., Hao, Y., … Chen, R. (2016). NONCODE 2016: An informative and valuable data source of long non-coding RNAs. *Nucleic Acids Research, 44*(D1), D203–D208. https://doi.org/10.1093/nar/gkv1252.

Zhao, J., Sun, B. K., Erwin, J. A., Song, J. J., & Lee, J. T. (2008). Polycomb proteins targeted by a short repeat RNA to the mouse X chromosome. *Science, 322*(5902), 750–756. https://doi.org/10.1126/science.1163045.

Repetitive elements, heteromorphisms, and copy number variants

19

Thomas Liehr

*Jena University Hospital, Friedrich Schiller University, Institute of Human Genetics,
Jena, Germany*

Chapter outline

Background

Thinking about humans, phenotypically visible polymorphisms coming to one's mind may be things like: (i) different hair, eye, or skin color, (ii) (in)ability to role the tongue, (iii) different shapes of toes or fingers, (iv) variation in body size, or (v) in intelligence. These, as well as other phenotypically visible polymorphisms in animals, like the black panther variant, or many different color variants of ladybugs, have a genetic basis (Liehr, 2020). Thus, it is not surprising to find even more variation between individuals of a given species at the DNA level.

Accordingly, chromosomal heteromorphisms, repetitive elements, and copy number variants are some of the many possible designations describing the same basic phenomenon - an individual genome of a certain vertebrate species, even of the same gender, never looks to 100% the same (Liehr, 2014; Mkrtchyan et al., 2010).

Cytogenomics. https://doi.org/10.1016/B978-0-12-823579-9.00015-1

There are of course basic numbers of genes, euchromatic regions, and chromosomes, which are well defined, and more or less stable within a species. Still there is a well-known variation and variability in coding sequences of healthy individuals (denominated, for example, as different alleles); however, much more happens in noncoding sequences (Bessenyei et al., 2004; Liehr, 2014; Mkrtchyan et al., 2010). Even though DNA polymorphisms have been studied a lot in cytogenomic research (Liehr, 2021a, 2021b) during recent decades, still many things are not clear yet, e.g., the size of polymorphic DNA-blocks, all their exact locations especially in the human genome, and their relationships to each other.

Variation and variability are the rule rather than exceptions, and basically, those are important motors of evolution. However, especially in human genetics, cytogenomic variation was discovered and/or recognized by the scientific community at different time points during its history, and at different levels of resolution; the latter is meant in terms of DNA changes observed at base pair, kilobase pair, megabase pair, and/or chromosomal levels. Fig. 19.1 summarizes possible levels of variation in DNA; as humans are the most studied species, this chapter deals with this example only, if not indicated differently.

Interestingly, several times during the history of human genetics (Petermann et al., 2017), when a new level of variation was discovered, a wave of surprise went through the scientific community. Sometimes, even phenomena now considered as variants without any clinical impact were suggested to be correlated with patients' symptoms and syndromes (Liehr, 2014).

In Table 19.1 (as also included in Fig. 19.1) eight different types of polymorphism as observable in humans are defined and examples are given. In addition, it should be highlighted that there are overlaps between these eight types. Furthermore, the same type of DNA variant leading to a polymorphism without any meaning for the phenotype may also act in another DNA context as a disease-causing and/or phenotype-changing event. Finally, it is important to know that some polymorphisms may be mixed up with meaningful mutations, too, as also summarized in Table 19.1.

In the following, the eight different polymorphism-types from Table 19.1 are discussed in more detail, starting from the smallest to the largest in size. Furthermore, the abovementioned "phenotypically visible polymorphisms" plus "chromosomal and whole genomic variants" are included below; the latter are normally suggested not to be existent or even not being viable in humans, but in many other species.

Types of polymorphic DNA
Phenotypically visible polymorphisms

In humans, but also in the whole kingdom of life, phenotypically visible polymorphisms can regularly be found. Besides already mentioned color variants, such as the black panther and the "normal" leopard, or ladybug color variants, one may be reminded of phenotypically visible polymorphisms in human. Prominent examples are: (i) the hitchhiker's and the normal thumb, (ii) the ability or inability to roll the

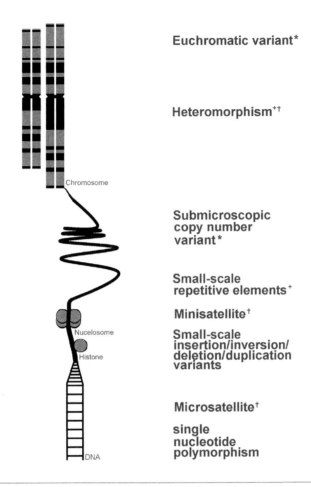

Euchromatic variant*

Heteromorphism[+†]

Chromosome

Submicroscopic copy number variant*

Small-scale repetitive elements[+]

Minisatellite[†]

Nucelosome

Histone

Small-scale insertion/inversion/deletion/duplication variants

Microsatellite[†]

single nucleotide polymorphism

DNA

FIG. 19.1

Eight genetic polymorphisms present at different resolution of the genome - as schematic drawing.

Eight genetic polymorphisms present at different resolution of the genome are depicted - starting from DNA double strand at the *bottom*, being spiralized more and more (via histones and nucleosomes to chromosomes). Single nucleotide polymorphism is the smallest variant; microsatellites, minisatellites, and chromosomal heteromorphism are, in part, expressions of the same phenomenon - here heterochromatic repetitive DNA with basic units between 1 and ~ 200 bp can be found - thus, they are marked in the figure by a cross symbol; small-scale insertions/inversions/deletions/duplications are in size between micro- and minisatellites; small-scale repetitive elements are, in part, overlapping with chromosomal heteromorphisms (thus both marked by a *plus sign*). Submicroscopic copy number variants and euchromatic variants are both nothing else than differently sized "copy number variants" (CNVs) and, accordingly, both marked with an *asterisk* in the figure.

Table 19.1 Eight genetic polymorphisms present at different resolution of the genome - as table.

The genetic polymorphisms are listed and detailed for a short definition, example(s) with and without clinical meaning and which phenomena they may be mixed up with.

Genetic polymorphism	Definition	Example(s) without clinical meaning	Example(s) with clinical meaning	To be mixed up with
Single nucleotide polymorphism	One base pair (ex) change	All as "benign" classified SNPs in genetic databases	Sickle cell anemia based on a point mutation	All as "variant of unknown significance" classified corresponding rearrangements in genetic database
			All as "pathogenic" classified corresponding SNPs in genetic databases	ABO-blood group system thought to have no clinical impact - disproven lately risk for heart attack
Microsatellite[a]	1 to ~ 10bp repeats at pericentromere and telomere	6bp repeats telomere sequence 5'-TTAGGG-3'	3 bp repeats in Huntington's chorea	None yet reported
Small-scale insertion/ inversion/deletion/ duplication	1–50bp in size	All as "benign" classified corresponding rearrangements in genetic databases	All as "pathogenic" classified corresponding rearrangements in genetic databases	All as "variant of unknown significance" classified corresponding rearrangements in genetic databases
Minisatellite[a]	10 to > 100bp repeats	Located at pericentromere and subtelomere	~ 5 human diseases associated with minisatellites	Acc. to integration site these elements may be harmless or harmful
Small-scale repetitive elements	Gain or loss of heterochromatic material, visible or invisible in microscope; repeats of ~ 0.1 to 8kb	Long interspersed nuclear elements (LINEs) - 6–8kb	> 50 human diseases associated with SINEs	Acc. to integration site these elements may be harmless or harmful

Submicroscopic copy number variant	Gain, loss or inversion of euchromatic material, not visible in light microscope, but in fluorescence microscope	Short interspersed nuclear elements (SINEs) – 0.1–0.4 kb Alpha-satellites (~ 170bp) at centromeres organized in higher order repeats and visible in microscope Amplification of 8q21.2	CNVs are associated with diseases	Oncogene amplification
Chromosomal heteromorphism	Gain, loss or inversion of heterochromatic MATERIAL, visible in microscope	13ph +	Submicroscopic inversions may predispose for MMDs 1q12 amplification is suggested to play a role in schizophrenia	MDDs E.g. der(13)t(6;13)(p22.2;p12)
Euchromatic variant	Gain or loss of euchromatic material, visible in microscope	Amplification of 8p23.1p23.1	None yet reported	8p23.1 duplication syndrome

[a] Micro- and minisatellites can be summarized as variable number of tandem repeats (VNTRs).

tongue, or (iii) different shapes of toes. Even though these features have been known for thousands of years, their genetics and the question of why they are there (i.e., if the one or the other feature may be advantageous) are far from being resolved. Some, like hitchhiker's thumb and rolling of the tongue, seem to have simple autosomal inheritance. Still, at least tongue-rolling has been shown to be inherited in a much more complex, obviously multifactorial way (McDonald & Boyd, 2020). Nonetheless, it is common sense that the genetic basis for most phenotypically visible polymorphisms is single nucleotide polymorphisms (SNPs). As an SNP is nothing else than a base pair exchange in a given gene, on molecular level there is no principal difference on DNA level between an SNP and a mutation in a disease-causing gene. A well-known example is the autosomal-recessively inherited sickle-cell-anemia, being due to a typical point mutation in the *HBB*-gene in chromosome 11p15.5; accordingly the beta (hemo)globin protein, consisting of 147 amino acids has, instead of glutamic acid, a valine in position six (Liehr, 2020).

Single nucleotide polymorphisms

Discussing phenotypically visible polymorphism in the previous paragraph, SNPs were touched on. Within the genome of two human beings, SNPs are, on average, different in more than 3 million positions; in addition, new SNPs are acquired during everyone's lifetime in each cell - accordingly, even monozygotic twins are not genetically identical (Bessenyei et al., 2004). The polymorphisms leading to different blood groups of the ABO blood group system are well-known; between one and four different SNPs distinguish the blood groups from each other (Storry & Olsson, 2009). While until recently no phenotypes were associated with the ABO blood group system, actual studies have now shown that there are indeed lower genetic risks for thrombotic vascular and coronary artery disease associated with blood group O compared to all others, especially group A (Chen et al., 2016).

As summarized in Bessenyei et al. (2004), the borderline between a single nucleotide exchange being (i) a meaningless SNP, (ii) a potentially meaningful, phenotype-changing alteration, and (iii) an adverse mutation cannot be defined clearly (Bessenyei et al., 2004). In accordance with this, all modern human genetic diagnostics based on sequencing approaches distinguish five different classifications for reporting of sequence variants (Richards et al., 2015). An SNP can be:

- clearly benign: this means, for example, there is enough evidence due to hundreds of reference cases that such a change/such an SNP does not lead to any monogenetic disease;
- likely benign: this can be an SNP that has never been reported before, but all prediction programs for protein sequence including splicing effects show no evidence for an adverse effect;
- likely adverse: such an SNP can, for example, also not have been previously seen; however, the abovementioned prediction programs show a clear deleterious effect on protein sequence, like an early stop-codon;

- clearly adverse: this may be an SNP having been seen at least twice before in a corresponding patient, and prediction programs show a clear effect on protein sequence; or
- if an SNP does not fit in any of the four previous classes, comprehensively defined in the "Standards and Guidelines for the Interpretation of Sequence Variants," use of the designation "variant of uncertain significance" (VOUS) is recommended (Richards et al., 2015).

Indeed, the criteria to classify sequence variants into the mentioned five groups are, besides for SNPs, also applied for small deletions, duplications, and inversions - see below. Still, it must be stressed that this way to interpret such data is only valid for monogenetic diseases. An influence of such alterations on a multigenic condition cannot be excluded and is yet hard to classify.

Most likely, it will need at least one decade until the majority of human SNPs can be attributed to the two classifications "clearly benign" or "clearly pathogenic," and it will most probably be much longer until it can be resolved which role specific SNPs play in multigenic disorders (see also Ungelenk, 2021).

Microsatellites

Repetitive DNA constitutes up to 75% of the human genome; this is noncoding DNA, which may be partially transcribed into RNA, but not to proteins. Microsatellites, as well as minisatellites, small-scale repetitive elements, and chromosomal heteromorphisms (see below) are built up by such repetitive DNA (Liehr, 2014). Different approaches by which these different polymorphic DNA types can be characterized are the major reason why these four groups have been defined. This means they were grouped according to their sizes and some special DNA sequence features.

Microsatellites are 1 to ~ 10 bp repeats, predominantly localized in (peri)centromeric and (sub)telomeric regions. A prominent microsatellite example is the 6 bp repeat of the telomeric sequence 5′-TTAGGG-3′. It may form different repeat lengths with interindividual, intercellular, and interchromosomal differences. In addition, in humans, telomere length is age-associated, while lab mice have extremely large terminal telomeric repeat blocks, hardly affected by aging (Goyns & Lavery, 2000). However, the telomeric repeats in the somatic cells shrink in length with the age of the human individual.

Besides being polymorphic in humans (within a certain frame with respect to its copy numbers), telomeres are important to keep chromosomes stable. They are the biomarkers that tell the cell: "This is my own DNA; my defense mechanisms against free viral DNA in the nucleus must not be activated." In addition, shorter telomeric repeats are present as so-called interstitial telomeric sequences (ITS) in the middle of chromosome-arms in many genomes. They are thought to be remains from evolution, highlighting places of chromosome-end-fusions in the past (Liehr et al., 2017).

Apart from their (peri)centric and (sub)telomeric locations, 1 to ~ 10 bp repeats are present, with species-specific distribution along all chromosomes. Thus, using

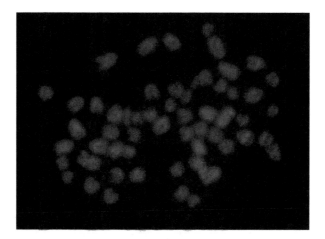

FIG. 19.2

Microsatellites used in molecular cytogenetic research.
Metaphase plate of fish species *Osteoglossum bicirrhosum* (Teleostei, Osteoglossiformes, Osteoglossidae), highlighting the chromosomal distribution of (CA)n microsatellite motif among 56 chromosomes.

Figure kindly provided by Prof. Marcelo de Bello Cioffi, Sao Carlos, Brazil.

the cytogenomic approach of molecular cytogenetics (Liehr, 2021c), such repeats can be characterized by fluorescence in-situ hybridization to learn about species evolution and/or to discriminate species by their specific microsatellite patterns (Sassi et al., 2019) (Fig. 19.2).

Clinical genetics and forensics also take advantage of the fact that each human being has an (almost) unique microsatellite pattern. Thus, microsatellite analyses based on polymerase chain reaction is applied, for example, to study uniparental disomy (see also Eggermann, 2021; Harutyunyan & Hovhannisyan, 2021) and/or to perform paternity testing or to identify a perpetrator. As mentioned in Table 19.1, detection of 3 bp-repeats (trinucleotides) within or nearby specific genes can also play a role in human inherited diseases like Huntington's chorea. This and other so-called trinucleotide-repeat diseases (TREDs) are caused by genes comprising a certain number of trinucleotide-repeats in each healthy individual. During meiosis, these trinucleotide repeats may be amplified or reduced in size. When they exceed a certain number, this leads to a structurally altered gene-product and in consequence to a disease (Liehr, 2020). This means microsatellites are, in most cases, polymorphic, but in rare cases, if being amplified in too many copies, lead to clinical problems.

Small-scale insertions/inversions/deletions/duplications

Changes in the DNA sequence, like single nucleotide deletions, insertions, duplications, or substitutions are summarized as SNPs (see above). All comparable changes plus small inversions affecting two to some dozen base pairs are called "small-scale insertions/inversions/deletions/duplications," the sizes of which are determined by the technical peculiarities of the applied next generation sequencing platform and the corresponding bioinformatics pipelines (Bisht & Avarello, 2021; Boschiero et al., 2015; Ungelenk, 2021).

The interpretation problem, i.e., "Which of these changes is a polymorphism and which a mutation?" is equivalent to that described for SNPs. If such a variant cannot be attributed as being clearly pathogen or clearly benign in human genetics, it is recommended to reevaluate the alteration bioinformatically after after 2 years at the earliest. The underlying suggestion is to have more comparable cases over time, and thus to come to a more meaningful interpretation later than at present (Richards et al., 2015).

In population genetics of humans (Posth et al., 2018), as well as other species (Souza-Shibatta et al., 2018), SNP analysis is an indispensable cytogenomic tool.

Minisatellites

Minisatellites are nothing else than microsatellites (see above), only larger, with 10–100 bp repeats. Thus, these two types can also be summarized as VNTRs (variable number of tandem repeats). Minisatellites, like microsatellites, are predominantly located at pericentric and subtelomeric regions, but can also be found everywhere else in the genome; according to their integration site/localization, presence of such a minisatellite may be harmless or harmful. Yet only five human diseases associated with minisatellites are reported; like in TREDs minisatellites are exclusively deleterious in those cases where their copy numbers exceed a certain threshold (Boschiero et al., 2015). VNTRs are thus again an example where the same phenomenon on a DNA level can be a meaningless variant or a deleterious event, depending on its position, as previously discussed for SNPs and small-scale insertions/inversions/deletions/duplications. As for these other examples, the majority of minisatellite variations are harmless, and only a few are associated with diseases in humans.

Small-scale repetitive elements

Small-scale repetitive elements may be gain, loss, or insertion of heterochromatic DNA, which is constituted of 0.1 to 8 kb repeats. Normally, these elements are present in a few to many copies, but still invisible in light microscopy, i.e., they are below 5 Mb in size.

A major part of these inserts in genomes are of (retro)viral origin; they are a normal component of eukaryotic genomes (Holmes, 2011). Cytogenetically (Weise & Liehr, 2021a) invisible in humans are the so-called "long interspersed nuclear elements" (LINEs), which have a basic length of 6–8 kb, and the "short interspersed nuclear elements" (SINEs), being 0.1–0.4 kb in unit length. LINEs are formed by a group of retrotransposons making up about 21% of the human genome, the majority of which are truncated; active ones can still double and integrate at new sites in the genome (Bertuzzi et al., 2020). SINEs, another abundant class of retrotransposons, compose about 13% of the human genome (SINEs, 2020).

According to their integration sites, these elements may be harmless or harmful. For LINEs, "there are nearly 100 examples of known diseases caused by retroelement insertions, including some types of cancer and neurological disorders. Correlation between LINE mobilization and oncogenesis has been reported for epithelial cell cancer (carcinoma). Hypomethylation of LINES is associated with chromosomal instability and altered gene expression and is found in various cancer cell types in various tissues types" (LINES, 2020). "SINE loci are generally transcriptionally repressed in somatic cells but can be robustly induced upon infection with multiple DNA viruses" (Tucker & Glaunsinger, 2017). More than 50 human diseases are associated with SINEs (SINEs, 2020).

Human LINEs and SINEs are, in the majority of cases, something easy to consider as polymorph DNA. Again, a feature in common with previously discussed variant DNAs is that it is a matter of location and circumstances (like for SINEs the presence or absence of specific viruses) if these conditions lead to problems.

Another form of polymorphic nuclear DNA present in all eukaryotes is "polymorphic mitochondrial insertions" (NumtS). These are an expression of ongoing integration of mitochondrial DNA into the nucleus of the eukaryontic cell. In humans, > 1000 NumtS are already known (Dayama et al., 2014). The mitochondrial genome is ~ 16.5 kb in size and NumtS are normally shorter and can be arranged in arrays. Normally, NumtS are too small to be seen in cytogenetics. However, recently an exceptional case with an insertion of a cytogenetically visible NumtS-block in chromosome 14 has been reported in a healthy carrier (Harutyunyan et al., 2020).

A completely different kind of "small-scale repetitive elements" is the so-called "satellite DNAs." They are stretches of DNA of ~ 170 bp in length, which can be found along the whole chromosome; however, they are concentrated around the centromeres (see Fig. 19.1), and there especially, they form so-called higher-order repeats (HORs) and are easily visible in light microscopy. Even though they have been known of for decades and characterized and denominated in much detail, the majority of these satellite DNAs, making up about 5% of the human genome, are not included in the genome browsers. While the interspersed alpha-, ß-, and other satellite DNAs cannot be seen in cytogenetics, they are clearly visible as heterochromatic blocks in karyotyping and in molecular cytogenetics (Liehr, 2021c). It is known that some of these satellite DNAs are transcribed into mRNA; however, their role is as yet unclear. It is also unknown if and what influence gain or loss of these DNA stretches (variations may be up to several tens of Mb in size) may have for the individual. It is considered still that such changes are without any consequences, but this seems to be unlikely (Garrido-Ramos, 2017; Ichikawa & Saitoh, 2021; Liehr, 2014).

Submicroscopic copy number variants (CNVs)

In 2004, "the picture of the human genome was remarkably changed and extended by a new kind of polymorphism, called large-scale copy-number variations (LCV), or submicroscopic copy number variations (CNVs). These were found by DNA microarray technology and include hundreds of previously undetected structural variants in the human genome such as deletions, gains and inversions. Surprisingly, CNVs can have sizes of ten to several hundred thousand base pairs and are located in euchromatic regions all over the genome" (Weise et al., 2008; Weise & Liehr, 2021b). Corresponding data are now collected in the "database of genomic variants" (http://dgv.tcag.ca/dgv/app/home) (see also Iafrate et al., 2004; Sebat et al., 2004). As has been reported above for NumtS, normally these CNVs are too small to be seen on a chromosomal level. However, they can be visualized by molecular cytogenetics and even used as markers to distinguish single chromosomes (Weise et al., 2008). Furthermore, in human cytogenetic routine diagnostics, an altered banding pattern in a clinically normal person can sometimes be attributed to cytogenetically visible CNVs. If such large CNVs are found, more often they are called "euchromatic variants" (see below); one cytogenetically euchromatic variant recently discovered is localized in 8q21.2 (Manvelyan et al., 2011).

At this point of this chapter it is not surprising that, for CNVs, nowadays the majority of them are considered to be harmless and meaningless; however, a few of them are associated with disease. If a CNV region encompasses at least one copy number-sensitive gene (i.e., for its correct function it is important to be present in exactly two copies, and not in one or three), and is flanked by repetitive elements, this region may easily be affected by loss or duplication during meiosis. Such DNA constitutions lead to so-called microdeletion/microduplication diseases (MDD) (Weise et al., 2012). Accordingly, in human genetic diagnostics it is important to understand the difference between a harmless CNV, and a region being copy number variant and potentially connected with an MDD. Furthermore, inborn and altered during aging harmless CNVs (Mkrtchyan et al., 2010) must be clearly distinguished from acquired cancer related gene-amplifications, being present as double minutes and/or homogeneously staining regions (Shimizu, 2009).

Chromosomal heteromorphisms

A chromosomal heteromorphism is an alteration in a genome, which can be repeatedly found within a certain percentage of a population and is clearly visible in light microscopy. Similar to that described above for micro- and minisatellites, small-scale insertions/inversions/deletions/duplications, and CNVs, chromosomal heteromorphisms can also be gain, loss, or inversion of heterochromatic material. The latter is mainly concentrated in the (peri)centric regions and, in humans, also at the end of the long arm of the male Y-chromosome and at the short arms of the acrocentrics (chromosomes 13, 14, 15, 21, and 22). In other, even closely related species like chimpanzees, such heteromorphic DNA is completely different on DNA-level and differently distributed among the chromosomes than in human (Fig. 19.3).

In such heteromorphic DNA micro- and minisatellites, alpha-, ß-, and other satellite DNAs, often organized in HORs, can be found. The many different possible human chromosomal heteromorphisms are summarized elsewhere (Liehr, 2014; Liehr T. Cases with Heteromorphisms, 2020). An example is the enlargement of a chromosome 13 short arm, denominated 13ph +, which can in rare cases be mixed up with

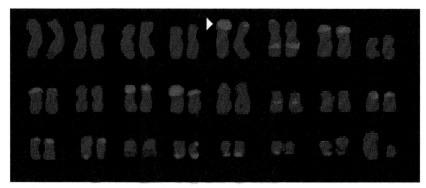

FIG. 19.3

Heteromorphic DNA in the chromosomes of *Pan troglodytes*.
A glass-needle-based microdissection derived DNA-probe *(red label)* from heterochromatic DNA localized at the tip of *Pan troglodytes* (PTR) chromosome 5p *(arrowhead)*, was hybridized back to a male metaphase of PTR *(blue counterstain)*. The probe gave signals on both chromosomes 5, highlighting a chromosomal heteromorphism for this region as larger and smaller signal. This not yet molecular genetically characterized probe had heterochromatic homologous regions at the terminal regions of PTR chromosomes 6p, 8p, 10p, 11p, 12p, 13p, 16p, 17p, 18p, 18q, 20p, 20q, 21q, 22q, and 23q, and interstitially in 6q and 14q. Using this probe in a fluorescence in situ hybridization experiment in humans, no signals could be obtained (results not shown).

cryptic imbalanced translocations (Trifonov et al., 2003). While in the early development of cytogenetic diagnostics such chromosomal heteromorphims were wrongly associated with human diseases (Liehr, 2014), recent work again suggests, for example, that 1q12-amplification plays a role in schizophrenia susceptibility (Ershova et al., 2020).

Euchromatic variants

Comparable to chromosomal heteromorphisms, euchromatic variants can be gain or loss of euchromatic material, visible by microscope. Basically, each euchromatic variant is a special appearance of an abovementioned submicroscopic CNV. In humans, these regions are localized in 4p16.1, 8p23.1, 8q21.2, 9p12 ~ p11.2-p13, 9q12-q21.12 ~ q13-q21.12, 15q11.2, and 16p11.2. Some of these harmless variants can be cytogenetically mixed up with meaningful ones, like the amplification of 8p23.1p23.1 with 8p23.1 duplication syndrome (Liehr, 2014).

Chromosomal and whole genomic variants

The abovementioned types of polymorphisms are best studied in humans. All of them also exist in other species, including animals (vertebrates and invertebrates) and plants. Two additional polymorphisms considered most often in literature to be nonexistent in humans need to be added to the list shown in Table 19.1: the "chromosomal variant" and the "whole genomic variant."

The "chromosomal variant," or more precisely, the "polymorphism in chromosome numbers," can be found in many species. Here it is about supernumerary "B" chromosomes (Bs), which

> *"are extra karyotype units in addition to A chromosomes and are found in some fungi and thousands of animals and plant species. Bs are uniquely characterized due to their non-Mendelian inheritance. A classical concept based on cytogenetics and genetics is that Bs are selfish and abundant with DNA repeats and transposons, and in most cases, they do not carry any function"*

(Ahmad & Martins, 2019). Within a given population, Bs may be present in different copy numbers and variants. No clear effect of their presence or absence for an individual of the population can be observed. However,

> *"the modern view states that Bs are enriched with genes for many significant biological functions, including but not limited to the interesting set of genes related to cell cycle and chromosome structure. Furthermore, the presence of Bs could favor genomic rearrangements and influence the nuclear environment affecting the function of other chromatin regions. Thus, Bs might play a key function in driving their transmission and maintenance inside the cell, as well as offer an extra genomic compartment for evolution"*

(Ahmad & Martins, 2019). In humans, the existence of Bs is under discussion. Some of the so-called small supernumerary marker chromosomes could be candidates for human Bs (Liehr et al., 2008).

In most vertebrates, a diploid genome is the norm. Some vertebrates can also be observed as tri- or tetraploid variants; some insects in the haploid state are male and in diploid condition female. These examples show that "whole genomic variants" are principally possible; in insects within one species, while in vertebrates and plants polyploidization normally leads to new species (Spangenberg et al., 2020). In humans 100% of tri- and tetraploid individuals are aborted prenatally. Still, in mosaic with a normal cell line, such conditions may even be found in clinically normal persons (Natera-De Benito et al., 2014). Thus, this may also be considered as a mosaic polymorphic condition.

Insights and conclusions

As outlined above, genetic polymorphisms, sometimes leading to phenotypically polymorphisms or susceptibilities, are the rule rather than the exception. It is remarkable that even alleged polymorphisms, like in the ABO blood group system, turned out to have some influence on the phenotype. Accordingly, it may be not too great a stretch to state that the future will definitely hold many surprises for cytogenomic research. While euchromatic variants are studied, registered, and published intensively, this is not the case for heterochromatic variants. As it has already been shown that satellite DNAs are - especially in early embryogenesis and in tumors - part of highly expressed noncoding RNA of cells, they should finally receive more attention, also in connection with present concepts on chromosomes (Daban, 2021) and nuclear architecture (Liehr, 2021d).

References

Ahmad, S., & Martins, C. (2019). The modern view of B chromosomes under the impact of high scale omics analyses. *Cell*, *8*(2), 156.

Bertuzzi, M., Tang, D., Calligaris, R., Vlachouli, C., Finaurini, S., Sanges, R., … Gustincich, S. (2020). A human minisatellite hosts an alternative transcription start site for NPRL3 driving its expression in a repeat number-dependent manner. *Human Mutation*, *41*(4), 807–824. https://doi.org/10.1002/humu.23974.

Bessenyei, B., Márka, M., Urbán, L., Zeher, M., & Semsei, I. (2004). Single nucleotide polymorphisms: Aging and diseases. *Biogerontology*, *5*(5), 291–303. https://doi.org/10.1007/s10522-004-2567-y.

Bisht, P., & Avarello, M. D. M. (2021). Molecular combing solutions to characterize replication kinetics and genome rearrangements. In T. Liehr (Ed.), *Cytogenomics* (pp. 47–71). Academic Press. Chapter 5 (in this book).

Boschiero, C., Gheyas, A. A., Ralph, H. K., Eory, L., Paton, B., Kuo, R., … Burt, D.W. (2015). Detection and characterization of small insertion and deletion genetic variants in modern layer chicken genomes. *BMC Genomics*, *16*(1), 23250. https://doi.org/10.1186/s12864-015-1711-1.

Chen, Z., Yang, S. H., Xu, H., & Li, J. J. (2016). ABO blood group system and the coronary artery disease: An updated systematic review and meta-analysis. *Scientific Reports*, *6*. https://doi.org/10.1038/srep23250, 23250.

Daban, J.-R. (2021). Multilayer organization of chromosomes. In T. Liehr (Ed.), *Cytogenomics* (pp. 267–296). Academic Press. Chapter 13 (in this book).

Dayama, G., Emery, S. B., Kidd, J. M., & Mills, R. E. (2014). The genomic landscape of polymorphic human nuclear mitochondrial insertions. *Nucleic Acids Research*, *42*(20), 12640–12649. https://doi.org/10.1093/nar/gku1038.

Eggermann, T. (2021). Epigenetics. In T. Liehr (Ed.), *Cytogenomics* (pp. 389–401). Academic Press. Chapter 20 (in this book).

Ershova, E. S., Malinovskaya, E. M., Golimbet, V. E., Lezheiko, T. V., Zakharova, N. V., Shmarina, G. V., … Kostyuk, S.V. (2020). Copy number variations of satellite III (1q12) and ribosomal repeats in health and schizophrenia. *Schizophrenia Research*, *223*(1), 199–212. https://doi.org/10.1016/j.schres.2020.07.022.

Garrido-Ramos, M. A. (2017). Satellite DNA: An evolving topic. *Genes*, *8*(9), 230. https://doi.org/10.3390/genes8090230.

Goyns, M. H., & Lavery, W. L. (2000). Telomerase and mammalian ageing: A critical appraisal. *Mechanisms of Ageing and Development*, *114*(2), 69–77. https://doi.org/10.1016/S0047-6374(00)00095-6.

Harutyunyan, T., Al-Rikabi, A., Sargsyan, A., Hovhannisyan, G., Aroutiounian, R., & Liehr, T. (2020). Doxorubicin-induced translocation of mtdna into the nuclear genome of human lymphocytes detected using a molecular-cytogenetic approach. *International Journal of Molecular Sciences*, *21*(20), 1–11. https://doi.org/10.3390/ijms21207690.

Harutyunyan, T., & Hovhannisyan, G. (2021). Approaches for studying epigenetic aspects of the human genome. In T. Liehr (Ed.), *Cytogenomics* (pp. 155–209). Academic Press. Chapter 10 (in this book).

Holmes, E. C. (2011). The evolution of endogenous viral elements. *Cell Host & Microbe*, *10*(4), 368–377. https://doi.org/10.1016/j.chom.2011.09.002.

Iafrate, A. J., Feuk, L., Rivera, M. N., Listewnik, M. L., Donahoe, P. K., Qi, Y., … Lee, C. (2004). Detection of large-scale variation in the human genome. *Nature Genetics*, *36*(9), 949–951. https://doi.org/10.1038/ng1416.

Ichikawa, Y., & Saitoh, N. (2021). Shaping of genome by long noncoding RNAs. In T. Liehr (Ed.), *Cytogenomics* (pp. 357–372). Academic Press. Chapter 18 (in this book).

Liehr, T. (2014). *Benign & pathological chromosomal imbalances; microscopic and submicroscopic copy number variations (CNVs) in genetics and counseling*. Academic Press.

Liehr, T. (2020). *Human genetics—Edition 2020: A basic training package*. Epubli. 2020.

Liehr T. Cases with Heteromorphisms. (2020). http://cs-tl.de/DB/CA/HCM/0-Start.html. (Accessed 28 October 2020).

Liehr, T. (2021a). A definition for cytogenomics - Which also may be called chromosomics. In T. Liehr (Ed.), *Cytogenomics* (pp. 1–7). Academic Press. Chapter 1 (in this book).

Liehr, T. (2021b). Overview of currently available approaches used in cytogenomics. In T. Liehr (Ed.), *Cytogenomics* (pp. 11–24). Academic Press. Chapter 2 (in this book).

Liehr, T. (2021c). Molecular cytogenetics. In T. Liehr (Ed.), *Cytogenomics* (pp. 35–45). Academic Press. Chapter 4 (in this book).

Liehr, T. (2021d). Nuclear architecture. In T. Liehr (Ed.), *Cytogenomics* (pp. 297–305). Academic Press. Chapter 14 (in this book).

Liehr, T., Buleu, O., Karamysheva, T., Bugrov, A., & Rubtsov, N. (2017). New insights into Phasmatodea chromosomes. *Genes (Basel)*, *8*(11), 327. https://doi.org/10.3390/genes8110327.

Liehr, T., Mrasek, K., Kosyakova, N., Ogilvie, C., Vermeesch, J., Trifonov, V., & Rubtsov, N. (2008). Small supernumerary marker chromosomes (sSMC) in humans; are there B chromosomes hidden among them. *Molecular Cytogenetics*, *1*(1), 12. https://doi.org/10.1186/1755-8166-1-12.

LINES. (2020). https://en.wikipedia.org/wiki/Long_interspersed_nuclear_element.

Manvelyan, M., Cremer, F. W., Lancé, J., Kläs, C., Kelbova, C., Ramel, H., … Liehr, T. (2011). New cytogenetically visible copy number variant in region 8q21.2. *Molecular Cytogenetics*, *4*(1), 1. https://doi.org/10.1186/1755-8166-4-1.

McDonald, J. H., & Boyd, K. (2020). *Myths of human genetics: Tongue rolling*. http://udel.edu/~mcdonald/mythtongueroll.html.

Mkrtchyan, H., Gross, M., Hinreiner, S., Polytiko, A., Manvelyan, M., Mrasek, K., … Weise, A. (2010). The human genome puzzle—The role of copy number variation in somatic mosaicism. *Current Genomics*, *11*(6), 426–431. https://doi.org/10.2174/138920210793176047.

Natera-De Benito, D., Poo, P., Gean, E., Vicente-Villa, A., García-Cazorla, A., & Fons-Estupiña, M. (2014). Mosaicismo diploide/triploide: un fenotipo variable, pero característico [Diploid/triploid mosaicism: A variable but characteristic phenotype]. *Revista de Neurologia*, *594*(4), 158–163.

Petermann, H., Harper, P., & Doetz, S. (Eds.). (2017). *History of human genetics; Aspects of its development and global perspectives* Springer.

Posth, C., Nakatsuka, N., Lazaridis, I., Skoglund, P., Mallick, S., Lamnidis, T. C., … Reich, D. (2018). Reconstructing the deep population history of central and South America. *Cell*, *175*(5), 1185–1197.e22. https://doi.org/10.1016/j.cell.2018.10.027.

Richards, S., Aziz, N., Bale, S., Bick, D., Das, S., Gastier-Foster, J., … Rehm, H.L. (2015). Standards and guidelines for the interpretation of sequence variants: A joint consensus recommendation of the American College of Medical Genetics and Genomics and the Association for Molecular Pathology. *Genetics in Medicine*, *17*(5), 405–424. https://doi.org/10.1038/gim.2015.30.

Sassi, F. D. M. C., Oliveira, E. A. D., Bertollo, L. A. C., Nirchio, M., Hatanaka, T., Marinho, M. M. F., … Cioffi, M.D. (2019). Chromosomal evolution and evolutionary relationships of lebiasina species (Characiformes, lebiasinidae). *International Journal of Molecular Sciences*, *20*(12), 2944. https://doi.org/10.3390/ijms20122944.

Sebat, J., Lakshmi, B., Troge, J., Alexander, J., Young, J., Lundin, P., … Wigler, M. (2004). Large-scale copy number polymorphism in the human genome. *Science*, *305*(5683), 525–528. https://doi.org/10.1126/science.1098918.

Shimizu, N. (2009). Extrachromosomal double minutes and chromosomal homogeneously staining regions as probes for chromosome research. *Cytogenetic and Genome Research*, *124*(3–4), 312–326. https://doi.org/10.1159/000218135.

SINEs. (2020). https://en.wikipedia.org/wiki/Short_interspersed_nuclear_element.

Souza-Shibatta, L., Kotelok-Diniz, T., Ferreira, D. G., Shibatta, O. A., Sofia, S. H., de Assumpção, L., … Makrakis, M.C. (2018). Genetic diversity of the endangered neotropical cichlid fish (*Gymnogeophagus setequedas*) in Brazil. *Frontiers in Genetics*, *9*, 13. https://doi.org/10.3389/fgene.2018.00013.

Spangenberg, V., Arakelyan, M., Cioffi, M. D. B., Liehr, T., Al-Rikabi, A., Martynova, E., … Kolomiets, O. (2020). Cytogenetic mechanisms of unisexuality in rock lizards. *Scientific Reports*, *10*(1), 8697. https://doi.org/10.1038/s41598-020-65686-7.

Storry, J. R., & Olsson, M. L. (2009). The ABO blood group system revisited: A review and update. *Immunohematology*, *25*(2), 48–59.

Trifonov, V., Seidel, J., Starke, H., Martina, P., Beensen, V., Ziegler, M., … Liehr, T. (2003). Enlarged chromosome 13 p-arm hiding a cryptic partial trisomy 6p22.2-pter. *Prenatal Diagnosis*, *23*(5), 427–430. https://doi.org/10.1002/pd.595.

Tucker, J. M., & Glaunsinger, B. A. (2017). Host noncoding retrotransposons induced by DNA viruses: A SINE of infection? *Journal of Virology*, *91*(23). https://doi.org/10.1128/JVI.00982-17, e00982-17.

Ungelenk, M. (2021). Sequencing approaches. In T. Liehr (Ed.), *Cytogenomics* (pp. 87–122). Academic Press. Chapter 7 (in this book).

Weise, A., Gross, M., Mrasek, K., Mkrtchyan, H., Horsthemke, B., Jonsrud, C., … Liehr, T. (2008). Parental-origin-determination fluorescence in situ hybridization distiguishes homologous human chromosomes on a single-cell level. *International Journal of Molecular Medicine*, *21*(2), 189–200.

Weise, A., & Liehr, T. (2021a). Cytogenetics. In T. Liehr (Ed.), *Cytogenomics* (pp. 25–34). Academic Press. Chapter 3 (in this book).

Weise, A., & Liehr, T. (2021b). Molecular karyotyping. In T. Liehr (Ed.), *Cytogenomics* (pp. 73–85). Academic Press. Chapter 6 (in this book).

Weise, A., Mrasek, K., Klein, E., Mulatinho, M., Llerena, J. C., Hardekopf, D., … Liehr, T. (2012). Microdeletion and microduplication syndromes. *Journal of Histochemistry and Cytochemistry*, *60*(5), 346–358. https://doi.org/10.1369/0022155412440001.

Epigenetics

20

Thomas Eggermann

Uniklinik RWTH Aachen, Institut für Humangenetik, Aachen, Germany

Chapter outline

Introduction

With the first description of heterochromatin regions as compactly stained chromosomal subsegments and the discrimination between heterochromatin and euchromatin in the last century, a major fundamental level of chromatin organization was identified (see also Liehr, 2021a). The cellular functions of chromosomal organization levels afterwards became increasingly obvious with the development of a broad repertoire of methods to analyze genomes within a field later called cytogenomics (Liehr, 2021b, 2021c). Meanwhile, it is out of question that genome organization is closely related to gene expression. In turn, it is not surprising that molecular levels of DNA and chromatin organization are closely linked to epigenetics (see as well Harutyunyan & Hovhannisyan, 2021), the latter being a basic mechanism for gene expression. Additionally, proper epigenetic signature is also a prerequisite for genome integrity, as shown for the role of histone H4 acetylation in human oocytes (H4K12ac) (Baranov & Kuznetzova, 2021; Van Den Berg et al., 2011). In contrast to genomic variation affecting DNA sequence, the term epigenetics refers to heritable molecular changes that do not involve the order of the nucleotides within the DNA backbone. These alterations often have an impact on activity and expression of genes and comprise modifications

Cytogenomics. https://doi.org/10.1016/B978-0-12-823579-9.00016-3

of DNA methylation (noncoding), RNA (Ichikawa & Saitoh, 2021), histone modifications as well as local and genome-wide chromatin organization (Liehr, 2021d). Epigenetic signatures and their functional consequences contribute to the normal development of an organism, but environmental factors also influence the epigenetic constitution of an organism. Epigenetic marks can be transmitted through cell divisions, and due to their dynamics they play a major role in the regulation of cellular differentiation. In addition, transgenerational inheritance of epigenetic patterns has also been demonstrated. A specific mode of epigenetic regulation in higher mammals and some other organisms which serve as a model to decipher epigenetic mechanisms is genomic imprinting. This epigenetic phenomenon allows the cell to discriminate between the parental origin of alleles, and thereby results in a monoallelic expression either of the maternal or the paternal gene copy. It is estimated that genomic imprinting regulates the expression of 1%–2% of human protein-coding genes in a parent-of-origin-specific manner (for review Patten et al., 2016). The monoallelic expression of a specific parental allele contributes to the balanced transcription of key elements (Tucci et al., 2019), basically involved in development, growth, and metabolism. It is therefore not surprising that molecular disturbances cause congenital disorders - the so-called imprinting disorders - showing common clinical features, e.g., disturbed growth, mental retardation, and tumorigenesis (Soellner et al., 2017).

Epigenetic regulation and chromatin organization

A key factor in epigenetic regulation is chromatin organization, which has a direct and indirect impact on correct development by mediating the function of genomic elements in normal metabolism and propagation of epigenetic cell memory. This chromatin organization (see also Liehr, 2021d), visible as heterochromatin and euchromatin in banding cytogenetic analysis (Weise & Liehr, 2021), is based on the wrapping of DNA around an octamer core of histone proteins (so-called nucleosome), and these histone proteins exhibit numerous posttranslational modifications, like methylation and acetylation. The combination of these histone modifications affects nucleosome density, and thereby the transcription activity of the wrapped DNA (Fig. 20.1) (Rao et al., 2014; Sawan et al., 2008; Tan et al., 2018; Zheng et al., 2016). The density of DNA wrapping by nucleosomes is a major regulation factor of its transcription activity. The dynamic modification of histone proteins with an impact on local chromatin organization and on the whole genome organization thereby allows a dynamic and flexible transcription during development.

Epigenetic landscape during development

Correct progression of development from fertilization to maturity (Baranov & Kuznetzova, 2021; Pellestor et al., 2021) is ensured by epigenetic mechanisms. (Harutyunyan & Hovhannisyan, 2021) It starts after fertilization when the control of development passes from maternal to fetal genome (Eckersley-Maslin et al., 2018; Tong et al., 2000). This transition from maternal to zygotic control

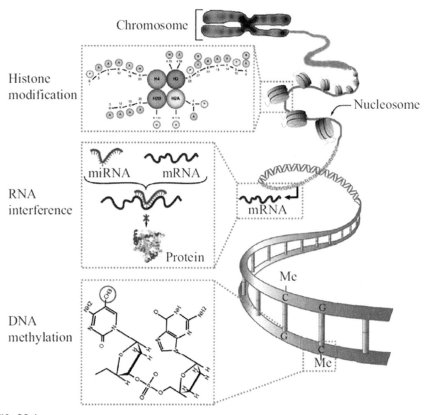

FIG. 20.1

Molecular levels of epigenetic regulation. On DNA level, a major DNA modification with an impact on gene expression is DNA methylation. Gene expression, as well as DNA methylation and chromatin organization are furthermore influenced by (noncoding) RNAs, e.g., by silencing of messenger RNAs or secondary structures affecting the setting of DNA methylation. The posttranslational histone modifications have a broad effect on the regulation of epigenetically imprinted regions, as well (see text).

From Sawan, C., Vaissière, T., Murr, R., & Herceg, Z. (2008). Epigenetic drivers and genetic passengers on the road to cancer. Mutation Research – Fundamental and Molecular Mechanisms of Mutagenesis, 642(1–2), 1–13. https://doi.org/10.1016/j.mrfmmm.2008.03.002.

of development goes along with molecular alterations on different levels of epigenetic regulation.

- On **DNA level**, the early embryo undergoes a wave of extensive demethylation of the fifth carbon of cytosine (5mC) from zygote to blastocyst, and methylation is re-established during gastrulation (for review Messerschmidt et al., 2014). Imprinted genes and clusters are excluded from these de- and re-methylation waves, but all methylation processes are prone to disturbances of enzymes of the DNA methylation machinery. This machinery comprises factors which catalyze the demethylation, either actively in case of the paternal genome, or passively for maternal genome by protection from demethylation. In the early embryo, several of the factors involved in these processes are organized in the subcortical maternal complex (SCMC; see the following section) (Monk et al., 2017) and mediate the demethylation of 5mC on DNA level directly (e.g., TET3, Messerschmidt et al., 2014) or indirectly, by histone modification and altered chromatin structure (e.g., binding of Stella/DPPA3 at the maternal histone H3 dimethylation of residue Lys9: H3K9me2, Nakamura et al., 2012). During this progression, maternal RNAs and proteins encoded by genes of the SCMC and others are depleted and the embryonic genome is activated. In particular, the activation of the zygotic genome is a highly complex process and in humans, it starts with a minor transcriptional wave from the zygote to the four-cell embryo, and a major transcriptional burst until the end of the eight-cell stage.

- Corresponding to DNA methylation and transcription, **posttranslational alterations of the histones** are dynamic as well during development and thereby influence the constitution of nucleosomes and chromatin organization. In the preimplantation embryo, trimethylation histone H3 residue Lys4 (H3K4me3) has been identified to be associated with increased transcription activity, whereas the modification H3K27me3 is linked to chromatin inactivation (Liu et al., 2016). Open chromatin during this stage is additionally obtained by acetylation of H3K27 (H3K27ac) (Dahl et al., 2016). Evidence for a functional link between DNA methylation and posttranslational histone modification is provided by the observation that DNA methylation protects large chromatin (>10 kb) from acquiring H3K4me3 (Hanna et al., 2018).

- The posttranscriptional alterations and the high expression activity are reflected in embryonic development by an increased **chromatin accessibility** (Lu et al., 2016). This does not only affect promotor regions, resulting in higher expression levels, but also other parts of the genome with probable, but currently unknown function for embryonic development. However, this chromatin accessibility is reduced, but dynamic in later developmental stages, and thereby so far inactive genes can become active, and vice versa.

- The **higher order genome organization** consists of discrete topologically associating domains (TADs) which comprise chromatin stretches of 400–500 kb in length (Yumiceba et al., 2021). In turn, these TADs are organized into ~5 Mb regions and differentiated into A and B compartments, alternating along the chromosome and corresponding to transcriptionally active and inactive domains. Long-range chromatin interaction is generally more frequent within a TAD than across TAD boundaries and within a compartment. Finally, the position of chromatin in the nucleus is the third higher order organization level of the genome; B compartments tend to mediate the interaction between chromatin and nuclear lamina (van Steensel & Belmont, 2017). However, in early embryogenesis (in the preimplantation stage), the latter has not yet been observed. The TAD organization is profoundly changed during oocyte maturation, and in meiosis II oocyte, TADs and compartments are absent

(Du et al., 2017; Ke et al., 2017), whereas chromatin in sperms is structured (Du et al., 2017). After fertilization, the genome becomes increasingly structured by TADs and compartments, corresponding to the gene expression shift silent B compartments turn to active A ones.

The maternal and the paternal genomes are functionally disparate in higher mammals

The functional disparity of the parental genomes in higher mammals becomes obvious by the phenomenon of genomic imprinting, but it is also reflected by the different structural genome organization in gametes and in zygote. Maternal and paternal genomes differ by volume, compactness (sperm DNA is loaded with protamines), and TAD compartment strength (Du et al., 2017; Flyamer et al., 2017; Ke et al., 2017). However, in local chromatin structure, differences are not obvious. Imprinted genes are exempted from this, as they differ in chromatin accessibility in addition to different DNA methylation patterns (Liu et al., 2016). Differences between the two genomes can be observed in posttranscriptional histone modifications: whereas in the paternal genome H3K27me3 signatures are practically lost after fertilization and re-established in case of silenced genes, H3K27me3 in the oocyte can be transmitted to the genome in the zygote depending on its functional position (Zheng & Dean, 2009). The comprehensive remodeling of expression, histone modification, and chromatin organization during development requires a tightly concerted interaction of a huge number of factors and pathways. Among others, these include cell cycle regulators, DNA (de)methylation machinery enzymes, DNA damage repair, histone modification, chromatin remodeling (for review Eckersley-Maslin et al., 2018), and the subcortical maternal complex (SCMC). Maternal effect mutations in genes encoding SCMC components and their clinical consequences illustrate the close functional links between progression of embryonic development and imprinting.

Disturbances of the subcortical maternal complex

The close functional association between epigenetics and chromosomal segregation in meiosis has recently been demonstrated by the identification of so-called maternal effect mutations, i.e., pathogenic variants in genes encoding factors of the SCMC. The SCMC is a large multimeric protein complex formed in the mature mammalian oocyte and is localized at its periphery. Its components are exclusively expressed from the maternal genome in oocytes and early embryos, and then degraded in further embryonic development without compensation. Several functions of the SCMC in oocyte and early embryo progression have been suggested (for review Monk et al., 2017) and these comprise meiotic spindle formation and epigenetic reprogramming of the zygote (for review Bebbere et al., 2016). Accordingly, pathogenic variants in SCMC genes in the mother cause aneuploidy and disturbed epigenetic programming in the offspring (Table 20.1) and are generally associated with an increased risk for

Table 20.1 Compilation of maternal effect variants in the currently known members of the human SCMC, and their clinical consequences.

SCMC member (alternative name)	Aneuploidy	Disturbed imprinting	Miscarriages/ pregnancy loss	Number of cases	Reference (see in text below for detailed information)
NLRP2	Yes	Yes	Yes	5	1; 2
				1	
NLRP5 (MATER)		Yes	Yes	5	3
		Yes	NR	2	4
NRLP7		NA	Yes	~70	For review: 5
		Yes	NR	3	For review: 6
	Yes	Yes	Yes	5	1; 7; 8; 9
KHDC3L (c6orf221, ECAT1, FILIA)		Yes	Yes	>6	For review: 5
PADI6	Yes	NA	Yes	3	10; 11
TLE6		Yes	NR	4	1
		NA	NR	3	12
OOEP (floped)		Yes	NR	1	1

References for this table: 1 = (Begemann et al., 2018); 2 = (Meyer et al., 2009); 3 = (Docherty et al., 2015); 4 = (Soellner et al., 2019); 5 = (Nguyen et al., 2018); 6 = (Sanchez-Delgado et al., 2016); 7 = (Caliebe et al., 2014); 8 = (Soellner et al., 2017); 9 = (Deveault et al., 2009); 10 = (Xu et al., 2016); 11 = (Qian et al., 2018); 12 = (Alazami et al., 2015).

reproductive failure (for review Elbracht et al., 2020). The mechanisms behind the functional link between proper chromosomal segregation and epigenetic marking are not yet fully understood, but in mice and human, disruption of components of the SCMC have been shown to be associated with developmental arrest at the two- or four-cell stage (Qian et al., 2018; Xu et al., 2016; Tashiro et al., 2010). Maternal null mice embryos for *Khdc3l* (Filia) exhibit defective zygote spindle assembly and chromosomal alignment, resulting in mitotic delay and aneuploidies (Zheng & Dean, 2009). In the process of epigenetic programming, the SCMC members were suggested to be essential to ensure the correct cellular localization and nuclear translocation of epigenetic factors (Sanchez-Delgado et al., 2016).

The imprinted regions in 11p15.5 as an example of the interaction between local chromatin regulation and local chromatin organization

The chromosomal region 11p15.5 harbors two imprinting centers (*H19/IGF2*:IG-DMR/IC1 and *KCNQ1OT1*:TSS-DMR/IC2) which regulate the monoallelic expression of several imprinted genes. Disturbances of the 11p15.5 imprinted gene clusters are associated with opposite growth disorders as Beckwith-Wiedemann syndrome (BWS; OMIM #130650) and Silver Russell syndrome (SRS; OMIM #180860), and comprise both genomic alterations (copy number variants/CNVs, point mutations, uniparental disomy), as well as methylation defects (epimutations) (for review Brioude et al., 2018; Wakeling et al., 2017). Whereas patients with BWS exhibit a loss of methylation (LOM) of the centromeric IC2, gain of methylation (GOM) of IC1, paternal uniparental disomy of 11p15.5 (upd(11)pat) or copy number variants in 11p15.5, SRS patients almost exclusively show LOM of the telomeric IC1. As illustrated in Fig. 20.2 for IC1, the parent-of-origin-specific patterns of DNA methylation are linked to differences in chromatin organization (Monk et al., 2019; Rao et al., 2014; Tan et al., 2018). Thereby, methylation of IC1 on the paternal allele does not only block *H19* expression but brings specific enhancer motifs close to the *IGF2* promotor and thereby activates *IGF2* expression. In the case of SRS and BWS, the aberrant methylation of the IC1 results in a switch of chromatin organization in the sense of a "maternalization" of the paternal allele (SRS) or a "paternalization" of the maternal allele (BWS). In fact, the physiological cause for the major molecular alterations in both disorders (IC2 LOM in BWS and IC1 LOM in SRS) are basically unknown, but increasing numbers of reports on cases with small and rare structural variants within the two 11p15.5 regions significantly contribute to the understanding of the regulative elements of the IC1 and IC2. For the regulation of the IC1, the relevance of the correct spatial order of binding sites for the CCCTC-binding factor (CTCF) and other elements involved in regulation and expression of the neighbored imprinted genes has been shown (e.g., Beygo et al., 2013; Kraft et al., 2019; Sparago et al., 2018). In these cases, deletions of different sizes and localization within the CTCF-binding region of IC1 allowed the identification of binding sites relevant for proper gene expression, and it can be assumed that these disturbances do not only

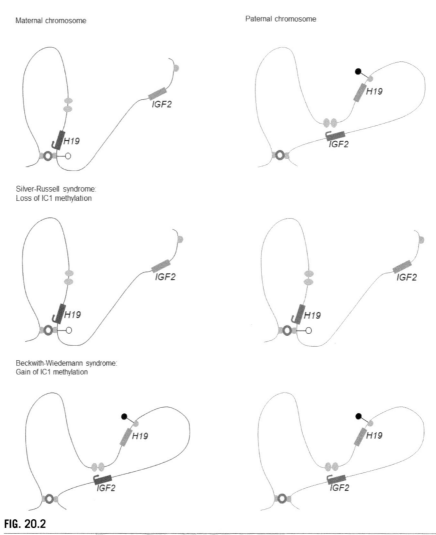

Maternal chromosome

Paternal chromosome

IGF2

H19

H19

IGF2

Silver-Russell syndrome:
Loss of IC1 methylation

IGF2

H19

IGF2

H19

Beckwith-Wiedemann syndrome:
Gain of IC1 methylation

H19

IGF2

H19

IGF2

FIG. 20.2

Parent-of-origin-specific patterns of DNA methylation are linked to differences in chromatin organization. Based on results from chromatin capture studies, alternative chromatin loops of the maternal and the paternal IC1 allele have been proposed. On the maternal allele, the loop formation is linked to the unmethylated H19/IGF2:IG-DMR close to the H19 promotor region. The unmethylated status allows the expression of H19. Furthermore, the loop formation prevents the spatial approach of two enhancer elements needed for the expression of IGF2. On the paternal chromosome 11p15.5, the loop is formed by a region close to the IGF2 promotor, whereas the loop formation at the H19/IGF2:IG-DMR is blocked. The spatial arrangement of the enhancer elements and the IGF2 promotor region results in expression of IGF2. The functional consequences of disturbed imprinting patterns of the IC1 are illustrated for SRS where the paternal chromosome is "maternalized" with silencing of IGF2, and for BWS, the maternal chromosome becomes "paternalized" (chromatin regions are not to scale).

affect the binding sites but also the organization of corresponding chromatin, and thereby its accessibility for binding factors. Within IC2, several imbalances have been identified, too. Some of them provide for the contribution of noncoding and coding RNAs in imprinted regions concerning their proper regulation and interaction with chromatin organization. An example is genomic variants affecting the *KCNQ1* gene. Those are associated with the LongQT 1 syndrome, but in the case of maternal transmission, they additionally cause LOM of IC2 and a BWS phenotype in case of functionally truncating *KCNQ1* variants. Mouse studies have indicated that transcription across ICs is a prerequisite for establishing methylation imprints in the maternal germline (Chotalia et al., 2009; Singh et al., 2017). Since IC2 is located within the transcriptional unit of the *KCNQ1*, it has been hypothesized that maternally inherited truncated *KCNQ1* variants result in LOM of maternal IC2 methylation in BWS (Beygo et al., 2019; Eßinger et al., 2020; Valente et al., 2019).

Conclusion and outlook

Due to the broadness and complexity of the functional relations between epigenetic regulation and chromatin organization, this chapter can only provide excerpt insights in this fascinating field. However, independent from the genomic region, interactions between DNA sequence, chromatin organization, and epigenetic mechanisms are similar along the whole genome, and are based on three molecular levels: (i) DNA modifications, (ii) RNA interference, and (iii) histone modifications. Additionally, higher order structures contribute to the proper realization of epigenetic information. In summary, these examples highlight the relevant role of cytogenomic research in deciphering epigenetic regulation mechanisms. With the implementation of next and third generation sequencing techniques, as well as low-input single-cell methods, these molecular levels can now be optimally addressed. Nucleic acid modifications can be profiled on single-cell level, simultaneously with the analysis of chromatin organization and transcriptome. Thereby, spatial and temporal relationships between the different levels of epigenetic and chromatin organization will be uncovered comprehensively.

References

Alazami, A. M., Awad, S. M., Coskun, S., Al-Hassan, S., Hijazi, H., Abdulwahab, F. M., … Alkuraya, F.S. (2015). TLE6 mutation causes the earliest known human embryonic lethality. *Genome Biology, 16*(1), 240. https://doi.org/10.1186/s13059-015-0792-0.

Baranov, V. S., & Kuznetzova, T. V. (2021). Nuclear stability in early embryo. Chromosomal aberrations. In T. Liehr (Ed.), *Cytogenomics* (pp. 307–325). Academic Press. Chapter 15 (in this book).

Bebbere, D., Masala, L., Albertini, D. F., & Ledda, S. (2016). The subcortical maternal complex: Multiple functions for one biological structure? *Journal of Assisted Reproduction and Genetics, 33*(11), 1431–1438. https://doi.org/10.1007/s10815-016-0788-z.

Begemann, M., Rezwan, F., Beygo, J., Docherty, L., Kolarova, J., Schroeder, C., … Mackay, D. (2018). Maternal variants in NLRP and other maternal effect proteins are associated with multilocus imprinting disturbance in offspring. *Journal of Medical Genetics, 55*(7), 497–504. https://doi.org/10.1136/jmedgenet-2017-105190.

Beygo, J., Bürger, J., Strom, T. M., Kaya, S., & Buiting, K. (2019). Disruption of KCNQ1 prevents methylation of the ICR2 and supports the hypothesis that its transcription is necessary for imprint establishment. *European Journal of Human Genetics*, *27*(6), 903–908. https://doi.org/10.1038/s41431-019-0365-x.

Beygo, J., Citro, V., Sparago, A., De Crescenzo, A., Cerrato, F., Heitmann, M., … Riccio, A. (2013). The molecular function and clinical phenotype of partial deletions of the IGF2/H19 imprinting control region depends on the spatial arrangement of the remaining CTCF-binding sites. *Human Molecular Genetics*, *22*(3), 544–557. https://doi.org/10.1093/hmg/dds465.

Brioude, F., Kalish, J., Mussa, A., Foster, A., Bliek, J., Ferrero, G., … Maher, E. (2018). Clinical and molecular diagnosis, screening and management of Beckwith-Wiedemann syndrome: An international consensus statement. *Nature Reviews Endocrinology*, *14*(4), 229–249. https://doi.org/10.1038/nrendo.2017.166.

Caliebe, A., Richter, J., Ammerpohl, O., Kanber, D., Beygo, J., Bens, S., … Siebert, R. (2014). A familial disorder of altered DNA-methylation. *Journal of Medical Genetics*, *51*(6), 407–412. https://doi.org/10.1136/jmedgenet-2013-102149.

Chotalia, M., Smallwood, S. A., Ruf, N., Dawson, C., Lucifero, D., Frontera, M., … Kelsey, G. (2009). Transcription is required for establishment of germline methylation marks at imprinted genes. *Genes and Development*, *23*(1), 105–117. https://doi.org/10.1101/gad.495809.

Dahl, J., Jung, I., Aanes, H., Greggains, G., Manaf, A., Lerdrup, M., … Klungland, A. (2016). Broad histone H3K4me3 domains in mouse oocytes modulate maternal-to-zygotic transition. *Nature*, *537*(7621), 548–552. https://doi.org/10.1038/nature19360.

Deveault, C., Qian, J., Chebaro, W., Ao, A., Gilbert, L., Mehio, A., … Slim, R. (2009). NLRP7 mutations in women with diploid androgenetic and triploid moles: A proposed mechanism for mole formation. *Human Molecular Genetics*, *18*(5), 888–897. https://doi.org/10.1093/hmg/ddn418.

Docherty, L. E., Rezwan, F. I., Poole, R. L., Turner, C. L., Kivuva, E., Maher, E. R., … Mackay, D.J. (2015). Mutations in NLRP5 are associated with reproductive wastage and multilocus imprinting disorders in humans. *Nature Communications*, *6*(1), 8086. https://doi.org/10.1038/ncomms9086.

Du, Z., Zheng, H., Huang, B., Ma, R., Wu, J., Zhang, X., … Xie, W. (2017). Allelic reprogramming of 3D chromatin architecture during early mammalian development. *Nature*, *547*(7662), 232–235. https://doi.org/10.1038/nature23263.

Eßinger, C., Karch, S., Moog, U., Fekete, G., Lengyel, A., Pinti, E., … Begemann, M. (2020). Frequency of KCNQ1 variants causing loss of methylation of imprinting centre 2 in Beckwith-Wiedemann syndrome. *Clinical Epigenetics*, *12*(1), 63. https://doi.org/10.1186/s13148-020-00856-y.

Eckersley-Maslin, M. A., Alda-Catalinas, C., & Reik, W. (2018). Dynamics of the epigenetic landscape during the maternal-to-zygotic transition. *Nature Reviews Molecular Cell Biology*, *19*(7), 436–450. https://doi.org/10.1038/s41580-018-0008-z.

Elbracht, M., Mackay, D., Begemann, M., Kagan, K. O., & Eggermann, T. (2020). Disturbed genomic imprinting and its relevance for human reproduction: Causes and clinical consequences. *Human Reproduction Update*, *26*(2), 197–213. https://doi.org/10.1093/humupd/dmz045.

Flyamer, I. M., Gassler, J., Imakaev, M., Brandão, H. B., Ulianov, S. V., Abdennur, N., … Tachibana-Konwalski, K. (2017). Single-nucleus Hi-C reveals unique chromatin reorganization at oocyte-to-zygote transition. *Nature*, *544*(7648), 110–114. https://doi.org/10.1038/nature21711.

Hanna, C. W., Taudt, A., Huang, J., Gahurova, L., Kranz, A., Andrews, S., ... Kelsey, G. (2018). MLL2 conveys transcription-independent H3K4 trimethylation in oocytes. *Nature Structural and Molecular Biology*, *25*(1), 73–82. https://doi.org/10.1038/s41594-017-0013-5.

Harutyunyan, T., & Hovhannisyan, G. (2021). Approaches for studying epigenetic aspects of the human genome. In T. Liehr (Ed.), *Cytogenomics* (pp. 155–209). Academic Press. Chapter 10 (in this book).

Ichikawa, Y., & Saitoh, N. (2021). Shaping of genome by long noncoding RNAs. In T. Liehr (Ed.), *Cytogenomics* (pp. 357–372). Academic Press. Chapter 18 (in this book).

Ke, Y., Xu, Y., Chen, X., Feng, S., Liu, Z., Sun, Y., ... Liu, J. (2017). 3D chromatin structures of mature gametes and structural reprogramming during mammalian embryogenesis. *Cell*, *170*(2), 367–381.e20. https://doi.org/10.1016/j.cell.2017.06.029.

Kraft, F., Wesseler, K., Begemann, M., Kurth, I., Elbracht, M., & Eggermann, T. (2019). Novel familial distal imprinting centre 1 (11p15.5) deletion provides further insights in imprinting regulation. *Clinical Epigenetics*, *11*(1), 30. https://doi.org/10.1186/s13148-019-0629-x.

Liehr, T. (2021a). Repetitive elements, heteromorphisms, and copy number variants. In T. Liehr (Ed.), *Cytogenomics* (pp. 373–388). Academic Press. Chapter 19 (in this book).

Liehr, T. (2021b). A definition for cytogenomics - Which also may be called chromosomics. In T. Liehr (Ed.), *Cytogenomics* (pp. 1–7). Academic Press. Chapter 1 (in this book).

Liehr, T. (2021c). Overview of currently available approaches used in cytogenomics. In T. Liehr (Ed.), *Cytogenomics* (pp. 11–24). Academic Press. Chapter 2 (in this book).

Liehr, T. (2021d). Nuclear architecture. In T. Liehr (Ed.), *Cytogenomics* (pp. 297–305). Academic Press. Chapter 14 (in this book).

Liu, X., Wang, C., Liu, W., Li, J., Li, C., Kou, X., ... Gao, S. (2016). Distinct features of H3K4me3 and H3K27me3 chromatin domains in pre-implantation embryos. *Nature*, *537*(7621), 558–562. https://doi.org/10.1038/nature19362.

Lu, F., Liu, Y., Inoue, A., Suzuki, T., Zhao, K., & Zhang, Y. (2016). Establishing chromatin regulatory landscape during mouse preimplantation development. *Cell*, *165*(6), 1375–1388. https://doi.org/10.1016/j.cell.2016.05.050.

Messerschmidt, D. M., Knowles, B. B., & Solter, D. (2014). DNA methylation dynamics during epigenetic reprogramming in the germline and preimplantation embryos. *Genes and Development*, *28*(8), 812–828. https://doi.org/10.1101/gad.234294.113.

Meyer, E., Lim, D., Pasha, S., Tee, L. J., Rahman, F., Yates, J. R. W., ... Maher, E.R. (2009). Germline mutation in NLRP2 (NALP2) in a familial imprinting disorder (Beckwith-Wiedemann syndrome). *PLoS Genetics*, *5*(3). https://doi.org/10.1371/journal.pgen.1000423, e1000423.

Monk, D., Mackay, D. J. G., Eggermann, T., Maher, E. R., & Riccio, A. (2019). Genomic imprinting disorders: Lessons on how genome, epigenome and environment interact. *Nature Reviews Genetics*, *20*(4), 235–248. https://doi.org/10.1038/s41576-018-0092-0.

Monk, D., Sanchez-Delgado, M., & Fisher, R. (2017). NLRPS, the subcortical maternal complex and genomic imprinting. *Reproduction*, *154*(6), R161–R170. https://doi.org/10.1530/REP-17-0465.

Nakamura, T., Liu, Y. J., Nakashima, H., Umehara, H., Inoue, K., Matoba, S., ... Nakano, T. (2012). PGC7 binds histone H3K9me2 to protect against conversion of 5mC to 5hmC in early embryos. *Nature*, *486*(7403), 415–419. https://doi.org/10.1038/nature11093.

Nguyen, N. M. P., Khawajkie, Y., Mechtouf, N., Rezaei, M., Breguet, M., Kurvinen, E., ... Slim, R. (2018). The genetics of recurrent hydatidiform moles: New insights and lessons from a comprehensive analysis of 113 patients. *Modern Pathology*, *31*(7), 1116–1130. https://doi.org/10.1038/s41379-018-0031-9.

Patten, M., Cowley, M., Oakey, R., & Feil, R. (2016). Regulatory links between imprinted genes: Evolutionary predictions and consequences. *Proceedings of the Biological Sciences*, *283*(1824), 20152760. https://doi.org/10.1098/rspb.2015.2760.

Pellestor, F., Gaillard, J.-B., Schneider, A., Puechberty, J., & Gatinois, V. (2021). Chromoanagenesis phenomena and their formation mechanisms. In T. Liehr (Ed.), *Cytogenomics* (pp. 213–245). Academic Press. Chapter 11 (in this book).

Qian, J., Nguyen, N. M. P., Rezaei, M., Huang, B., Tao, Y., Zhang, X., … Slim, R. (2018). Biallelic PADI6 variants linking infertility, miscarriages, and hydatidiform moles. *European Journal of Human Genetics*, *26*(7), 1007–1013. https://doi.org/10.1038/s41431-018-0141-3.

Rao, S., Huntley, M. H., Durand, N. C., Stamenova, E. K., Bochkov, I. D., Robinson, J. T., … Aiden, E.L. (2014). A 3D map of the human genome at kilobase resolution reveals principles of chromatin looping. *Cell*, *159*(7), 1665–1680. https://doi.org/10.1016/j.cell.2014.11.021.

Sanchez-Delgado, M., Riccio, A., Eggermann, T., Maher, E. R., Lapunzina, P., Mackay, D., & Monk, D. (2016). Causes and consequences of multi-locus imprinting disturbances in humans. *Trends in Genetics*, *32*(7), 444–455. https://doi.org/10.1016/j.tig.2016.05.001.

Sawan, C., Vaissière, T., Murr, R., & Herceg, Z. (2008). Epigenetic drivers and genetic passengers on the road to cancer. *Mutation Research, Fundamental and Molecular Mechanisms of Mutagenesis*, *642*(1–2), 1–13. https://doi.org/10.1016/j.mrfmmm.2008.03.002.

Singh, V. B., Sribenja, S., Wilson, K. E., Attwood, K. M., Hillman, J. C., Pathak, S., & Higgins, M. J. (2017). Blocked transcription through KvDMR1 results in absence of methylation and gene silencing resembling Beckwith-Wiedemann syndrome. *Development (Cambridge, England)*, *144*(10), 1820–1830. https://doi.org/10.1242/dev.145136.

Soellner, L., Begemann, M., Mackay, D. J. G., Grønskov, K., Tümer, Z., Maher, E. R., … Eggermann, T. (2017). Recent advances in imprinting disorders. *Clinical Genetics*, *91*(1), 3–13. https://doi.org/10.1111/cge.12827.

Soellner, L., Kraft, F., Sauer, S., Begemann, M., Kurth, I., Elbracht, M., & Eggermann, T. (2019). Search for cis-acting factors and maternal effect variants in Silver-Russell patients with ICR1 hypomethylation and their mothers. *European Journal of Human Genetics*, *27*(1), 42–48. https://doi.org/10.1038/s41431-018-0269-1.

Sparago, A., Cerrato, F., & Riccio, A. (2018). Is ZFP57 binding to H19/IGF2: IG-DMR affected in Silver-Russell syndrome? *Clinical Epigenetics*, *10*(1), 23. https://doi.org/10.1186/s13148-018-0454-7.

van Steensel, B., & Belmont, A. S. (2017). Lamina-associated domains: Links with chromosome architecture, heterochromatin, and gene repression. *Cell*, *169*(5), 780–791. https://doi.org/10.1016/j.cell.2017.04.022.

Tan, L., Xing, D., Chang, C. H., Li, H., & Xie, X. S. (2018). Three-dimensional genome structures of single diploid human cells. *Science*, *361*(6405), 924–928. https://doi.org/10.1126/science.aat5641.

Tashiro, F., Kanai-Azuma, M., Miyazaki, S., Kato, M., Tanaka, T., Toyoda, S., … Miyazaki, J.I. (2010). Maternal-effect gene Ces5/Ooep/Moep19/Floped is essential for oocyte cytoplasmic lattice formation and embryonic development at the maternal-zygotic stage transition. *Genes to Cells*, *15*(8), 813–828. https://doi.org/10.1111/j.1365-2443.2010.01420.x.

Tong, Z. B., Gold, L., Pfeifer, K. E., Dorward, H., Lee, E., Bondy, C. A., … Nelson, L.M. (2000). Mater, a maternal effect gene required for early embryonic development in mice. *Nature Genetics*, *26*(3), 267–268. https://doi.org/10.1038/81547.

Tucci, V., Isles, A., Kelsey, G., Ferguson-Smith, A., & Erice Imprinting Group. (2019). Genomic imprinting and physiological processes in mammals. *Cell*, *176*(5), 952–965. https://doi.org/10.1016/j.cell.2019.01.043.

Valente, F., Sparago, A., Freschi, A., Hill-Harfe, K., Maas, S., Frints, S., … Cerrato, F. (2019). Transcription alterations of KCNQ1 associated with imprinted methylation defects in the Beckwith–Wiedemann locus. *Genetics in Medicine*, *21*(8), 1808–1820. https://doi.org/10.1038/s41436-018-0416-7.

Van Den Berg, I. M., Eleveld, C., Van Der Hoeven, M., Birnie, E., Steegers, E. A. P., Galjaard, R. J., … Van Doorninck, J.H. (2011). Defective deacetylation of histone 4 K12 in human oocytes is associated with advanced maternal age and chromosome misalignment. *Human Reproduction*, *26*(5), 1181–1190. https://doi.org/10.1093/humrep/der030.

Wakeling, E., Brioude, F., Lokulo-Sodipe, O., O'Connell, S., Salem, J., Bliek, J., … Netchine, I. (2017). Diagnosis and management of silver-Russell syndrome: First international consensus statement. *Nature Reviews Endocrinology*, *13*(2), 105–124. https://doi.org/10.1038/nrendo.2016.138.

Weise, A., & Liehr, T. (2021). Cytogenetics. In T. Liehr (Ed.), *Cytogenomics* (pp. 25–34). Academic Press. Chapter 3 (in this book).

Xu, Y., Shi, Y., Fu, J., Yu, M., Feng, R., Sang, Q., … Wang, L. (2016). Mutations in PADI6 cause female infertility characterized by early embryonic arrest. *American Journal of Human Genetics*, *99*(3), 744–752. https://doi.org/10.1016/j.ajhg.2016.06.024.

Yumiceba, V., Souto Melo, U., & Spielmann, M. (2021). 3D cytogenomics: Structural variation in the three-dimensional genome. In T. Liehr (Ed.), *Cytogenomics* (pp. 247–266). Academic Press. Chapter 12 (in this book).

Zheng, P., & Dean, J. (2009). Role of filia, a maternal effect gene, in maintaining euploidy during cleavage-stage mouse embryogenesis. *Proceedings of the National Academy of Sciences of the United States of America*, *106*(18), 7473–7478. https://doi.org/10.1073/pnas.0900519106.

Zheng, H., Huang, B., Zhang, B., Xiang, Y., Du, Z., Xu, Q., … Xie, W. (2016). Resetting epigenetic memory by reprogramming of histone modifications in mammals. *Molecular Cell*, *63*(6), 1066–1079. https://doi.org/10.1016/j.molcel.2016.08.032.

Subject Index

Note: Page numbers followed by *f* indicate figures and *t* indicate tables.

403